4

Developments in Petroleum Science, 13

enhanced oil recovery

Developments in Petroleum Science, 13

enhanced oil recovery

Proceedings of the third European Symposium on Enhanced Oil Recovery,
held in Bournemouth, U.K., September 21—23, 1981

Edited by

F. JOHN FAYERS

Atomic Energy Establishment, Winfrith, Dorchester, England

ELSEVIER SCIENTIFIC PUBLISHING COMPANY
AMSTERDAM — OXFORD — NEW YORK   1981

ELSEVIER SCIENTIFIC PUBLISHING COMPANY
Molenwerf 1
P.O. Box 211, 1000 AE Amsterdam, The Netherlands

Distributors for the United States and Canada:

ELSEVIER/NORTH-HOLLAND INC.
52, Vanderbilt Avenue
New York, N.Y. 10017

ISBN 0-444-42033-9 (Vol. 13)
ISBN 0-444-41625-0 (Series)

Printed in The Netherlands

TABLE OF CONTENTS

MISCIBLE GAS DISPLACEMENT

NUMERICAL METHODS

EXPERIMENTAL TECHNIQUES

THERMAL RECOVERY METHODS

UNITED STATES RESEARCH PROGRAMME

FOREWORD

This residential symposium is the third in the series of symposia which have been held on the subject of enhanced oil recovery in the United Kingdom; the other two being held at Britannic House of BP in London in May 1977, and at Heriot-Watt University in Edinburgh in July 1978.

Since 1977, when the first symposium was held in London, the annual production and the number of fields in operation in the UK sector of the North Sea has roughly doubled and it is perhaps right to re-iterate the remarks of the Chairman of the organising committee of the first meeting. He said that, "There is an urgent need to decide which enhanced oil recovery techniques are suitable for use in the North Sea. Once this decision is made, the selected R&D goals should be vigorously pursued, leading, hopefully, to the development of specific tailor-made techniques effective in the individual fields in the North Sea area".

Although these remarks are still valid today, in the intervening period throughout Europe significant progress has been made. We have seen an increase in the number of pilot field experiments undertaken by the oil industry, an increase in the research work carried out at universities, research institutes and oil company laboratories. A number of Government programmes have been initiated or expanded. Against this background of an increased R&D activity, some significant, albeit tentative, steps in the application of enhanced oil recovery offshore have been taken.

The continuing increase in the price of oil over the past few years renders the timing of the present symposium to be particularly relevant to the question of improvement in oil recovery in all the sectors of the North Sea and for the provision of future supplies of energy to Europe. The occasion of the present conference provides an international forum for research workers in enhanced oil recovery to exchange information and to develop an increased awareness of the research studies currently being pursued elsewhere. It is hoped that new directions for research, applicable to the European Continental Shelf, may become apparent and the future adoption of enhanced oil recovery techniques in this area advanced.

This volume is a collection of the papers to be presented and discussed at the Symposium.

F J FAYERS
Chairman, Organising Committee

September 1981

FUNDAMENTAL ASPECTS OF SURFACTANT-POLYMER FLOODING PROCESS

D. O. SHAH

*Department of Chemical Engineering and Anesthesiology,
University of Florida, Gainesville, Florida 32611*

ABSTRACT

Surfactant-polymer flooding process offers a promising approach to recover additional oil from the water flooded reservoirs which may contain as much as 70% of original oil-in-place. The capillary number, which determines the microscopic displacement efficiency of oil, can be increased by 3 to 4 orders of magnitude by reducing the interfacial tension (IFT) of oil ganglia below 10^{-3} dynes/cm. Conceptual events involved in the mobilization and displacement of oil ganglia are described including the role of ultralow interfacial tension, the role of interfacial viscosity in coalescence of oil ganglia and formation of the oil bank, the propagation of the oil bank, the surfactant-polymer incompatibility, the formation and flow of emulsions in porous media, the role of wettability as well as the influence of surface charge density of the rock/fluid interface and oil-brine interface in oil displacement efficiency. It is shown that there are two regions of ultra-low IFT; 1) in the low surfactant concentration (0.1-0.2%) and the other in the high surfactant concentration region (2.0%-10.0%). In the low concentration systems, the ultra-low interfacial tension occurs when the aqueous phase of the surfactant solution is about the apparent critical micelle concentration. And, the salinity is at the critical electrolyte concentration for the coacervation process. The migration of surfactant from the aqueous phase to the oil phase via coacervation process appears to be responsible for the ultralow interfacial tension.

In high surfactant concentration systems, a middle phase microemulsion in equilibrium with excess oil and brine forms in a narrow salinity range. The salinity at which equal volumes of oil and brine are solubilized in the middle phase microemulsion is referred to as the optimal salinity of the system. At the optimal salinity, the interfacial tension at both interfaces is equal. Evidence is presented that the middle phase microemulsion at the optimal salinity is a water external microemulsion formed due to coacervation and subsequent phase separation of micelles from the aqueous phase. The optimal salinity can be shifted to a desired value by varying the structure and concentration of alcohol. The shift in optimal salinity can be correlated with the brine solubility of the alcohol used in a given surfactant formulation. It was further observed that the optimal salinity increases with the oil chain length. In order to form middle phase microemulsions at very high salinity, ethoxylated surfactants or alcohols can be incorporated into a surfactant formulation which can shift the optimal salinity to as high as 32% NaCl concentration. Such high salinity formulations consisting of petroleum sulfonates and ethoxylated sulfonates are relatively insensitive to divalent cations.

The coalescence rate or the phase separation time was minimum at optimal salinity. It was also observed that the apparent viscosity was minimal at the optimal salinity for the flow of microemulsions through porous media. The rate of flattening of an oil drop in a surfactant formulation increases strikingly in the presence of alcohol. It appears that the presence of alcohol promotes the mass transfer of surfactant from the aqueous phase to the interface. The addition of alcohol also promotes the coalescence of oil drops, presumably due to a decrease in the interfacial viscosity.

The surfactant-polymer incompatibility can lead to a phase separation of a surfactant and polymer even in the absence of oil. In the presence of oil, the formation of middle phase microemulsion is promoted by the presence of polymer in the aqueous phase. The surfactant-polymer incompatibility is explained in terms of excluded volume effects and the maximization of solvent for polymer molecules.

Some novel concepts for surfactant-polymer flooding process have been discussed including the use of two different surfactant slugs, two different polymer slugs, salinity gradient design and the injection of an oil bank to promote oil recovery.

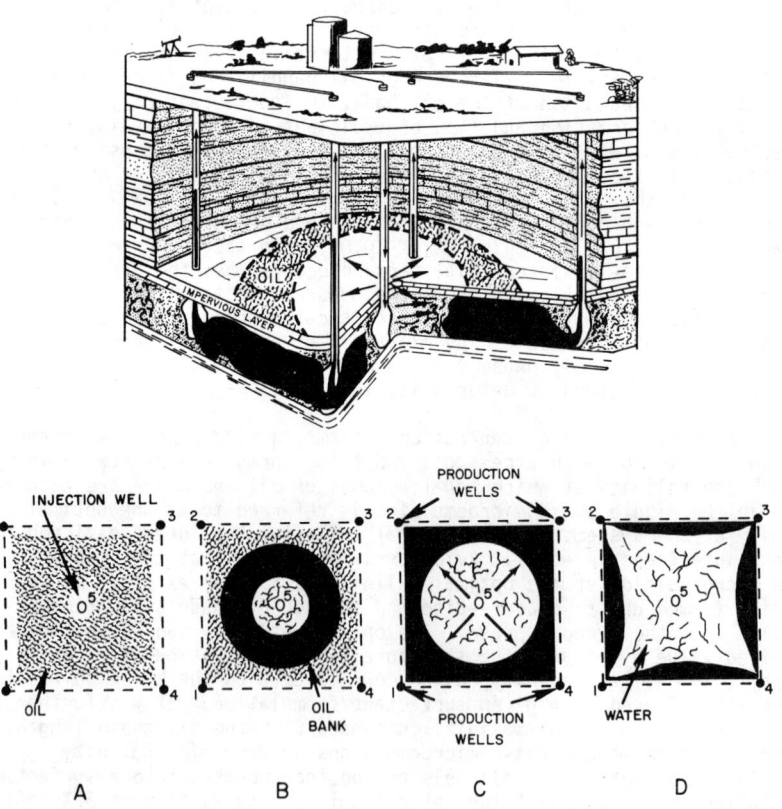

Fig. 1   Schematic diagram of an oil reservoir and the displacement of oil by water or chemical flooding.

A. INTRODUCTION

It is well recognized that the energy consumption per capita and the standard of living of a society are interrelated. Among various sources of energy, fossil fuels or crude oils play an important role in providing the energy supply of the world. It also serves as a raw material for feed stocks in chemical industry. In view of the worldwide energy crisis, the importance of enhanced oil recovery to increase the supply of crude oil is obvious and various enhanced oil recovery processes have been proposed and tested both on a laboratory scale and in the field. For heavy oils, thermal processes have been used extensively whereas for light oils, chemical processes such as polymer flooding, caustic flooding, miscible flooding and surfactant-polymer flooding have attracted great interest. The major research findings in the enhanced oil recovery area have been reported in recent literature and the symposia proceedings of various conferences during the last decade (1-11). The present paper focuses on the fundamental aspects of the surfactant-polymer flooding process and related phenomena.

Figure 1 schematically shows a three-dimensional view of a petroleum reservoir.

At the end of water-flooding, the oil that remains in the reservoir is believed to be in the form of oil ganglia trapped in the pore structure of the rock as shown in Figure 1A. These oil ganglia are entrapped due to capillary forces. However, if a surfactant solution is injected to lower the interfacial tension of the oil ganglia from its value of 20-30 dynes/cm to 10^{-3} dynes/cm, the oil ganglia can be mobilized and can move through narrow necks of the pores. Such mobilized oil ganglia form an oil bank as shown in Figure 1B. Figures 1C and 1D schematically show the oil bank approaching the production well and the subsequent breakthrough of the drive water. Figure 2 schematically illustrates a two-dimensional view of the surfactant-polymer flooding process.

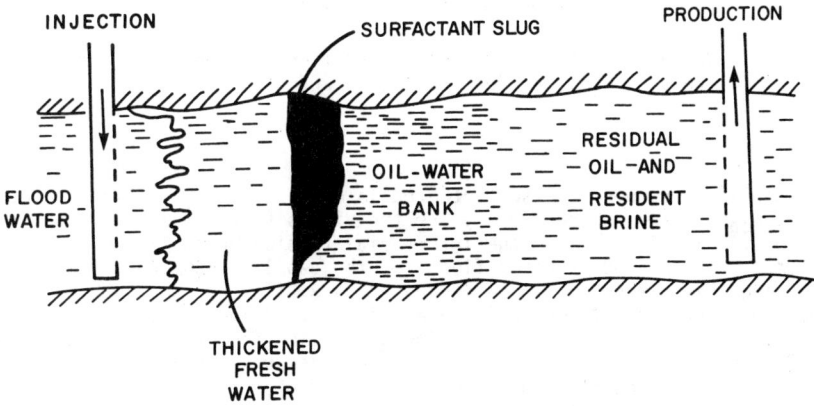

Fig. 2   Schematic diagram of the surfactant-polymer flooding process.

The surfactant slug is followed by a polymer slug for a proper mobility control of the process.

4

B. CAPILLARY NUMBER AND CONCEPTUAL ASPECTS OF THE PROCESS

Recently, in an excellent review article, Taber (12) has summarized various emperical dimensionless numbers proposed by several investigators to correlate the oil displacement efficiency in porous media. Figure 3 shows such a correlation reported by Foster (13) between the capillary number and residual oil in porous media.

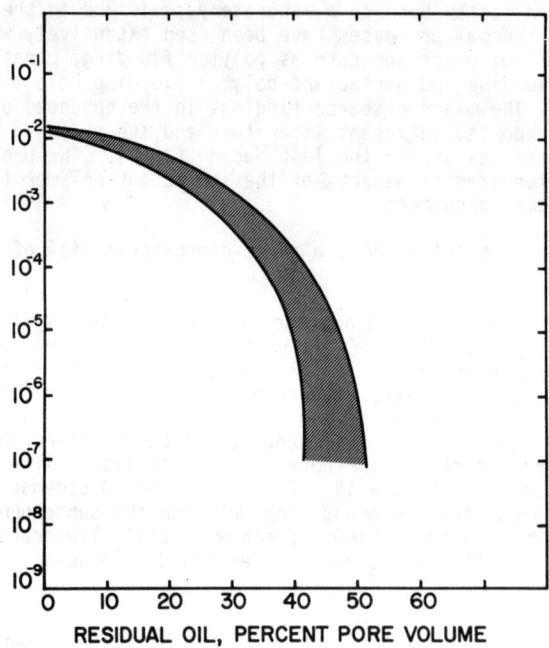

Fig. 3   Dependence of residual oil saturation on Capillary Number
(Foster, W.R., J. Pet. Tech., p. 206, Feb. 1973).

The capillary number represents the ratio of viscous to capillary forces (i.e. $N_{ca} = \mu v/\sigma\phi$ where μ and v are the viscosity and velocity of the displacing fluid, σ is the interfacial tension and ϕ is the pore volume). At the end of water flooding, the capillary number is around 10^{-6} and this number has to be increased by 3 to 4 orders of magnitude for tertiary oil recovery processes (14). Under practical reservoir conditions, the reduction in interfacial tension from a high value of 20 or 30 dynes/cm to 10^{-3} or 10^{-4} dynes/cm offers such a possibility. Therefore, the main function of the surfactant is to produce such an ultra-low interfacial tension at the oil ganglia/surfactant formulation interface. Figure 4 schematically shows the role of ultralow interfacial tension in promoting the mobilization of oil ganglia in porous media. Subsequently, the displaced oil ganglia must coalesce to form an oil bank. For this a very low interfacial viscosity is desirable (Figure 5). It is known that high interfacial viscosity results in the formatin of stable emulsion (15).

5

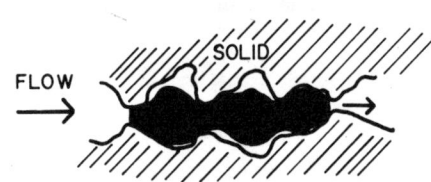

FOR THE MOVEMENT OF OIL
THROUGH NARROW NECK OF
PORES, A <u>VERY LOW OIL/WATER</u>
<u>INTERFACIAL TENSION</u> IS
DESIRABLE ≈ .OOI DYNES/CM

Fig. 4   Schematic diagram of the role of low interfacial tension in the
         surfactant-polymer flooding process.

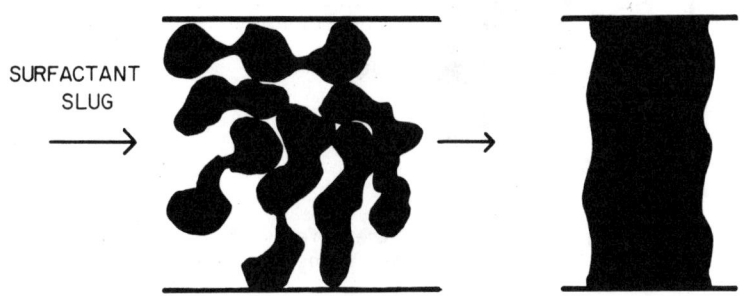

DISPLACED OIL GANGLIA MUST COALESCE TO FORM
A CONTINUOUS OIL BANK: FOR THIS A <u>VERY LOW</u>
<u>INTERFACIAL VISCOSITY</u> IS DESIRABLE

Fig. 5   Schematic diagram of the role of low interfacial viscosity in
         the surfactant-polymer flooding process.

Once an oil bank is formed in the porous medium, it has to be propagated through the porous medium without increasing the entrapment of
oil at the trailing edge of the oil bank. As shown in Figure 6, the
maintenance of ultralow interfacial tension at the oil bank/surfactant/
slug interface is essential for minimizing the entrapment of the oil in
the porous medium whereas the leading edge will coalesce with the oil
ganglia.

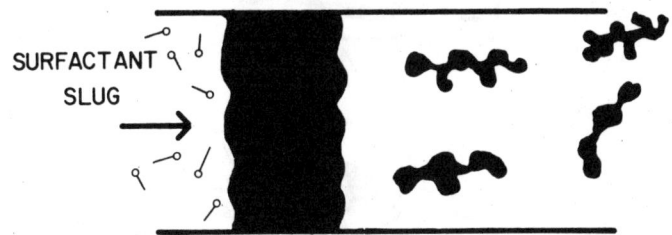

COALESCENCE OF OIL GANGLIA WITH OIL
BANK CAUSES FURTHER DISPLACEMENT OF OIL

Fig. 6   Schematic diagram of the role of coalescence of oil ganglia in
         the surfactant-polymer flooding process.

Figure 7 schematically illustrates the movement of the oil bank, surfactant slug and the mobility control polymer slug in the porous medium.

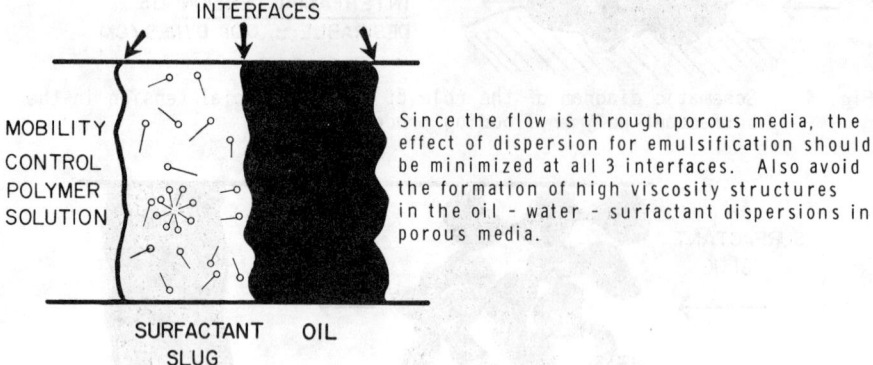

INTERFACES

MOBILITY CONTROL POLYMER SOLUTION

Since the flow is through porous media, the effect of dispersion for emulsification should be minimized at all 3 interfaces. Also avoid the formation of high viscosity structures in the oil - water - surfactant dispersions in porous media.

SURFACTANT SLUG   OIL

Fig. 7   Schematic illustration of the effects of dispersion and emulsi-fication between the various slugs during the surfactant-poly-mer flooding process.

Since the flow through the porous medium causes dispersion of these fluids, emulsions will be formed at the oil bank/surfactant slug inter-face and a mixed surfactant-polymer zone will occur at the surfactant-polymer solution interface. High viscosity structures at both these interfaces should be avoided in order to improve the efficiency of the process. The mass transfer of surfactant to the oil bank can influence the magnitude of interfacial tension (16). Trushenski (17) has shown that surfactant-polymer incompatibility leading to a phase separation of surfactant and polymer strikingly reduces the efficiency of the process.

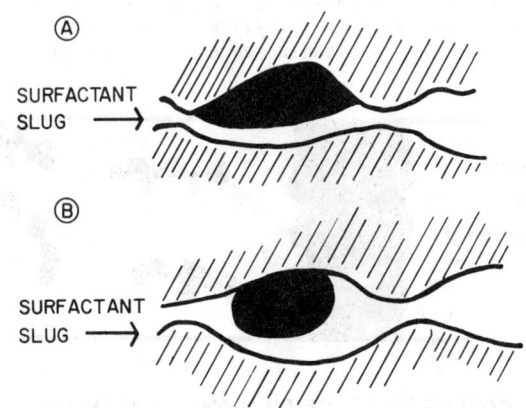

Ⓐ

SURFACTANT SLUG ⟶

Ⓑ

SURFACTANT SLUG ⟶

PROPER CHOICE OF SURFACTANT CAN CHANGE Ⓐ TO Ⓑ

Fig. 8   The role of wettability and contact angle on oil displacement.

Figure 8 schematically illustrates the role of wettability of solid surface on the oil ganglia.

The choice of surfactant can influence the wettability of the rock surface to oil and brine (12).

Another parameter that we have found (18, 19) that influences the interfacial tension and interfacial viscosity and oil recovery is the surface charge at the oil-brine as well as rock-brine interfaces. We found that a high surface charge density leads to a lower interfacial tension, lower interfacial viscosity and higher oil recovery (Figure 9).

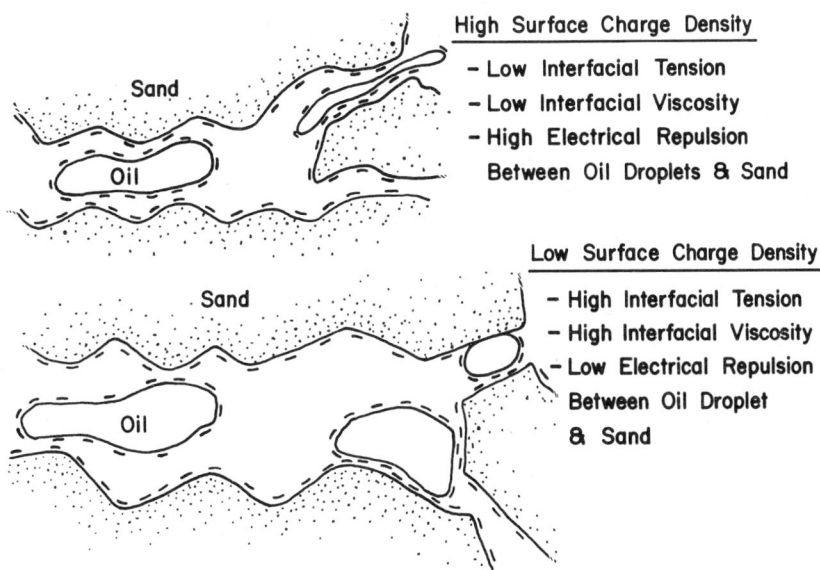

Fig. 9   Schematic diagram of the role of surface charge in the oil displacement process.

The conceptual processes described in Figures 3 to 9 are supported by the results of our studies described in the following sections.

C. LOW SURFACTANT CONCENTRATION SYSTEMS

Figure 10 shows the interfacial tension as a function of surfactant concentration in a dodecane/brine system.

It is evident that there are two regions of ultra-low interfacial tension (IFT). At low surfactant concentrations, the system appears to be a two-phase system, namely, oil and brine in equilibrium with each other, whereas at high surfactant concentration systems (around 4 to 8% surfactant concentration), a middle phase microemulsion exists in equilibrium with excess oil and brine. The formation of middle phase microemulsion and related phenomena will be discussed in section D.

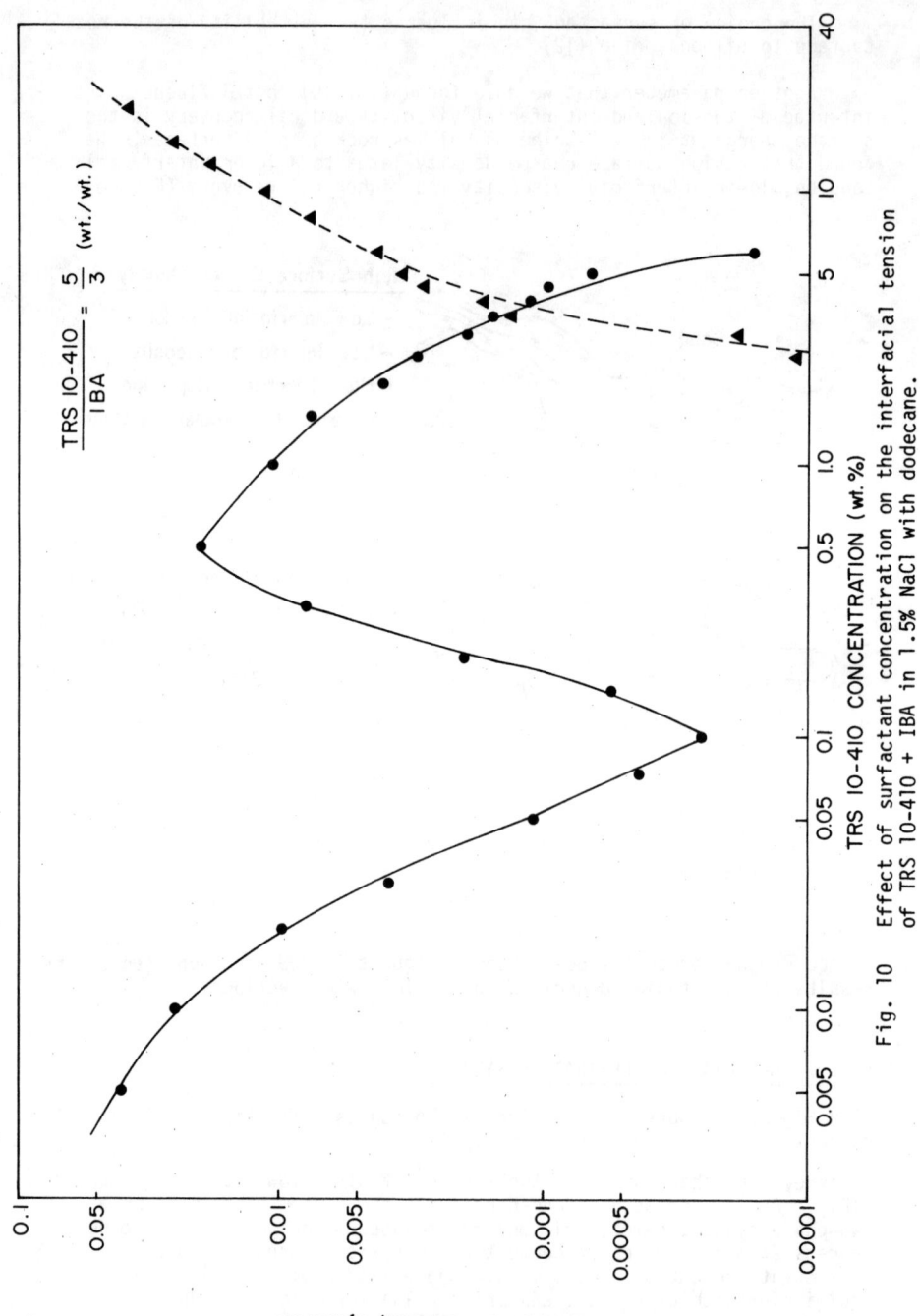

Fig. 10   Effect of surfactant concentration on the interfacial tension
of TRS 10-410 + IBA in 1.5% NaCl with dodecane.

For low surfactant concentration systems, we have shown that the ultralow IFT occurs when surfactant molecules migrate from the aqueous phase to the oil phase (19-21). Figure 11 shows the interfacial tension and the partition coefficient of a surfactant in an octane/brine system. The ultra low IFT occurred around a partition coefficient of unity in this system (19,20). However, it should be emphasized that since the partition coefficient changes abruptly in this region the exact value of partition coefficient can vary significantly around ultralow IFT. We believe that a reasonable conclusion is that lowering of interfacial tension is observed when micelles leave the aqueous phase due to coacervation process (19-23).

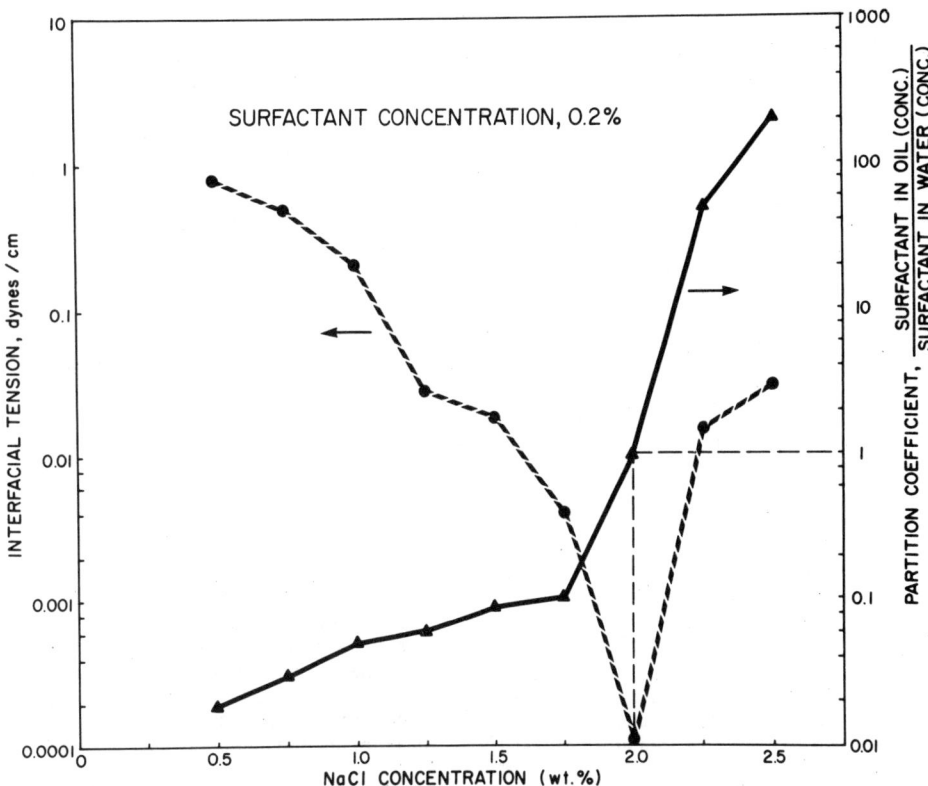

Fig. 11   Effect of added electrolyte on interfacial tension and surfactant partition coefficient of the system 0.2%TRS 10-80 + brine + octane.

Since commercial petroleum sulfonates involve a distribution of molecular weights and isomeric structures we also investigated the interfacial tension using isomerically pure sulfonates. Figure 12 shows the IFT behavior as a function of salinity, oil chain length and surfactant concentration using petroleum sulfonates (TRS 10-80 or TRS 10-410 and an isomerically pure surfactant UT-1). It is evident that both the salinity and oil chain length effects were similar for both these classes of sur-

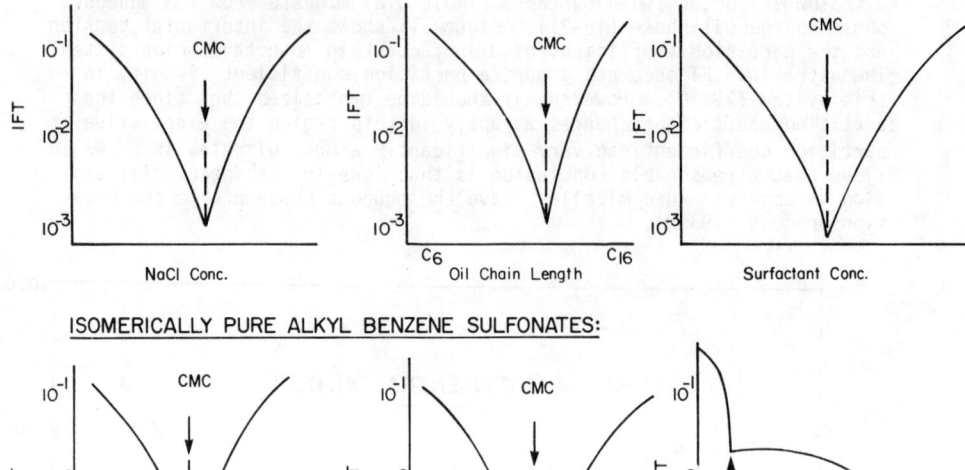

Fig. 12   Schematic diagram of the effect of salt concentration, oil
chain length and surfactant concentration on the interfacial
tension of pure and impure alkyl benzene sulfonates.

factants, namely, there is a specific salinity and specific oil chain
length where we obtain an ultralow IFT. However, the effect of surfactant concentration on IFT was different for commercial and isomerically
pure surfactants. For low surfactant concentration systems, we also observed that the ultra low IFT appears when the aqueous phase is at the
apparent cmc for the surfactant remaining in the aqueous phase. These
conclusions were in aggreement with osmotic pressure, light scattering
and spectroscopic measurements on the equilibrated aqueous phase (22).

Figure 13 is a generalized diagram showing the IFT, phase behavior
and the two critical electrolyte concentrations for both pure and commercial surfactants at low as well as high surfactant concentrations.
By direct analysis of surfactant concentrations in each phase, we found
(21) that the salinity at which surfactant molecules leave the aqueous
phase is lower than the salinity at which they enter the oil phase.
Thus, we define two critical electrolyte concentrations, namely, CEC_1,
and CEC_2, to represent the electrolyte concentrations at which the surfactant concentration begins to decrease in the aqueous phase and begin
to increase in the oil phase respectively. We further observed that the
minimum interfacial tension occurs in the vicinity of the first critical
electrolyte concentration. In between CEC_1 and CEC_2, the surfactant
may precipitate or may form a coacervate phase below the aqueous phase or
in between the aqueous and the oil phase depending upon its density (21).

In low concentration systems, it is possible that an extremely small volume of middle phase may exist between the oil and brine phases even though it may not be visible. This suggestion is in agreement with observation that the volume of the middle phase microemulsion increases linearly with the surfactant concentration and the straight line passes through the origin (24). It should be emphasized that the general behavior and inter-relationship shown in Figure 13 is valid for both commercial and isomerically pure surfactants (21).

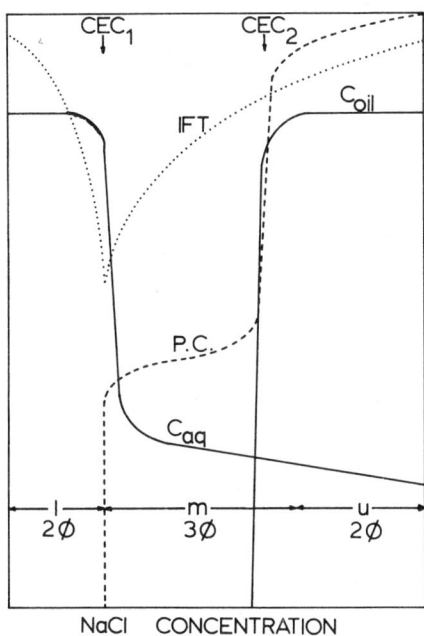

Fig. 13   Generalized diagram of the effect of salt concentration on surfactant partitioning, phase behavior and interfacial tension.

Figure 14 shows the effect of oil chain length on CEC_1 and CEC_2 in Aerosol OT/brine/oil systems.   It is evident that the CEC_1 increases with oil chain length until it reaches the critical oil chain length (C_{11}) above which the value of CEC_1 remains constant.   On the other hand, CEC_2 continues to increase with the oil chain length. Interestingly, we observed that the ultralow IFT only occurs for oil chain lengths below the critical oil chain length ($< C_{11}$), whereas the interfacial tension remains high for oils above the critical oil chain length (21).

We propose that all the oils below the critical oil chain length are able to solubilize in the micelles whereas the oils having chain length above the critical oil chain length are unable to solubilize in the micellar solution. Thus, it appears that solubilization of oil within the micelles is an important requirement for producing ultralow IFT. From our extensive studies on interfacial tension and partitioning of surfactant in relation to many parameters, we have proposed the following 5 necessary conditions to achieve ultralow IFT's.

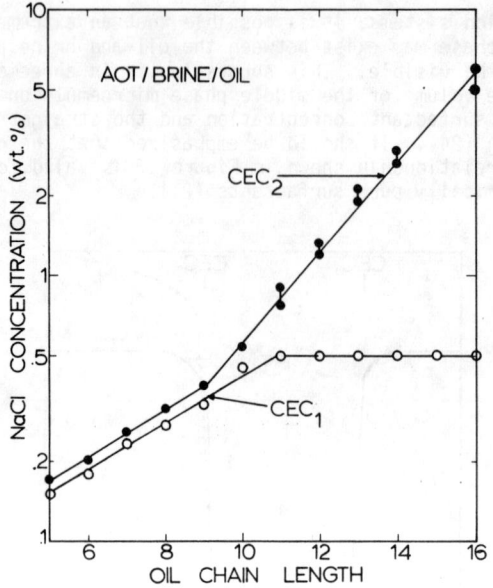

Fig. 14 Effect of oil chain length on the first and second critical electrolyte concentrations of Aerosol OT.

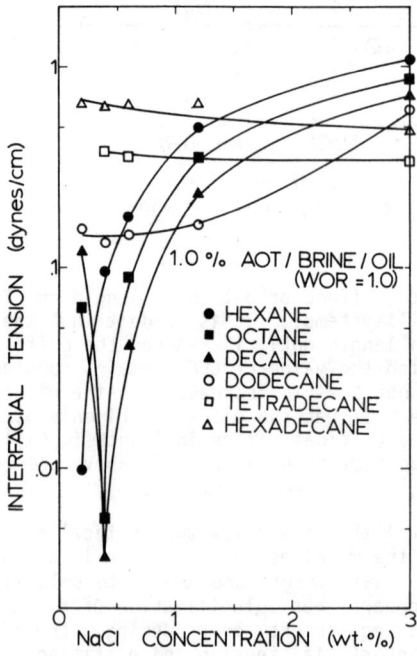

Fig. 15 Effect of oil chain length on the interfacial tension of the systems 1.0% AOT/brine/oil.

1) The total surfactant concentration should be appreciably above the apparent cmc in the aqueous phase.

2) The surfactant should be soluble in both the aqueous and the hydro-carbon phase.

3) Micelles in the aqueous phase should be able to solubilize oil from the hydrocarbon phase.

4) The aqueous phase salinity should be near the first critical electrolyte concentration (CEC_1).

5) There should be a large slope in the surfactant partition coefficient curve in the region of the ultralow IFT. (i.e. a steep partition coefficient curve for surfactant).

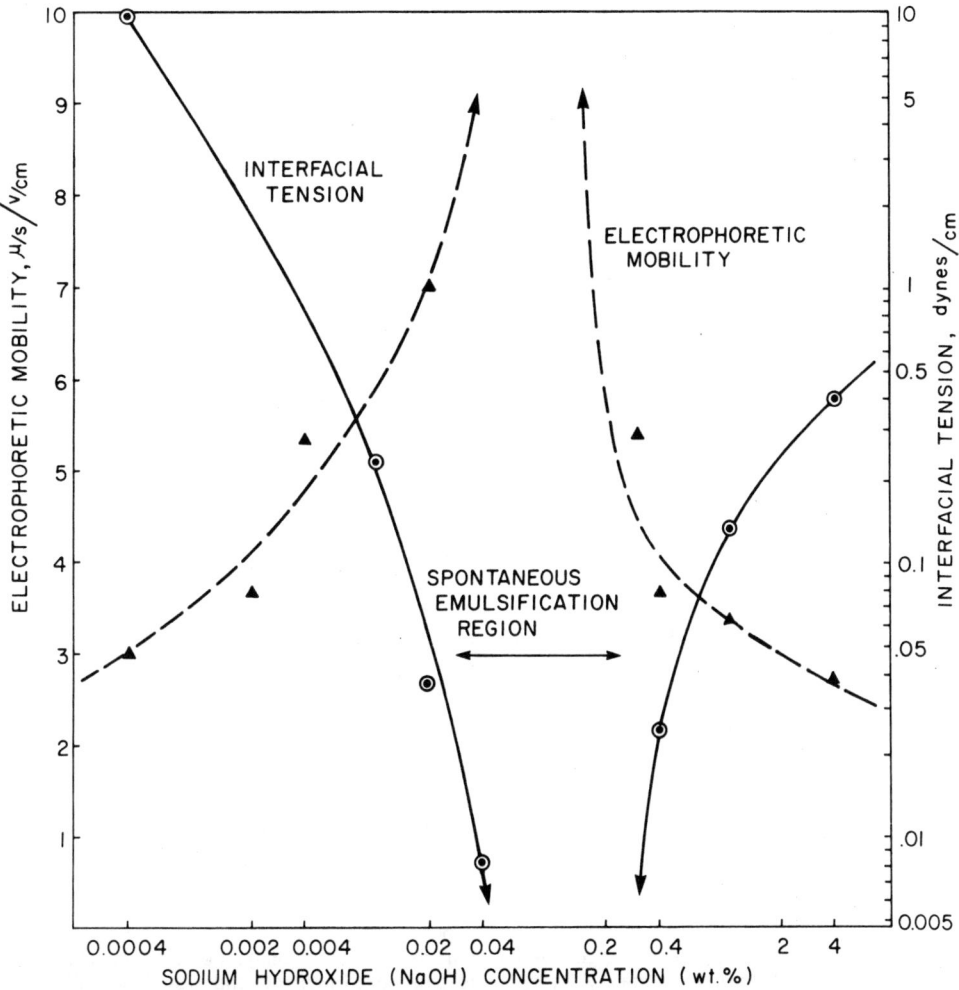

Fig. 16 A correlation between interfacial tension and electrophoretic mobility for crude oil-NaOH solutions.

14

Figure 16 shows the correlation of interfacial tension with electrophoretic mobility in crude oil/caustic systems (18,19,25,26). We have
observed for several crude oils that the ultra low IFT occurs in the region where the electrophoretic mobility is maximum. This suggests that
the maximum in surface charge density coincides with the minimum in interfacial tension. This correlation was also observed for the effect of
salinity and surfactant concentration (19). Figure 17 schematically
illustrates 3 components of the interfacial tension, namely, 1) surface
concentration of the surfactant, 2) surface charge density, and 3) mutual
solubilization of oil and brine. We propose that by adjusting any of
these variables one can influence the magnitude of interfacial tension.

Using the conceptual approach shown in Figure 17, we were able to
broaden and lower the magnitude of interfacial tension as well as increase the salt tolerance limit of the surfactant formulation.

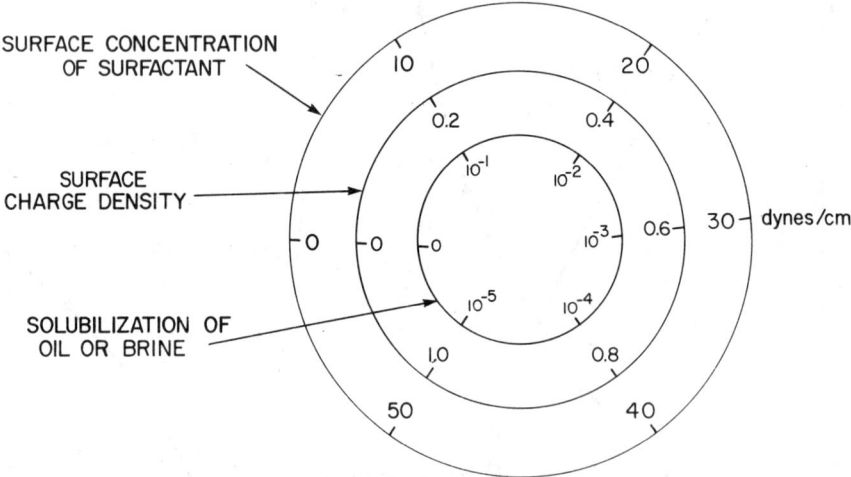

Fig. 17   A schematic illustration of the factors affecting the magnitude
of the interfacial tension.

Figure 18 shows the interfacial tension of a petroleum sulfonate
TRS 10-410/n-octane/brine system when gradually the petroleum sulfonate
is replaced with an ethoxylated phosphate ester (Klearfac AA-270).

The Klearfac AA-270 containing a phosphate group possesses two ionic
oxygen atoms and hence can generate a high surface charge density at the
interface. This presumably is responsible for lowering the magnitude of
IFT and broadening the salinity range over which the ultralow IFT occurs
for the mixed surfactant systems (27).

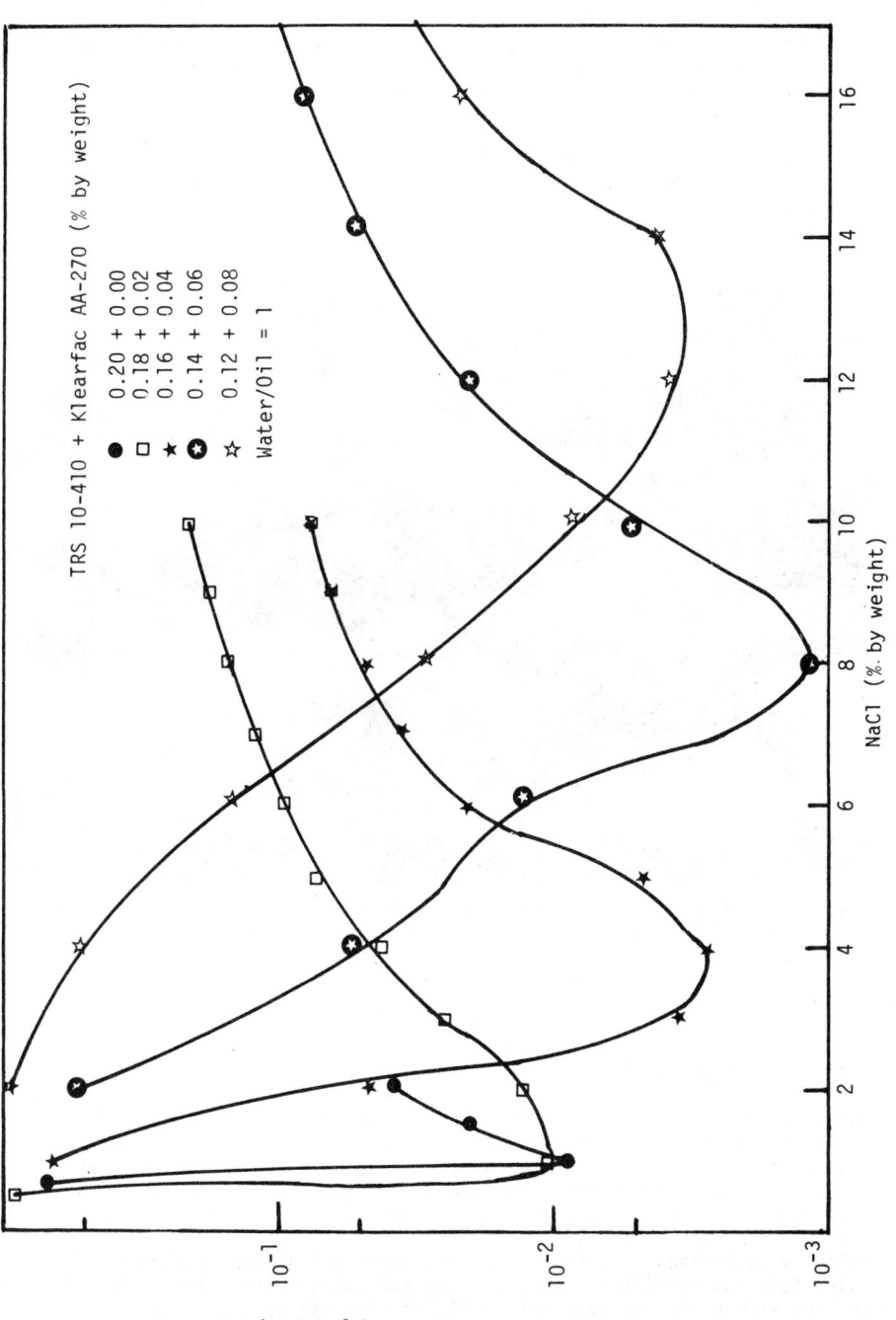

Fig. 18   An illustration of the reduction and broadening of the minimum in interfacial tension by the addition of Klearfac AA-270.

D. HIGH SURFACTANT CONCENTRATION SYSTEMS AND THE OPTIMAL SALINITY

Many surfactant formulations exhibit extreme flow or static birefringence in a given salinity range or in a given temperature range. Often these optically anisotropic formulations exhibit ultralow IFT with oil. The microstructure of such birefringent formulations should be of interest in understanding the changes in molecular associations occurring in these systems.

Figures 19-21 illustrate the microstructure of a birefringent surfactant formulation consisting of 5% TRS 10-410 + 3% isobutanol and 2% NaCl brine.

Fig. 19   Freeze-fracture electron micrograph of the anisotropic system
5% TRS 10-410 + 3% Isobutanol + 2% NaCl (8550 x).

The freeze-fracture electron microscopic technique used to obtain these pictures is believed to preserve the microstructure of the samples due to the very rapid cooling rate (24). These electron micrographs clearly indicate that the birefringent formulations consist of bubbles filled with brine and separated from each other by a thin surfactant membrane. Figure 21 clearly shows the structure of this membrane consisting of several thin layers. The dimension of each layer is close to a surfactant bilayer (approximately 70A). Therefore, it appears that when the salinity is increased in the surfactant formulation, the surfactant molecules form the multilayer structure while keeping their polar groups in contact with brine and form such cells or foamlike stable structure. We have called these structures birefringent cellular fluids (24).

Fig. 20   Freeze-fracture electron micrograph of the above system at
18,000X.

Fig. 21   Freeze-fracture electron micrograph of the above system at
30,000X.

Figure 22 shows the similarity between coacervation of a micellar solution in the absence of oil and the formation of a middle phase microemulsion in the presence of oil. The lower part of Figure 22 shows the transition of a birefringent surfactant formulation to an isotropic coacervate phase upon addition of salt. On the other hand, the same formulation in the presence of an equal volume of dodecane shows the formation of lower phase, middle phase and upper phase microemulsions. We propose that the middle phase microemulsion is similar to the coacervated phase containing some solubilized oil. Additional studies in support of these models have been reported elsewhere (21, 23, 24).

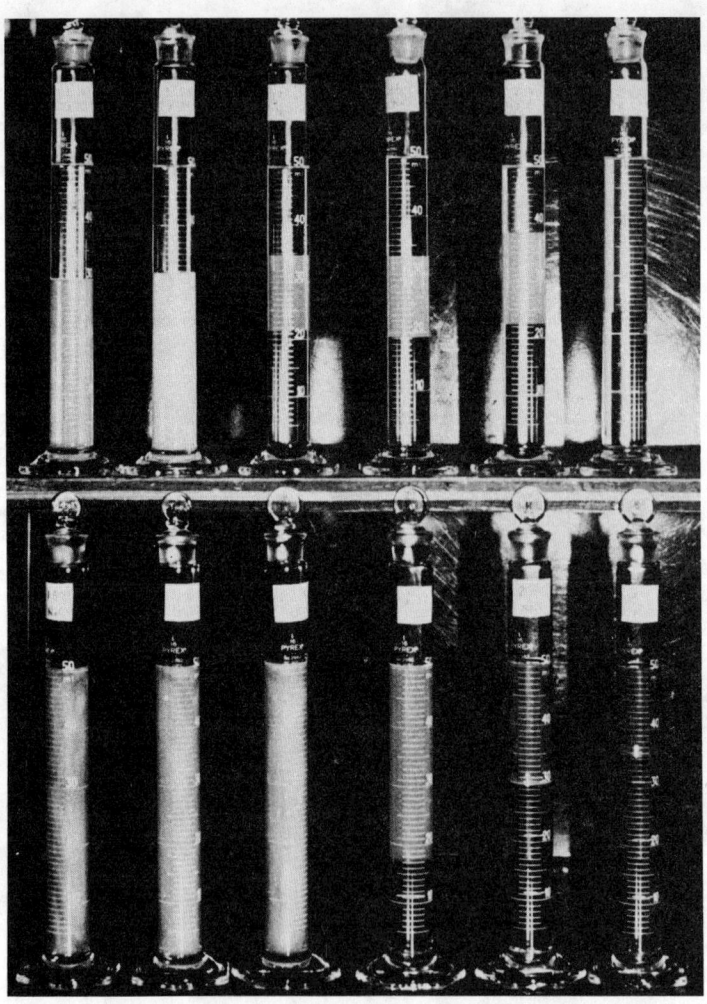

Fig. 22   A comparison of coacervation in aqueous solution with middle phase formation in surfactant/oil/brine/alcohol systems.

Figure 23 schematically shows the mechanism of formation of middle phase microemulsions as salinity is increased.

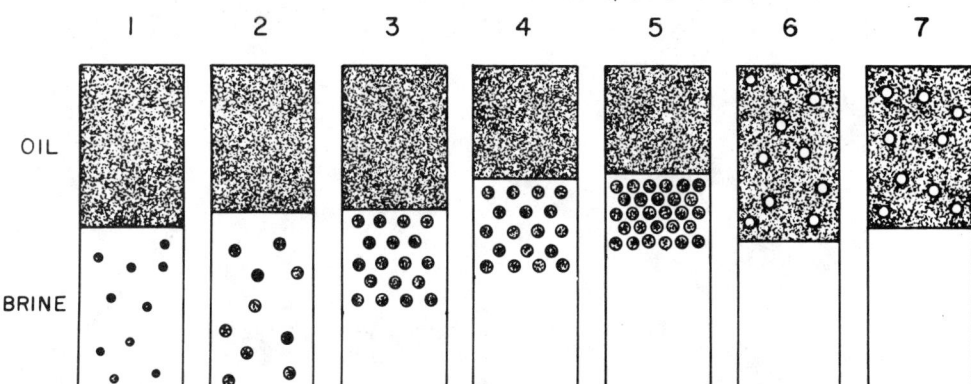

⊕   oil swollen micelles (microdroplets of oil)

○   reverse micelles (microdroplets of water)

Fig. 23   A schematic illustration of the l→m→u transition for the TRS 10-410/Isobutanol/Oil/Brine System.

As one increases the salinity, the cmc decreases, the aggregation number of the micelles increases and the solubilization of oil within micelles increases. The compression of the electrical double layer around micelles will occur, hence reducing the repulsive forces between the micelles. Thus the reduction in the repulsive forces and increase in the attractive forces between the micelles will bring the micelles closer and ultimately lead to a separation of a micelle rich phase forming the middle phase microemulsion. Upon further increase in salinity, the solubilization of oil in this middle phase increases whereas that of brine decreases. The magnitude of interfacial tension of the middle phase with oil or brine depends upon the extent of solubilization of oil and brine in the middle phase. In general, the higher the solubilization of oil or brine in the middle phase microemulsion, the lower is the interfacial tension with respect to these excess phases (28). The salinity at which equal volumes of oil and brine are solubilized in the middle phase microemulsion is referred to as optimal salinity for the surfactant-oil-brine systems under given physical chemical conditions (29, 30).

Figure 24 shows the freeze-fracture electron micrograph of a middle phase microemulsion formed in the system extensively studied by Reed and Healy (28-30).
It clearly shows the discrete spherical structures embedded in a continuous aqueous phase consistent with the mechanism proposed in Figure 23. It should be pointed out that other investigators (40-47) have proposed the possibility of bicontinuous structure or the coexistance of oil external and water external microemulsions in the middle phase. In very high surfactant concentration systems, (15-20%) the existence of anamalous structure which are neither conventional water external or oil external microemulsions have been proposed to account for some unusual properties of these systems (43-46).

Figure 25 shows that the transition l→m→u is not only achieved by increasing the salinity but is also possible by changing any of the other 8 variables.

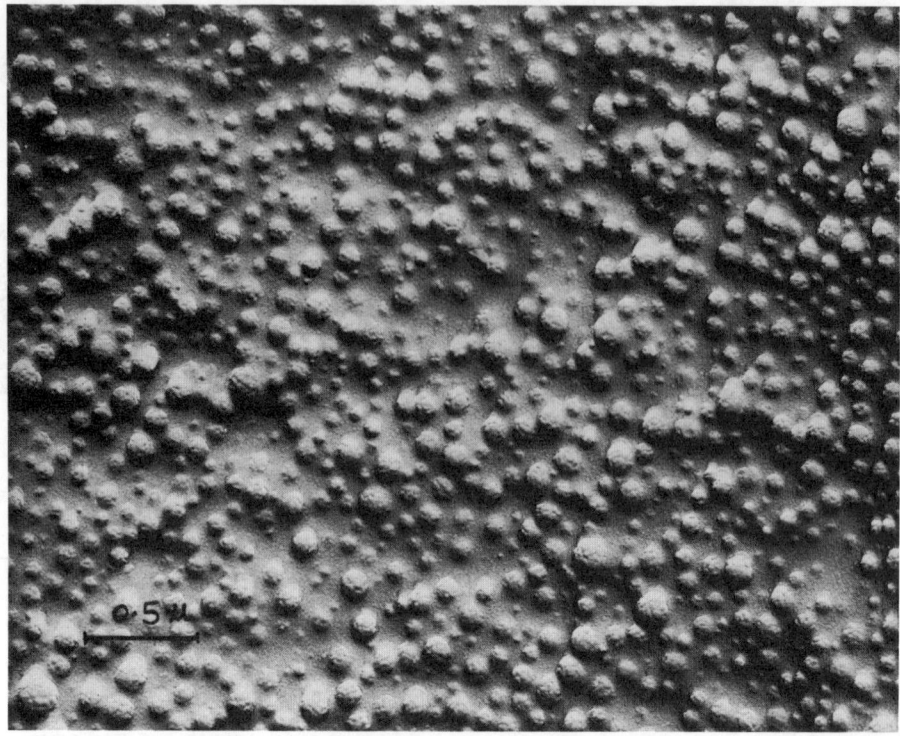

Fig. 24 Freeze-fracture electron micrograph of the middle phase of the
Exxon system at the optimal salinity.

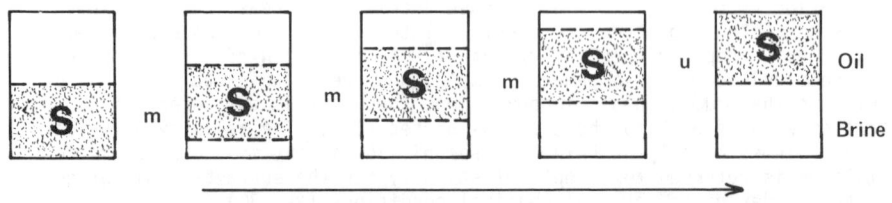

Parameter Increasing

The transition l ⟶ m ⟶ u occurs by:

1. Increasing Salinity
2. Decreasing oil chain length
3. Increasing alcohol concentration (C_4, C_5, C_6)
4. Decreasing temperature
5. Increasing total surfactant concentration
6. Increasing brine/oil ratio
7. Increasing surfactant solution/oil ratio
8. **Increasing molecular weight of surfactant**

Fig. 25 Schematic illustration of the factors influencing the l → m → u
transition in surfactant/oil/brine/alcohol systems.

Thus, by choice of a suitable parameter, one can obtain the transition in the structure of these microemulsions. At optimal salinity, the partition coefficient of surfactant in the excess oil and brine phases is found to be near unity and the interfacial tension between the excess oil and excess brine is also minimum (19).

The importance of the optimal salinity concept for enhanced oil recovery is shown in the data illustrated in Figure 26.

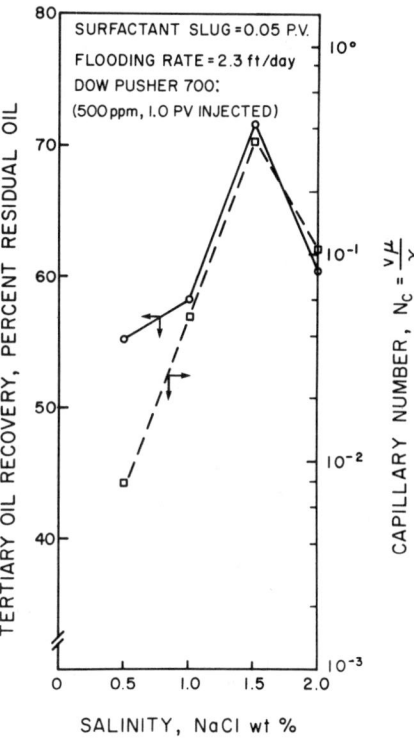

Fig. 26   Effect of salinity on the capillary number and tertiary oil recovery in sand packs.

It is evident that the oil recovery is maximum at optimal salinity for the systems reported here. An excellent correlation between the capillary number and oil recovery is also evident from Figure 26 (48). In view of this observation, the surfactant formulation for a practical application should be designed such that the reservoir salinity becomes the optimal salinity under the given reservoir conditions.

Figure 27 shows the effect of oil chain length on optimal salinity of the TRS 10-410 + isobutanol systems (49) and the corresponding interfacial tension at the optimal salinity for each oil chain length. It was observed that as the oil chain length increases, the optimal salinity increases and the volume of the middle phase decreases. The range over which the middle phase microemulsion exists also increases as the oil

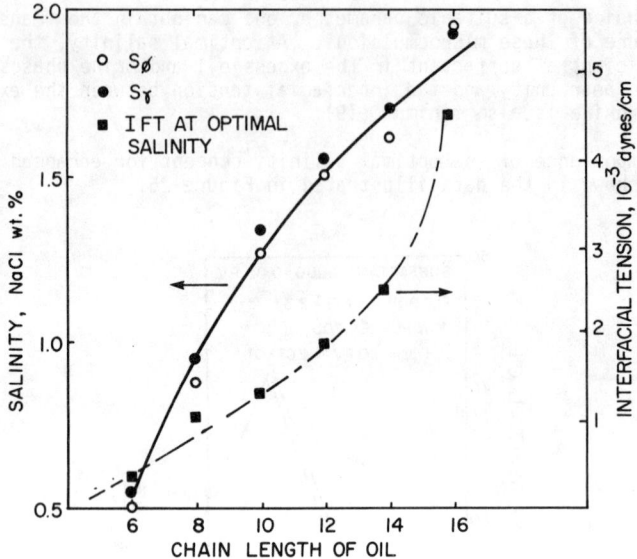

Fig. 27   Effect of oil chain length on the optimal salinity and interfacial tension at the optimal salinity.

chain length increases. It should be pointed out that from extensive
studies on mixed alkanes, the concept of Equivalent Alkane Carbon Number
(EACN) has been proposed to correlate the interfacial tension of pure
alkanes with those of the mixtures (50). Many light crude oils have been
simulated by the mixtures of pure hydrocarbons (51). Most light oils or
the EACN for most light crude oils were found to be between C_7 and
C_{11}.

Figure 28 shows the correlation of optimal salinity in the presence
of various alcohols with their solubility in brine.
Figure 28 summarizes the data obtained by three research groups (49,52,
53). It is interesting that the optimal salinity of a given oil and surfactant formulation lies near the intersection of the brine solubility.
This correlation suggests that if one determines the optimal salinity in
the presence of 2 or 3 alcohols, one can predict the optimal salinity in
the presence of other alcohols from their brine solubility data. This is
a very useful correlation and eliminates the time consuming and laborious
procedure of obtaining the optimal salinity in the presence of each alcohol.

E. <u>TRANSIENT PROCESSES</u>

There are several transient processes, such as the formation and
coalescence of drops as well as their flow through porous media, that are
likely to occur in the surfactant-polymer flooding process. Figure 29
shows the coalescence or phase separation time of handshaken and sonicated macroemulsions as a function of salinity.

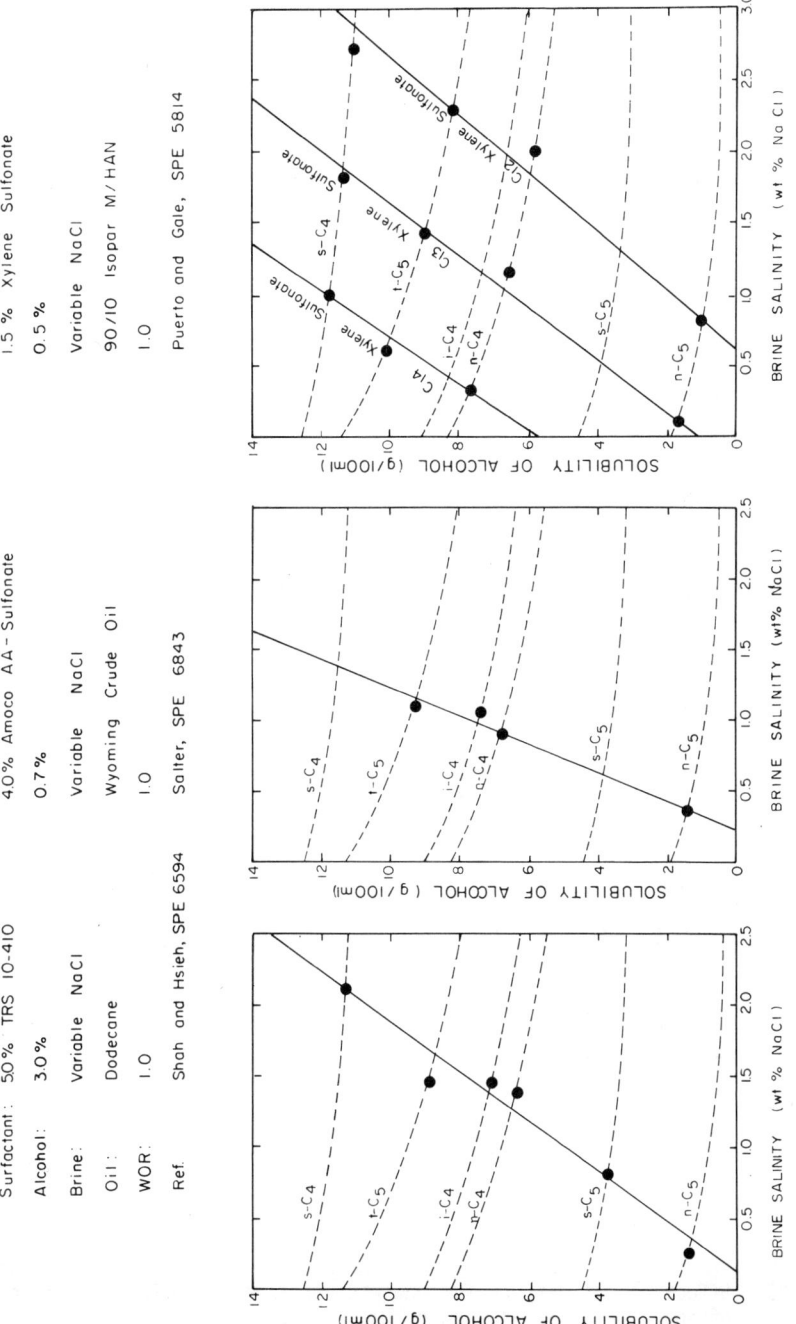

Fig. 28   A correlation of optimal salinity in the presence of various alcohols with their solubility in brine.

24

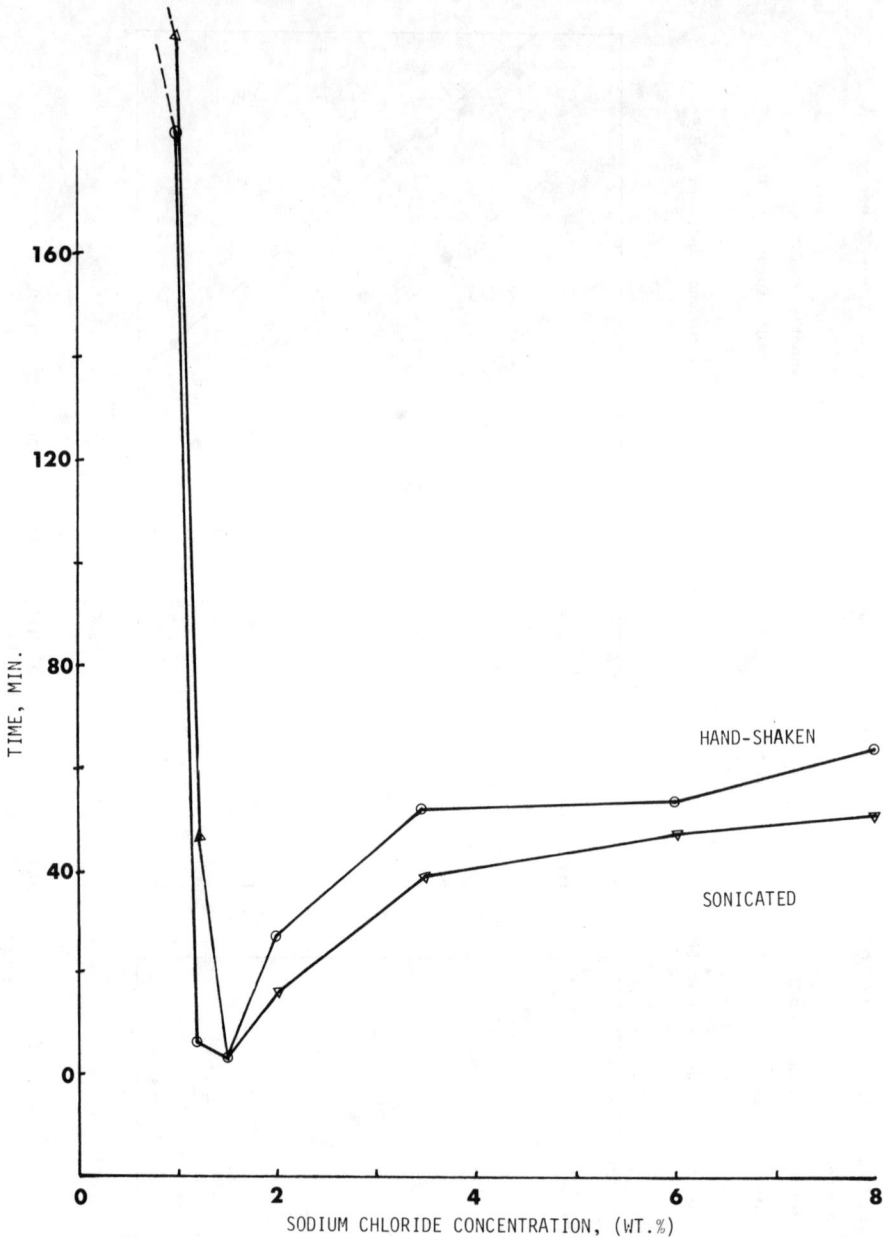

Fig. 29   Effect of salinity on the phase separation or coalescence rate
of sonicated and hand-shaken emulsions.

It is obvious that minimal phase separation time or the fastest coalescence rate occurs at the optimal salinity (54). The rapid coalescence
could contribute significantly to the formation of an oil bank from the
mobilized oil ganglia. This also suggest that at the optimal salinity
the interfacial viscosity must be very low to promote the rapid coalescence.

Figure 30 shows the pressure drop across a porous medium when emulsions prepared at various salinities flow through it. It is evident that
the minimum pressure drop occurs at and around the optimal salinity of
the surfactant formulation. This also suggests that the interfacial tension is an important factor influencing the pressure drop across porous
media (54).

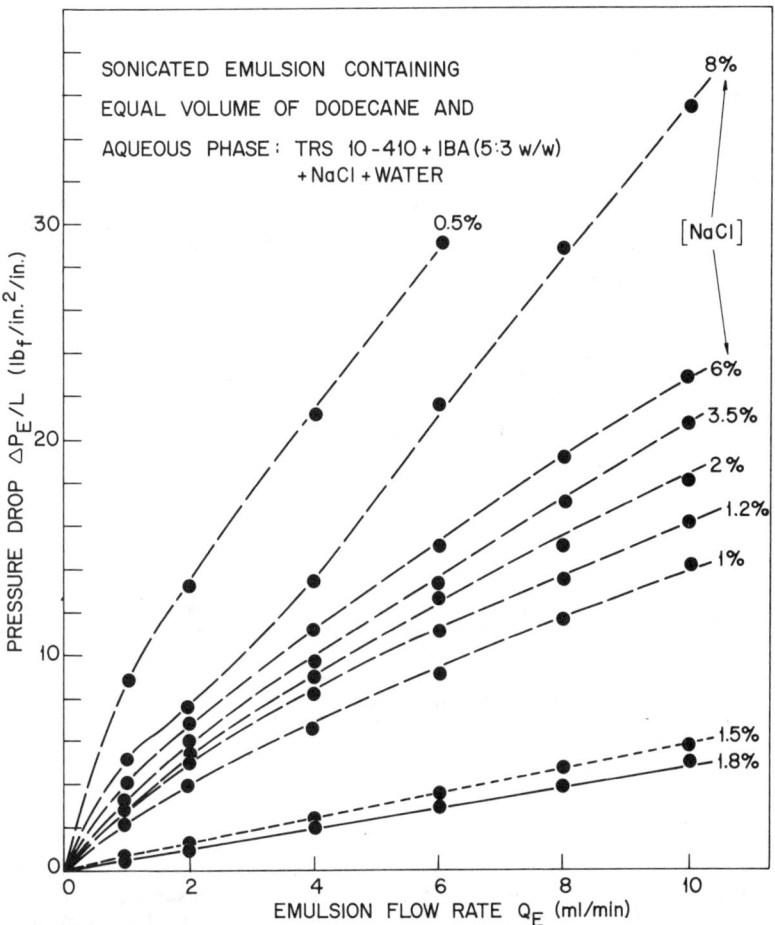

Fig. 30 Effect of salinity on the pressure drop-flow rate curves of
sonicated emulsions.

Figure 31 shows a very interesting and important correlation between the coalescence rate in emulsions and the apparent viscosity in the flow through porous media. The minimum apparent viscosity for the flow of emulsions in porous media coincides with minimum phase separation time at the optimal salinity.

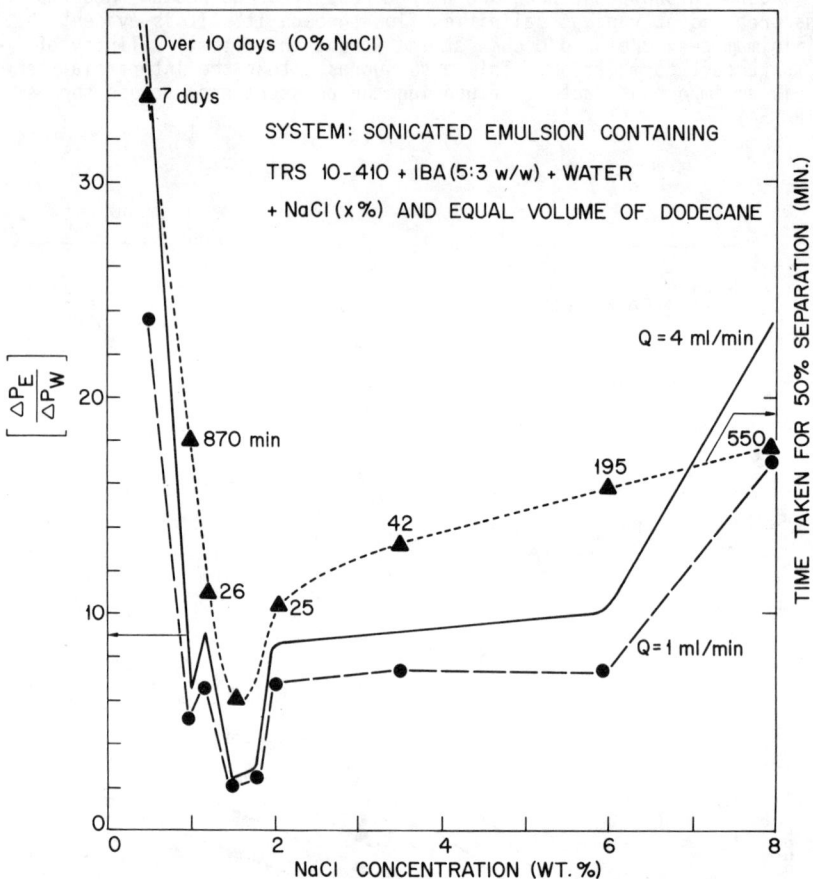

Fig. 31 A correlation between the apparent viscosity and coalescence rate of sonicated emulsions.

This correlation between the phenomena occurring in porous media and outside the porous medium allows us to use coalescence measurements as a screening criterion for many surfactant formulations for their possible behavior in porous media. It is likely that a rapidly coalescing emulsion will give a lower apparent viscosity for the flow in porous media (54).

Figure 32 summarizes all the phenomena occurring at the optimal salinity in relation to enhanced oil recovery by surfactant-polymer flooding.

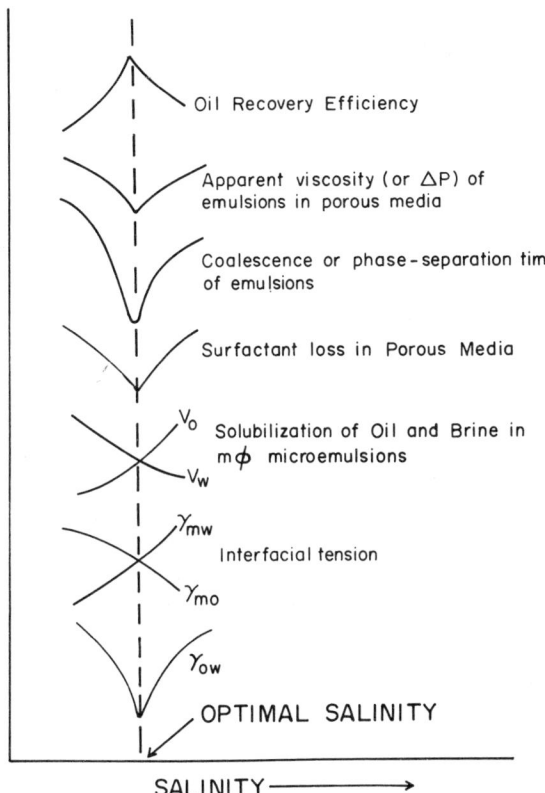

Fig. 32   A summary of various phenomena occurring at the optimal salinity in relation to enhanced oil recovery by surfactant-polymer
flooding.

It is evident that the maximum in oil recovery efficiency correlates well with transient and equilibrium properties of surfactant-oil-brine systems. In our preliminary studies, we have found that the surfactant loss in porous media is also minimum at the optimal salinity presumably due to reduction in the entrapment process for the surfactant phase. Therefore, the maximum in oil recovery at optimal salinity might be a combined effect of all these processes taking place at the optimal salinity.

Since optimal salinity leads to favorable conditions for optimal oil recovery, one would like to design approaches to alter the optimal salinity of a given surfactant formulation (55-57). Figure 33 shows the optimal salinity of a mixed surfactant formulation consisting of a petroleum sulfonate and ethoxylated sulfonate (EOR-200).

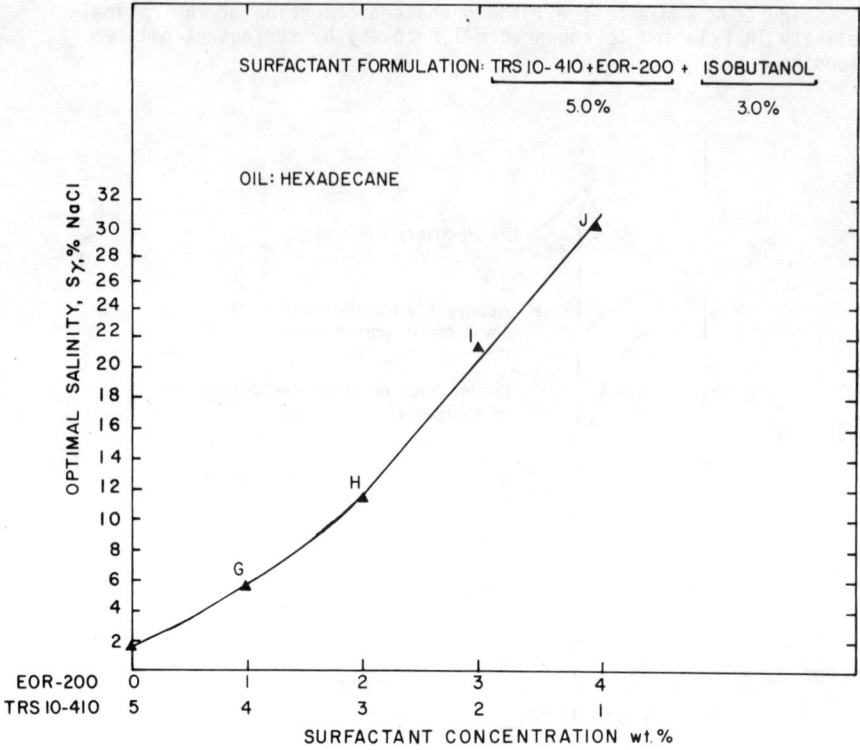

Fig. 33. Increase in the optimal salinity of surfactant formulation by addition of EOR-200.

As one replaces petroleum sulfonates with the ethoxylated sulfonate the optimal salinity increases and can reach as high as 32% NaCl brine. Interestingly, these formulations when equilibrated with oil produced middle phase microemulsions having very low interfacial tension. Thus, the mixed surfactant formulations containing petroleum sulfonates and ethoxylated sulfonates or alcohol are promising candidates for high salinity formulations (55,56).

Figure 34 shows the shape of an oil drop upon contacting a surfactant formulation consisting of 0.05% TRS 10-80 in 1% NaCl. It is evident that as surfactant molecules migrate from the aqueous phase to the interface and subsequently to the oil phase the interfacial tension decreases and the spherical drop gradually flattens out. This flattening

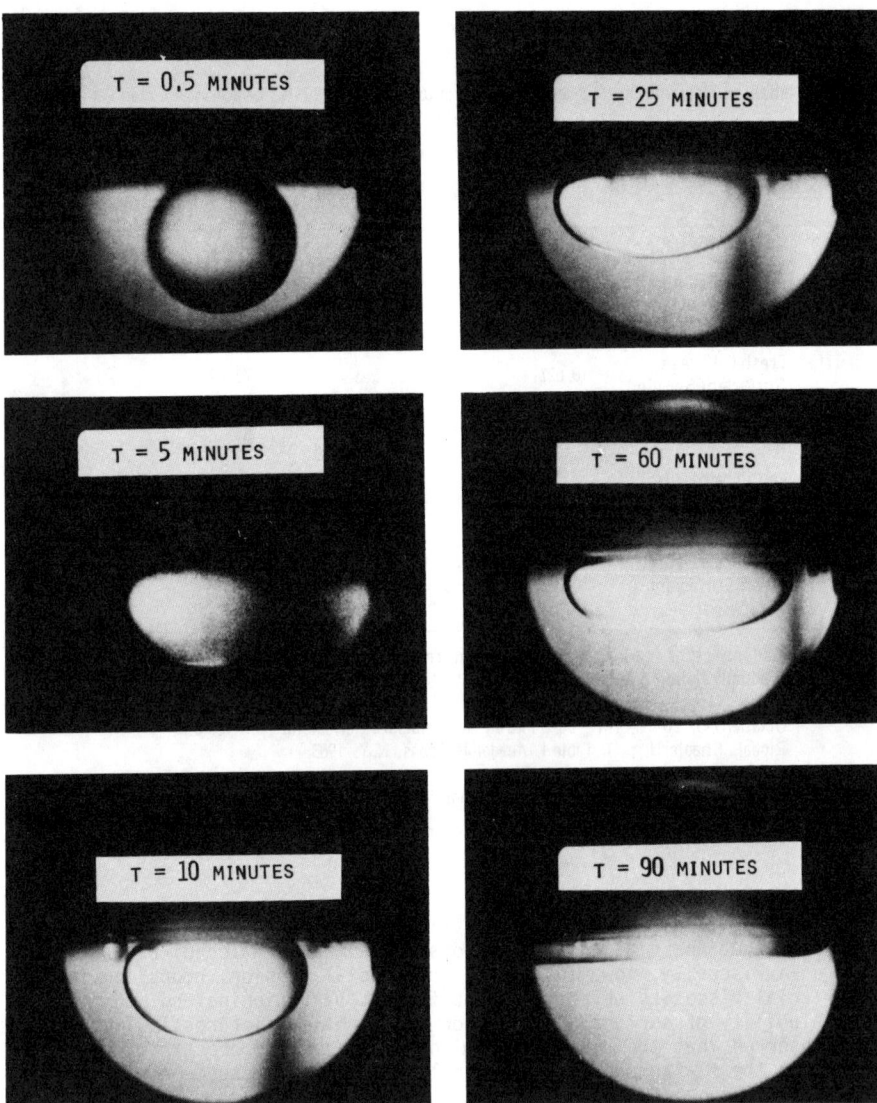

Fig. 34. An illustration of the drop flattening phenomenon for a drop of
octane in an equilibrated solutrion of 0.05% TRS 10-80 in 1%
NaCl.

time reflects the rate at which molecules accumulate at the oil-brine
interface. As shown in Table 1, there is a good correlation between the
flattening time, IFT and the oil recovery. The reduction in the flattening time leads to favorable oil recovery efficiency (16,48).

TABLE 1

IFT, Flattening Time and Oil Recovery Efficiency of 0.05% TRS 10-80 in 1% NaCl vs. n-Octane at 25°C

SYSTEM		IFT (mN/m)	FLATTENING TIME* (seconds)	OIL RECOVERY+ (% OIP)
I.	Fresh Oil/1% NaCl	≈50.8**	∞	61-63
II.	Fresh Oil/Equili-brated Surfactant Solution	0.731	6600	44-52
III.	Fresh Oil/Fresh Surfactant Solution	0.627	480	75-77
IV.	Equilibrated Oil/% NaCl	0.121	900	83
V.	Equilibrated Oil/ Equilibrated Surfac-tant Solution	0.0267	240	94
VI.	Equilibrated Oil/ Fresh Surfactant Solution	0.00209	15	-

*Flattening time is defined as the time required for the n-octane drop to gradually flatten out.

**Octane/H_2O, 20°C, IFT = 50.8 mN/m, "Interfacial Phenomena", Davies and Rideal, Chapter I, p. 17 Table I, Academic Press, N.Y. 1963.

+Sandpack dimensions: 1.06" dia. x 7" long; Permeability= 3 darcy; flow rate: 2.3 ft./day.

In general, a surfactant formulation for enhanced oil recovery in-cludes a short chain alcohol. The possible effect of alcohol can be the change in viscosity, lowering of the interfacial tension, reduction in interfacial viscosity or change in surfactant partitioning and modifying the solubility of surfactant in oil or brine phase. Interestingly, we have observed that the presence of alcohol has a much more striking effect on the flattening time of an oil drop in the presence of a surfac-tant formulation. As shown in Table 2 it compares the many interfacial properties, flattening time and oil recovery efficiency in the presence and absence of alcohol (16). It is evident that the flattening time de-creases strikingly in the presence of alcohol suggesting that the alcohol promotes the mass transfer to the interface and a rapid reduction in the magnitude of the interfacial tension.

There are also time dependent changes in the surface properties of a surfactant formulation. This include the chemical degradation (58,59), or changes in the aggregation process of micelles (60). Several investi-gators have shown that the interfacial tension changes with time (61). We have also shown that using several physical techniques that molecular association also changes with time leading to the aging effects of the surfactant formulation (58). The aging processes may occur over a period of months or years.

Table 2

The Effect of IBA on Flattening Time,
IFT, IFV, Partition Coefficient, and Oil
Displacement Efficiency

SYSTEM	0.1% TRS 10-410 in 1.5% NaCl vs. n-Dodecane	0.1% TRS 10-410 + 0.06% IBA in 1.5% NaCl vs. n-Dodecane	0.05% TRS 10-80 in 1% NaCl vs. n-Octane	0.05% TRS 10-80 + 0.04% IBA in 1% NaCl vs. n-Octane
Run	S100-48	S100-43	S100-02	S100-44
Flattening Time	90 sec	< 1 sec	420 sec	< 1 sec
IFT (dynes/cm)	0.086	0.088	0.025	0.024
Interfacial Viscosity (s.p.)	0.096	0.086	0.023	0.018
Partition Coefficient	0.010	0.009	0.3	1.36
Secondary Recovery				
By Brine Flooding	–	–	61.2%	60.08%
By Surfactant Soln Flooding	84.37%	98.32%	60%	91%
Tertiary Recovery	–	–	0	76.84%
Final Oil Saturation	11.73%	1.28%	30%	5.36%

*All displacement experiments are carried out with nonequilibrated systems in sand packs at 25°C; Dimensions and flow rates same as given in Table 2.

Secondary and tertiary oil recovery values are percent of oil-in-place, whereas final oil saturation is percent of total pore volume.

32

F. SURFACTANT-POLYMER INCOMPATIBILITY

Trushenski (17) has shown that surfactant-polymer incompatibility can lead to a considerable reduction in the efficiency of the process. The surfactant-polymer incompatibility manifests itself as a phase separation and alteration of the viscosity of the separated phases. The entrapment of the high viscosity phase will effectively remove that component from the flooding process. The mixing of the surfactant and polymer in the porous medium occurs due to both dispersion effects as well as excluded volume effects for the flow of polymer molecules in porous media.

Figure 35 shows the effect of mixing surfactant and a polymer solution in the absence of oil.

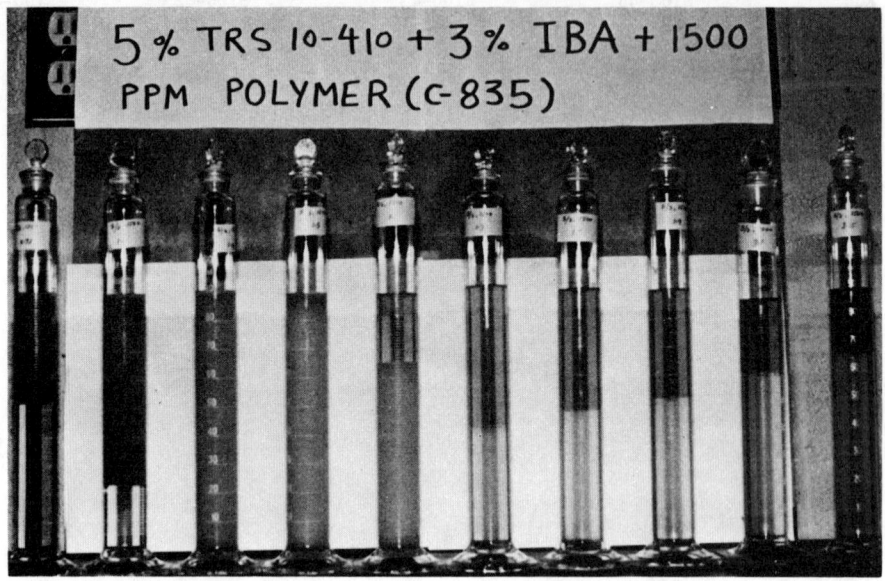

Fig. 35. Effect of addition of polymer on the phase behavior of aqueous surfactant solutions.

It is evident that there are two regions of phase separation, one at low salinity and the other at high salinity separated by a metastable colloidal dispersion. We refer to the separation at the lower salinity as region 1 and those at high salinity as region 2. The separation of a surfactant-rich phase in region 2 is similar to that in coacervation process, whereas the separation of micelles in region 1 is induced by the presence of polymer molecules. The surfactant-polymer incompatibility shows up strikingly in the formation of region 1 (62).

The addition of polymer to an oil/brine/surfactant/alcohol system shows that the formation of middle phase microemulsion is promoted by the presence of polymer (Figure 36). However, the transition middle phase to upper phase microemulsion is not influenced at all by the presence of polymer. We have further shown (62,63) that the optimal salinity is not significantly influenced by the presence of polymer in the oil/brine/surfactant/alcohol system.

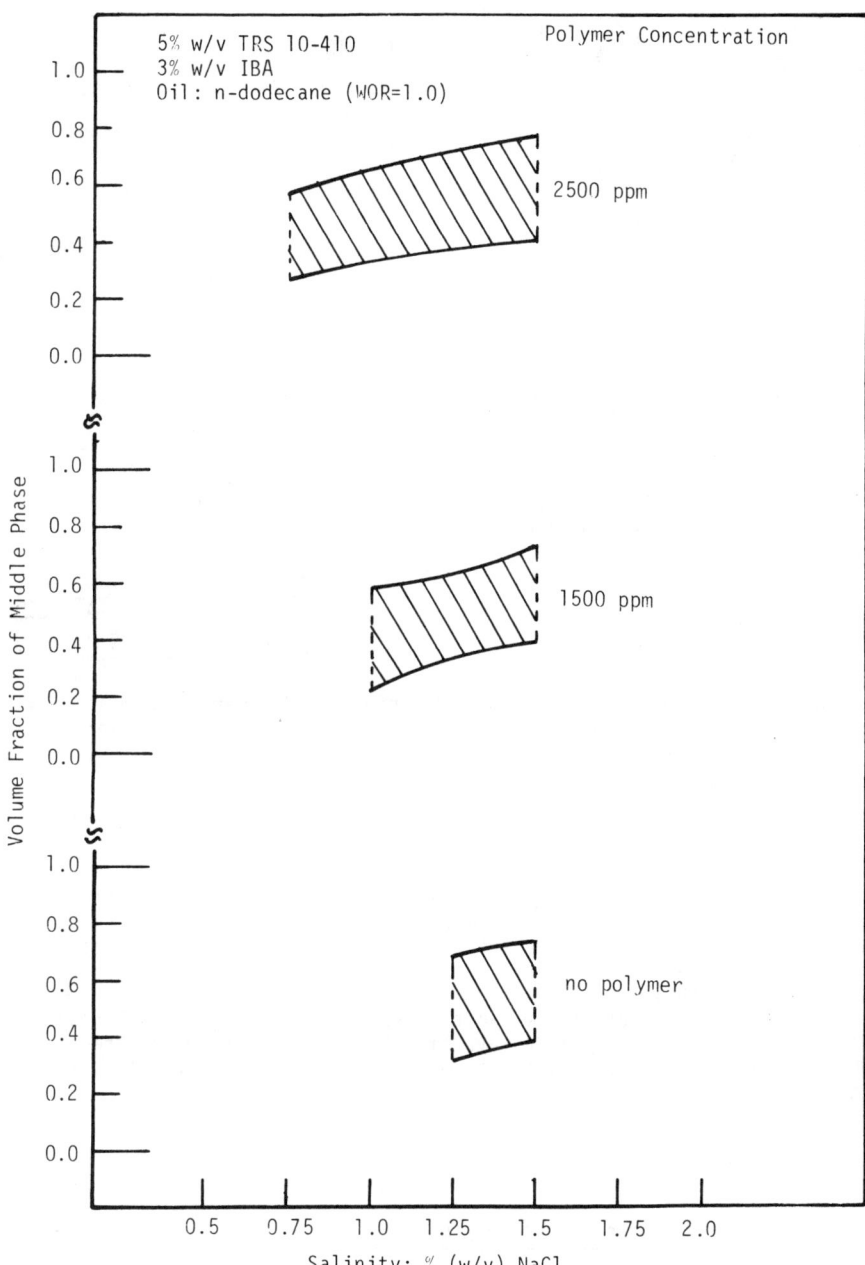

Fig. 36. Effect of polymer concentration on the salinity range for formation of middle-phase microemulsion.

Figure 37 shows the schematic explanation of the surfactant polymer incompatibility and concomittant phase separation.

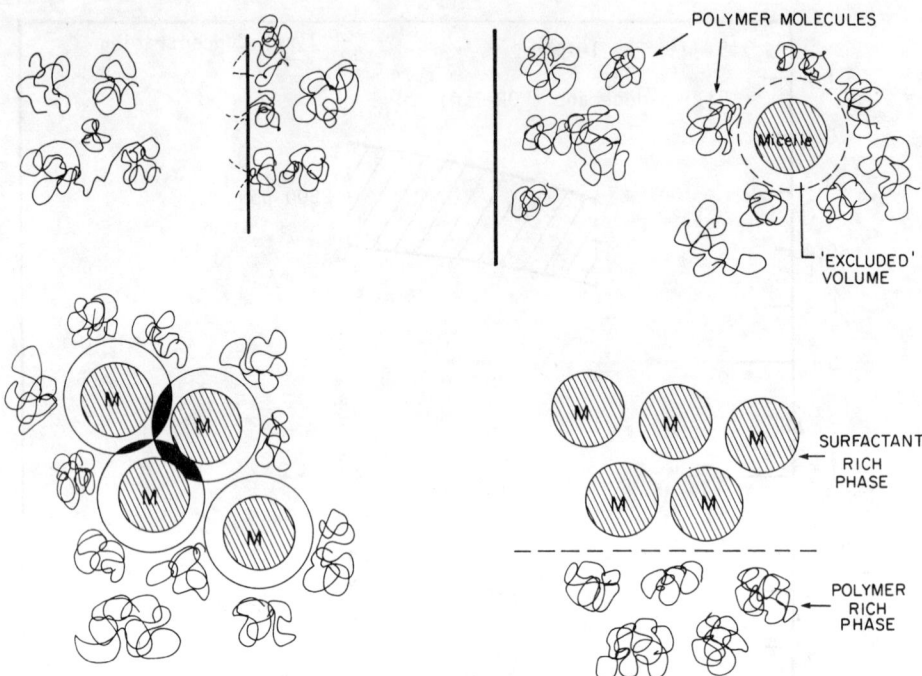

Fig. 37. Schematic illustration of surfactant-polymer incompatibility
leading to phase separation in mixed surfactant-polymer systems.

We propose that around each micelle there is a region of solvent that is
excluded to polymer molecules. However, when these micelles approach
each other there is overlapping of this excluded region. Thus if all
micelles separate out then the excluded region diminishes due to the
overlap of the shell and more solvent becomes available for the polymer
molecules. This effect is similar to what is called polymer depletion
stabilization (61). Thus in a way, this is similar to osmotic effect
where the polymer molecule tends to maximize the solvent for all possible
configurations.

G. NEW CONCEPTS IN EOR PROCESS DESIGN

The formation of an oil bank is a very important event in the surfactant-polymer flooding process. This was established from studies on
the injection of an artificial oil bank followed by the surfactant formulation that can produce ultralow IFT with the injected oil. We found
that the oil recovery increased substantially and the residual oil saturation decreased with the injection of oil bank as compared to the same
studies carried out in the absence of an injected oil bank (48).

Figure 38 schematically shows that the oil bank formation and its propagation is analogous to "snowball effect".

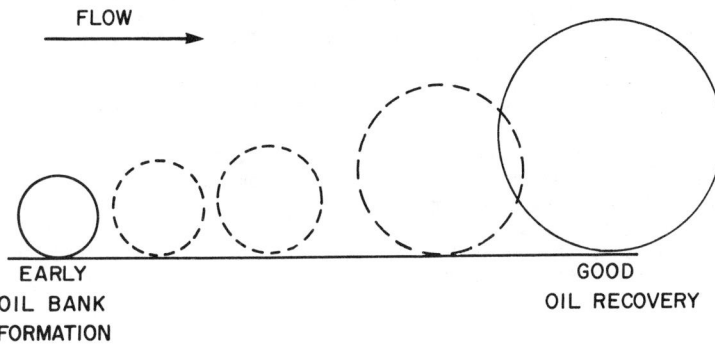

FLOW

EARLY
OIL BANK
FORMATION

GOOD
OIL RECOVERY

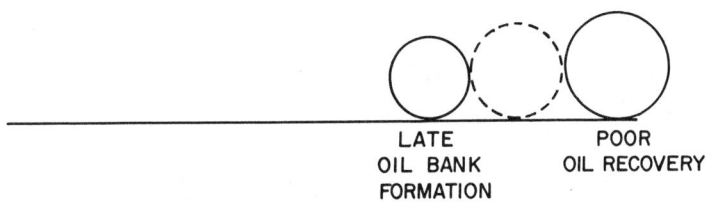

LATE
OIL BANK
FORMATION

POOR
OIL RECOVERY

NO OIL BANK FORMATION

NO
OIL RECOVERY

Fig. 38.   Schematic illustration of the injection of an oil bank and the subsequent "snowball effect" in enhanced oil recovery.

If an early oil bank forms as it moves through the porous medium, it accumulates additional oil ganglia resulting in an excellent oil recovery whereas a late oil bank formation will result in a poor oil recovery.

We have shown that the salinity of polymer solution is far more important than the salinity of connate water (23). When the salinity of polymer solution was at the optimal salinity of the surfactant formulation, high oil recovery efficiency was obtained over a wide range of connate water salinities. Evidence showed that the phase behavior of the surfactant slug in porous media is largely determined by the salinity of the polymer solution (65). For better mobility control and minimal surfactant loss a two-slug design of a surfactant formulation was employed (23). In this design, the first surfactant slug has an optimal salinity close to the connate water salinity and the second surfactant slug has a much lower optimal salinity. The polymer solution salinity is made equal to the optimal salinity of the second surfactant slug. With this design, high oil recovery in berea cores can be obtained even in the presence of high salinity (6% NaCl + 1% calcium chloride) connate water.

The optimal salinity concept is further extended to include the effect of mobility control and surfactant dispersion and entrappment in porous media (65). The proposed salinity shock design of mobility polymer solutions employs two slugs of polymer solutions in which the first polymer slug is at the optimal salinity of the preceeding surfactant formulation and the second polymer slug is at a much lower salinity.

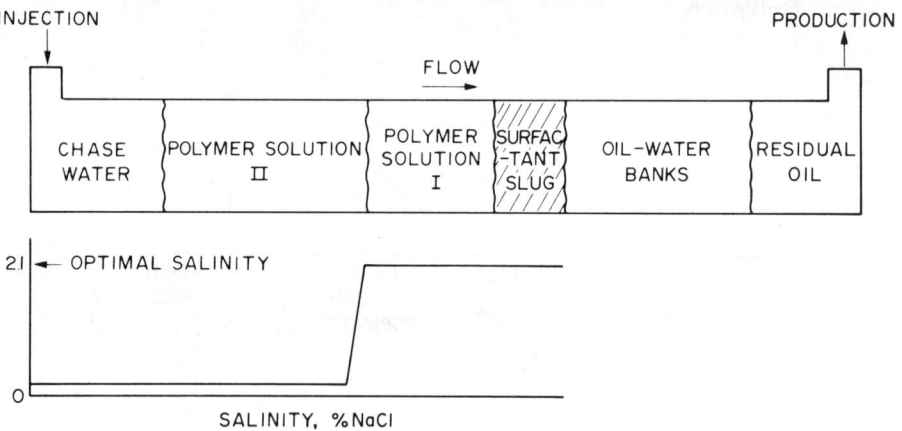

Fig. 39. Schematic representation of the graded-salinity design of polymer buffer solution for enhanced oil recovery.

With this unique design high oil recovery and high surfactant recovery can be obtained for soluble oil flooding in sandpacks, while the polymer consumption can be greatly reduced.

Figure 40 schematically shows our results obtained using the salinity shock design. The optimal salinity for the surfactant formulation used was 2.1% NaCl and the reservoir brine was 3% NaCl plus 1% calcium chloride. By the use of two polymer slugs we were able to obtain in berea cores 88% tertiary oil recovery and 48% surfactant recovery.

For aqueous micellar-polymer flooding with crude oil in Berea cores, it has been shown (66-69) that a contrast salinity design of the preflush micellar-polymer flooding process may produce a better oil recovery than that obtained from a constant salinity process. In the contrast salinity design, the salinity of the preflush water is made higher while the salinity of the polymer solution is made lower than the optimal salinity of the surfactant formulation. The rationale of this design is that an optimal salinity profile can be established in the vicinity of the surfactant slug upon mixing of the injected fluids in the porous medium.

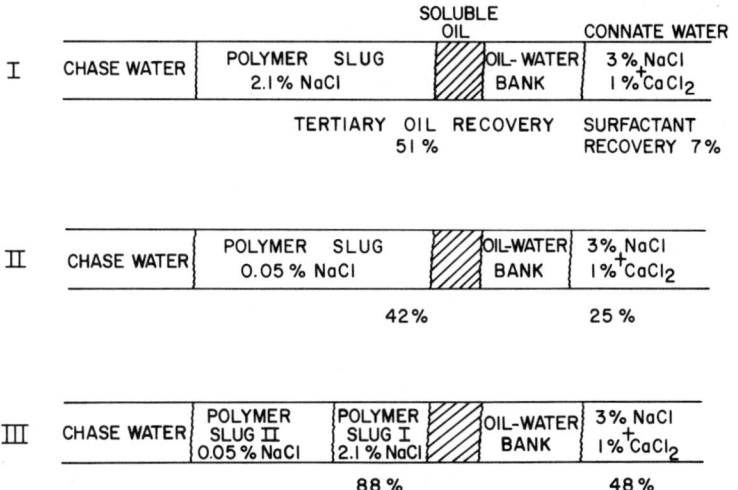

Fig. 40. The effect of salinity shock of polymer buffer solution an oil displacement efficiency and surfactant loss.

It is hoped that the experimental results presented in this paper contribute in a small way to increasing our understanding of phenomena occurring in porous media. It should be emphasized that results we have obtained using laboratory scale experiments are neither conducted nor intended to be extrapolated to the oil field processes. It is recognized that the processes occuring in petroleum reservoirs are far more complex than those that we can design and control using a laboratory setup.

ACKNOWLEDGEMENTS

The author wishes to express his sincere thanks and appreciation to the National Science Foundation - RANN, ERDA and the Department of Energy (Grant No.· DE-AC1979BC10075) and the consortium of the following Industrial Associates for their generous support of the University of Florida Enhanced Oil Recovery Research Program during the past seven years: 1) Alberta Research Council, Canada, 2) American Cyanamid Co., 3) Amoco Production Co., 4) Atlantic-Richfield Co., 5) BASF-Wyandotte Co., 6) British Petroleum Co., England, 7) Calgon Corp., 8) Cities Service Oil Co., 9) Continental Oil Co., 10) Ethyl Corp., 11) Exxon Production Research Co., 12) Getty Oil Co., 13) Gulf Research and Development Co., 14) Marathon Oil Co., 15) Mobil Research and Development Co., 16) Nalco Chemical Co., 17) Phillips Petroleum Co., 18) Shell Development Co., 19) Standard Oil of Ohio Co., 20) Stepan Chemical Co., 21) Sun Oil Chemical Co. 22) Texaco, Inc., 23) Union Carbide Corp., 24) Union Oil Co., 25) Westvaco, Inc., 26) Witco Chemical Co., and the University of Florida. He also wishes to convey his sincere thanks to his colleagues in Chemical Engineering, Petroleum Engineering and Institute for Energy Studies of Stanford University for their collaboration during his stay at Stanford University.

REFERENCES

1. "Improved Oil Recovery by Surfactant and Polymer Flooding", D.O. Shah
    and R.S. Schechter eds., Acad. Press, Inc., N.Y. (1977).

2. "Surface Phenomena in Enhanced Oil Recovery", D.O. Shah, ed.
    Proceedings of the Symposium on Enhanced Oil Recovery, Stockholm,
    Sweden, Aug. 20-25, 1979, Plenum Publishing Company, N.Y. (in
    press).

3. "SPE Improved Oil Recovery Reprints," Proceedings of Symposium on
    Improved Oil Recovery, Tulsa, OK., April 16-19, 1972.

4. "Proceedings of the SPE-AIME Symposium on Improved Oil Recovery",
    Tulsa, OK, April 22-24, 1974.

5. "Proceedings of the SPE-AIME Symposium on Improved Oil Recovery",
    Tulsa, OK, March 22-24, 1976.

6. "Proceedings of the SPE-AIME Symposium on Improved Oil Recovery",
    Tulsa, OK, April 16-19, 1978.

7. "Proceedings of the SPE International Symposium on Oilfield and
    Geothermal Chemistry," Houston, TX, Jan. 22-24, 1979.

8. "Proceedings of the SPE-AIME Symposium on Improved Oil Recovery",
    Tulsa, OK, April 20-23, 1980.

9. "Proceedings of the European Symposium on Enhanced Oil Recovery,
    Edinburgh, Scotland, July 5-7, 1978.

10. "Proceedings of the 1980 SPE Internatinal Symposium on Oilfield and
    Geothermal Chemistry, Stanford, CA., May 28-30, 1980.

11. "Secondary and Tertiary Oil Recovery Processes," Interstate Oil
    Compact Commission, Oklahoma City, OK (1974).

12. Taber, J.J., "Research on Enhanced Oil Recovery: Past, Present and
    Future" in "Surface Phenomena in Enhanced Oil Recovery", pp. 13-53,
    D.O. Shah, ed., Plenum Publishing Co., N.Y. (in press).

13. Foster, W.R., J. Pet. Tech., 25, 205 (1973).

14. Bansal, V.K. and Shah, D.O., in "Micellization, Solubilization, and
    Microemulsions", Vol. I pp. 87-113, K.L. Mittal, ed., Plenum Press,
    N.Y. (1977).

15. Pasquarelli, C.H. and Wasan, D.T., in "Surface Phenomena in Enhanced
    Oil Recovery", pp. 237-248, D.O. Shah, ed., Proceedings of the Sym-
    posium on Enhanced Oil Recovery, Stockholm, Sweden, Aug. 20-25,
    1979, Plenum Publishing Company, NY. (in press).

16. Chiang, M.Y. and Shah, D.O., "The Effect of Alcohol on Surfactant
    Mass Transfer across the Oil/Brine Interface and Related Phenome-
    na," SPE 8988, presented at the SPE 5th Intl. Sympo. on Oilfield
    and Geothermal Chemistry, Stanford, CA, May 28-30, 1980.

17. Trushenski, S.P., in "Improved Oil Recovery by Surfactant and Polymer Flooding", D.O. Shah and R.S. Schechter, eds., Acad. Press, Inc., N.Y., (1977).

18. Chiang, M.Y., Chan, K.S. and Shah, D.O., J. Can. Pet. Tech., 17(4), 1 (1978).

19. Chan, K.S. Ph.D. Dissertation, University of Florida (1978).

20. Wade, W.H., Vasquez, E., Salager, J.L., El-Emory, M., Koukounis, C. and Schechter, R.S., in "Solution Chemistry of Surfactants", Vol. II, pp. 801-816, K.L. Mittal, ed., Plenum Press, N.Y. (1979).

21. Noronha, J.C., Ph.D. Dissertation, University of Florida (1980).

22. Chan, K.S. and Shah, D.O., J. Disp. Sci., Tech. 1(1), 55 (1980).

23. Chou, S.I., Ph.D. Dissertation, University of Florida, (1980).

24. Hsieh, W.C., Ph.D. Dissertation, University of Florida (1977).

25. Bansal, V.K., Chan, K.S. McCallough, R. and Shah, D.O., J. Can. Pet. Tech., 17(1), 1 (1978).

26. Shah, D.O., Chan, K.S. and Giordano, R.M., in "Solution Chemistry of Surfactants", Vol. I pp. 391-406, K.L. Mittal, ed., Plenum Press, N.Y. (1977).

27. Rashid, S.N., in "University of Florida Research on Surfactant-Polymer Oil Recovery Systems - Annual Report" pp. I1-I4, Dec. 1979.

28. Robbins, M.L., "Theory of the Phase Behavior of Microemulsions", SPE 5839, presented at the SPE Improved Oil Recovery Symposium, Tulsa, OK March 22-24 (1976).

29. Healy, R.N. and Reed, R.L., SPE J., 491 (Oct. 1974).

30. Healy, R.N. and Reed, R.L., SPE J., 147 (June 1976).

31. Shah, D.O. and Hamlin, R.M. Jr., Science, 171, 483 (1971)

32. Shah, D.O., Tamjeedi, A., Falco, J.W. and Walker, R.D. Jr., AIChE J., 18(6) 1116 (1972).

33. Falco, J.W., Walker, R.D., Jr. and Shah, D.O., AIChE J., 20(3), 510 (1974).

34. Shah, D.O., Walker, R.D., Hsieh, W.C., Shah, N.J. Dwivedi, S., Nelander, J., Pepinsky, R. and Deamer, D.W., SPE 5815 presented at the SPE Improved Oil Recovery Symposium, Tulsa, OK, March 22-24, 1976.

35. Bansal, V.K., Chinnaswamy, K., Ramachandran, C. and Shah, D.O., J. Colloid Interface Science, 72(3), 524 (1979).

36. Bansal, V.K., Shah, D.O. and O'Connell, J.P., J. Colloid Interface Science, 75(2), 462 (1980)

37. Chou, S.I. and Shah, D.O., J. Colloid Interface Sci., 78(1), 249 (1980).

38. Chou, S.I. and Shah, D.O., J. Colloid Interface Sci., 80(1), 49 (1981).

39. Chou, S.I. and Shah, D.O., J. Colloid Interface Sci., 80(2), 311 (1981).

40. Scriven, L.E., Nature, 263, 123 (1976).

41. Scriven, L.E., in "Micellization, Solubilization and Microemulsions", ed. K.L. Mittal, Vol. II, pp. 877, Plenum Press, N.Y. (1977).

42. Ramachandran, C., Vijayan, S. and Shah, D.O., J. Phys. Chem., 84, 1561 (1980).

43. Hwan, R., Miller, C.A. and Fort, T. Jr., J. Colloid Interface Sci., 68, 221 (1979).

44. Shinoda, K. J. Colloid Interface Sci., 24, 4 (1967).

45. Shinoda, K. and Saito, H., J. Colloid Interface Sci., 26, 70 (1968).

46. Friberg, S., Lapczynska, I. and Gilberg, G., J. Colloid Interface Sci., 56, 19 (1976).

47. Miller, C.A., Hwan, R., Benton, W.J. and Fort, T. Jr., J. Colloid Interface Sci., 61(3), 554 (1977).

48. Chiang, M.Y., Ph.D. Dissertation, University of Florida (1978).

49. Hsieh, W.C. and Shah, D.O., "The Effect of Chain Length of Oil and Alcohol as well as Surfactant to Alcohol Ratio on the Solubilization, Phase Behavior and Interfacial Tension of Oil/Brine/Surfactant/Alcohol Systems",SPE 6594, presented at the 1977 SPE-AIME Intl. Symposium on Oilfield and Geothermal Chemistry, LaJolla, CA, June 27-28, 1977.

50. Morgan, J.C., Schechter, R.S. and Wade, W.H., in "Improved Oil Recovery by Surfactant and Polymer Flooding", D.O. Shah and R.S. Schechter, eds., Acad. Press, Inc., N.Y. (1977).

51. Cash, R.L., Cayias, J.L., Fournier, G., Jacobson, J.K., Schares, T., Schechter, R.S. and Wade, W.H., "Modeling Crude Oils for Low Interfacial Tension", SPE 5813, presented at the SPE Symposium on Improved Oil Recovery, Tulsa, OK, March 22-24, 1979.

52. Satter, S.J., "The Influence of Type and Amount of Alcohol on Surfactant-Oil-Brine Phase Behavior and Properties," SPE 6843, presented at the 52nd Annual Fall Conference and Exhibition of SPE-AIME, Denver Co., Oct. 9-12, 1977.

53. Puerto, M.C. and Gale, W.W., "Estimation of Optimal Salinity and Solubilization Parameters for Alkyl Orthoxylene Sulfonate Mixtures", SPE 5814, presented at the SPE Improved Oil Recovery Symposium, Tulsa, OK March 22-24, 1976.

54. Vijayan, S., Ramachandran, C., Doshi, H. and D.O. Shah, in "Surface Phenomena in Enhanced Oil Recovery", D.O. Shah ed., pp. 327-376, Plenum Publishing Co., N.Y. (in press).

55. Bansal, V.K. and Shah, D.O., J. Colloid Interface Sci., 65(3), 451 (1978).

56. Bansal, V.K. and Shah, D.O., SPE J., 167 (June, 1978).

57. Bansal, V.K. and Shah, D.O., J. Am. Oil Chemists Soc., 55 (3), 367 (1978).

58. Vijayan, S., Ramachandran, C., and Shah, D.O., J. Am. Oil Chemists Soc., 58(4), 566 (1981).

59. Vijayan, S., Ramachandran, C. and Shah, D.O., J. Am. Oil Chemists Soc., 58(6), 746 (1981).

60. Vijayan, S., Ramachandran, C. and Doshi, H., "University of Florida Research on Chemical Oil Recovery Systems Semi-Annual Report", pp. B11-B57, June, 1978.

61. Cash, R.L., Cayias, J.L., Hayes, M., McAllister, D.J. Schares, T. and Wade, W.H., J. Pet. Tech., 985 (Sept. 1976).

62. Desai, N.N., in "University of Florida Research on Surfactant-Polymer Oil Recovery Systems-Annual Report", pp. I27-I49, Dec. 1979.

63. Desai, N.N., in "University of Florida Research on Surfactant-Polymer Oil Recovery Systems-Annual Report", pp. I35-I-48, Dec. 1980.

64. Hesselink, F. Th. and Faber, M.J., in "Surface Phenomena in Enhanced Oil Recovery," D.O. Shah, ed., pp. 861-869, Plenum Publishing Co., N.Y. (in press).

65. Chou, S.I. and Shah, D.O., in "Surface Phenomena in Enhanced Oil Recovery", D.O. Shah, ed., pp. 843-860, Plenum Publishing Co., N.Y. (in press)

66. Paul, G.W. and Froning, H.R., J. Pet. Tech., 25, 957 (1973).

67. Gupta, S.P. and Trushenski, S.P., SPE J., 19, 116 (1979).

68. Nelson, R.C., "The Salinity Requirement Diagram-A Useful Tool in Chemical Flooding Research Development", SPE 8824, presented at the SPE Improved Oil Recovery Symposium, Tulsa, OK, April 20-23, 1980.

69. Hirasaki, G.J., Van Damselaar, H.R. and Nelson, R.C., "Evaluation of the Salinity Gradient Concept in Surfactant Flooding", SPE 8825, presented at the SPE Improved Oil Recovery Symposium, Tulsa, OK, April 20-23, 1980.

SURFACTANTS FOR ENHANCED OIL RECOVERY PROCESSES IN HIGH-SALINITY SYSTEMS – PRODUCT SELECTION AND EVALUATION –

M. H. AKSTINAT

Institute of Petroleum Engineering,
Technical University of Clausthal, West Germany

ABSTRACT

Under consideration of high-salinity brines and different compositions of crude oils a standard test program for the rapid selection of surfactants was developed. By means of this program for testing surface-active products approximately 12 00 surfactants of different classes have been tested for their suitability as agents for enhanced oil recovery (EOR). From the results of these investigations,systematic correlations between chemical structure, product class and oil composition (naphthenic, aromatic, paraffinic) can be derived.
Especially the properties of mixed surfactants (anionic-nonionic) are discussed in detail ,with respect to temperature dependent interfacial phenomena, sorption processes and chromatographic separation. The research work led to the surfactant classes of the Ethercarboximethylates and Ethersulfonates.

INTRODUCTION

In the so-called primary oil recovery phase (natural flow), the yield is generally only 10-20 % of the original oil in place (OOIP). By secondary measures, e.g., water flooding, the yield can be increased to 20-30 % OOIP. By application of enhanced oil recovery (EOR) methods, e.g., surfactant flooding, an additional 10-20 % increase in yield is possible, depending on the conditions of the oil field and the type of oil. For less complicated reservoir conditions (low reservoir temperature, low salinity of brine)surfactants for EOR processes are already known. The aim of these investigations is to find suitable surfactants for application in reservoirs with
- high salinity of the reservoir brine ($100 < c_s < 250$ g.l$^{.1}$ salt concentration) and
- elevated reservoir temperatures ($40 < \theta < 80^{o}C$)

Moreover,the products selected should be economical to use and technically simple to handle /11/.
The working concept can be characterized by the following aspects:
- Preservation of the reservoir properties
- Development of appropriate rapid test methods:
  Because of the multitude of products available, rapid test programs must be developed, in order to permit the selection of suitable products in the shortest possible time
- Consideration of boundary conditions for specific groups of oil reservoirs:

The characteristic data pertaining to a specific group of reservoirs must be evaluated in order to provide a representative
survey of the boundary conditions required for both the rapid
test methods and the preliminary flooding experiments.
- Conclusions with the most general applicability:
  Since there should be no restriction to one oil reservoir only,
  the results should be of the most general possible validity,
  and surfactant solutions with a broad range of application should
  be sought. Hence, maximal demands should be imposed on the flooding media, in order to ensure test results which are applicable to as many oil reservoirs as possible.

TEST PROGRAM/STANDARDIZATION

Reproducible test conditions are always required for investigating surfactants, in order that various products be appraised and
compared. Such conditions can be fulfilled only by model systems,
since the properties of real systems (reservoir water, crude oil,
 reservoir rock) are usually subject to variations.
Hence a test program requires the designing of model systems, in
which as many parameters of real systems as possible are considered. Since surfactants are generally dissolved in water, the total
salinity and the composition of reservoir brine are of utmost importance for the selection of suitable surfactants.
Based on several hundred chemical analyses of water samples from
German oil reservoirs, a classification into three brine categories was possible:
- type AM (low salinity) TDS $\leqslant 10$ g.l^{-1}
- type BM (intermediate salinity) 10 g.$l^{-1} \leqslant$ TDS < 165 g.l^{-1}
- type CM (highly saline) TDS $\geqslant 165$ g.l^{-1}

Observed significant characteristics of highly saline reservoir
brines are that
- many salts occur in the dissolved state at concentrations exceeding their usual solubility products
- all water samples contain heavy metal ions, such as Fe^{2+}
- the pH-values of brines BM and CM lie in a relatively acidic
  range (3,0-6,5)
- all brines show comparatively high sulfate contents
- all brines contain large quantities of Ca^{2+} and Mg^{2+} ions and
  lower concentrations of Sr^{2+} and Ba^{2+}.

Trace elements ($\geqslant 10$ mg.l^{-1}) were not considered. At reservoir pressure between 50 and 100 bar and reservoir temperatures between 40
and 80°C about 4 g of CO_2 dissolves in 100 g of water. However,
above 6 bar the pH-value of water containing CO_2 already tends toward a constant value of 3,3 /15/, which probably also dominates
in most reservoir brines of type CM.
For the surfactant investigations a statistical composition was
ascertained for a highly saline model reservoir water CM (table 1).
This standardized brine CM was employed for all subsequent tests,
unless otherwise indicated.
The primary screening criteria of surfactants for EOR processes
in highly saline systems may be listed as follows /10/:
- solubility in reservoir brine (TDS > 165 g.l^{-1})
- long-term stability in the temperature range of 30-80°C
- low interfacial tensions in the system brine/crude oil
  ($\gamma < 1$ mN.m^{-1})

Table 1: Composition of synthetic reservoir brine CM

Salt	Concentrations in mg.l^{-1}
NaCl	165 000
$CaCl_2$. $6H_2O$	49 349 (25 000)*
$MgCl_2$. $6H_2O$	12 810 (6 000)*
KCl	750
KBr	400
KJ	20
LiCl	100
NH_4Cl	350
$SrCl_2$. $6H_2O$	1 681 (1 000)*
$BaCl_2$. $2H_2O$	58 (50)*
$NaHCO_3$	650
Na_2SO_4 . $10H_2O$	680 (300)*
CO_2 injected for at least 1 h	
$NaCO_3$. $4H_2O$	523 (250)*
$FeSO_4$. $9H_2O$	366 (200)*

TDS: 200 070 mg.l^{-1}

*The concentration in parentheses refers to the quantity without water of crystallization

Based on these fundamental requirements a standard test program for surfactants was developed (see fig. 1).

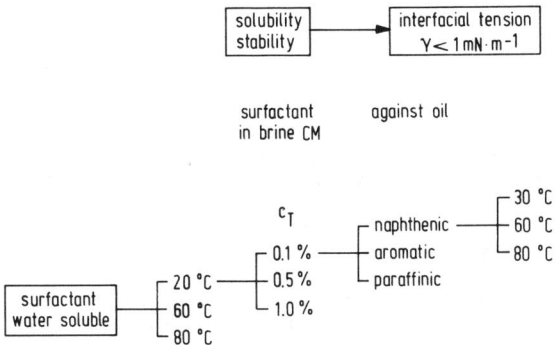

Figure 1: Test program for water-soluble surfactants

Besides the aforementioned criteria suitable surfactants should show further,
- low adsorption on reservoir rock
- favourable partition coefficients and
- broad range of application.

It is also necessary for the determination of the parameters mentioned already to standardize the crude oil and the reservoir rock.
Since it was impossible to find a representative crude oil, and
since the influence of the typical basic oils should be determined too, three oils (PAE-laboratory of German Shell AG, Hamburg)
were used as standards for the investigations:
- paraffinic oil P (683/76)
- naphthenic oil N (655/76)
- aromatic oil A   (671/76)
As a standard for the reservoir rock (adsorption measurements) a
pure quartz sand, type P 0,03-0,15 (V. Busch, Schnaittenbach/West
Germany), was used.

PRIOR CONSIDERATIONS /16/

Which path should the chemist   take prior to testing to find effective surfactants for enhanced oil recovery? The following systematic procedure seems to be a successful way:
- fundamental selection criteria must be identified and considered (from the standpoint of application technology and economics)
- commercially available surfactants should be tested by classes.

First the generally known range of surfactants should be arranged
according to functional groups (fig. 2).

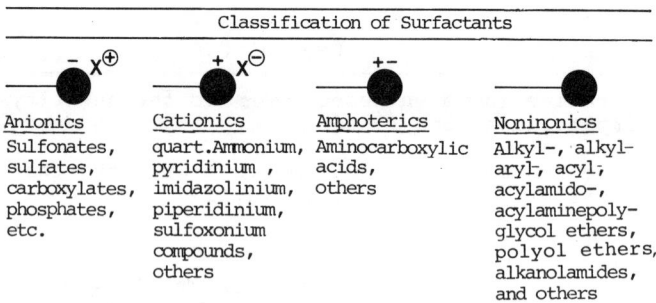

Classification of Surfactants

Anionics	Cationics	Amphoterics	Noninonics
Sulfonates, sulfates, carboxylates, phosphates, etc.	quart.Ammonium, pyridinium , imidazolinium, piperidinium, sulfoxonium compounds, others	Aminocarboxylic acids, others	Alkyl-, alkyl-aryl-, acyl-, acylamido-, acylaminepoly-glycol ethers, polyol ethers, alkanolamides, and others

Figure 2: Surfactant classes

However, this arrangement,without further information from the
user,still does not offer an adequate basis for the selection of
surfactants for EOR-processes.
The first important criterion may be the commercial availability
of surfactants with consideration of the allowable costs of oil
recovery. This will already limit the number of surfactants that
can be considered for EOR processes.
There remain about 14 important and commercially readily available
surfactant groups (table 2).

FUNDAMENTAL SELECTION CRITERIA FOR SURFACTANTS
PRIOR TO TESTING

The following parameters must be considered in selecting surfactants for EOR processes prior to testing:

Table 2: Commercially available surfactant groups

Structure		Designation
Ionic Surfactants		
$R-CH_2-COONa$	$R=C_{10-16}$	Na-salts of fatty acids
$R-C_6H_4-SO_3Na$	$R=C_{11-13}$	Alkylbenzenesulfonate
$\underset{R}{\overset{R}{>}}CH-SO_3Na$	$R+R'=C_{11-17}$	Alkane sulfonate
$R-CH_2-CH=CH-CH_2-SO_3Na +$ $R'-CH_2-CH-(CH_2)n-CH_2-SO_3Na$ $\quad\quad\quad\overset{|}{OH}$	$R=C_{10-14}$	α-olefinsulfonate Hydroxyalkanesulfonate as byproduct
$R-CH-C\overset{\nearrow O}{\underset{\searrow OCH_3}{}}$ $\overset{|}{SO_3Na}$	$R=C_{14-16}$	α-Sulfo fatty acid ester
$R-CH_2-O-SO_3Na$	$R=C_{11-17}$	Fatty alcohol sulfate
$\underset{R}{\overset{R'}{>}}CH-O-(C_2H_4O)_2-SO_3Na$	1. $R'=H,R=C_{11-13}$ 2. $R+R'=C_{10-14}$	Fatty alcohol ether sulfate*
$R-CH_2-O-(C_2H_4O)n-CH_2COONa$	$R=C_{11-13}$ $n = 4,5$	Fatty alcohol ethoxylate acetate**
$\underset{R^2}{\overset{R^1}{>}}\overset{+}{N}\underset{R^4}{\overset{R^3}{<}} \quad Cl^\ominus$		Quaternary ammonium salts
NonIonic Surfactants		
$\underset{R}{\overset{R'}{>}}C-O-(C_2H_4O)n-H$	1. $R=C_{8-18}$ bei $R'=H$ $\quad n=3-15$ 2. $R+R'=C_{10-14}$ $\quad n=3-12$	Primary or secondary alcohol ethoxylates
$R-C_6H_4-O-(C_2H_4O)n-H$	$R=C_{6-12}$ $n=7-10$	Alkylphenol ethoxylates
$\overset{CH_3}{\underset{\overset{|}{CH_3}}{C_{12}H_{25}-\overset{|}{N}\rightarrow O}}$		Amine oxides
Ampholytic Surfactants		
$R^1-\overset{R^2}{\underset{R^3}{\overset{|}{N}}}-(CH_2)_3-SO_3^\ominus$		Sulfobetains
$R^1-\overset{R^2}{\underset{R^3}{\overset{|}{N}}}-CH_2-COO^\ominus$		Betains

* Fatty alcohol polyethylene glycol ether sulfate, Na-salt
** Fatty alcohol polyethylene glycol ether carboxymethylate, Na-salt

- type of crude (paraffinic, naphthenic, aromatic or mixed type)
- colloidal chemistry of crude (content of asphaltenes, resins, etc.)
- adsorption phenomena (composition of reservoir rock)
- characteristics of reservoir environment (pH, temperature, wetting conditions, salinity)
- diffusion phenomena (rapid diffusion to the O/W-interface).

The importance of the individual parameters can vary greatly depending on the conditions of application, and cannot be generalized. For this reason, two complex parameters will be discussed in detail.

DIFFUSION PHENOMENA

It is known from studies on O/W systems that the diffusion is dependent, among others, on the structure of surfactants /1/. Yet, the diffusion coefficient of surfactants itself is of decisive importance. In general, the following relationships apply:

- the diffusion coefficient decreases with increasing degree of
  alkoxylation
- the diffusion coefficient increases with increasing concentration of surfactant (up to c.m.c.)
- the diffusion coefficient is directly proportional to temperature
- the diffusion coefficient for dissolved surfactants is inversely
  proportional to the viscosity of the solvent
- branched block copolymers diffuse more readily than long-chain
  linear types.

The structure of the polyether chains of synthetic surfactants can
also be of importance for diffusion processes. It is well known
that polyether chains, depending on the degree of alkoxylation,
can exist in the so-called zig-zag form or in the meander form (see
fig. 3 /2/).

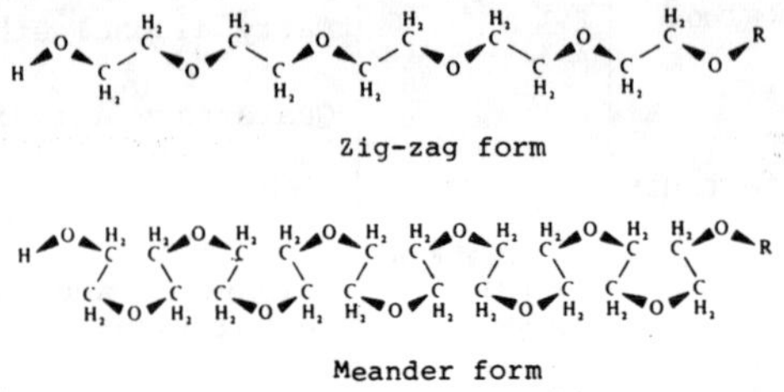

Zig-zag form

Meander form

Figure 3: Shapes of polyether chains /2/

With increasing EO number, the width/length coefficient of the nonionics increases, and diffusion coefficient thus decreases. By
blocking the polyether oxygen for hydration as a result of $O{\to}CH_2$
dipole forces, a change can also occur in the cloud points, the
critical micelle formation concentration (c.m.c.), and thus the
interfacial activity or solubility behavior /3/.

ADSORPTION

For the question of adsorption phenomena as a function of surfactant structure or reservoir rock, numerous findings are of importance /1, 9,12, 14/.

GENERAL CRITERIA

Surfactants

- Amphiphatic surfactants are readily adsorbed on hydrophobic rock
  surfaces, depending on their structure
- The greater the solubility of a surfactant, the smaller is its
  adsorption (greatest adsorption of surfactant occurs in high-
  salinity water because of diminished solubility)
- With increasing temperature and viscosity of the solvent adsorption decreases
- With increasing surfactant concentration adsorption increases

AIM: Low total adsorption but high rate of adsorption up to the sa-
     turation concentration

Reservoir system

- Hydrophilic easily water-wettable rocks: quartz, clay
- Hydrophobic, poorly water-wettable rocks: carbonates

SPECIAL CRITERIA

Surfactants

- Total adsorption decreases with increasing molecular mass of surfactant (the total area accessible to adsorption becomes smaller)
- Nonionic surfactants /9/ are adsorbed mostly in unimolecular layers, adsorption decreases with increasing EO degree, but adsorption increases with increasing length of the hydrocarbon chain; derivatives with an aliphatic hydrocarbon chain are more strongly adsorbed than derivatives with an aromatic hydrocarbon chain
- Ionic surfactants are adsorbed for the most part in polymolecular layers (cationics: about 250 layers).
- Limiting concentration: synthetic surfactants (0.05-0.07 %) $\ll$ natural surfactants (0.25 %).

Reservoir rock

- Clay: adsorption of unsaturated hydrocarbons (in part polymerized)>aromatics >naphthenates >alkanes. Cationics $\gg$nonionics > anionics.
- Silicates: slight adsorption of nonionics (oil-wetted > water-wetted), adsorption increases with temperature; strong adsorption of cationics on quartz (lowered by addition of nonionics).

The increased adsorption of cationics, independent of the reservoir rock, is thus clearly evident. As a guide, values of about $0.5 \cdot 10^{-4}$ mg/cm^2 can be given for the admissible adsorption on quartz surfaces.

PHYSICOCHEMICAL PROPERTIES OF SURFACTANTS

Prior to testing, a few generally-known rules and some empirical data from the chemistry of surfactants can be used:

Good wetting action: C_8-C_{12} range ($<C_{10}$: small micelles, low surface activity). Branched and solvatable groups should lie close to the centre of the molecule.

Hydrophilic character: 3 CH$_2$ groups $\triangleq$ 1 OH-group
$-\overset{\text{O}}{\underset{}{\text{C}}}$-NH-group $\triangleq$ -O$^{\underline{2}}$group (hydrogen bridging)

$\langle\bigcirc\rangle\!\!\rightarrow \triangleq$ 3 CH$_2$-groups

Solubility: n/3 EO → beginning water solubility
n/2 EO → medium water solubility
1-1,5 n EO → good water solubility
(n = number of carbon atoms in hydrophobic chain)
Solubility decreases with rising temperature (→ cloud point/through dehydration and increasing electrolyte content (see fig. 4 /7/).

HLB value:
W/O emulsifiers	3-6
Wetting agents	7-9
O/W emulsifiers	8-12
O/W dispersing agents, W/O demulsifiers, solubilizing agents	/13-18/

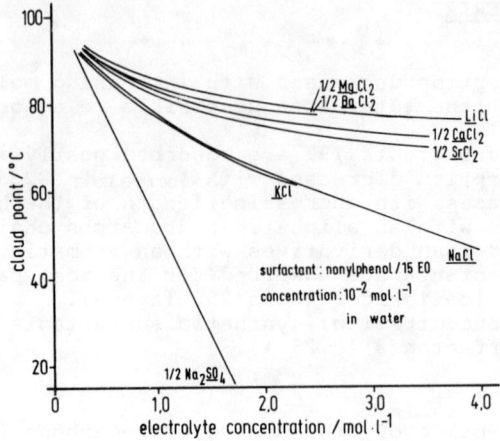

<u>Figure 4:</u> Effect of electrolyte concentration and type on cloud point TP /7/

On the basis of this preliminary information ,it is now already possible to get the most important requirements on surfactants for EOR processes /4, 5/:
- Enrichment at the interface
- Formation of oriented monolayers
- Permanent lowering of interfacial tension in the system oil/water to <1 mN.m^{-1} at low surfactant concentration /13/
- Tendency to micelle formation
- Partial oil solubility
- Stabilization of O/W emulsions
- Solubility or dispersability in highly saline formation water
- Long-term stability (1-2 years) under reservoir conditions
- Low adsorption on reservoir rock
- Low cost coupled with high effectiveness

A list of possible building stones available commercially for the synthesis of surfactants is given in table 3. These considerations then lead to classes of promising products, which in part should exhibit very strong interfacial activity and are described in the US-literature as effective for EOR processes (see table 4 and 5).

These known surfactants are suitable primarily for low salinities (1 % NaCl with about 100-200 ppm Ca^{2+} and Mg^{2+} only).Without a polyether chain with sufficient dispersing power, however, the solubility in high salinity systems (15-25 % NaCl, 20 000-40 000 ppm Ca^{2+} and Mg^{2+}) is for the most part too low or the electrolyte sensitivity to alkaline earth ions too high. Even in the case of polyethoxylates the electrolyte content of the reservoir water can lower the cloud point strongly (fig. 4) and thus can bring about a decreased solubility as well as increased adsorption and a partial passage of the surfactant into the oil phase /6/. In general the interfacial activity of the anionics is likewise reduced strongly in water with a high electrolyte content /5/.
Frequently also surfactant mixtures for EOR processes have been described and applied.
There remains unclear the question of chromatographic phenomena in the use of complicated surfactant mixtures in reservoir, in which the quite different components of the mixture can exhibit completely different rates of migration.

Table 3: Possible building stones for surfactants

Surfactant building stones

———— α-olefins
oligomeric alkenes
fatty acids (saturated and unsaturated) and
   derivatives, natural oils
alkanols (alfols, oxo-alcohols, fatty alcohols)
alkylaromatics
isoalkylphenols
alkylamines (fatty amines)
polyalkylene glycol ethers
polybutylene oxide (polypropylene oxide)

———● SO_3, ($\bigcirc \rightarrow SO_3$), $ClSO_3H$, H_2NSO_3H, Na_2SO_3, $NaHSO_3$

, $HOC_2H_4SO_3Na$, $(CH_2)_3 \begin{smallmatrix} SO_2 \\ | \\ O \end{smallmatrix}$, $(CH_2)_2 \begin{smallmatrix} SO_2 \\ \\ C \\ \| \\ O \end{smallmatrix} O$

, H_3PO_4 (P_2O_5), $(CH_3O)_2SO_2$ $(N^\oplus)$

H_2O_2 (N), $ClCH_2CO_2H$, (HNO_3)
      $\downarrow$
      O

(Formaldehyde, epichlorohydrin, $RO-CH_2CH-CH_2$,

aliphatic oligoamines, polyols, etc.)

Table 4: Anionic surfactants for EOR processes (international
literature)

Anionic surfactants

Chemical constitution	Designation	Structural type
$R^1-CH-CO_2R^2$, SO_3H (Na)	α-Sulfo fatty acid esters	
$R-CH_2-CH-R'-COONa$, OSO_3Na	OH fatty acid sulfates	
$R-CON\begin{smallmatrix}R^1\\R_2\end{smallmatrix}$, SO_4Na	Sulfated amide oils	
$C_{12}H_{25}-\bigcirc-O-\bigcirc$, SO_3Na SO_3Na	Didecyldiphenyl ether disulfonates	
$R-\overset{H}{\underset{OH}{C}}-C_2H_4-\overset{H}{N}-CH_2SO_3Na$	Hydroxyalkylaminosulfonic acids	
$R\overset{O}{\overset{\|}{\square}}N-\bigcirc-SO_3Na$, R	Alkenylsuccin-N-(alkyl)-phenylimidesulfonates	
$R-N-R'-(OC_2H_4)_xOSO_3Na$, R	Dialkylamino polyether sulfates	
$R-CH_2\overset{H_2O}{\underset{O}{\underset{SO_3}{\diagup}}}\begin{smallmatrix}R-CH_2-\overset{OH}{CH}-CH_2-CH_2-SO_3H\\R-CH=CH-CH_2-CH_2-SO_3H\\R-CH_2-CH=CH-CH_2-SO_3H\end{smallmatrix}$	Alkenyl-, OH-Alkane sulfonates	
$R-\bigcirc-O(C_nH_{2n}O)_xR$, SO_3Na (TEXACO)	Sulfates of iosalkylphenyl polyether sulfonates	
$HN^\oplus\begin{smallmatrix}C_2H_4O-\overset{O}{\overset{\|}{C}}-R\\C_2H_4O-\overset{O}{\overset{\|}{C}}-R\\C_2H_4(OC_2H_4)_yOC-\overset{H}{\overset{\|}{C}}-CH_2-CO_2Na\\O SO_3^\ominus\end{smallmatrix}$	Bisfatty acid esters of triethanolamine polyglycol ether sulfocarboxylates	
$R-O-(CH_2-CH_2-O)_n-\overset{O}{\overset{\|}{P}}-ONa$, ONa	Alkanol polyether phosphates	

Table 5: Amphoteric surfactants for EOR processes (international literature)

Amphoteric Surfactants

Chemical constitution	Structural type
$C_{12}H_{25} \cdot \overset{CH_3}{\underset{CH_3}{N}} + CH_2 - \langle \rangle - SO_3^-$	Sulfobetains
$H_{25-37}C_{12-18} - \overset{CH_3}{\underset{CH_3}{\overset{\oplus}{N}}} - (CH_2)_3 - SO_3^{\ominus}$	Sulfobetains
$R - \overset{R'}{\underset{R'}{\overset{\oplus}{N}}} CH_2 - \overset{H}{\underset{OH}{C}} - CH_2 - SO_3^{\ominus}$	Sulfobetains
$H_{25-37}C_{12-18} - \overset{CH_3}{\underset{CH_3}{\overset{\oplus}{N}}} - CH_2 - COO^{\ominus}$	Betain
$R - \langle N \rangle$, $HO-CH_2-CH_2 \diagdown {}_N \diagup CH_2 - COO^-$	Alkylimidazol- iniumbetains
$H_{2n+1} - C_n - CO - NH - (CH_2)_m^{\oplus} - \overset{CH_3}{\underset{CH_3}{N}} - CH_2 - COO^{\ominus}$	Amidoalkylbe- tains

Otherwise it is probably possible to increase, by the use of such surfactant mixtures, the packing density at the interface and thus the degree of wetting (see fig. 5 /10/); further, also the formation of mixed micelles is possible (fig. 6 /4/).

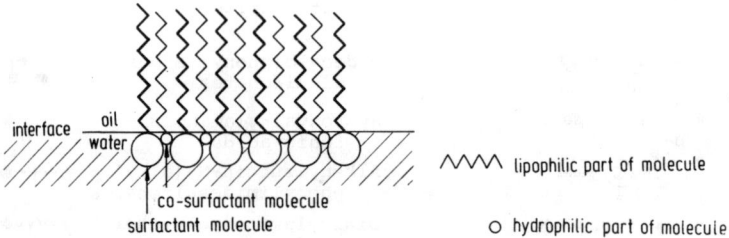

Figure 5: Increased packing density by surfactant mixtures at O/W interfaces

When pure nonionics are used, such as isoalkylphenol ethoxylates, attainment of satisfactory interfacial activities demands a higher degree of alkoxylation. These products are not electrolyte-sensitive and have a good solubility in brine (a rule of thumb is that at 50°C, only nonionics with a degree of ethoxylation n of 10 or more are soluble in 10 % NaCl).

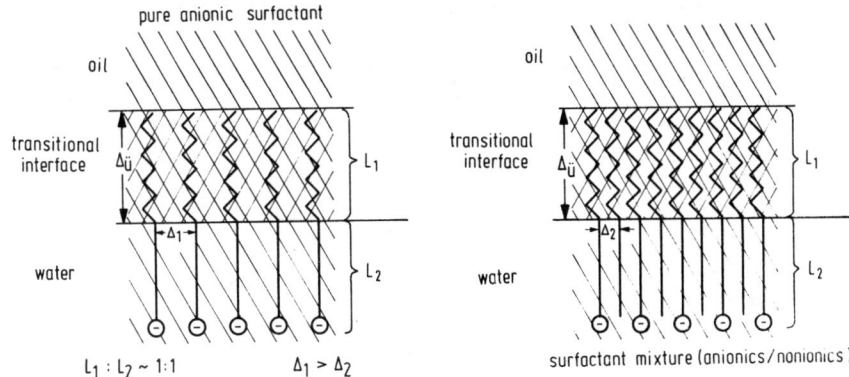

Figure 6: a) packing density of pure anionic surfactant at interface
b) packing density of anionic-nonionic surfactant at
interface

EXPERIMENTAL

Under consideration of as many selection criteria, physico-chemical properties and possibilities of surfactant synthesis as possible more than 1 200 surfactants were tested for their applicability to EOR processes. The screening of the surfactants was
carried out according to the rapid screening program already introduced.
Further tests on a surfactant were proposed only, if it had passed the screening program.

SOLUBILITY IN BRINE CM

All expectations on the solubility of surfactants in high-salinity
brines were confirmed in all respects. Especially the surfactants
with polyether chains and anionic groups have shown good solubilities up to the mark. Table 6 presents some typical products,
which were selected on the basis of the above-described solubility criteria.

INTERFACIAL ACTIVITY

During the measurements of the interfacial activity (Lecomte du
Noüy) a strong dependence of the interfacial tension on the temperature and salinity was established in the system oil/water without addition of any surfactants (see fig. 7).
The experiments have shown that
- all standard oils are characterized by a typical interfacial tension - relationship
- interfacial tension depends strongly on the oil composition
- naphthenic oil shows the highest values of interfacial tension
against high-salinity brines
- in general an increase in salinity is accompanied by a decrease in
interfacial tension/A minimum in interfacial tension will be
passed.
The results are summarized in figure 7.

Table 6: Some surfactants with good solubility, tested in high-
salinity brine

Chem. constitution	Designation	Structural type
i–C$_9$H$_{19}$–⟨O⟩–O(C$_2$H$_4$O)$_{4-4}$CH$_2$CO$_2$Na	Isoalkylphenylpolyether acetates	—⌇●
i–C$_9$H$_{19}$–⟨O⟩–O(C$_2$H$_4$O)$_{20}$SO$_3$Na 　　　i–C$_9$H$_{19}$	Diisoalkylphenylpolyether sulfates	⟩⌇●
C$_{13/15}$H$_{27/31}$(OC$_2$H$_4$)$_5$OSO$_3$Na	Alkylpolyether sulfates	—⌇●
R–C(=O)–N⟨(C$_2$H$_4$O)$_x$ SO$_3$Na / (C$_2$H$_4$O)$_x$ SO$_3$Na (2 x = 5)	Acylamidopolyether sulfates	●
(R–C(=O)–OC$_2$H$_4$–N⟨(C$_2$H$_4$O)$_y$ SO$_3$Na / (C$_2$H$_4$O)$_y$ SO$_3$Na) (2 y + 1 = 2 x)	(Esteramine polyether sulfates)	∼
i–C$_{12}$H$_{25}$–⟨O⟩–O–CH$_2$–CH(OC$_n$H$_{2n}$)$_x$OR'–CH$_2$–(OC$_n$H$_{2n}$)$_x$ OR' 　　SO$_3$Na x = 0 – 20 R' = H, SO$_3$Na, 　 –CH$_2$COONa n = 2, 3	(Sulfone/sulfate-isoalkyl-phenylpolyethoxyglycerol ether)	●⌇●
⁺+⟨O⟩–(OC$_2$H$_4$)$_{20}$OH	Ditert.-alkylphenyl poly-ethers	⟩⌇
i–C$_9$H$_{19}$–⟨O⟩–(OC$_2$H$_4$)$_{20}$OH	Isoalkylphenylpolyethers	—⌇

If surfactants were added to the standard oils, characteristic curves resulted for the function interfacial tension γ = f (surfactant concentration c_T).
The shape of the curves depend on
- type of oil (composition)
- temperature
- salinity
- type of surfactant and concentration of surfactant.

Typical diagrams of surfactant mixtures (anionics-nonionics) are
presented in figure 8 a/b.
If the interfacial tension reach values of < 1 mN.m^{-1}, the accuracy of the method of Lecomte du Noüy is no longer sufficient. Further tests on "successful" surfactants makes the application of
a spinning-drop-tensiometer (SITE) necessary.
A "successful" surfactant must comply with the following criterion:
The interfacial tension of a successful surfactant must be <
1 mN.m^{-1} against all three standard oils in the temperature range
of 30-80°C.
On the basis of this criterion a general temperature interfacial
tension-relationship was derived for the surfactants tested (see
fig. 9).
By this it was evident, that anionics (with an optimal content of
nonionics) will exhibit the lowest interfacial tension.

TEMPERATURE STABILITY

For the investigation of the temperature stability, surfactant
solutions of various concentrations in brine CM were kept for

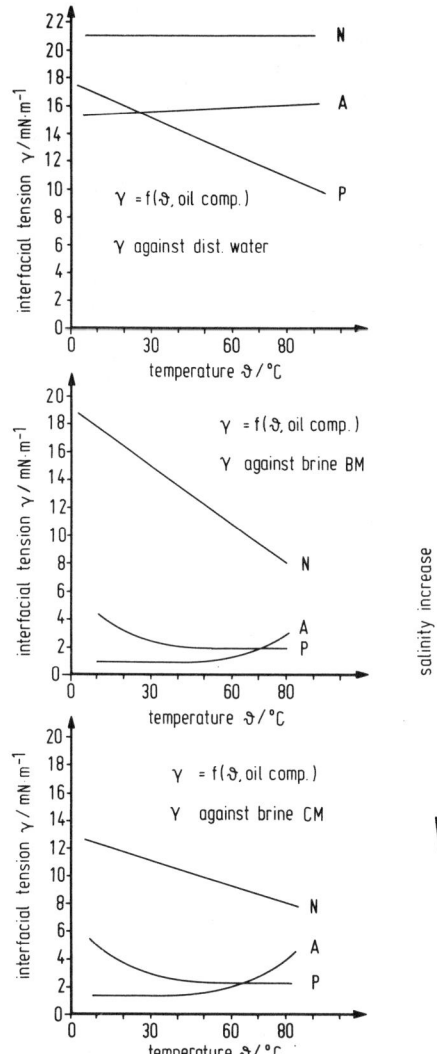

N naphthenic oil
A aromatic oil
P paraffinic oil

Figure 7 : Interfacial tension as a function of temperature, salinity and oil composition (naphthenic, aromatic, paraffinic)

6 month at a temperature of 80°C.
After this time the interfacial activity was compared to that of a standard solution.
On the basis of these experiments, the following statement was possible:
Ether phosphates → ether sulfates → ether carboximethylates → ether sulfonates

arrow in direction of increasing stability

SPECIAL FINDINGS

Polyether sulfates, -carboximethylates and -sulfonates of the following structure are particularly suitable for EOR processes:
$R(OC_2H_4)_x Y^- Me^+$ (Fig.10).

Numerous surfactants with especially low values of the interfacial tension may be classified as mixed surfactants (Mischtenside) (anionic/nonionic). The composition of the mixed surfactant is usually governed by the manufacturing process or the degree of conversion. In this respect it was observed that a degree of conversion of 50 to 80 per cent nonionic to anionic surfactant gives rise to particularly favourable surfactant properties. A typical homologue distribution for such a surfactant is shown in figure 11 .

The following important data and research results are worth mentioning:
- The distribution curve for the surfactant homologs (alkoxilates) should be as broad as possible (more polydisperse) i.e., alkali-catalyzed alkoxilation (not Lewis-acid catalyzed).
- The degree of alkoxilation n must be adjusted according to the crude type (and Y^-). General rules are
  paraffinic crudes: n = 4 ± 2 EO   HLB: 8-10
  naphthenic crudes: n = 6 ± 2 EO   (partial oil solubility)

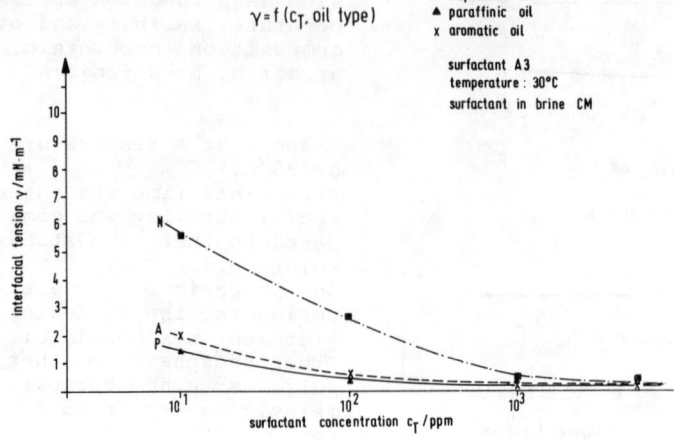

$\gamma = f(c_T, \text{oil type})$

■ naphthenic oil
▲ paraffinic oil
x aromatic oil

surfactant A3
temperature : 30°C
surfactant in brine CM

Figure 8 a:
Interfacial ten-
sion as a func-
tion of oil com-
position, sur-
factant concen-
tration and tem-
perature (30°C)
surfactant:
$C_{12/14}$-fatty
alkohol-polygly-
colether-(4,5
EO)-carboxme-
thylate, Na-salt

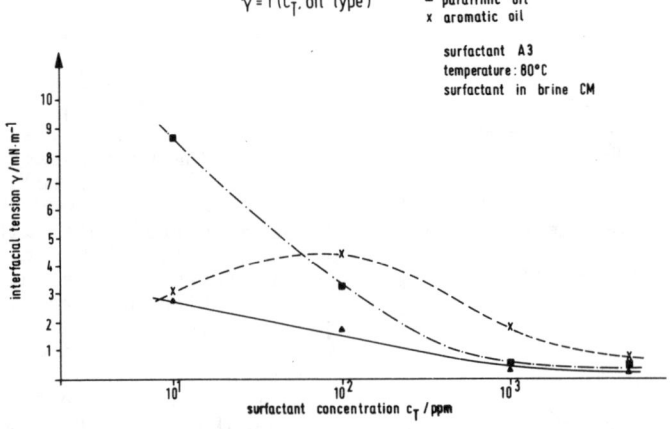

$\gamma = f(c_T, \text{oil type})$

■ naphthenic oil
▲ paraffinic oil
x aromatic oil

surfactant A3
temperature : 80°C
surfactant in brine CM

Figure 8 b:
same as fig.
7a/temperature
80°C

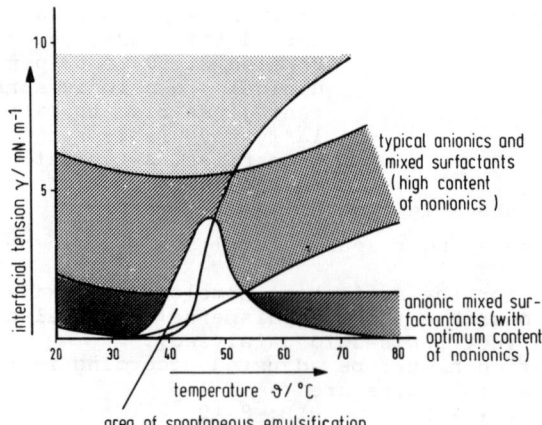

typical anionics and
mixed surfactants
(high content
of nonionics)

anionic mixed sur-
factantants (with
optimum content
of nonionics)

area of spontaneous emulsification

Figure 9 : Temperature/
interfacial tension-re-
lationship for some im-
portant surfactant groups

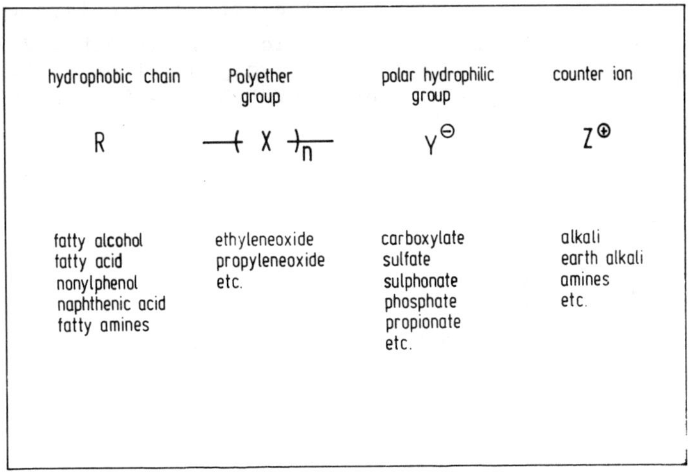

hydrophobic chain	Polyether group	polar hydrophilic group	counter ion
R	$\xrightarrow{}\!\!\left(\!X\!\right)_{\!n}\!\!\xrightarrow{}$	$Y^{\ominus}$	$Z^{\oplus}$
fatty alcohol	ethyleneoxide	carboxylate	alkali
fatty acid	propyleneoxide	sulfate	earth alkali
nonylphenol	etc.	sulphonate	amines
naphthenic acid		phosphate	etc.
fatty amines		propionate	
		etc.	

<u>Figure10:</u> Surfactants suitable for EOR processes

<u>Quantitative analysis of i-nonylphenolpolyglycolether-(6EO)-carboxymethylate, Na-salt</u>

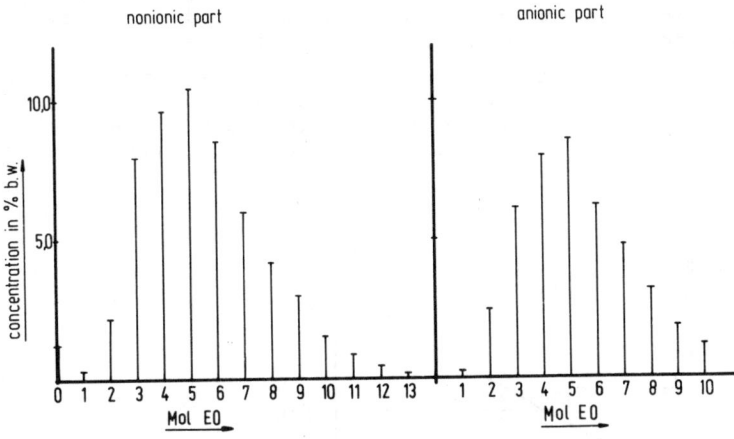

Figure 11: Quantitative analysis of i-nonylphenol-polyglycol-
           ether (6 EO) - carboximethylate, Na salt, by HPLC

aromatic crudes: n = 8 $\pm$ 2 EO

(Propoxilates are in general less effective, as are EO-PO adducts)
- The hydrophobic chain R must be tailored with precision.
  paraffinic crudes: $C_{14} \pm 4$ (saturated, unbranched)
  naphthenic crudes: $C_{12} \pm 2$
  aromatic crudes: alkylaromatics (iso-C_{8-12}-alkyl).
- Cation (Z^+) : Na^+, K^+, N^+R_4 or NH_4^+

- The length of the hydrophobic and hydrophilic molecular parts
  should be roughly in the 1 : 1 ratio (partial oil solubility).

Example: $C_{12/14}$-alkylpolyglycol ether sulfate-(4,5 EO), Na-salt
(hydrophobic: hydrophilic chain length = 2,2 nm: 2,3
nm)or for nonylphenolpolyglycol ether sulfate-(4 EO),
Na-salt (hydrophobic: hydrophilic chain length =
2,0: 2,1 nm (see fig. 12).

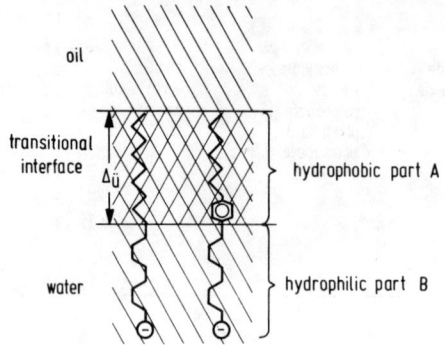

Figure 12 : Optimal chain ratio of suitable surfactants for EOR processes

ratio at optimum A:B-1:1

- The degree of conversion of nonionics into sulfates, carboximethylates, sulfonates, etc., should be 50-80 % (mixed surfactant formation from nonionics and anionics).

The caracteristic behaviour of anionic-nonionic mixed surfactants with temperature (minimum of interfacial tension) can be explained with the help of the phase diagram for such systems (see figure 13), whereby the occurrence of a miscibility gap is decisive.

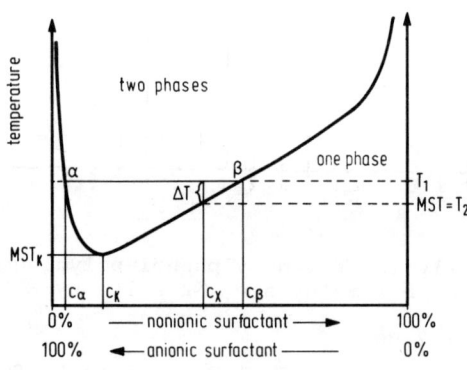

Figure 13 : Schematic phase diagram for mixed surfactant (anionic-nonionic) with miscibility gap

MST_K = lower critical micell - splitting - temperature

c_K = critical splitting - concentration of mixed micelles at MST_k

c_X = composition of mixed micelles

$MST=T_2$ = splitting - temperature of mixed micelles

ΔT = $T_1 - T_2$

α, β = coexistent phases with concentrations c_α and c_β

A separation into water-/oil-soluble surfactants occurs when the
mixed micelles formed from anionics & nonionics reach the micelle splitting
temperature (MST). When the MST is exceeded, various interesting
phenomena may be observed (see figure 14); these are accompanied
by transport processes at the interfaces.

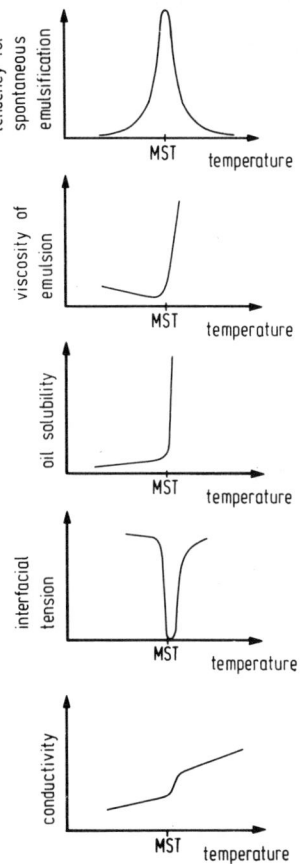

Figure 14: Phenomena at
the micelle-splitting
temperature (MST) for mixed
surfactant (nonionic-an-
ionic)

CONCLUSIONS

With the technological possibilities taken into consideration, and
with the help of a rapid test procedure, it was possible to select surfactants suited for EOR processes in high-salinity systems
from a large number of products. The selected surfactants are anionics and belong to the classes of polyglycolethercarboximethy-
lates and polyglycolethersulfonates. As a result of the manufacturing process, these products may be classified as mixed surfactants (nonionic-anionic). Since mixed micelles are formed, these
products possess special temperature-dependent properties which
are interesting for EOR processes.
In the long term, tailor-made products, especially surfactant mixtures or mixed surfactants, offer special promise from the economic point of view.

Nomenclature

c.m.c.	=	critical micelle formation concentration
c_s	=	salinity/concentration of salts dissolved; $g.l^{-1}$
EO	=	ethylene oxide
EOR	=	enhanced oil recovery (tertiary oil recovery phase)
HLB	=	hydrophilic/lipophilic balance
HC	=	hydrocarbons
OOIP	=	original oil in place, %
O/W	=	oil/water
PO	=	propylene oxide
ppm	=	parts per million
TDS	=	total dissolved solids, %
W/O	=	water-in-oil
θ	=	temperature, $^{\circ}C$

Abbreviations for figures

c_T	=	surfactant concentration in ppm or %
Δ_1, Δ_2	=	distance between surfactant molecules at interface, nm
$\Delta_{\ddot{u}}$	=	thickness of transitional interface
L_1	=	length of hydrophilic chain, nm
L_2	=	length of hydrophobic chain, nm
γ	=	interfacial tension, $mN.m-1$

LITERATURE

1	Babalyan, G.A.:	Physicochemical processes in oil production, "Publishing House "Nedra", Moscow, 1974 (in Russian)
2	Rösch, M.:	The configuration of the polyethyleneoxide chain of nonionic surfactants (part 1 & 2) (in German) Tenside Detergents 8 (1971), pp. 302-313 Tenside Detergents 9 (1972), pp. 23-28
3	Schönfeldt, N.:	"Grenzflächenaktive Ethylenoxid -Addukte" (Interface-Active Ethylene Oxide Adducts), Wiss. Verlags GmbH, Stuttgart
	Schick, M.J.:	"Nonionics Surfactants", Marcel Dekker, Inc., New York, 1967/ Chapter 22
4	Akstinat, M.H.:	Viscous flooding media for tertiary oil recovery in highly saline systems - selection criteria, testing methods and experimental results (in German) Ph. D. thesis, TU Clausthal 1978
5	Gutscho, S.J.:	"Surfactants and Sequestrants", Noyes Data Corp., Park Ridge, N.J. , 1977,
6	Balzer, D.; Kosswig, K.:	The phase-inversion-temperature as a criteria for selection of surfactants for EOR (in German) Tenside Detergents 16 (1979), pp. 256 - 261
7	Schick, M.J.:	Surface films of nonionic detergents I. Surface tension study J. Coll. Sci. 17 (1962),pp. 801-813
8	Crook, E.H.; Fordyce, D.B.; Trebbi, G.F.:	Molecular weight distribution of nonionic surfactants/II. Partition coefficients of normal distribution and homogeneous p, t - Octylphenoxyethoxiethanols (OPEs) J. Coll. Sci. 20 (1965), pp. 191-204
9	Kravchenko, J.J.:	Effect of temperature on the adsorption of nonionic surface-active substances on solid adsorbents Coll. J. USRR 33 (1971), pp. 379-381
10	Akstinat, M.H.:	Surface-active agents for tertiary oil recovery: selection criteria and selection methods (in German) Tenside Detergents 14 (1977), pp. 57-63

11   Rieckmann, M.:   Tertiary oil recovery methods (in
                                   German)
                                   Erdöl-Erdgas-Z. 91 (1975), pp. 348-
                                   359

12   Rudi, V.P.;   Influence of surfactants on the pro-
     Sobkiv, E.R.:   perties of clays (in Russian)
                                   Kolloid Zh. 28 (1966), pp. 119-122

13   Cash, R.L. et al.:   Surfactant aging: a possible detriment
                                   to tertiary oil recovery
                                   50. SPE of AIME Ann. Fall Mtg., 28.3.-
                                   1.10.1975, Dallas/Tx.
                                   SPE-Paper 5564

14   Trogus, F.J. et al.   Adsorption of mixed surfactant systems
                                   52. SPE of AIME Ann.Fall Techn. Conf.
                                   & Exh., 9.-12.10.1977, Denver/Col.
                                   SPE-Paper 6845

15   Wright, C.C.:   The use of Carbon Dioxide in water-
                                   floods
                                   API Prod. Div. Pacific Coast Distr.
                                   Mtg., 21.-23.5.1963, Los Angeles
                                   Preprint 801-39 k

16   Oppenländer, K.;   Surfactants for enhanced oil recovery
     Akstinat, M.H.;   in high-salinity systems - criteria
     Murtada, H.:   for the surfactant selection and appli-
                                   cation
                                   Tenside Detergents 17 (1980), pp. 57-
                                   67

PRELIMINARY STUDIES OF THE BEHAVIOUR OF SOME COMMERCIALLY AVAILABLE SURFACTANTS IN HYDROCARBON–BRINE–MINERAL SYSTEMS

C. ANDREWS, N. M. COLLEY and R. THAVER

British Gas Corporation, London Research Station

ABSTRACT

Some commercial surfactants have been studied with a view to their usefulness for enhanced oil recovery applications.   The following aspects of their behaviour have been assessed.

1.   Their interfacial tension behaviour with crude oil and pure alkanes.

2.   The variation of phase inversion temperature with different variables.

3.   Their adsorption onto rock surfaces

The interfacial tensions were measured by the spinning drop technique.   As the temperature varies, the interfacial tension of a surfactant–brine– hydrocarbon mixture passes through a minimum.   Some surfactants have given interfacial tensions approaching 10^{-3} dynes cm^{-1}.

We have found:

1.   The phase inversion temperature decreases with increasing salinity, the hydrocarbon and the surfactant concentration and composition remaining constant.

2.   For constant salinity and surfactant concentration phase inversion temperature increases with increasing equivalent alkane carbon number.

3.   The phase inversion temperature increases with ethylene oxide content of the surfactant, salinity and hydrocarbon remaining constant.

4.   The phase inversion temperature decreases with increasing lipophilic alcohol content of the systems.

5.   Static adsorption tests on reservoir rock show Langmuir adsorption isotherms.

Introduction

London Research Station, the corporate laboratory of British Gas became involved in enhanced oil recovery after an invitation by the Department of Energy to take part in its research programme coordinated by A.E.E. Winfrith.   After a review of information available to us on the reservoirs operated by British Gas Corporation we decided that our resources would be most usefully employed studying micellar/polymer and miscible flooding. This paper describes the work we have performed so far to identify commercially available surfactants with interfacial tensions-lowering properties to suit the conditions prevailing in our reservoirs, and to assess their sensitivity to changes in reservoir variables, lack of sensitivity being a desirable (but attainable?) ideal. Measurements of phase inversion temperature (PIT), interfacial tensions and adsorption onto mineral surfaces have been made.

The Reservoir

Conditions in the target reservoir are similar to those listed below:

Oil type E.A.C.N. (Reservoir)		7-8
(Stock tank)		10
Temperature		43°C
Formation water		90,000 mgNaCl/litre
		1,300 mgCa/litre
		500 mgMg/litre
Flood water (sea water)		30,000 mgNaCl/litre
		400 mgCa/litre
		1,200 mgMg/litre

Chemicals

Surfactants. Samples of the surfactants listed below were obtained from Hoeschst AG.

Anionics: Hostapal* BV., an alkylaryl polyglycol ether sulphate - Na salt.

    (7 ethylene oxide (e.o.)units, 50% w/w active).

Surfactant A straight chain alkyl phenol ether acetate, 4 ethylene oxide units.

Surfactant B "   "   "   "   "   " 6 ethylene oxide units.

Non ionics: Sapogenate* T80 tri-butyl phenylpolyglycol ether(8 ethylene oxide units.)
T100 "   "   "   "   "(10 ethylene oxide units.)

    T110 "   "   "   "   " (11   ethylene oxide units.)
    T130 "   "   "   "   " (13   ethylene oxide units.)
        Arkopal* N060 Nonyl phenylpolyglycol ether (6 ethylene oxide units.)
(All 100% active).

Hydrocarbons used in this work were specified to be greater than 99% pure.

---

* Hostapal, Sapogenate and Arkopal are trade marks of Hoechst AG.

Phase Inversion Temperature

The phase inversion temperature, PIT, of a hydrocarbon/brine/
surfactant system indicates the existence of a minimum interfacial tension at that
temperature. Since the lowering of interfacial tension is a requirement for the
mobilisation of oil trapped in constricted capillaries and all oil reservoirs are
essentially isothermal, PIT represents a useful parameter for the selection of
surfactant for a given reservoir. For nonionic surfactants below the PIT the
surfactant partitions preferentially into the aqueous phase and the emulsion formed
between the two phases is predominately 'oil-in-water'. Above the PIT, it
partitions mainly into the oleic phase and forms a 'water-in-oil' emulsion (Balzer
and Kosswig,1979). Balzer and Kosswig (1979) have carried out some parametric
studies of PIT with a range of anionic carboxy methylated nonyl phenol ethyoxylate
surfactants. They found:

1. PIT increases with increasing equivalent alkane carbon number (EACN) of the oil
and that aromatic hydrocarbons show very low values of EACN.   Mixtures of aromatic
compounds and alkanes give intermediate PITs.

2. PIT increases with decreasing salinity.

3. PIT increases with increasing number of ethylene oxide groups in ethoxylated
surfactants

We have extended this work to nonionic surfactants in studying the following
variables on PIT. The effects of these variables must be considered if a surfactant
flood is to maintain its oil mobilising properties as it passes through the
reservoir. The parameters studied in this work are:

1. Oil type expressed as EACN.
2. Surfactant type and concentration
3. Salinity
4. Co-surfactant type and concentration
5. Phase ratio
6. Number of ethylene oxide units in surfactant molecule.

1. The EACN for a given reservoir oil should be constant. The EACN of our
reservoir crude has been assessed from measured EACN of stock tank crude and
calculated from a well stream analysis.

2. The concentration of surfactant at some point away from the injection well is
likely to change because of adsorption onto the reservoir rock surfaces. Adsorption
measurements are therefore important.

3. The salinity of the brine in contact with residual crude in a waterflooded
reservoir may vary from pure injection water to pure formation water.

4. Co-surfactant effectiveness may change with concentration and type.

5. Variable oil/brine ratios will occur in a reservoir as a flood proceeds. This
must be taken into account when performing laboratory tests.

6. The hydrophilic/lipophylic balance of a surfactant will depend upon its ethylene
oxide content,(Shinoda 1965). Commercial surfactants are usually assigned a nominal
ethylene oxide content, but actually contain a distribution of e.o. chain lengths.
If PIT is dependent upon the number of e.o. units, then selective adsorbtion by
reservoir rock will change PIT.

Interfacial Tension

It is necessary to augment the data obtained from PIT measurements. The inversion of emulsions occurs over a small temperature range. For this to occur with minimum energy an interfacial tension minimum is implied. A typical plot of I.F.T. against temperature is sketched in Fig.1.

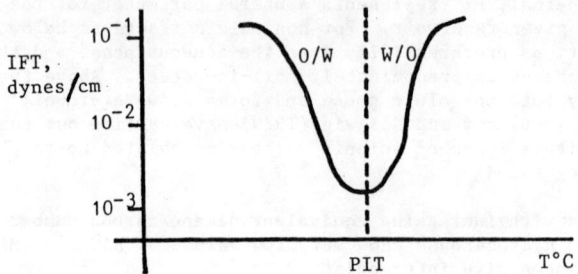

IFT, dynes/cm

Figure 1

Measurements have been made to determine the way IFT changes with temperature.

Methods

Phase Inversion Temperature was determined by means of electrical conductivity measurements (Balzer and Kosswig 1979). For an oil-in- water emulsion with a non-ionic surfactant initially below the PIT the conductivity slowly increased with temperature but fell rapidly as the emulsions inverts and the aqueous phase became discontinuous and therefore non-conducting. Figure 2 shows a typical curve. More than one 'minimum' may occur for impure surfactants.

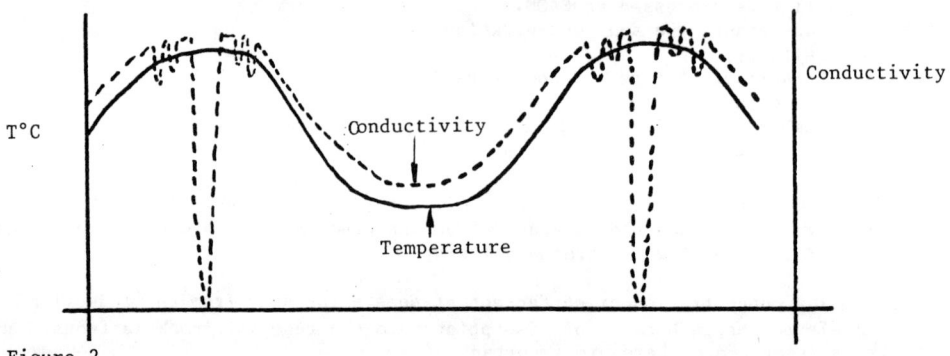

Figure 2

Adsorption

The investigation consisted of a series of experiments to measure A, the adsorptive capacity of the reservoir rock for surfactant material. The method used was based on that of Somasundaran & Hannah (1979). The method of analysis for surface-active material was the titration procedure of Reid et al., (1967).

Interfacial Tension

Measurements were made at ambient pressure with the University of Texas spinning drop tensiometer.

TESTS, RESULTS AND DISCUSSION

1. Variation of PIT with EACN and determination of EACN value of the phenyl and cyclohexyl groups.

PITs were determined on the following mixtures at the phase ratios stated (brine/oil).

Surfactant	Concentration g /litre	Brine	Hydrocarbons	Phase Ratio brine/oil	Results
A	10	Seawater	n-alkanes C7-C10	5:1	Fig 3 line A
A+B	5 of each	Seawater	n-heptane-toluene mixtures EACN 4 to 7	5:1	Fig 3 line B
T80	10	Seawater	N-alkanes C_6 to C_{11} methylcyclohexane	5:1	Fig 4 line A
T80	As above after storage for 3 months at room temperature				Fig 4 line B
T80	10	30g /NaCl/ litre	n-alkanes C_6-C_{11}	4:1	Fig 5 line A
T80	50	" "	" "	"	Fig 5 line B
T100	50	" "	" "	"	Fig 5 line C
T100	10	" "	" "	"	Fig 5 line D
T100	50	" "	butyl cyclohexane	4:1	Table 1
"	"	" "	phenyl heptane	"	Table 1
"	"	" "	phenyl octane	"	Table 1

Results and Discussion

The results presented in Figs 3 to 5 show a linear relationship between EACN and PIT over the EACN range studied.

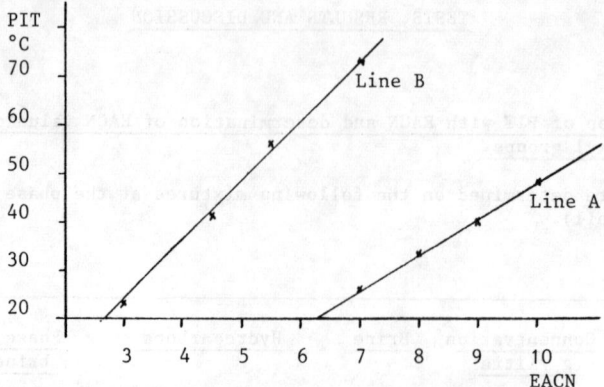

Figure 3: Variation of PIT with EACN for
two different surfactant solutions.

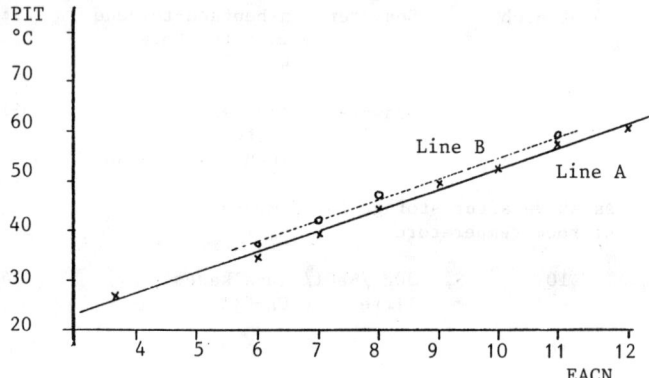

Figure 4: Variation of PIT with EACN
for 10g /1 Sapogenate T80 in seawater.

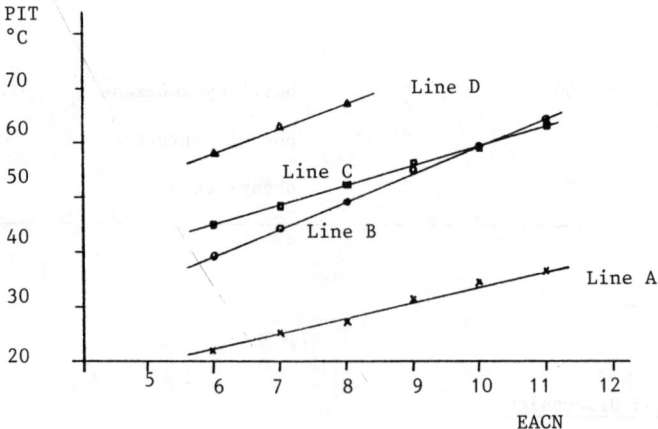

Figure 5: Variation of PIT with EACN; pure n-alkanes,
30 g /litre NaCl

The EACN of butyl cyclohexane was determined relative to the Fig 5 line B and line
C, phenyl heptane and phenyl octane were determined relative to Fig 5 line C.   The
values found are listed in Table 1.

TABLE I

	EACN found		Assigned EACN of ring
butylcyclohexane	6.5 (Fig 5 Line C)	6.75 fig 5 line B	2.6
phenyl heptane	5.0 (" " ")	-	-2.0
phenyl octane	6.0 (" " ")	-	-2.0

Thus we are able to assign an EACN of 3.6 to methyl cyclohexane.   This may enable
us to dilute stocktank crude with hydrocarbons with rings to obtain mixtures of
hydrocarbon at ambient pressures having EACNs more representative of reservoir
crudes.
A shift of PIT of 2 to 3°C was observed on storage at room temperature for 3
months of the stock 50 gms T80/litre solution from which line B fig 4 was obtained.
 No further change was observed on storage for a further 5 months, nor on a freshly
prepared solution stored at 40°C for 3 weeks.

At the higher concentration of T80 the dependance of PIT on EACN is reduced.   The
affect is not as marked in the case of T100.

The difference in slopes shown between lines A and B, Fig 3 suggest the possibility
of modifying the sensitivity of a system to change in temperature by the addition of
another surfactant.   Differential adsorption within a reservoir could cause
problems in practice.

PIT can change rapidly with EACN. Rates of change of PIT of up to 14°C/EACN unit
(line B Fig 3) are possible and the slope can change with surfactant concentration.
Rates of change of PIT as low as 3°C/EACN unit are possible (line A Fig 3).

2.   Variation of PIT with surfactant concentration

Tests   The effect of increasing surfactant concentration was studied on the
following mixtures.

Surfactant	Concentration g /litre	Brine g NaCl/1	Hydrocarbon	Phase ratio brine/oil	Results
T80	Various 10 to 70	30	heptane	4:1	Fig 6 Curve 1A
T80	" "	"	octane	4:1	Fig 6 Curve 1B
T100	" "	"	heptane	4:1	Fig 6 Curve 2A
T100	" "	"	octane	4:1	Fig 6 Curve 2B

Results and Discussion

The results are shown in Fig 6.   Both surfactants exhibit a non-linear
relationship, with PIT increasing with decreasing surfactant concentration.   This
is in agreement with the work of Shinoda and Arai (1964).   The rate of change of
PIT is lower at high surfactant concentrations which indicates that a high
concentration flood could be less susceptible to concentration changes.

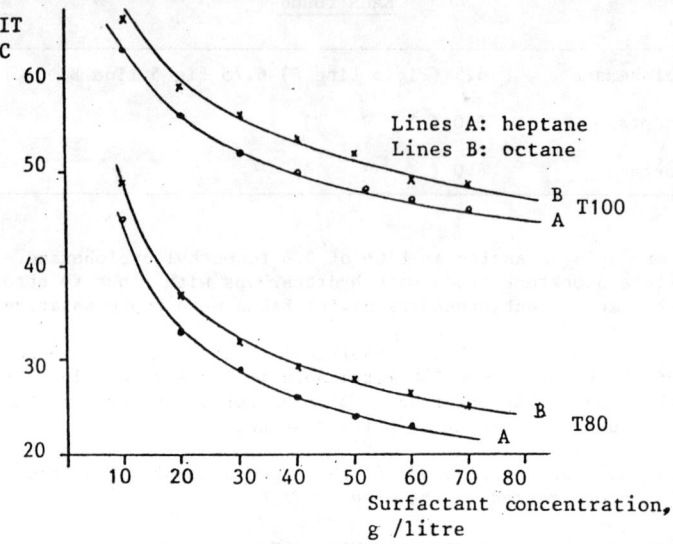

Figure 6:   Variation of PIT with concentration
of Sapogenate T80 and T100.

3.   Variation of PIT with Salinity

Tests   PIT's were determined on the following mixtures.

Surfactant	Concentration g /litre	Brine	Hydrocarbon	Phase ratio brine/oil	Results
T80	10	NaCl only (various concentrations)	Stock Tank Crude EACN 10	5:1	Fig 7 line A
T80	10	"	hexane	4:1	Fig 7 line B
T80	10	"	heptane	4:1	Fig 7 line C
T80	10	"	octane	4:1	Fig 7 line D

Results and discussion

The results are shown in Fig 7. For all hydrocarbons tested, the rate of change of PIT with salinity is independent of the hydrocarbon used. The decrease in PIT with increasing salinity is to be expected as the surfactant partitions more readily into the oleic phase as salinity increases (Knickerbocker et al, 1979).

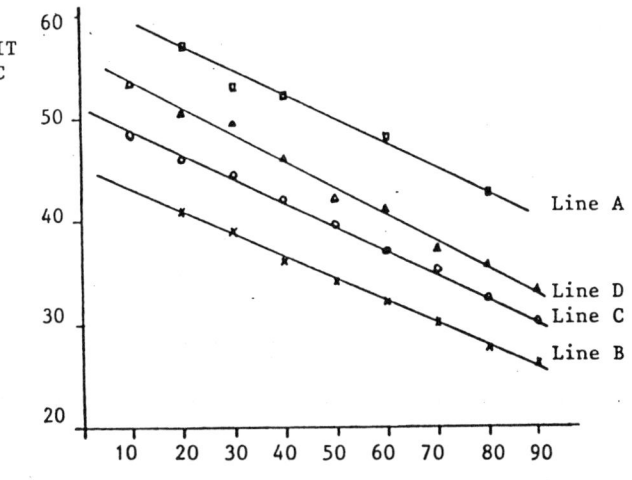

Figure 7: Variation of PIT with brine salinity; 10 g /litre Sapogenate T80 solution.

4. Variation of PIT with Alcohol (Cosurfactant) type and concentration

Tests   Measurements were made on 50 g T100/litre brine. Brine concentration was 30 g /NaCl/litre and oil EACN 7.5 at a phase ratio of 4:1.

Alcohols studied were:
(a) iso-butanol (lipophilic)
(b) iso pentanol("   )
(c) isopropanol (hydrophilic)

Results and discussion.

The results shown in Fig 8 agree with trends predicted in the literature, (Knickerbocker et al, 1979), in that increasing the concentration of a lipophilic alcohol (lines A and B) will increase the partitioning of the surfactant into the oleic phase and tend to lower the PIT. The hydrophilic alcohol (line C) has the opposite affect but less pronounced. Increasing the concentration of hydrophilic alcohols has the opposite effect on PIT as increasing surfactant concentration. If an alcohol has to be used as a viscosity modifier then hydrophilic alcohols may be more manageable with respect to their effect on PIT than lipophilic alcohols.

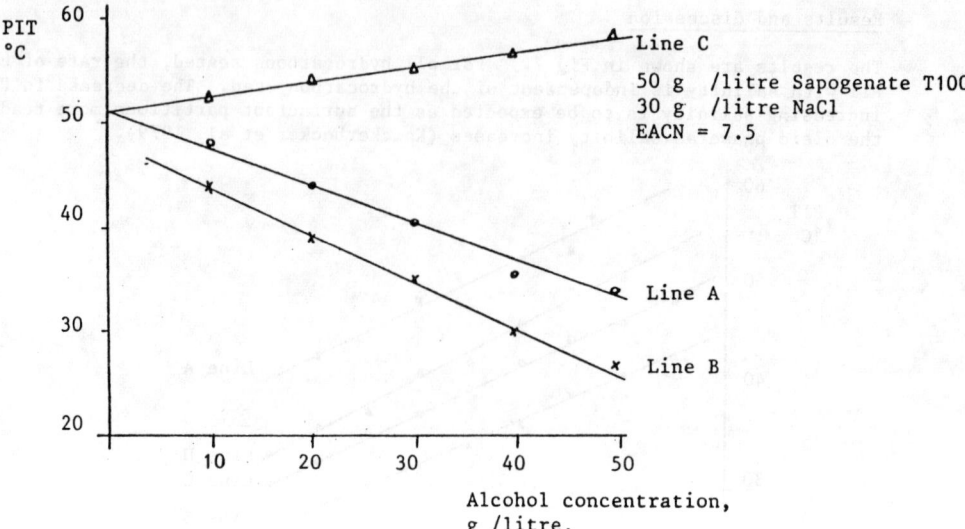

Figure 8: Variation of PIT with alcohol concentration

5. Variation of PIT with brine/oil phase ratio

Tests   PITs were measured on the following mixtures to find out if PIT varied with phase ratio.

Surfactant	Concentration	Brine	Hydrocarbon	Results
T80	50 g /litre	30 g Nacl /litre	n heptane	Fig 9 line A
T80	10 g /litre	" " "	n hexane	Fig 9 line B
T80	" " "	" " "	n heptane	Fig 9 line C
T80	" " "	" " "	n octane	Fig. 9 line D

Results and Discussions

The possibility that PIT would depend upon phase ratio was indicated when PITs obtained from the measurements in the spinning drop tensiometer did not correspond exactly to those made by the conductivity measurements.

The results show that PIT increases as the proportion of oleic phase increases. This is contrary to the findings of Balzer and Kosswig (1979) who reported smaller changes with the opposite slope.   Arai (1965) reports the same effect as Balzer and Kosswig (op.cit.)

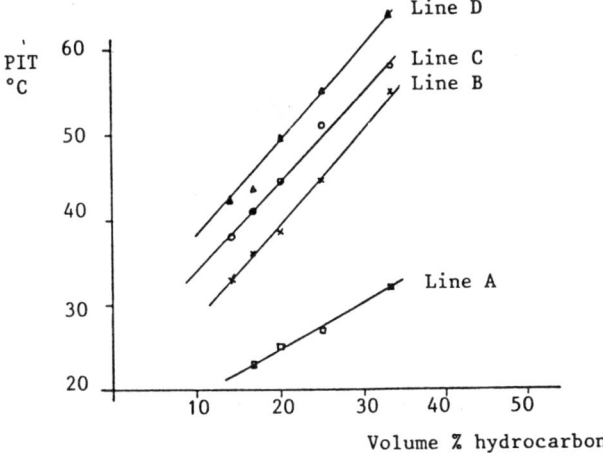

Figure 9:   Variation of PIT with phase ratio.

6.   Variation of PIT with Ethylene Oxide (eo) content of surfactant

Tests   PITs were measured using a mixture of pure normal alkanes, EACN 7.5
with 30 g NaCl/litre brine containing 50 g of surfactant/litre.
The surfactants used were Sapogenate T80,T100,T110 and T130 which
contain (nominally) 8, 10, 11 and 13 ethylene oxide units
respectively.   Intermediate eo contents were obtained from mixtures
of the adjacent surfactants as supplied and were calculated on a molar
basis.

Results and Discussion

The results presented in Fig 10 show a linear relationship between PIT and the
number of eo units per molecule. The deviation from linearity above eo = 11 is
probably due to evaporation of the hydrocarbon during the test.

The findings are in agreement with those of Bourell et al,(1980).   The change
in PIT is explained by the increased hydrophilic properties with increased eo
content (Shinoda, 1965).

The Arkopal series of surfactants probably exhibits a similar trend but only
two have been tried i.e. NO60 (6 eo's) and NO80(8 eo's).   These gave PITs of
about 3⁰C and 70-75⁰C respectively at a concentration of 50 g /litre in
the same brine/hydrocarbon system.   The effect of the number of eo units is
greater with the Arkopal series than the Sapogenates.

Where a surfactant contains a spectrum of eo contents, selective adsorption by
the reservoir rock may change its effective eo value and thus affect the PIT of
the system.

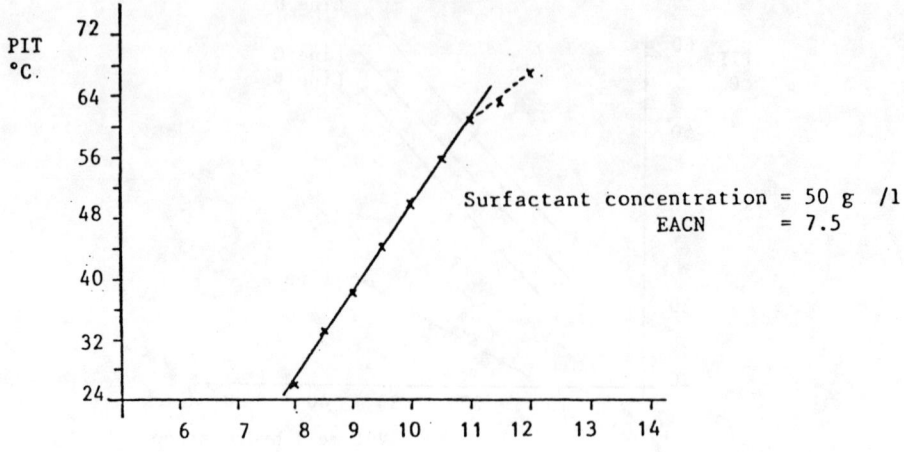

Surfactant concentration = 50 g /1
EACN = 7.5

Surfactant ethylene oxide number.

Figure 10: Variation of PIT with surfactant
ethylene oxide number

7. Variation of IFT with temperature

Tests   Interfacial tension measurements were made between the upper and lower
phases obtained from mixtures whose PIT's had been determined in an attempt to
confirm the presence of an IFT minimum at the PIT.

Tests were performed with the following mixtures.

Surfactants	Concentration	Brine	Hydrocarbons	Phase ratio	PIT	Results
T80	10 g /litre	30 gm NAC1/ litre	n hexane	4:1	39	Fig 11
T80	" " "	" " "	n octane	4:1	49.5	Fig 12
NO60	" " "	" " "	Crude EACN =10	5:1	23.5	Fig 13

Results and Discussion

The results obtained are shown in Figs 11 to 13. The PITs obtained from
conductivity measurements are included in the above table. Repeat determinations of
I.F.T. were usually found to agree within $\pm$ 5%.

Figs 11 and 12 indicate that a minimum does occur at the PIT but that more than one
'minimum' can occur.   This is supported by conductivity traces made during PIT
measurements and is probably due to a proportion of surfactant having a different
number of eo units than the stated nominal value.

The equipment available only allowed for the transfer of phases into the tensiometer
at room temperature.   Measurements were made at various temperatures after heating

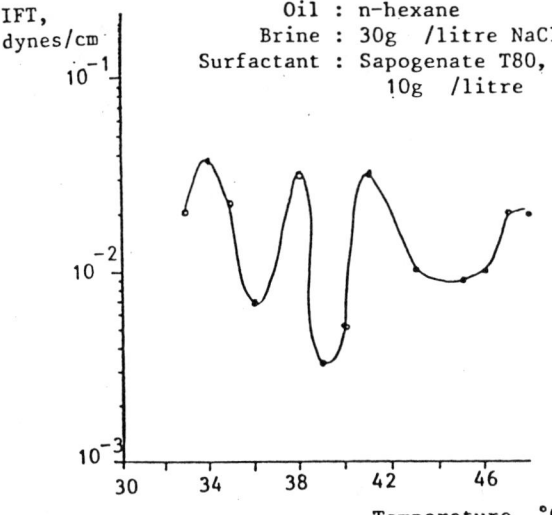

Figure 11: Variation of IFT with temperature.

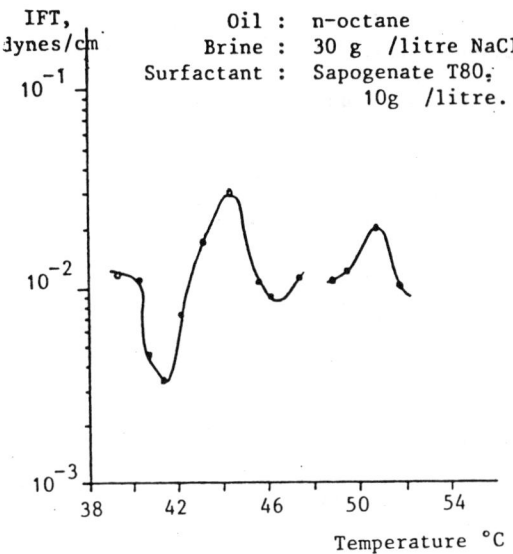

Figure 12: Variation of IFT with temperature.

from room temperature.   As one increases the temperature of the sample tube a third
phase (microemulsion) begins to develop as the equilibrium is disturbed.   In order
to make meaningful measurements the middle phase is separated from the remaining
oil drop. This was achieved with some difficulty especially in the case of
colourless oils.

Ideally, equilibration, sampling and measurements should be carried out at the same
temperature.

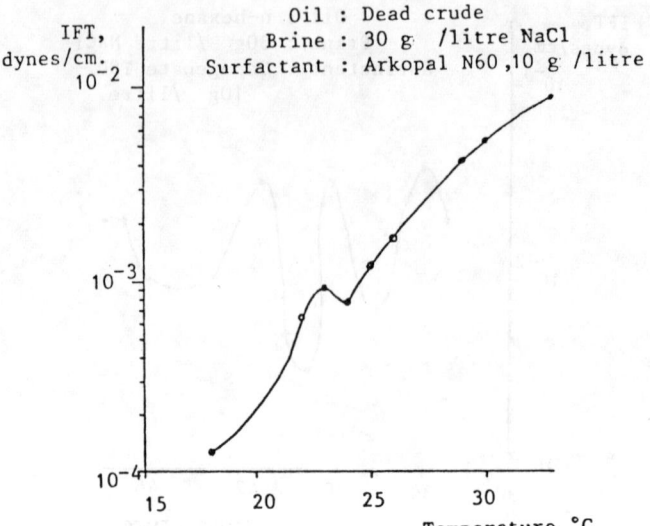

Oil : Dead crude
Brine : 30 g /litre NaCl
Surfactant : Arkopal N60 ,10 g /litre

Figure 13: Variation of IFT with temperature

Adsorption of Surfactants

Tests This section describes the results obtained for adsorption of Hostapal BV on reservoir material in various states of disaggregation.   Although work on Hostapal BV was terminated, (because optimal salinity falls outside the range of our interest), the results showed some of the limitations of static adsorption tests.

Samples of reservoir rock were taken from cores and crushed in a ball mill until the powder passed 180μ sieve.   25 g of sample were taken and equilibrated with 50 cm^3 of various concentrations of Hostapal BV in distilled water or sea-water.   The suspension was stirred constantly for 4 hours 40°C.   The aqueous portion ws then decanted and centrifuged for 30 minutes, by which time, the supernatant liquid was clear.   Analysis of this liquid then gave the remaining concentration of Hostapal BV.   Hence the amount abstracted by the solids was calculated.

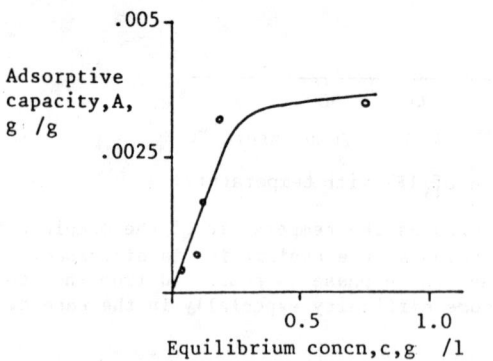

Figure 14: Adsorption of Hostapal BV onto reservoir rock, Sample 1.

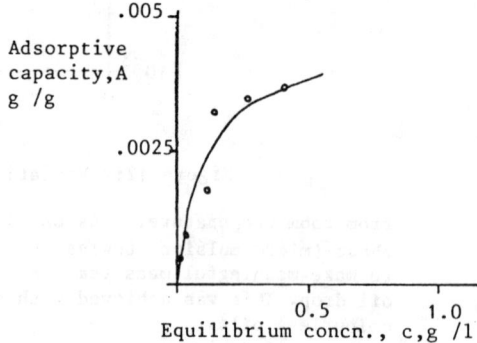

Figure 15: Adsorption of Hostapal BV onto reservoir rock, Sample 2.

Results and Discussion

The tests performed are listed below and the results presented in Figs 14 to 18. Table 2 shows a typical sets of results.

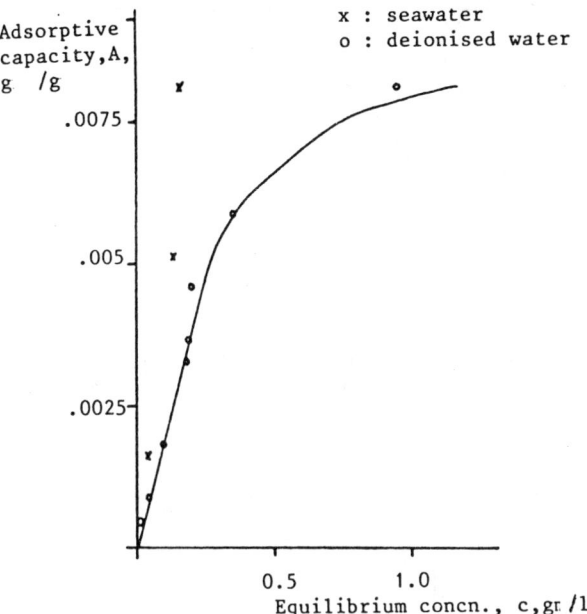

Figure 16:   Adsorption of Hostapal BV onto reservoir rock, Sample 3.

TABLE 2 (SAMPLE 3)

Initial surfactant concentration g /litre	Final (equilibrium) surfactant concentration,C g /litre	Adsorptive capacity A, g surfactant/g rock
0.25	0.0078	0.00048
0.50	0.033	0.00092
1.0	0.098	0.0018
1.75	0.164	0.0033
2.0	0.184	0.0036
2.5	0.193	0.0046
3.3	0.352	0.0059
5.0	0.957	0.0080

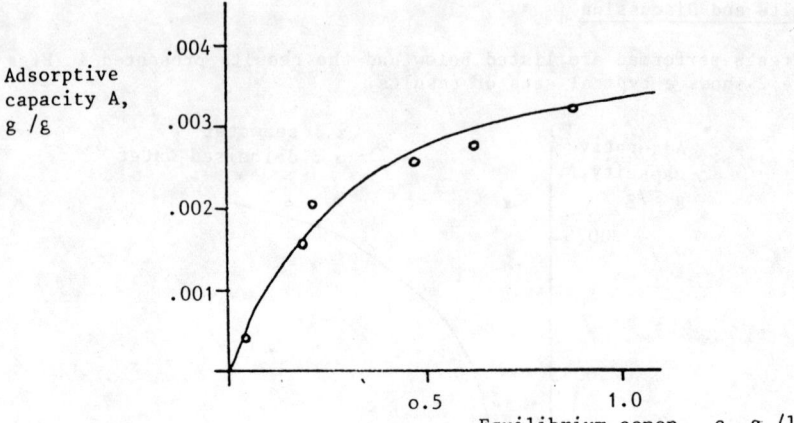

Figure 17: Adsorption of Hostapal BV onto
reservoir rock, sample 4.

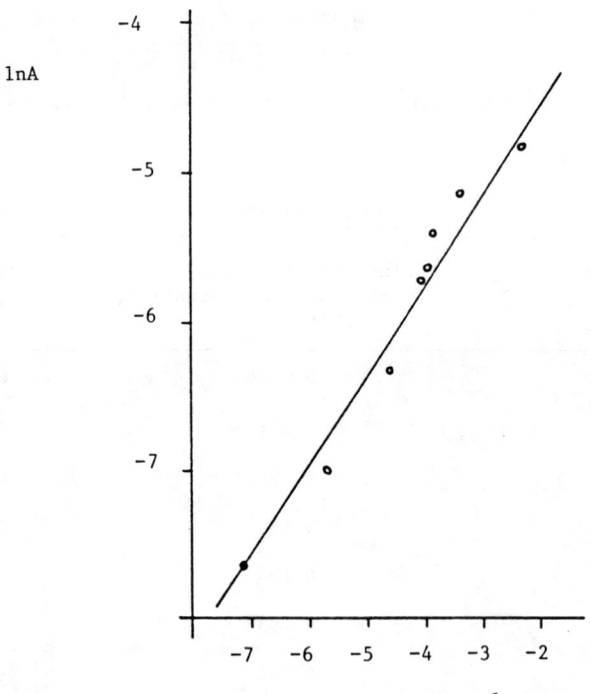

Figure 18: Data from Figure 16 plotted logarithmically

All the figures show the tendency for A to tend towards a constant for a given sample as the equilibrium concentration increases. Portions of Sample 3 (Fig 16) were also equilibrated with the surfactant in seawater and the results appear to show a much higher adsorptive capacity. Sample 3 has similar permeability and porosity characteristics to the other samples. The curves have the same general form as the classical adsorption isotherms.

$$A = Kc$$

where

A= adsorptive capacity
c= final concentration
K and n are constants

or $\ln A = \ln K + \ln C$

The data in Table 2 (sample 3) are plotted as ln A against ln C in Fig 18.

There are indications in Fig 14 and 15 that adsorption may be proceeding in layers.

A qualitative test of the effect of particle size on the equilibrium adsorption of surfactant was performed in a similar manner to those described above. The results obtained are shown in Table 3.

Table 3

	A g /g
<180µ	0.0036
'fine powder	0.0028
'coarse grains'	0.0022
cm size pieces	0.0002

This would appear to limit the usefulness of static adsorption and calculation of a 'worst case' total adsorption capacity of a reservoir. More useful data would be obtained from core flood experiments.

CONCLUSIONS

Phase inversion temperature can serve as a guide to the conditions under which a non ionic surfactant will give an interfacial tension minimum.

Using the linear relationship between EACN and PIT it is possible to assign an EACN to; stock tanks crudes, and aliphatic and aliphatic cyclic hydrocarbons. It should then be possible to use cyclic hydrocarbons to lower the EACN of stock tank crude to that of reservoir crude for use in partitioning and phase studies at ambient pressure.

If the way in which the parameters which affect surfactant properties change during the course of a flood can be assessed it should be possible to design a flood which will maintain it's properties. This, however, requires detailed knowledge of the reservoir.

Acknowledgements

We would like to thank the British Gas Corporation for their permission to publish this paper.

BIBLIOGRAPHY

Balzar, D, and Kosswig, K.; The phase inversion temperature as a criterion for the selection of survace active agents in the tertiary production of mineral oil.

Tenside Det. 16 (1979), 5, pp 256-261.

Shinoda, K.; The comparison between the PIT system and the HLB - value system to emulsifier selection.

Comptes rendus du 5 eme Congress International de la Detergence, Barcelona, (1965), pp 275-283.

Somasundaran, P. and Hannah, S.; Adsorption of sulphonates on reservoir rocks.

Soc. Pet. Eng. Jour. August 979, pp 221-232.

Reid, V.W., Longman, G.F. and Heinerth, E.; Determinaton of anionic active detergents by two-phase titration.

Tenside Det. 4 (1967), pp 292-304.

Shinoda, K. and Arai, M. The correlation between phase inversion temperature in emulsion and clould point in solution of nonionic emulsifier.

Jour. Phys. Chem. 68 (1964), 12, pp 3485-3490.

Knickerbocker, B.M., Pesheck, C.V., Scriven, L.E. and Davis, H.T. Phase behaviour of alcohol-hydrocarbon-brine mixtures.

Bourrel, M., Salager, J.L., Schechter, R.S. and Wade, W.H. A correlation for phase behaviour of nonionic surfactants.

Jour.Colloid Interface Sci. 75 (1980), 2, pp 451-461.

THE PROVISION OF LABORATORY DATA
FOR EOR SIMULATION

C. E. BROWN and G. O. LANGLEY

*Petroleum Engineering Branch, Exploration and Production
Division, BP Research Centre, Sunbury-on-Thames*

Abstract

     Laboratory core tests are important in the development and
assessment of EOR processes. It is vital that the core data obtained
characterise the physical processes relevant to the field, and are
appropriate to their use in field simulators. The key parameter of
relative permeability is discussed, and its extension to low tension
immiscible displacement assessed. The current status of these concepts
is discussed with reference to oil slug propagation along a core.

1.   Introduction

     Our overall aim is to obtain data from laboratory tests on core
samples, to use these data for the prediction and analysis of field
trial performance, and ultimately for the prediction of full field
performance.

     There are many problems on the way from core tests to reservoir
performance prediction, one of the biggest being reservoir description
and identification of reservoir heterogeneity. In this paper we will
limit our discussions to the topic of laboratory core data, and how this
may be used to examine the physical processes involved in oil recovery.
We will work on the principal that if successful & efficient displacement
cannot be obtained from a core sample, then there will be little
chance of obtaining success on the field scale. We will consider the
theoretical aspects associated with oil bank propagation, and how the
predictions are affected by the relative permeability input data. A
brief discussion of methods of assessing potentially useful surfactant
systems and possible artifacts associated with core tests on the
laboratory scale is included. In this paper we will discuss low tension,
immiscible floods only. Miscible flooding will not be considered.

2.   Laboratory core waterflood tests

     Before considering low tension flooding, we will review the
more conventional waterflood case, since this often forms the basis
for enhanced oil recovery methods. Waterflood tests on core samples
can be used to gain information on the efficiency of displacement of
oil by water from actual reservoir rock. Laboratory displacement
tests are usually confined to one dimension.

For strongly water-wet rocks the efficiency of displacement is good in the sense that the displacement appears piston-like; practically all of the oil is recovered before breakthrough of the flooding water. However, residual oil is trapped behind the waterflood front as insular globules of oil usually occupying the larger pore spaces (1). This residual oil level is dependent on the initial oil saturation; the relationship being influenced by pore structure. The scope for tertiary oil recovery may be high in this case.

For strongly oil-wet rock the displacement does not appear to be efficient. Early breakthrough of water often occurs, and large amounts of oil are recovered after water breakthrough, at high fractional flow of water. Residual oil is less well defined in that oil is by-passed rather than trapped and occupies smaller pores and surface grooves. However, oil may continue to be produced until very low oil saturation is obtained (2). The scope for tertiary oil recovery may be low in this case. The contrast in the oil-wet and water-wet case is shown in Figure 1.

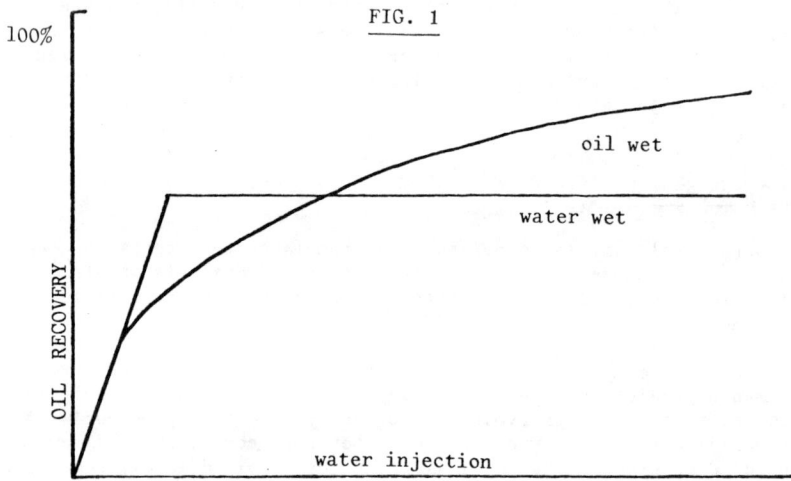

FIG. 1

In order to predict waterflood performance, it would be convenient to generate a data set which could be used for the prediction of flood performance and which is a unique property of the rock. The concept of relative permeability was introduced to fulfil this role, which has now become central to conventional numerical simulation of oil recovery processes.

3. Relative Permeability

The basic concept of relative permeability is limited to movement of two or more continuous phases in the same direction through porous media at a steady saturation level. It assumes that it is valid to extend the Darcy equation for single phase flow to the multiphase case.

The concept is most likely to hold in the case of a sample which has initially 100% saturation of the wetting phase and where the saturation of the non-wetting phase is increasing. Relative permeability data can be generated either by a steady-state method (where both phases are injected simultaneously into the core sample and permeability measurements are made at a steady-state fractional flow), or by analysing flood data (e.g. using the Johnson, Bossler method (3) based on the Buckley-Leverett theory (4)).

Some validation of the relative permeability concept comes in two ways:

1)   The steady-state and flood derived data are similar

2)   Data produced from floods using different viscosity ratios give data which lie on the same curve, even though the recovery performance may be significantly different.

Figure 2 shows typical relative permeability curves for an oil flood of a water-wet rock. It is noted from Figure 2 that the relative permeability to oil increases quite rapidly, and that to water drops rapidly. The sum of the relative permeabilities throughout their range is very much less than unity indicating interference between the phases.

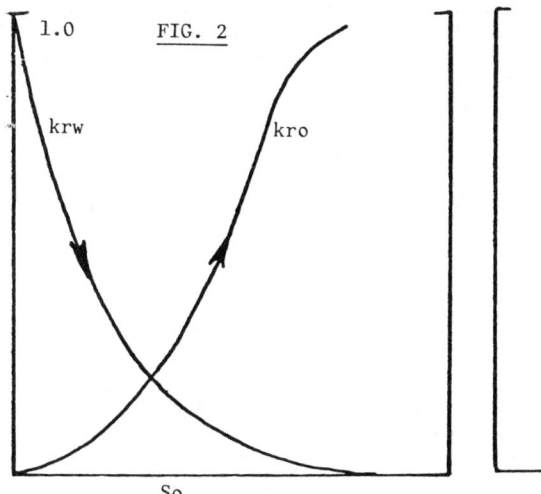

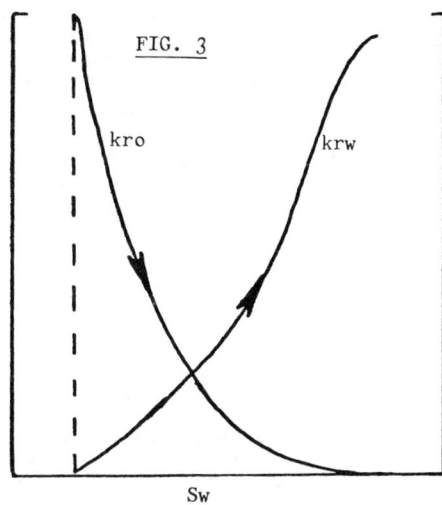

3.1   Waterflooding - the oil-wet case

Parallel arguments apply to a waterflood of an oil-wet rock. Figure 3 shows typical relative permeability curves for the oil-wet case.

Provided that capillary dispersion and end-effect problems are not involved, the waterflood performance (and hence the relative permeability obtained from waterflood data), does not appear to depend on flood rate (5). Moreover, provided that the phases remain continuous and immiscible then, in concept, the relative permeabilities should not depend on interfacial tension.

It is often suggested that the relative permeability curves will change shape as the interfacial tension is lowered, tending towards straight lines where the sum of the relative permeabilities is unity at all saturations (6). This view often results from the study of miscible or partially miscible displacement tests where diffusion processes act to distribute the fluid components equally over the pore structure of the rock. Diffusion processes have the effect of improving the displacement efficiency. Theoretically, the flood efficiency can be improved by straightening the oil relative permeability curve to reduce the rate of change of fractional flow of water. The resulting relative permeability curves for these miscible displacements are pseudo-functions with little predictive capacity. Steady-state tests using miscible systems are likely to produce straight line relative permeability data since a mixture of the components slows in all conductive flow channels.

As mentioned earlier, an alternative viewpoint for immiscible systems is that the oil drainage relative permeability curves will not change as interfacial tension is reduced. This would require that the oil recovery performance is independent of interfacial tension. Support for this view comes from the fact that increasing the capillary number (VL/γ) has little effect (assuming constant viscosities) on recovery performance, provided that capillary dispersion and end-effects are negligible (5). In addition, for oil-wet systems residual oil saturation is not a well defined quantity; in any case it is very low, provided that enough water has passed through the rock, and that end-effects are negligible. Reduction of interfacial tension will therefore have little effect on residual oil.

The conceptual reasoning for the argument that relative permeability does not change as the capillary number increases is that the displacing fluid will preferentially occupy pore channels with higher flow capacity, irrespective of interfacial tension. More experimental work is needed before definite conclusions can be drawn as to which relative permeability behaviour is relevant to low tension immiscible displacement.

3.2 Waterflooding – the water-wet case

For the case of waterflooding of a water-wet rock, the basic concepts of relative permeability are not strictly adhered to; principally because the non-wetting oil phase does not remain in communication. As mentioned earlier, the recovery performance of a waterflood on a water-wet core usually appears piston-like. Therefore no relative permeability data can be generated using conventional analysis of the flood performance, apart from end-point data.

Relative permeability curves can be generated from steady-state tests, but the data can vary depending on the test method. A common method is to change the fractional flow in steps which results in a saturation front travelling down the sample. Perhaps a true steady-state test should avoid such saturation gradients in the core sample. This situation can be approached by gradually changing the fractional flow over a long period of time. Possible shapes of the relative permeability curves are shown in Figure 4. The residual oil values obtained by the two steady-state methods are not necessarily the same, and might not agree with that obtained from a flood.

The shape of the relative permeability curves may be considered to be academic in the water-wet case, since both sets of relative permeability curves shown in Figure 4 will probably predict plug displacement if the Buckley-Leverett theory is used. However, there may be differences of predicted flood performance when using coarse grid finite difference numerical models, but this is an artifact of the model and is usually overcome by empirically altering curves of type 1 to be more like type 2. The truth of the matter is that we do not have a satisfactory theory to combine viscous and capillary flow effects, and we usually resort to some form of pseudo functions to match observation.

The residual oil saturation obtained after a waterflood is a definite value, in that oil flow completely ceases (unlike the oil-wet case). The residual oil level is dependent on initial oil saturation and on flood rate. The distribution of residual oil is also dependent on flood rate (1). In addition, high flood rate can make a weakly water-wet rock appear oil-wet (7,8).

It is noted from Figure 4 that the relative permeability to water at residual oil saturation is low, indicating that trapped oil occupies the largest pore channels. From the point of view of a field flood, this can be advantageous since water mobility following the flood front is kept low, thus suppressing viscous fingering on the reservoir scale.

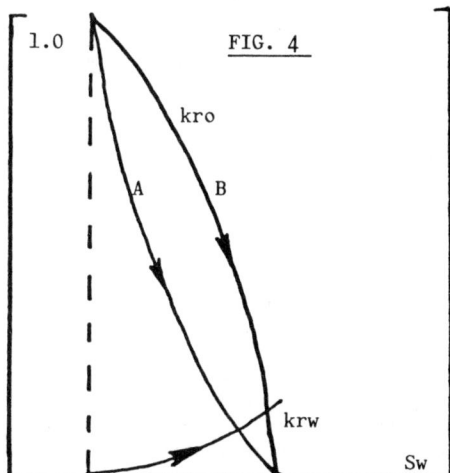

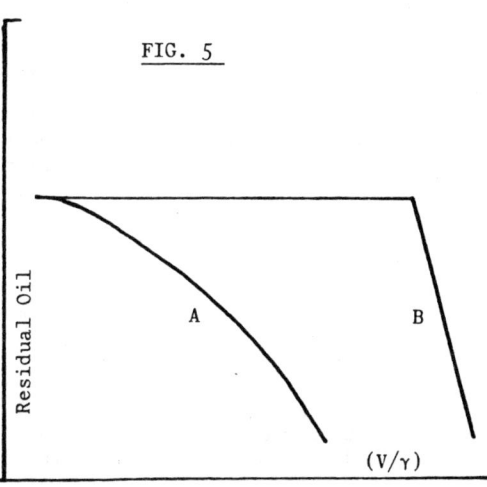

The residual oil saturation is also dependent on interfacial tension. The absence of imbibition in the case of very low interfacial tension can also make a water-wet sample look oil-wet. Considering the changes to the water-wet relative permeability curves as the interfacial tension is lowered, it is to be expected that significant changes will occur. Data obtained by the steady-state method (9) indicate that below an interfacial tension of 0.1 mN/m large changes in the water-wet curves occur. The tendancy is for the relative permeability curves to become straight lines, the sum of the relative permeabilities approaching unity at all saturations. As the interfacial tension is lowered during steady-state tests it is more likely that an emulsion of oil-in-water or water-in-oil will form.

As the interfacial tension decreases to very low levels the drop size
is likely to become extremely small. A situation will be approached
where the same fluid system is flowing in all conductive flow channels.
In this situation straight line relative permeability data might be
expected, although the basic concepts of relative permeability in fact
no longer apply.

Relative permeability data obtained from displacement tests have
shown somewhat different results (9). As the interfacial tension was
lowered the relative permeability to water increased and that to oil
decreased. At low interfacial tension (0.01 mN/m) the relative
permeability curves resembled oil-wet data. This supports the
proposal that oil drainage curves could be used as a first approximation for the back end of an oil bank even for a water-wet rock.

Curves of residual oil saturation versus V/γ (for constant
viscosity) can be generated by two methods:

i) Where each flood starts from the same initial oil
saturation (curve A, Figure 5).

ii) Having established a residual oil saturation at low values
of V/γ we can increase V/γ by increasing V or decreasing
γ. The residual oil saturation will eventually decrease
as trapped oil is mobilised (curve B, Figure 5).

4.   Application of laboratory data to enhanced recovery prediction

Reduction of residual oil saturation is a necessary but not
sufficient condition for a successful tertiary recovery flood. It may
be essential to develop and maintain an oil bank. Once an oil bank is
developed, it is the mobilised oil which collects and mobilises
residual oil at the front of an oil bank. Surfactant at the back of
the bank prevents retrapping of the oil.

We need to consider generation of secondary drainage relative
permeability data, starting from the residual oil saturation left after
primary imbibition, for application at the leading edge of the oil bank.
Curves of the type shown in Figure 6 can be generated from steady-state
tests. It has been suggested that the secondary drainage curves retrace
the primary imbibition curves (10, 11, 12) but this needs confirmation,
since it will depend on how the primary imbibition curves were obtained.
Prediction of oil flood performance obtained using steady-state relative
permeability data can be compared to oil flood data starting from
residual oil saturation after the primary waterflood.

If we work on the principal that the interfacial tension at the
back end of the oil slug is low enough to prevent retrapping of oil,
then as suggested earlier we might use oil-wet type relative permeability
curves to describe fractional flow at the back end of the bank. The
relative permeability set would then look like those shown in Figure 7.

Theoretical predictions using this type of data will be discussed
in the next section. Although many people have looked at the problem
of generating relevant relative permeability data for low tension floods
(13 - 21), there is still a need to obtain more data to clarify the
position.

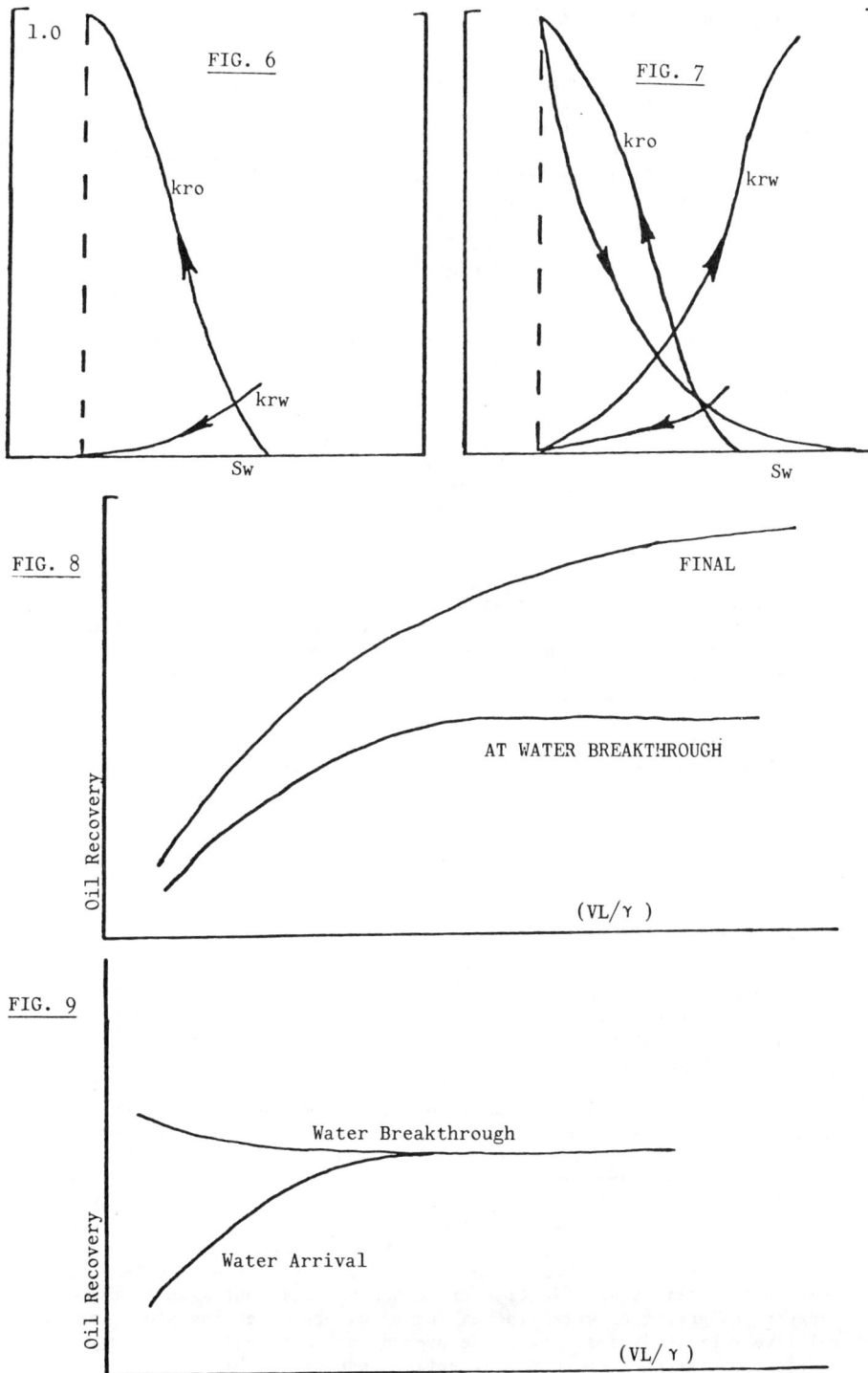

FIG. 6

FIG. 7

FIG. 8

FIG. 9

5.   OIL BANK PROPAGATION

In EOR processes the development and propagation of an oil bank is considered to be of importance;   certainly the correct analysis of such phenomena is crucial to core testing.   For highly complex EOR systems, analysis can only be carried out on the basis of assumptions which are difficult to validate, unless simplified systems can be studied.   Curiously little work has been reported which attempts to do this. A recent paper by Gladfelter and Gupta (22) is valuable, in that it sets some experimental evidence against which current views of oil bank propagation may be weighed. One of the key findings is that a region of increased oil saturation (an oil bank) can be generated from the fractional flow properties, without invoking other mechanisms. The additional feature proposed to explain this behaviour was hysteresis of the relative permeability curves.

The following outlines some features of oil bank propagation in two component, two phase (oil/water) displacement in cores as given by 1-D numerical simulation.   This can be compared with the experimental results and Buckley-Leverett analysis of Gladfelter and Gupta (22).   The test system simulated consists of a 45 cm core, divided into forty grid blocks, with an oil/water ratio of 3:1.   For each simulation run, a 0.0475 pore volume slug of oil was injected at high rate into the core, which was initially set at residual oil saturation.   The relative permeability hysteresis set first used was as shown as curves A and B in Figure 10(a), with the arrows indicating which limb corresponds to which directional change in water saturation.   The corresponding fractional flow curves are shown in Figure 10(b).   The corresponding oil slug behaviour as simulated is shown in Figure 11(a), showing a series of oil bank profiles as it progresses along the core.   The front of the oil bank progresses with a well defined front, but the size of the bank degrades as oil is 'lost' into the following tail.   Use of the same relative permeability set, but with the hysteresis directions reversed gives the behaviour shown in Figure 11(b);   in this the oil bank is not recognisably propagated since it is quickly dispersed.   This result highlights the sensitivity of oil bank propagation behaviour to the relative permeability curves.   They also show that numerical dispersion is not dominant in the present examples.

Following the procedure of Gladfelter and Gupta (22), injection of an oil and water mixture (in the present case, 25% oil) into a core at residual oil saturation gave the results depicted in Figure 11(c); a sharp oil front is formed, but without the presence of a bank of increased oil saturation as experimentally observed.   This result is as expected, since the numerical model moves incrementally up the relative permeability curves to the imposed injection composition;   it cannot overshoot this point, as can be achieved with   Rankine-Hugoniot shock front criteria as employed in a simple fractional flow analysis.   However, the principal of an oil bank stabilised by relative permeability hysteresis can be demonstrated by simulating injection of an oil bank followed by injection of an oil/water mixture.   This is shown in Figure 11(d). The importance of an oil bank induced by hysteresis effects extends not only to the valid identification of "enhanced" oil, but also to the correct assignment of water-increasing or water-decreasing steady-state relative permeabilities;   i.e. the presence of a transient bank may in fact involve changes in the direction reverse to that overall.

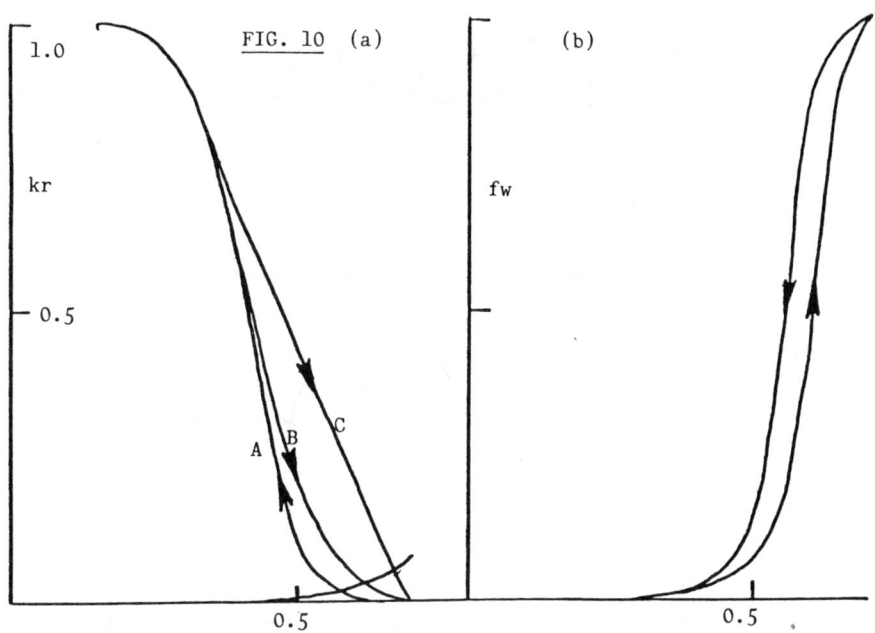

FIG. 10 (a)

kr

1.0

0.5

A B C

0.5

(b)

fw

0.5

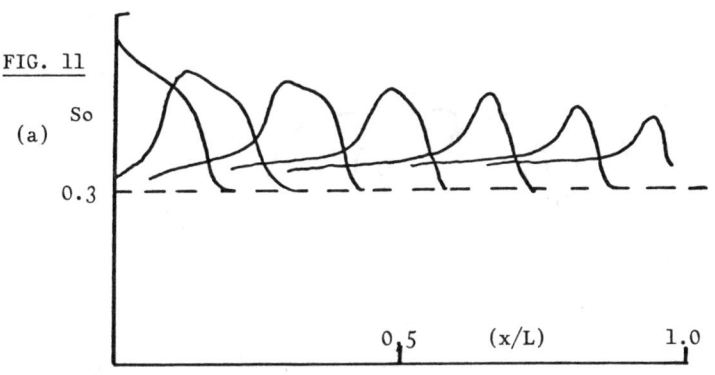

FIG. 11

(a) So

0.3

0.5 (x/L) 1.0

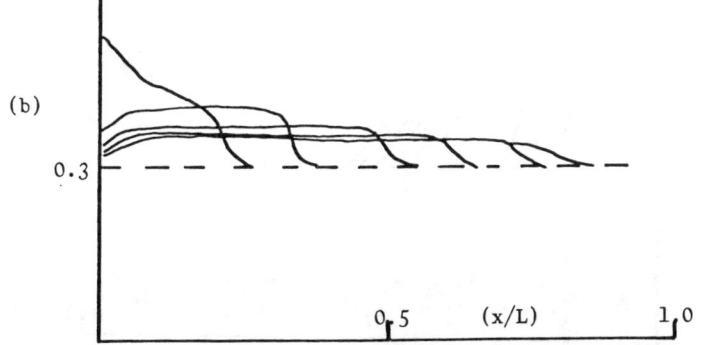

(b)

0.3

0.5 (x/L) 1.0

90

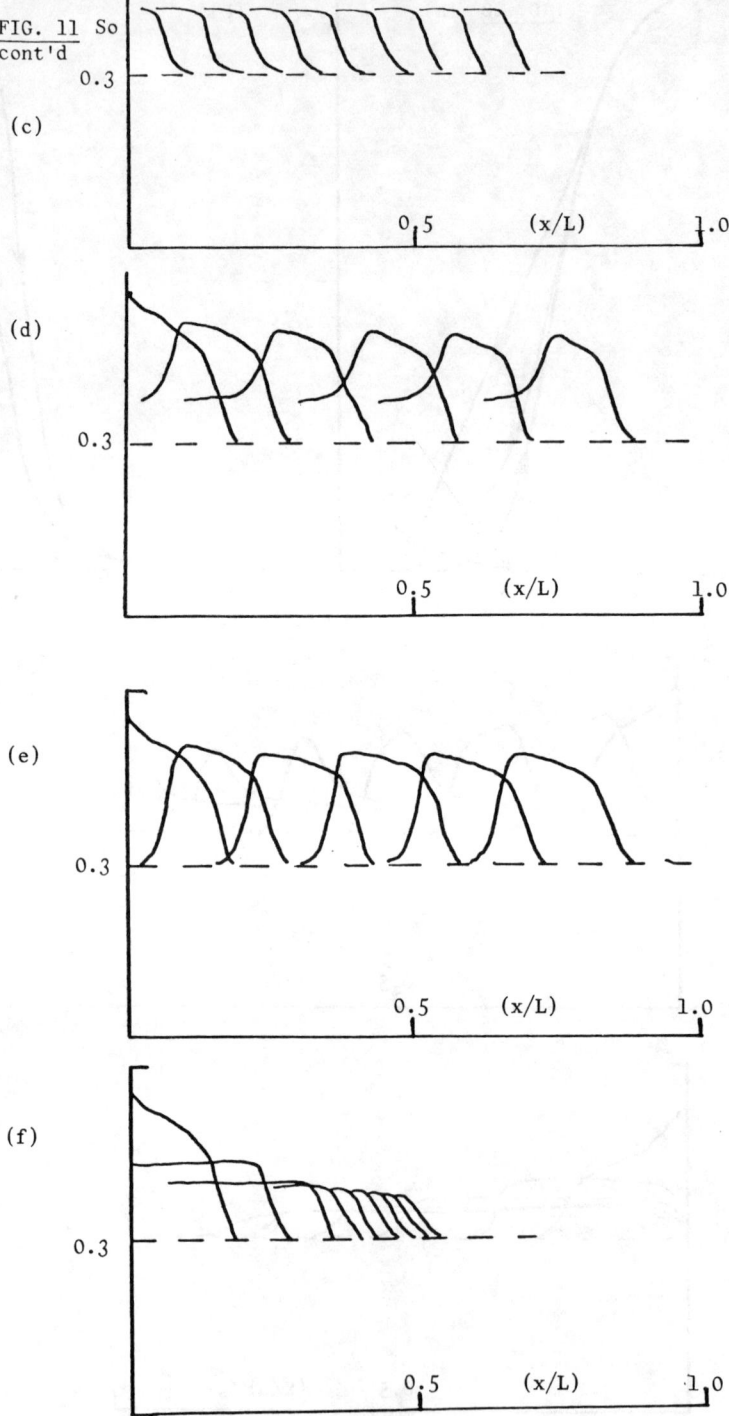

FIG. 11
cont'd

(c)

(d)

(e)

(f)

More drastic disparity between the relative permeability curves used at the leading and trailing fronts of an oil bank can of course give stabilised bank propagation without an oil/water mixture being subsequently injected. Figure 11(e) shows this, using curves A and C of Figure 10(a). The straightening of curve A to give curve C follows the commonly assumed functional change in the relative permeability curve induced by reducing the interfacial tension. However, experimental verification of the relative permeability behaviour in eor systems is still limited, and the assumed curves may not offer a good description of the physical processes involved. If the immiscible displacement of oil by surfactant acts as a high rate water flood as suggested earlier (since in both cases capillary forces are dominated by viscous forces) the aqueous phase will preferentially travel through the larger pore channels. This could result in a relative permeability behaviour as shown in Figure 7, with an increased krw, but with a kro curve falling below the prior curve at high oil saturations, i.e. the low tension flood does not have the imbibition which in a water-flood case ensures the oleic phase preferentially occupies the larger pores of a water-wet medium. The simulation results for this case (Fig. 11(f)) once more show degradation of an injected oil slug.

These simulation runs indicate the need to take proper account of relative permeability variations in the analysis and simulation of transient processes, and show that under certain conditions oil bank propagation will not occur. The use of assumed "ideal" relative permeability data (i.e. straight line extrapolations) may artificially predict stable bank propagation.

6. IRREVERSIBILITY AND HYSTERESIS IN CORE TESTS

In this section some modelling approaches to describe non-identity are outlined.

Core tests can display a wide range of irreversible characteristics. Jones and Rozelle (23) consider that irreversibility results from the common "S" shaped fractional flow curve if the conventional tangent constructions are applied to waterflooding and oil flood respectively. However, in considering the Buckley-Leverett theory, it can be noted that two solutions to the problem are to be expected, since the material balance may be applied for either the waterflood or the oilflooding directions. The conclusion that the direction appropriate to the particular stage of an EOR process must be used still holds. A more fundamental problem is whether the fractional flow curve itself is a unique function of saturation.

6. RELATIVE PERMEABILITY HYSTERESIS

6.1 Primary Hysteresis

The best documented "hysteresis" effects in relative permeability curves are associated with the irreversibility of the primary curves i.e. those curves which originate at initial conditions of complete saturation by a single phase. This hysteresis gives rise to residual (or irreducible) phase saturations. It is commonly assumed that essentially irreducible saturations can be established for both

wetting and non-wetting phases, although the distribution and properties
of these phases are by definition functions of wettability. Thus
it is usual to treat wetting and non-wetting phases asymmetrically.

The irreducible water saturation and the residual oil saturation
values are expected to be dependent on the initial saturation established
prior to reduction to residual. For strongly wetting phases, the
residual saturation tends to be ill-defined, so this effect is of little
consequence. For the non-wetting phase, Land (12) has proposed a semi-
empirical relationship which correlates initial and residual non-wetting
phase saturations; the Land relationship (taking water to be the wetting
phase and oil the non-wetting phase) is

$$(Sor^+)^{-1} - (Swi^+)^{-1} = C \qquad (1)$$

where $Sor^+ = Sor/(1- Swi)$

$Swi^+ = Soi/(1- Swi)$

with C being a constant for a given system.

Equations of this form have been used by Killough (24) to estimate
hysteresis sets descending from the primary non-wetting relative
permeability curve.

6.2 Secondary Hysteresis

The term secondary refers to those curves which start from the end
points of the primary curves. This region is (in principle) fully
accessible reversibly, that is in the saturation range Sw = Swi to (1-Sor).
In fact, it is only within a reversible range that the term hysteresis
can be truly applied. The characterisation of any hysteretic system can
prove complex, and for relative permeability curves the lack of experimental precision and the probable dependence of data on experimental
design (25) have so far prevented detailed analysis.

In the absence of a scientific approach, empirical relationships
are generally developed on the basis of plausibility and utility. It
may be questioned whether the initial formulation of the multiphase
flow relationships in the empirical Darcy-analogue form itself ensures
that all subsequent analysis can be no more than empirical.

A currently favoured parameterisation of relative permeability
curves uses a scaled power law relationship;

$$\left. \begin{array}{l} kro = kro(Swi)*(So^*)^{n_o} \\[2mm] krw = krw(Sor)*(Sw^*)^{n_w} \end{array} \right\} \qquad (2)$$

where

$So^* = (1-Sor-Sw)/(1-Swi-Sor)$

$Sw^* = (Sw-Swi)/(1-Swi-Sor)$

The secondary hysteresis can then be conveniently described by variation of the "n" exponents, as used by Evrenos and Comer (26) with an alternative different parameterisation scheme. A hysteresis parameter can be defined as

$$Hi \quad = \quad \left(\overset{+}{n}_o \Big/ \overset{+}{n}_w \right) \tag{3}$$

and similarly for the water exponents, where the superscript arrows denote increasing and decreasing water saturation. It can be noted that dependent on the value of Hi the hysteresis scan can occur in either direction, both types of behaviour being reported from experimental observations.

The variation of the relative permeability curves for an EOR process requires detailed description if a numerical simulation of the process is to be made. For surfactant flooding, the parameterised relative permeability curves are often modified in a systematic manner dependent on the new value of residual oil saturation (Sorc). Following this procedure with the hysteresis parameter, it is to be expected that the hysteresis becomes less pronounced as the residual

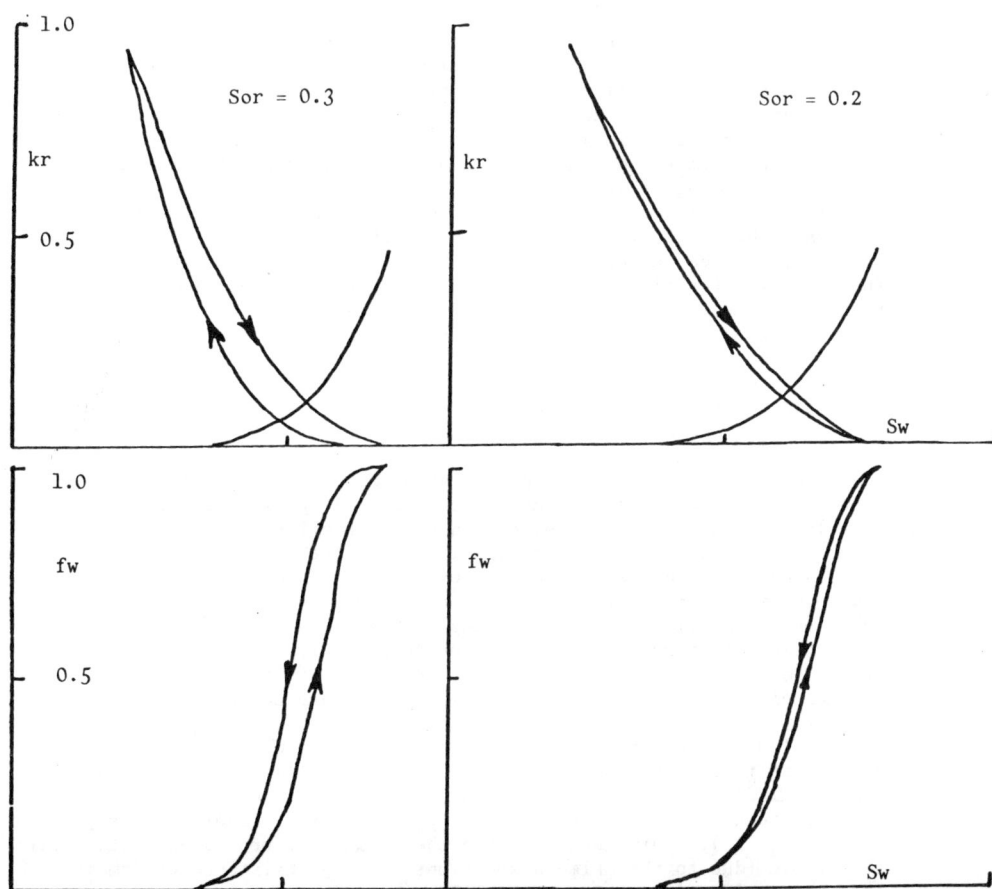

FIG. 12

oil saturation is decreased (since any conceptual model of the hysteresis mechanism centres on irreversibility following loss of hydraulic connectivity during phase trapping). Paralleling other dependencies on Sorc, a logarithmic relationship between Hi and Sorc can be used, or more simply a power law expression.

$$\frac{1 - Hi\ (Sorc)}{1 - Hi} = \left(\frac{Sorc}{Sor}\right)^a \tag{4}$$

Two of a family of relative permeability curves using this relationship is shown in Figure 12, with the dependence of the n_o exponent on Sorc being

$$\frac{1 - n_o\ (Sorc)}{1 - n_o} = \frac{Sorc}{Sor} \tag{5}$$

6.3   Scanning Hysteresis

Conventional oil recovery processes are usually described in terms of monotonic changes of saturation. Enhanced oil recovery processes aimed at mobilisation of post waterflood residual oil necessitate changes in the forms of the relative permeability curves and in the directions in which a process description scans them. In order to estimate how scanning from intermediate points of the secondary bounding curves will progress in the absence of definitive experimental evidence, parallels can be sought in other hysteretic systems. The other hysteretic process which is well established in petroleum engineering experience is the capillary pressure as a function of saturation. Since the same physical process gives rise both to relative permeability and to capillary pressure hysteresis, it is to be expected that mutually consistent descriptions should be possible. In order to compare the two sets of phenomena, the relative permeabilities need to be expressed as a single variable, and the fractional flow can be considered to be a state variable which offers more hope of comparability with other hysteresis phenomena.

The most direct consistency relationship between capillary pressure scans and fractional flow scans would be if one set were directly mapable onto the other. However, since the fractional flow set is rigorously bounded whereas the capillary pressure set is asymptotically bounded, this can at best be an approximation. In addition, the commonly made approximation is also present in the comparison of equilibrium capillary pressure values with dynamic systems. Attempts to correlate primary capillary pressure curves with fractional flow curves have not to date proven sufficiently satisfactory to enable mapping of scanning curves from capillary pressure to fractional flow data.

A less direct appeal to consistency can be made by application of similar functional forms to describe both capillary pressure and fractional flow hysteresis sets. The most general theory available to describe hysteresis systems is Independent Domain Theory, as propounded by Everett (27). Unfortunately, the theory has not proved quantitatively applicable (due to the prime assumptions i.e. the existence of domains and their independence, Topp and Miller (29)), but does provide a qualitative framework in which empirical relationships can be set.

In recent years a number of studies have been carried out in which unsaturated flow in porous media has been analysed numerically using empirical analytic hysteresis functions. However, these studies have concentrated on liquid/vapour systems of interest to hydrologists rather than on liquid/liquid systems and the hysteresis relationship examined has been between the saturation and flow potential. Applying an analogue of the empirical hysteresis function as recently used by Pickens and Gillham (29) to the fractional flow/saturation system gives

$$f_w = f_i \left\{ \frac{\cosh(S_w/S')^\alpha - (f_i - f_j)/(f_i+f_j)}{\cosh(S_w/S')^\alpha - (f_i - f_j)/(f_i+f_j)} \right\} \qquad (6)$$

where S', α, f_i, and f_j are fitting parameters which can be identified from sufficient data.

The functional form of this relationship can reproduce fractional flow curves, and scans within those curves which accord with the expected hysteretic behaviour as described by independent domain theory. However, this equation is not convenient in that the relative permeability hysteresis set cannot be explicitly separated from this representation.

A more flexible approach is to follow that which Killough (24) applied to capillary pressure hysteresis, by which scanning curves are described by a scaled combination of the bounding curves. Extending this to the relative permeability and fractional flow sets gives relationships of the type

$$\acute{kri} = \overset{+}{kri} - F(\overset{+}{kri} - \overset{+}{kri}) \qquad (7a)$$

$$\acute{fi} = \overset{+}{fi} - F(\overset{+}{fi} - \overset{+}{fi}) \qquad (7b)$$

where $F = \dfrac{(Si - \acute{Si} + \alpha)^{-1} - 1/\alpha}{(Simax - \acute{Si} + \alpha)^{-1} - 1/\alpha}$

where the subscript $\acute{}$ refers to a scanning curve from point Si, the superscript arrows refer to the bounding curves, and α is a fitting parameter. Examples are shown in Figure 13.

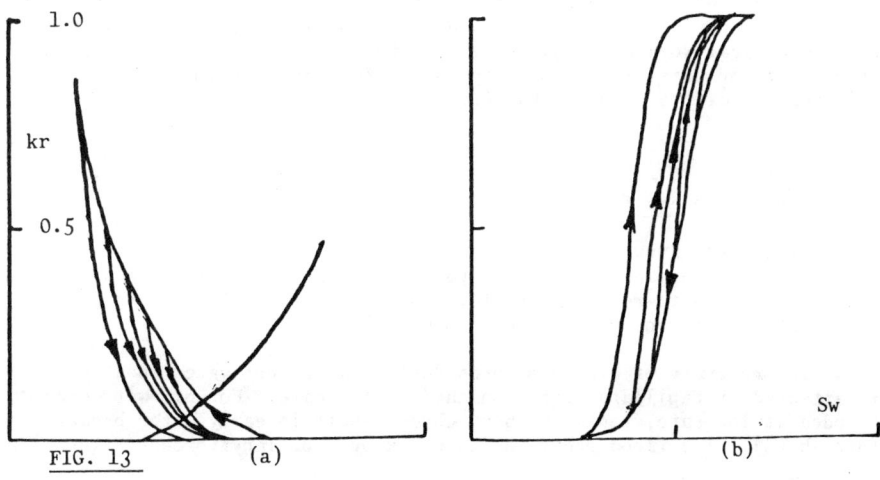

FIG. 13   (a)   (b)

The above discussion highlights the empirical way in which relative permeability hysteresis can be handled at present. For other important factors, such as core end effects and oil saturation residual to surfactant flooding, the position is even less well defined.

7. DISPLACEMENT TESTS FOR ASSESSING POTENTIALLY USEFUL SURFACTANT SYSTEMS

7.1 Screening tests

Any potentially useful surfactant system must be able to reduce interfacial tension between oil and water by a significant amount (e.g. down to 10^{-3} mN/m). Having found a surfactant capable of doing this, the next step is to conduct screening tests using glass bead or sand pack columns. If successful, the next step is to conduct core tests.

As we have discussed, the wetting conditions of a core sample play a large part in the displacement characteristics. It is therefore of considerable importance to match reservoir wetting conditions when conducting core tests. In order to do this we use preserved core material which is carefully prepared in order to change the surface wetting as little as possible. Our present procedure involves the flushing of live crude oil through the core at reservoir temperature.

The core is then left to "age" in contact with crude oil for several weeks. A low rate waterflood is then conducted (e.g. 1 ft/day advance rate) to establish a residual oil saturation. Before injecting surfactant solution, the brine flow rate is increased to ~ 200 ft/day. This is essential to ensure that oil left in the core is "trapped oil" and not oil retained by core-effects. Surfactant is then injected into the core at ~ 1 ft/day. Careful monitoring of the recovery performance enabled theoretical predictions to be tested. The theoretical work may show up reasons for lack of success in core flood tests or possible reasons why successful core tests may not necessarily lead to a successful field test.

7.2 Core scale artifacts

It is well known that laboratory tests on core samples can produce misleading information due to the small scale of the core plug in relation to reservoir scale. This is so even if the core plug represents a homogeneous reservoir rock.

The recovery of oil from cores during laboratory waterflood tests is affected by capillary forces on the sample scale. For oil-wet cores flooded at low rate, tests on short cores result in early water break-through (5). The flood front is dispersed by capillary forces (Figure 8).

If the core length or flood velocity is increased or the interfacial tension is decreased, then the oil recovery at water breakthrough is improved. If a low rate flood test is carried out on a core sample, e.g. to match reservoir type flow rates, then the oil recovery at water breakthrough may show an increase as the interfacial tension is decreased. However, on the reservoir scale, the reduction of interfacial tension would have no effect, according to the curves in Figure 8.

A similar situation arises in the case of residual oil saturation (i.e. after sufficient water has been passed through the core so that oil flow ceases). Capillary forces cause retention of oil at the outlet face of the core sample giving the impression of high residual oil saturation, particularly if the flood is carried out at rates representing reservoir flow rates. Reduction of interfacial tension can result in significant production of this end-effect oil which would not be representative of the reservoir scale.

In order to improve the situation, long cores are sometimes used to reduce the influence of end-effects. Small core plugs may be butted together to create a long core. Careful arrangement of the small plugs is necessary in order that relevant data is produced.

Turning to the water-wet case, the recovery of oil at water-breakthrough is less affected by sample length than in the oil-wet case (8). Water may arrive early at the outlet end of the sample but capillary forces delay water-breakthrough (Figure 9). The representation given in Figure 9 only applies over a small region of flow rate and interfacial tension, where capillary forces completely dominate fluid distribution on the pore scale. Above a certain flow velocity or below a certain interfacial tension, viscous forces will start to take over as the dominant factor controlling fluid distribution. Increase of sample length, however, will have no effect on fluid distribution on the pore scale.

The above discussion highlights the importance of careful interpretation of laboratory data if valid information is to be produced. In our discussions we have tended to consider the extreme case of oil-wet and water-wet rocks. In practice rocks may have a variety of wetting conditions. Indeed not all of the pore surface may have the same wetting. We have also considered an ideal situation where clays within the porous media do not effect the fluid flow characteristics. Adsorption of surfactants on to pore and clay surfaces has not been discussed. Full consideration of these interesting effects is outside the scope of this paper.

7.3 Simulation aspects

Preliminary analysis of core tests of EOR systems can be carried out with linear scaling relationships. Investigation of the sensivity to particular system properties in the optimisation of a system requires the use of more detailed simulation methods. In particular, the separation of core scale effects as discussed above demand careful study if the experimental results are to characterise the physical processes relevant to the field. In all cases the data must be considered in terms of its relevance to its intended use on input to a field simulator, with the attendant changes in scale which can only be treated mathematically. An additional role of the simulation of core floods is provided by the greater degree of experimental accessibility, and the ability to repeat laboratory work, as contrasted with the field situation. Despite the benefits of validation of a simulation model on the basis of laboratory results, there is no guarantee of success in the field (although, conversely, failure at the laboratory scale cannot easily be reconciled with confidence in subsequent field application). The simulation provides the language by which the dialogue between the laboratory and the field testing can proceed, and it is therefore important that it should be able to describe the real physical processes rather than aim at mere plausibility and utility.

CONCLUSIONS

1.   Despite its extensive use, the concept of relative permeability remains empirical. As with any empirical quantity, great care must be exercised when attempting to extend its area of applicability. In particular, the relative permeability functions which are appropriate to low tension immiscible displacement are as yet uncertain.

2.   The behaviour of an oil slug injected into a core provides a prototype to assess many of the assumptions which are present when core data are prepared for simulation of EOR processes. The availability of a correctly defined set of relative permeability curves, complete with hysteresis, irreversibility and system changes is of great importance to the correct modelling of an oil bank. Although in field use, such concepts may be of secondary importance to the heterogeneity, they cannot be ignored in the provision of data from core tests. Currently available simulation schemes are empirical; they require development and experimental validation.

3.   Numerical schemes which essentially scan relative permeability curves demand that these curves are defined over the full saturation interval. The sensivity to these curves of oil bank stability (and presumably all multi-front systems) is such that plausibility alone is too weak a criterion for acceptability. In addition, the behaviour of transient saturations associated with multiple shock fronts may not be adequately modelled.

Acknowledgements

Permission to publish this paper has been given by the British Petroleum Co. Ltd.

The authors wish to thank Mr. J.F. Berry and Dr. I. White who carried out the simulation work described in this paper.

List of Symbols

kr	relative permeability
S	fractional saturation
f	fractional flow
V	superficial velocity
γ	interfacial tension
L	core length
x	distance along core

a, C, F, fi, Hi, n, S^+, S^*, α,   - parameters, defined in the text.

Subscripts

o   oil
w   water
r   residual
i   irreducible

References

1. L.L. Handy and P. Datta   Fluid distributions during immiscible displacement in porous media. Soc. Pet. Eng. J, Sept. 1966 page 261.

2. R.A. Salathiel   Oil recovery by surface film drainage in mixed wettability rocks. SPE preprint 4104.

3. Johnson, E.F., Bossler, D.P. and Naumann, V.O. Calculation of relative permeability from displacement experiments. Trans AIME (1959) 216, 370-372.

4. Buckley, S.E. and Leverett, M.C.   Mechanisms of fluid displacement in sands. Trans AIME (1942) 146, 107 - 116.

5. L.A. Rapoport and W.J. Leas   Properties of linear waterfloods. Trans AIME (1953) 198, 139 - 148.

6. C. Bardon and D.G. Longeron   Influence of very low interfacial tension on relative permeability. Soc. Pet. Eng. J, Oct 1980 page 391.

7. F.F. Craig Jr. 1971. The reservoir engineering aspects of waterflooding. spe Monograph Vol. 3 H.L. Doherty Series. Page 24.

8. Kyte J.R. and Rapoport L.A.   Linear waterflood behaviour and end effects in water-wet porous media. Trans AIME (1958) 213, 423 - 426.

9. J.O. Amarefule and L.L. Handy 1981. The effect of interfacial tensions on relative oil-water relative permeabilities of consolidated porous media. SPE Preprint 9783

10. T.M. Geffen, W.W. Owens, D.R. Parrish and R.A. Morse Experimental investigation of factors affecting laboratory relative permeability measurements. Trans AIME (1951), vol. 192, page 99 - 110.

11. C.R. Sandberg, L.S. Gournay and R.F. Sippel   The effect of fluid-flow rate and viscosity on laboratory determinations of oil-water relative permeabilities. JPT, 1958, 36 - 43.

12. C.S. Land.   Comparison of calculated with experimental imbibition relative permeability Soc. Pet. Eng. J (Dec 1971) 419 - 425.

13. M.C. Leverett   Flow of oil-water mixtures through unconsolidated sands. Trans AIME (1939) 132, 149.

14. N. Mungan   Interfacial effects in immiscible liquid-liquid displacement on porous media. Soc. Pet. Eng. J (1966) 6, 247 - 253.

15. A.W. Talash 1976. Experimental and calculated relative permeability data for systems containing tension additives. SPE preprint 5810.

16. H.E. Gilliland and F.R. Conley.   Surfactant waterflooding
     Proc. of 9th World Pet. Congress, Tokyo May 11-16, 1975.

17. J.P. Batycky and F.G. McCaffery   Low interfacial tension
     displacement studies. paper 78-29-26, 29th Annual Tech.
     meeting of the Pet. Soc. of CIM, Calgary, Canada (June 13-16,
     1978).

18. H. Asar   Influence of interfacial tension on gas-oil
     relative permeability in gas-condensate systems. PhD dissertation
     University of Southern California (May 1980)

19. C.P. Thomas, W.K. Winter and P.D. Flemings 1977. Application of
     a general multiphase multicomponent chemical flood model to
     ternary, two-phase surfactant systems. SPE preprint 6727.

20. S.P. Gupta and S.P. Trushenski   Micellar flooding
     compositional effects on oil displacement. Soc. Pet. Eng. J
     April 1979 116 - 128.

21. G.A. Pope   The application of fractional flow theory to
     enhanced oil recovery. Soc. Pet. Eng. J. (June 1980) 191 - 202.

22. Gladfelter, R.E. and Gupta, S.P. Effects of Fractional Flow
     Hysteresis on Recovery of Tertiary Oil. SPE J (Dec. 1980)
     508 - 520.

23. Jones, S.C. and Rozelle W.O. Graphical Techniques for Determining
     Relative Permeability from Displacement Experiments. SPE J.
     (May 1978) 807 - 817.

24. Killough, J.E. Reservoir Simulation with History Dependent
     Saturation Functions. SPE J (Feb. 1976) 37 - 48.

25. Lin, C.Y. and Slattery, J.C. Three-Dimensional, Randomised
     Network Model for Two Phase Flow Through Porous Media.
     SPE 9803, 1981.

26. Evrenos, A.I. and Comer, A.G. Numerical Simulation of
     Hysteretic Flow in Porous Media. SPE 2693, 1969.

27. Everett, D.H. and Smith, F.W. A General Approach to Hysteresis.
     Trans. Faraday Soc. (1954) $\underline{50}$, 187 - 197.

28. Topp, G.C. and Miller, E.E., Hysteretic Moisture Characteristics
     and Hydraulic Conductivities for Glass Bead Media. Soil.
     Sci. Soc. Amer. Proc. (1966) $\underline{30}$, 156 - 162.

29. Pickens, J.F. and Gillham, R.W., Finite Element Analysis of
     Solute Transport Under Hysteretic Unsaturated Flow Conditions
     Water Resour. Res. (1980) $\underline{16}$, 1071 - 1078.

EXPERIMENTAL STUDY AND INTERPRETATION OF SURFACTANT RETENTION IN POROUS MEDIA

J. NOVOSAD

Petroleum Recovery Institute Calgary, Alberta, Canada T2L 2A6

ABSTRACT

Total retention of surfactants in a reservoir during chemical floods is probably one of the most important parameters in that it determines the economic success or failure of this enhanced oil recovery process. It is not, therefore, surprising that a substantial research effort has been devoted to laboratory evaluations of surfactant retention in porous media. Generally, the systems studied are complex as they contain a minimum of two liquid phases, and no less than five components: surfactant, cosurfactant, electroyte, water, and oil. The principal objective of this paper is to analyze and evaluate experimental procedures for determining surfactant adsorption and total surfactant retention. It is shown that the interpretion of experimental data is not straight forward, and that extreme caution must be exercised before any interpolation or extrapolation of adsorption or retention data is attempted.

Laboratory data on retention of pure sulfonate (Texas #1), petroleum sulfonate (TRS 10-80), and synthetic sulfonate (FA 400) are presented. These show the importance of experimental conditions as similar experiments may yield substantially different results when conditions are varied. Specifically, the effect of phase behavior on total surfactant retention is discussed and experimental procedures are outlined so that a differentiation can be made between losses of surfactant due to unfavorable phase behavior and those due to adsorption at the solid-liquid interface. An example of how experimental conditions may affect the measured values of surfactant losses is shown by the effect of surfactant slug size on the apparent level of retention and adsorption.

INTRODUCTION

Adsorption of surfactants considered for enhanced oil recovery applications has been studied extensively in the last few years[1-6] as it has been convincingly shown that it is possible to develop surfactant systems which displace oil from porous media almost completely when used in large quantities. Effective oil recovery by surfactants is not a question of principle but rather a question of economics. Since surfactants are more expensive than the crude oil, development of a practical enhanced oil recovery (EOR) technology depends on how much surfactant can be economically sacrificed in recovering additional crude oil from a reservoir. Therefore, it is quite clear why surfactant adsorption has always been considered critical to the success or failure of this EOR process.

It was recognized earlier that physico-chemical adsorption may be only one of a number of factors which contribute to total surfactant retention. Other physico-chemical mechanisms may include surfactant entrapment in an immobile oil phase[5], surfactant precipitation by divalent ions[6], surfactant precipitation caused by a separation of cosurfactant from surfactant[4], and surfactant precipitation due to chromatographic separation of different surfactant species[7]. When complications arising from ion-exchange phenomena usually involved in surfactant flooding are included, it should not come as a surprise that measured adsorption isotherms differ substantially from those previously observed for simpler surfactant systems.

A principal objective of this work is to evaluate the experimental techniques that can be used for measuring surfactant adsorption and to study experimentally two mechanisms responsible for surfactant retention. Specifically, an attempt is made to differentiate between the adsorption of surfactants at the solid-liquid interface and the retention of surfactants due to trapping in the immobile hydrocarbon phase which remains within the core following a surfactant flood.

MEASUREMENT OF ADSORPTION AT THE SOLID-LIQUID INTERFACE

Previous adsorption measurements of surfactants considered for enhanced oil recovery produced adsorption isotherms of unusual shapes with unexpected features. Primarily, an adsorption maximum has been observed when total surfactant retention has been plotted against the concentration of injected surfactant. Numerous explanations have been offered for these peaks; such as, a formation of mixed micells[6], the effects of "structure forming" and "structure breaking" cations[8], and the precipitation and consequent redissolution of divalent ions[2]. Which of these effects are responsible for the peaks in a particular situation and their relative importance is difficult to asses. However, it seems that, in view of the number of processes that are taking place simultaneously and the large number of components present in most of the systems studied, one should not expect smooth monotonically increasing isotherms that are patterned after adsorption isotherms for only two pure components. Also, it should be realized that most experimental procedures do not yield an amount of surfactant adsorbed but rather the surface excess.

It is shown next that an adsorption isotherm expressed in terms of the surface excess as a function of an equilibrium surfactant concentration must, by definition, contain a maximum if the data are measured over a sufficiently wide range of concentrations. It has been shown repeatedly that, for adsorption at the solid-liquid interface, the surface excess is the only consistently defined experimental variable which should be used in describing the preferential uptake of one component over another into the adsorbed layer[9]. The surface excess is defined by Equation (1)L:

$$n_i^e = \frac{n^o}{m}(x_i^o - x_i) \tag{1}$$

where, n^o = total mass of the liquid system (g),

x_i^o = relative concentration of component i in the bulk phase before adsorption takes place (fraction),

x_i = relative concentration of component i in the bulk phase after adsorption equilibrium is attained (fraction),

n_i^e = excess of mass of component i in the adsorbed phase per mass unit of adsorbent (g/g),

m = mass of adsorbent (g),

and an adsorption isotherm is defined as the surface excess dependence on equilibrium concentration of component i in the bulk phase:

$$n_i^e = f(x_i), \quad 0 < x_i < 1 \tag{2}$$

It should be clear from Equation (1) that the surface excess must be equal to zero for pure components ($x_i = 0$, $x_i = 1$) and, therefore, a non-zero adsorption isotherm must contain at least one peak. This applies to fully miscible systems in which adsorption isotherms are meaningful over the entire concentration range between pure component 1 and pure component 2. However, solubility limits of most surfactants in reservoir brines are reached at low concentrations, and measurements of adsorption above such concentration levels are meaningless as surfactant precipitation takes place. Therefore, a presence or an absence of a maximum in an adsorption isotherm is dependent upon surfactant solubility in a brine or other continuous phase.

Equation (1) applies also to multicomponent systems, however, such adsorption isotherms cannot be expressed graphically in a two-dimensional form without specifying a direction in which the adsorption surface (for three-component systems) is cut for viewing in two dimensions [4]. It is not practical to perform adsorption experiments with multicomponent systems in such a way that the adsorption surface is cut in a pre-determined way since it is not usually known apriori what the bulk composition of individual components is going to be after adsorption has taken place.

This problem is even more complicated when designing experiments with surfactant mixtures which are considered for surfactant flooding as individual components of the mixture are difficult to separate and, consequently, the whole mixture is usually treated as a single component. Even if it is assumed that a chosen analytical method can determine the sum of several surfactants accurately, it has been previously shown that isotherms for individual components may differ substantially from an overall isotherm.[4]

This is of importance for systems in which each component within the mixture serves a different function, such as either achieving an ultra low interfacial tension or improving the solubility of other components. The overall adsorption isotherm is then not suitable for predicting the system performance during the flood because depletion of individual components is not represented by the overall isotherm.

Treating the mixture as a single component brings additional uncertainty to adsorption experiments. As the individual components are adsorbed to different extents, their equilibrium concentrations become different from those originally present so that they will depend on specific experimental conditions such as the adsorbent/solution ratio. The apparent adsorption level may vary substantially from one experiment to another as equilibrium concentrations may be moving on the adsorption surface in different directions depending on the specific conditions of the experiment. These are the main reasons why experimental data from different laboratories are so difficult to compare and why a great deal of caution should be exercised in interpreting the shapes of isotherms which were determined in experiments in which the initial relative concentration of each individual component was held constant while the total concentration was varied.

METHODS FOR MEASUREMENT OF ADSORPTION ISOTHERMS

There are two distinctly different approaches for measuring adsorption at the solid-liquid interface. The first, a batch method, consists of measuring surfactant concentrations in the bulk phase before and after adsorption takes place, and adsorption is calculated from Equation (1). Since all measurements are performed in the bulk phase, the meaning of each variable in Equation (1) is without ambiguity and the calculated surface excess is a valid thermo-dynamic variable.

A main disadvantage of using the batch method lies in its poor accuracy at higher surfactant concentrations. The measured change in bulk surfactant concentration due to adsorption becomes small and adsorption is obtained by subtracting two numbers of similar size. It has been shown with surfactant systems considered for enhanced oil recovery that exceedingly accurate analytical methods are required for measurement of adsorption from solutions when surfactant concentrations exceeds one percent.

In the second method, a dynamic one, the retention of surfactants is determined from a flow-type experiment, and the losses of surfactant are calculated either from the delay of the surfactant breakthrough curve, if the amount of surfactant injected is so large that the effluent concentration reaches the injected concentration, or from the material balance when a smaller amount of surfactant is injected (Figure 1).

It should be realized that, if the surfactant concentration at the core outlet does not reach the injected concentration, then the adsorption determined from a material balance is attained over a concentration range that extends from the injected level at the core inlet to the maximum effluent concentrations measured at the core outlet. For example, if the injection of a 20% PV sur-factant slug results in a maximum outlet concentration corresponding to 10% of the injected concentration, the average surfactant concentration within the core would be approximately 55% of the injected level. A larger slug may give a maximum outlet concentration of 90% resulting in the average concentration within the core being about 95% of the injected concentration. Therefore, the retention values obtained from these two experiments would be related to different concentration regions, and the results would not be directly comparable.

A similar argument can be made when a different degree of separation between a cosurfactant and a surfactant occurs in experiments in which varying slug sizes are used. An example of such a separation is shown in Figure 2, and again retention results should be different since it has been shown previously that alcohol concentrations affect surfactant retention substantially.[10]

There are two additional considerations concerning measurements of adsorp-tion in displacement experiments in which an oil phase is present and when it is displaced from a core by a surfactant solution. In this case, adsorption cannot be determined directly from the delay of the breakthrough curve for two reasons. Firstly, the pore volume available to the surfactant is changing during the flood as oil is displaced. This makes the integration indicated in Figure 1(a) more difficult since the amount of oil recovered must be known at each point in the flood. Secondly, the possibility of surfactant being dis-tributed between the brine and hydrocarbon phases during the flood, in a way that differs from the injected solution, must be considered. When this occurs, the surfactant flow velocity becomes dependent not only on adsorption but also on two-phase flow characteristics. Unless the surfactant distribution is known precisely at each stage of the food, the delay in the breakthrough curve may be caused by either of the two phenomena, and a distinction between them cannot be made.

The previous paragraphs were intended to show that thermodynamically valid adsorption measurements can be best performed in batch experiments. However, since their sensitivity is often not sufficient for measurements of adsorption

from surfactant solutions of higher concentrations, displacement experiments
must be used for such systems. It should, however, be realized that such tests
measure averaged values of adsorption over concentrations that range from the
maximum effluent concentration to the injected concentration. The adsorption
values are also averaged over the range of cosurfactant/surfactant ratios which
depend on the specific characteristics of the surfactant system.

It follows that even minor variations in the displacement experiments can
produce substantially different results in terms of surfactant retention and
adsorption. Therefore, it is imperative that every adsorption measurement be
described in detail so that the results from different laboratories can be
realistically compared. This is not presently the case, as manifested in the
recent paper by Meyer and Salter[1] who surveyed the literature to determine the
effects of an oil presence on surfactant retention and found that an increase,
a decrease, or no effect had been observed. It is entirely possible that the
different experimental techniques and procedures could have been responsible
for such divergent results.

Another phenomenon affecting the retention of surfactants should be treated
separately. Surfactant solubility in reservoir fluids could possibly be grouped
within the phase behavior category but, since it may affect retention of
surfactants so substantially, it is discussed separately.

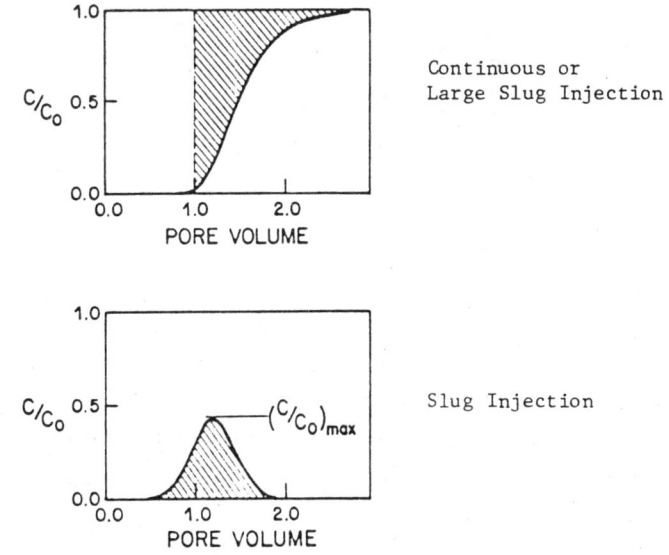

Figure 1: Determination of Surfactant Retention from
Displacement Experiments

It has been previously shown that there is an order of magnitude difference
in the retention of surfactants from dispersed solutions and from solutions in
which the surfactants are truly dissolved[4]. Even though most injected surfactant
solutions used in adsorption studies contain alcohols as cosurfactants in order
to keep surfactants fully dissolved, this may not be the case in the later stages
of a flood. Alcohols propagate through the porous mediua at different

velocities than surfactants because they distribute themselves between the oil and the brine differently than do the surfactants (Figure 2). Should this happen, it is likely that the surfactant loses its solubility in the brine, precipitates, and loses its ability to propagate through the core. This will result in an apparent increase in surfactant retention which cannot be easily distinguished from either adsorption or trapping in the hydrocarbon phase. Therefore, in most experiments described in this work, alcohols were used in excess quantities thereby eliminating or substantially reducing this possibility. In all cases, alcohol concentrations in the effluent were monitored so that a potential problem of poor surfactant solubility could be assessed at the end of each flood.

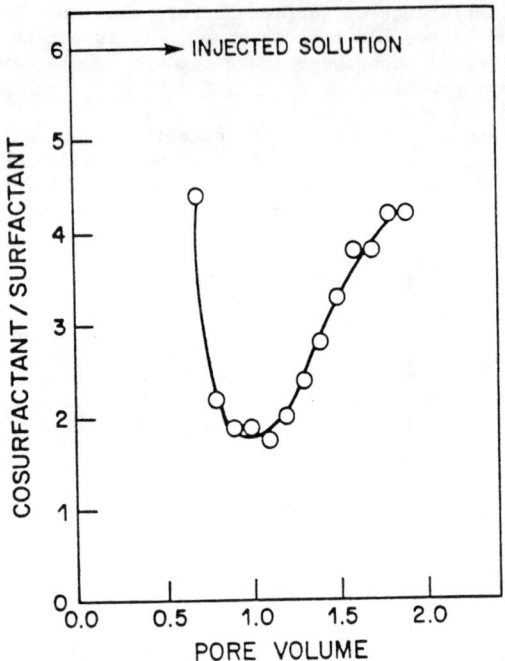

Figure 2: Cosurfactant/Surfactant Ratio at the Core
Outlet (100% PV Injection of 2% 1/6 Texas #1/
n-Propanol in 1.5% NaCl Brine)

EXPERIMENTAL

The main objective of this work is to determine surfactant retention in Berea cores with the main emphasis being to distinguish physico-chemical adsorption of surfactants from losses of surfactants due to trapping in the immobile hydrocarbon phase that is left in the core after a flood.

Chemicals

Three types of surfactants were used during the course of this study. TRS 10-80 served as an example of a commercial quality petroleum sulfonate which is produced by a direct sulfonation of petroleum-based feedstocks. PDM 337 was an example of a synthetic sulfonate, and Texas #1 was a well-defined pure surfactant, the structure of which was patterned after typical molecules found in petroleum-based feedstocks.

Texas #1 -   sodium salt of 8-phenyl-n-hexadecyl-p-sulfonate was obtained from Professor Wade of the University of Texas and has been used as received. According to Frances et al.[11] the purity of the sample exceeds 98%.

PDM 337 -   monoethanol amine salt of alkyl orthoxylene sulfonate supplied by Exxon Chemicals, Houston, Texas. According to the supplier, it is 84% active with a median alkyl chain size of around C_{12}. This surfactant was used as received.

TRS 10-80 -   petroleum sulfonate supplied by Witco Chemicals. Samples were desalted and deoiled according to the procedures described by Shah et al.[12]

Octane -   technical grade supplied by Phillips Petroleum Company.

Orthoxylene - boiling point range 143.5° - 144.5°C supplied by Matheson, Coleman and Bell Co.

Secondary Butyl alcohol - boiling point range 98° - 100°C, supplied by Eastman Kodak Company.

Brine   NaCl supplied by Fisher Scientific dissolved in deionized water.

Adsorbent -   Berea sandstone cores with air permeability range of 100 to 1,200 md, supplied by Cleveland Quarries.

Analytical Methods

The precision of analytical methods is of utmost importance in all studies of adsorption at the solid-liquid interface. The following methods have been extensively tested and employed for concentration determination in this work.

Surfactants - UV Spectrophotometry (Varian's Super Scan 3)
            - HPLC utilizing water-methanol-acetonitrile - sodium dihydro-phosphate solvents[13] (Waters Associates Instrument).

Octane/O-xylene/Secondary Butyl alcohol-GC employing Chromosorb V packing and the thermal conductivity detector (Hewlett-Packard Instrument).

Divalent Ions - Ionic Flame Spectrophometry (Perkin-Elmer Instrument)
             - Chelatometric titration

Experimental Procedures

Berea cores (2.5 x 2.5 cm^2 cross-section) were cut to 30 cm lengths and dried in a vacuum oven at 110°C for 24 hours. They were then saturated under vacuum with degassed brine, oil flooded to a connate water saturation, and then water-flooded to a residual oil saturation usually in the range of 30 to 35% of pore volume.

A surfactant slug was injected into the cores at residual oil saturation at constant rates of 2 ml/hour so that the apparent frontal advance rate of the fluid did not exceed 30 cm/day. In order to eliminate evaporative losses of volatile components, the outlet line was fed through a syringe needle piercing the septum of a collection tube. Synchronized movements of a fraction collector and the syringe needle were automated, thus allowing uninterrupted flooding in experiments lasting several days.

Surfactant floods were performed as follows. During the surfactant flood and the subsequent brine flood (no polymers or viscosity improving agents have been used in this work), the samples were collected at two-hour intervals which resulted in 5 to 10% of pore volume being collected in each sample. Effluent fluids were then analyzed for oil, brine, surfactant, and cosurfactant content. When the production of oil, surfactant, and cosurfactant ceased, several pore volumes of a hydrocarbon phase were injected into the core in an attempt to recover surfactants trapped in oil remaining in the core. Liquid produced by this hydrocarbon flood was analyzed for all components and recovered surfactants were considered to be surfactants trapped in the hydro- carbon phase during the surfactant flood. In some floods, octane was displaced by nonane or decane so that a complete displacement of residual oil could be verified and a material balance on oil closed.

After all surfactants trapped in the oil were displaced, the core was flooded with a strong solvent such as ethyl alcohol or isopropyl alcohol in a mixture with brine to remove all remaining surfactants from the core. This required injection of 5 to 10 pore volumes and the material balance on surfactant closed usually between 90 to 100% of injected surfactant. Surfactant removed from the core by alcohol solvents is considered to be surfactant adsorbed on the rock during the flood.

The flooding sequence described above allows a determination of the overall surfactant retention (i.e. the amount of surfactant lost during the flood) from the difference between the amounts of surfactant injected and recovered during the surfactant and subsequent brine injection. The hydrocarbon flood gives a amount of surfactant trapped in the oil phase due to unfavorable phase behavior, and the adsorbed surfactant recovered in the final solvent flood completes the material balance.

This procedure implicitly assumes that the hydrocarbon phase does not remove adsorbed surfactant from the core. This assumption was verified in the following way:

A 75% PV of 3% surfactant slug was injected in a brine-saturated core and followed with three additional pore volumes of brine. Surfactant loss was determined at 0.6 mg/g. Then, octane was continuously injected and an effluent was analyzed for surfactants. After more than 5 PV of throughput only 0.06 mg of surfactant per one gram of rock was recovered. This indicates that a minor amount of adsorbed surfactant can be recovered by the oil, and that the bulk of adsorbed surfactant will not be desorbed. However, even this small amount of adsorbed surfactant recovered by oil is sufficient to qualify this method for determination of trapped surfactant as qualitative.

In general, the best material balances were obtained in floods with TRS 10-80, and usually the most inaccurate results were obtained with PDM 337. It seems reasonable to suggest that a degree of surfactant solubility in alcohol solvents could explain this trend, however, no measurements of surfactant solu- bilities have been made.

In order to avoid experimental complications due to the possible precipitation of surfactants by divalent ions, sodium chloride brines were used throughout this study. Berea cores were preflushed with 5 to 7 pore volumes of sodium chloride brines in order to displace most of the exchangeable divalent ions. Even with these precautions, there is an increase in divalent cation concentration in the propagating surfactant slug (Figure 3). In our experiments, these levels have not exceeded 90 ppm. Separate phase behavior experiments indicated that such low divalent ion concentrations affected the phase behavior of surfactant solutions in that a minor shift toward upper phase microemulsions was noticed, but no surfactant precipitation was observed.

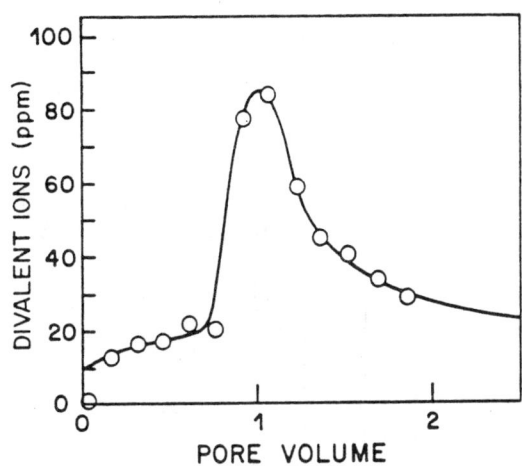

Figure 3: Divalent Ions Content in the Effluent (Injection of 75% PV of 2% 1/0.5 TRS 10-80/SBA in 1% NaCl)

It should be noted here that this procedure for differentiating trapped surfactant in the hydrocarbon phase from the adsorbed surfactant is not applicable to all situations. For example, in surfactant systems in which the surfactant distribution coefficient is not at extreme levels (i.e. $K = [(C_S)_{oil}/(C_S)_{brine}]$ tends to zero for lower phase microemulsions or $K \rightarrow \infty$ for upper phase microemulsions) the chase brine would bleed surfactant from the oil phase and no surfactant would ever be found trapped in the oil.

RESULTS AND DISCUSSION

Studies of oil recovery efficiency and surfactant retention indicate that better performing processes are usually accompanied by lower surfactant retention even though lower retention does not necessarily mean higher oil recovery.[14]

Since our experimental technique can distinguish between surfactant losses due to adsorption and losses due to unfavorable phase behavior, it was thought to be of interest to perform several series of similar experiments and then observe how these individual contributions to total surfactant retention are affected.

Effect of Cosurfactant on Surfactant Retention

It has been shown than, in systems containing no oil (i.e. systems containing only surfactant, cosurfactant, and brine), poor surfactant solubility may result in very high surfactant retention in Berea cores. An additional cosurfactant helped to dissolve the surfactant in the brine and the surfactant retention was reduced by one order of magnitude.[4] In systems containing oil, poor surfactant solubility may not result in surfactant molecule aggregation but may lead to a change in phase behavior in which case the surfactant dissolves in the upper hydrocarbon phase. In that case the surfactant retention would increase even though surfactant adsorption may either not change at all or may even decrease.

The PDM 337 surfactant with secondary butyl alcohol as a cosurfactant was selected for this part of the study. An increasing cosurfactant content makes the surfactant slightly more brine soluble and the phase behavior changes from an upper to a middle phase (Figure 4).

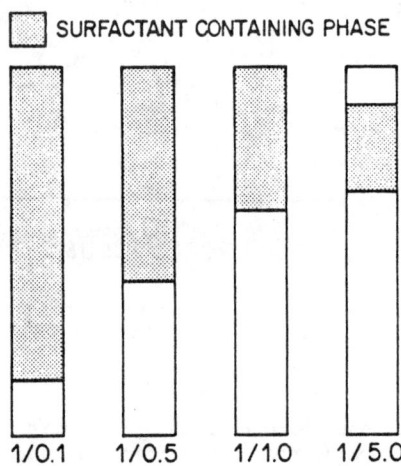

Figure 4: Phase Behavior of 3% PDM 337 Surfactant (80/20 volumetric ratio of 1.5% NaCl/ octane for different surfactant/secondary butyl alcohol ratios)

Surfactant/cosurfactant ratios of 1:0.1, 1:0.5, 1:1, and 1:5 were injected in four floods on Berea cores that had been waterflooded to residual oil saturations. The effluents were analyzed for surfactant, cosurfactant and oil content. Typical examples of the data collected are shown in Figures 5 to 7 and the results of these floods are summarized in Table 1. This series of floods clearly shows all of the difficulties which can be encountered when an attempt is made to compare adsorption data obtained from different displacement experiments.

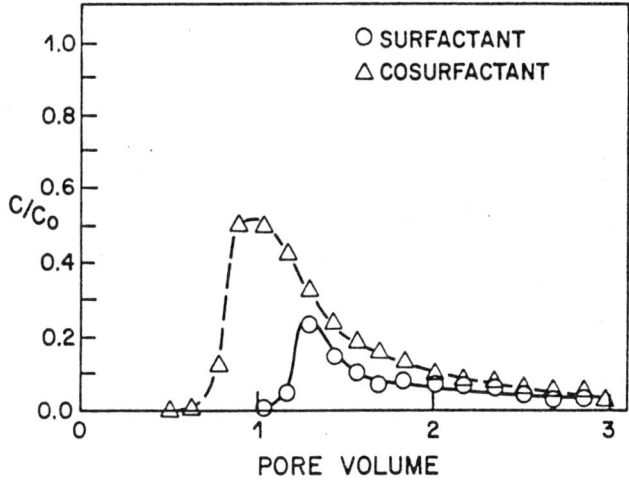

Figure 5: Surfactant and Cosurfactant Breakthrough Curves
(Flood 112: 50% PV Injection of 3%, 1:5 PDM 337/SBA
in 1.5% NaCl Brine)

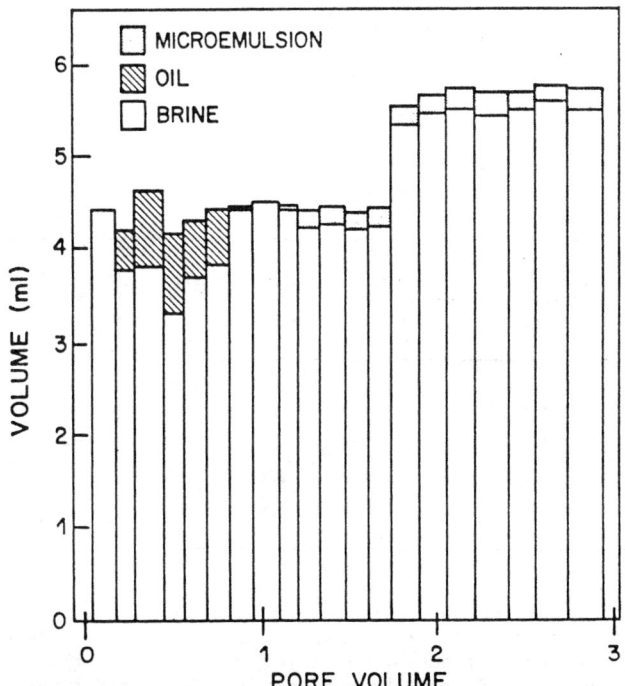

Figure 6: Effluent Phase Behavior (Flood 112)

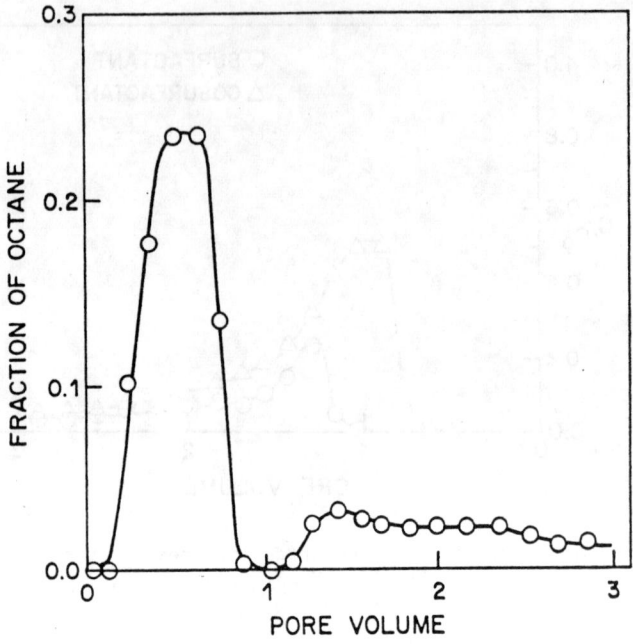

Figure 7: Fractional Flow of Oil (Flood 112)

Table 1: Summary of Flooding Results with 3% PDM 337
in 1.5% NaCl/Octane System (50% PV Injection)

Surfactant co-surfactant weight ratio	Surfactant Retention mg/g	Losses Due to Phase Behavior mg/g	Adsorption mg/g	$(c/c_o)_{max}$	Oil Recovery		
					% ROIP %	$(S_{or})_{final}$ %	$(f_{oil})_{max}$ %
1/0.1	1.2	-	1.2	0	52	15	17
1/0.5	1.2	-	1.2	.02	66	10	19
1/1	1.0	-	1.0	0.10	85	5	22
1/5	0.7	-	0.7	0.24	70	8	24

First, it should be noted that, in the floods with surfactant ratios of
1:0.1 and 1:0.5, essentially no surfactant is contained in the effluent. This
means that not enough surfactant was injected to satisfy the adsorption
capacity of the rock and that the surfaces near the end of the core are
probably not completely adsorbed with surfactant. Floods with the 1:1 and 1:5
surfactant/cosurfactant ratios have led to the production of some surfactant,

but the concentration peaks at the core outlets are substantially different from each other and, consequently, the adsorption values for the two different average surfactant concentrations are not directly comparable. Also, as Figure 8 shows, the normalized ratios of surfactant and cosurfactant concentrations are quite different for the two floods. Therefore, even though it may be tempting to suggest that there is enough data in Table 1 to ascertain the dependence of surfactant adsorption on alcohol content, a closer look shows that a comparison of surfactant adsorption for the four different systems cannot be made without conducting additional experiments.

The last three columns of Table 1 contain three indicators of the oil recovery of each surfactant flood. Results show that efficiency initially increases with cosurfactant content, however, the final flood performs less efficiently than the previous one. This confirms a conclusion reported previously that lower surfactant retention does not necessarily lead to the best oil recovery efficiency.[14]

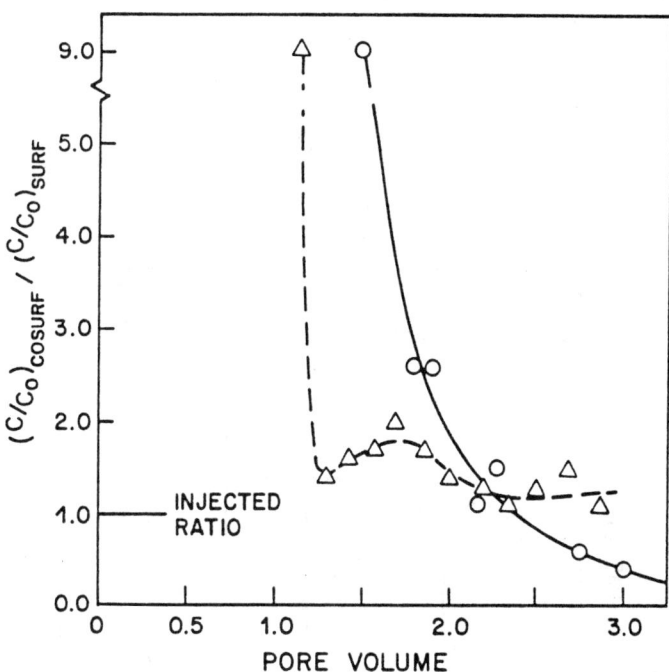

Figure 8: Normalized Cosurfactant/Surfactant Ratios at Core Outlets

The Effect of Slug Size on Surfactant Retention

A similar series of experiments was performed with TRS 10-80. The surfactant/cosurfactant ratio was varied from 1:0.5 to 1:10. Typical flooding results are shown in Figures 9 to 11, and Table 2 summarizes the data obtained in these five floods.

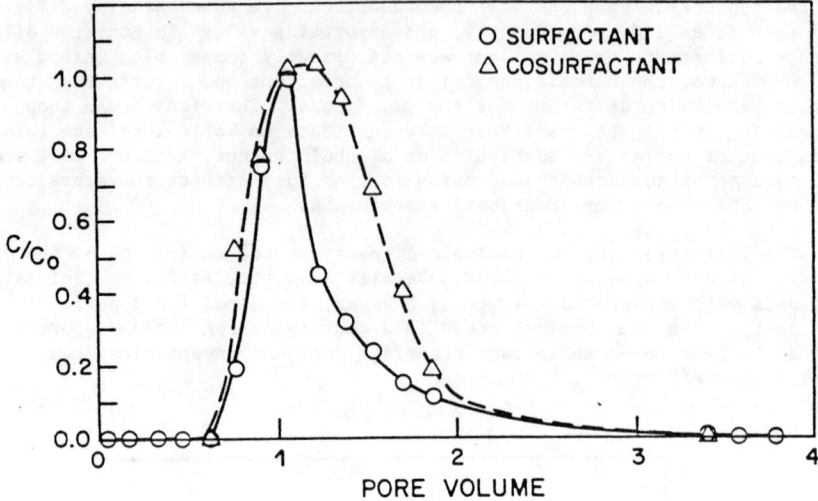

Figure 9:   Surfactant and Cosurfactant Breakthrough Curves
            (Flood 69:   75% PV Injection of 2%, 1:0.5 TRS 10-80/
            SBA in 1.0% NaCl)

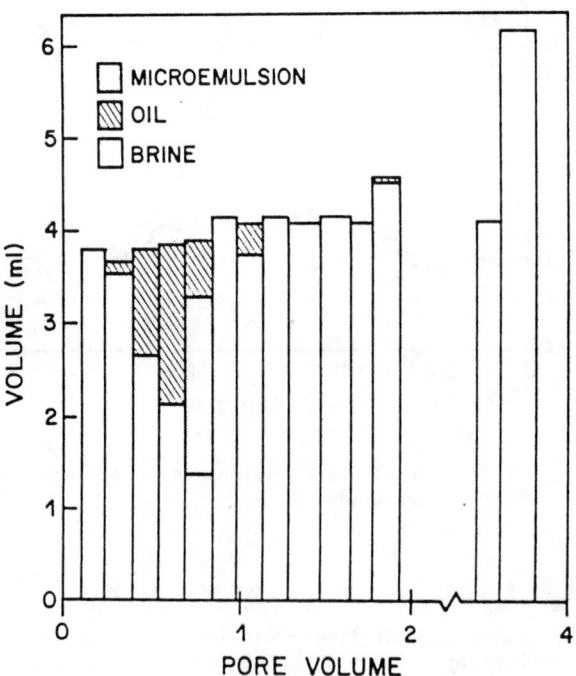

Figure 10: Effluent Phase Behavior

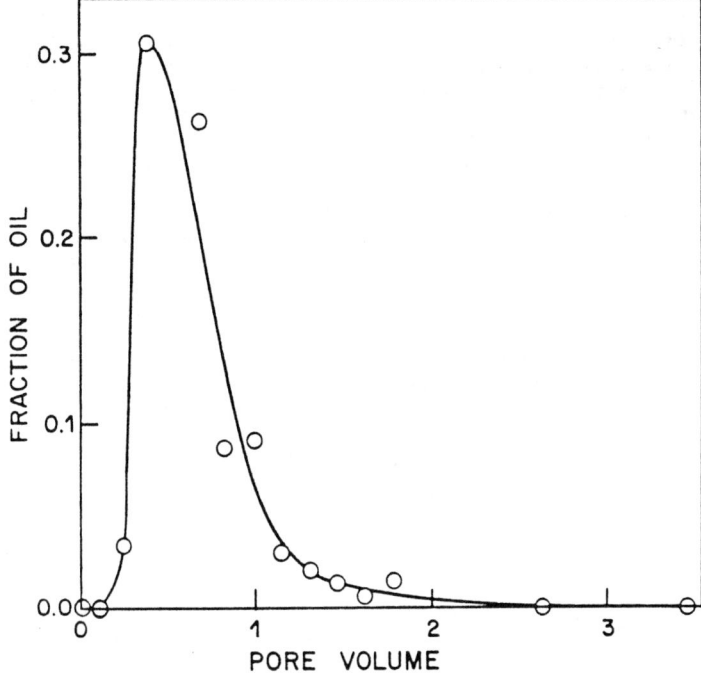

Figure 11: Fractional Flow of Oil (Flood 69)

Table 2: Summary of Flooding Results with 3% TRS 10-80/SBA
in 1.0% NaCl/Octane System

Surfactant co-surfactant	PV Injected %	Retention mg/g	Trapped Surfactant mg/g	Adsorption mg/g	$(c/c_o)_{max}$
1/10	80	0.35	0.2	.15	.50
1/5	94	0.50	0.4	0.10	0.95
1/3	150	0.2	-	0.15	1.0
1/3	75	0.3	0.1	0.20	0.85
1/1	75	0.5	-	0.48	0.8
1/0.5	75	0.6	-	0.55	1.0

116

It is interesting to compare the surfactant retention values observed in floods using a surfactant/cosurfactant ratio of 1:3 in which different size slugs of identical composition were injected. A 150% PV slug was sufficiently large to enable the effluent concentration to reach the level of the injected concentration. The injection volume in the other comparable flood was halved so that the effluent concentration reached only 85% of the injected concentration. While there is a difference in retention, the difference in adsorption is smaller. This apparent discrepancy can be explained in terms of the amount of oil trapped in the hydrocarbon phase. In the first flood, there is no trapped surfactant, while in the second about one-third of the surfactant loss is due to unfavorable phase behavior. This example shows clearly that information reflecting only overall surfactant retention may be very misleading.

Another series of experiments was performed with the pure Texas #1 surfactant. Figures 12 to 14 show an example of the experimental data and Table 3 presents a summary of the results. In this case, even though both overall retention and adsorption increase with increasing slug size, they do so at different rates. Again, it is the loss due to the phase behavior which is affected more by the size of the surfactant slug.

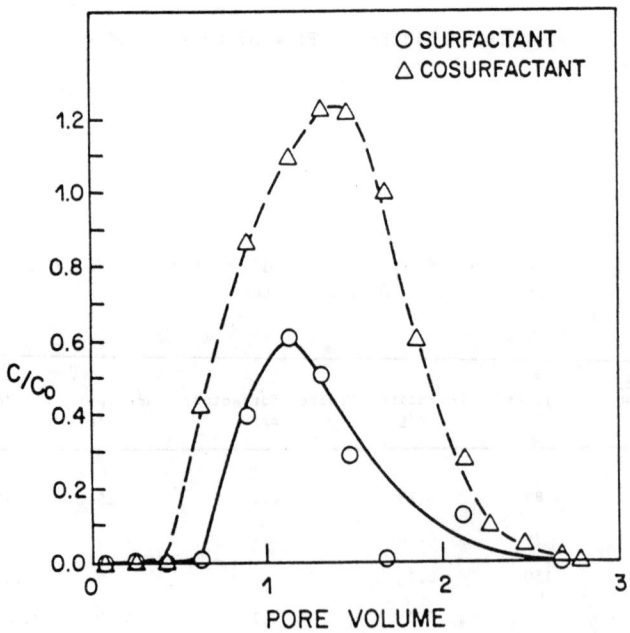

Figure 12: Surfactant and Cosurfactant Breakthrough Curves
(Flood 99: 100% PV Injection of 2%, 1:6 Texas #1/
n-Propanol in 1.5% NaCl)

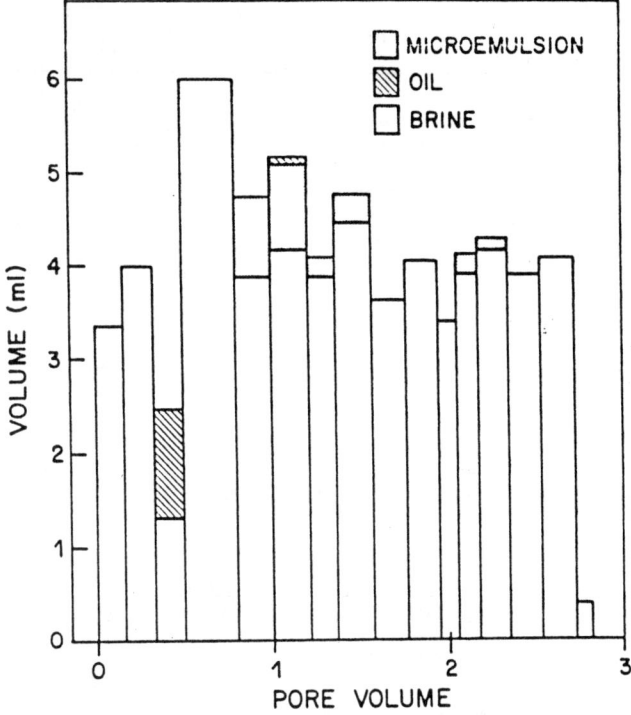

Figure 13: Effluent Phase Behavior (Flood 99)

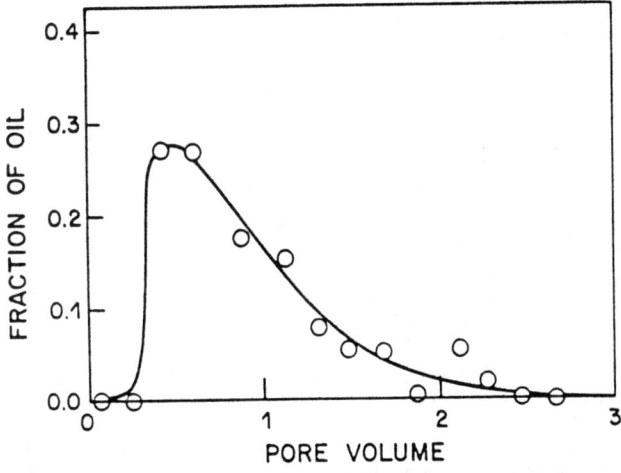

Figure 14: Fractional Flow of Oil (Flood 99)

Table 3: Summary of Flooding Experiments with 2%, 1:6
Texas #1/n-Propanol in 1.5% NaCl/Octane

PV Injected	Retention	Losses Due to Phase Behavior	Adsorption	$(c/c_o)_{max}$	Oil Recovery		
					%ROIP	$(S_{or})_{final}$	$(f_o)_{max}$
	mg/g	mg/g	mg/g		%	%	%
0.50	0.7	0.24	0.4	0.05	38	18	22
0.75	1.0	0.5	0.5	0.25	55	14	22
1.0	1.1	0.44	0.54	0.60	74	7	27

OTHER OBSERVATIONS

Experiments have been performed to evaluate the effect of cosurfactant presence within the chase brine on the retention of surfactants. Table 4 summarizes the results.

Table 4: Summary of Flooding Results with 3%, 1:1.75
PDM 337/SBA in 1.5% NaCl/Octane System

Flood #	Flood Description	Retention	Losses Due to Phase Behavior	Adsorption	$(c/c_o)_{max}$	Oil Recovery		
						% ROIP	$(S_{or})_{final}$	$(f_o)_{max}$
		mg/g	mg/g	mg/g		%	%	%
86A	No oil in the core	1.2	–	1.2	0.65	N/M		
85	Surfactant slug only	0.8	0.15	0.5	1.0	55	14	20
82	Surfactant followed by one PV of 3% SBA in brine	0.5	0.20	0.3	0.5	79	5	36
83	Same as 85 but at SLOWER INJECTION RATE	0.8	0.4	0.4	1.0	~67	~10	30

Flood 86A contained no oil, and adsorption of 1.2 mg/g was observed. Flood 85 contained oil at residual oil saturation and adsorption of 0.5 mg/g was determined. In addition, there was a loss of 0.15 mg/g surfactant due to phase behavior. The procedure used in Flood 82 was the same as for Flood 85 except that in Flood 82 the one PV of the brine that followed the surfactant slug contained 3% secondary butyl alcohol. As expected, the retention and adsorption levels are both lower, however, the amount of surfactant trapped in the oil phase did not change appreciably. The oil recovery was better as the final oil saturation is lowered from 14% PV in Flood 85 to 5% PV in Flood 82. Another interesting aspect observed in this experiment was the shape of the surfactant breakthrough curves (see Figure 15). Even though the floods were run at the same injection rates, the shape of the curve in Flood 82 gives the impression of a much higher level of dispersion than that in Flood 85. Several explanations are possible, but the limited data available do not allow for a unique interpretation and therefore none is offered. However, it is observed that data such as these should be of concern to people dealing with numerical models for chemical flooding since the data suggest that the chemical composition of the surfactant slug may substantially affect the apparent dispersion.

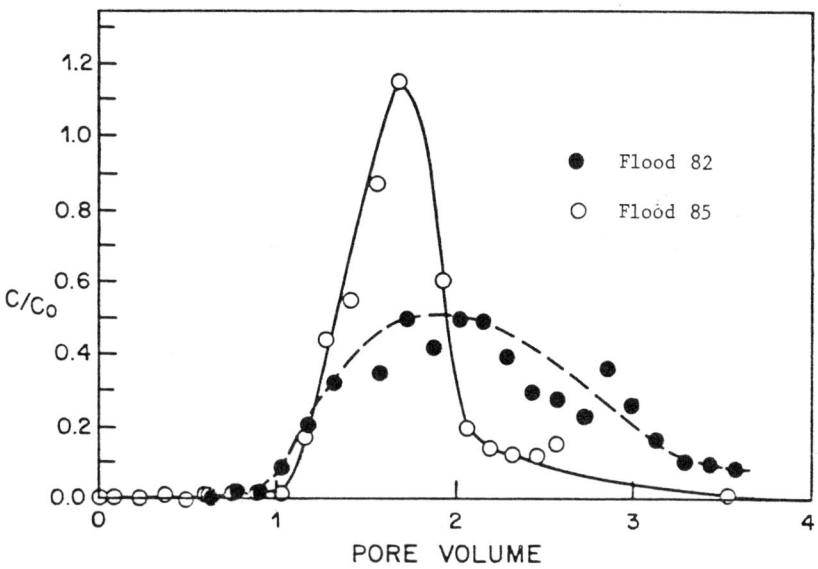

Figure 15:   Surfactant Breakthrough Curves

A recently published paper describing static adsorption experiments, among other results, indicated than an attainment of adsorption equilibrium required almost two weeks of contact between a surfactant solution and a solid adsorbent.[1] An attempt has been made to find out if similar phenomenon takes place during displacement tests in Berea cores. Therefore, Flood 85 was repeated but at an injection rate that was ten times lower and equal to an apparent frontal velocity of 3 cm/day. It took more than 10 days for the surfactant slug to propagate through the core. The oil recovery was better and an additional 4% PV of oil was recovered which is in agreement with the previously published data on this type of experiment.[15] The retention level was the same but the loss of surfactant by the phase trapping mechanism increased substantially while the adsorption loss was slightly lower. It therefore seems reasonable to suggest that the additional residence time for the surfactant in the core allowed it to be more concentrated in the oil phase, but than an increase in adsorption was not observed. It has been noted before that, for surfactant systems which are not at optimal formulation (i.e. not at a middle phase configuration), the time required for attainment of phase equilibrium may be substantial. Our experiments enable the suggestion that this process of surfactant redistribution among the phases may be more responsible for the time dependence of retention than is the slow attainment of adsorption equilibrium at the solid-liquid interface. This suggestion is supported by previously reported results on adsorption measurements in batch experiments in which, in the absence of oil, the adsorption always reached equilibrium within 24 hours.[4]

120

SUMMARY

Based upon more than one hundred displacement experiments with three types of surfactants in Berea cores, the following conclusions may be made:

1. Thermodynamically valid surfactant adsorption isotherms should be determined in batch experiments.

2. Displacement experiments yield surfactant retention values which involve averaging several variables. If any theory developed for adsorption is applied to retention data obtained from displacement experiments, the other causes of surfactant losses must be accounted for so only adsorption data are used.

3. Experimental procedures that permit differentiating between surfactant losses due to adsorption and those due to unfavorable phase behavior have been developed and tested.

4. Pure surfactant (Texas #1), synthetic sulfonate (PDM 337), and petroleum sulfonate (TRS 10-80) give comparable results for retention and adsorption in Berea cores.

5. Adsorption of surfactants can be reduced by the addition of low molecular weight alcohols (sec-butyl alcohol, n-propanol).

6. For the three surfactants studied, adsorption levels did not exceed 1.2 mg/g. If the overall retention is higher, surfactant losses due to unfavorable phase behavior or some other mechanism should be suspected.

ACKNOWLEDGEMENTS

The author wishes to acknowledge the assistance and dedication of Laurie Baxter and Gail Parker who performed the precise experiments necessary for this paper. Sincerely acknowledged are Bev Moore and Gail Donaldson for typing this manuscript.

REFERENCES

1. MEYERS, K. O. and SALTER, S. J.; "The Effect of Oil-Brine Ratio on Surfactant Adsorption from Microemulsion", paper SPE 8989 presented at the SPE 55th Annual Fall Meeting, Dallas, Texas (September 21-24, 1980).

2. CELIK, M. S., GOYAL, A., MANEV, E. and SOMASUNDURAN, P.; "The Role of Surfactant Precipitation and Redissolution in the Adsorption of Sulfonate on Minerals", paper SPE 8263 presented at the SPE 54th Annual Fall Meeting, Las Vegas, Nevada, (September 23-26, 1979).

3. KRUMRINE, P. H., CAMPBELL, T. C. and FALCONER, J. S.; "Surfactant Flooding I: The Effect of Alkaline Additives on IFT, Surfactant Adsorption, and Recovery Efficiency", paper SPE 8998 presented at the 5th Symposium on Oilfield and Geothermal Chemistry, Stanford, California (May 28-30, 1980).

4. NOVOSAD, J.; "Adsorption of Pure Surfactant and Petroleum Sulfonate at the Solid-Liquid Interface", Proceedings of the 3rd International Conference on Surface and Colloid Sciences held in Stockholm, Sweden, (August 20-25, 1979), Plenum Publishing, New York (1981).

5. GLOVER, C. J., PUERTO, M. C., MAERTER, J. M. and SANDVIK, E. I.; "Surfactant Phase Behavior and Retention in Porous Media", (June 1979) SPEJ 19, 183-193.

6. TROGUS, F. J., SCHECHTER, R. S. and WADE, W. H.; "A New Interpretation of Adsorption Maxima and Minima", (June 1979) J. Colloid Sci. 70, 293-305.

7. GALE, W. W. and SANDVIK, E. I.; "Tertiary Surfactant Flooding: Petroleum Sulfonate Composition - Efficacy Studies", (1973) SPEJ 13, 191-199.

8. SOMASUNDARAN, P. and HANNA, H. S.; "Adsorption of Sulfonates on Reservoir Rocks", paper SPE 7059 presented at the 5th Symposium on Improved Methods for Oil Recovery held in Tulsa, Oklahoma,(April 16-19, 1978).

9. SIRCAR, S., NOVOSAD, J. and MYERS, A. L.; "Adsorption from Liquid Mixtures on Solids: Thermodynamics of Excess Properties and Their Temperature Coefficients", (May 1972) I & EC Fundamentals 11, 249-254.

10. GILLILAND, H. E. and CONLEY, F. R.; "Surfactant Waterflooding".

11. FRANCES, E. I., DAVIS, H. T., MILLER, W. G. and SCRIVEN, L. E.; "Phase Behavior of a Pure Alkyl Aryl Sulfonate Surfactant", presented at the 175th ACS National Meeting, Anaheim, California (March 13-17, 1978).

12. SHAH, D. O. and WALKER, R. D.; "Research on Chemical Oil Recovery Systems", Semi-Annual Report, University of Florida, Gainesville (June 1977).

13. ZORNES, D. R., WILLHITE, G. P. and MICHNICK, M. J.; "An Experimental Investigation Into the Use of HPLC for the Determination of Petroleum Sulfonates", (June 1978) SPEJ 18, 207-218.

14. TRUSHENSKI, S. P., DAUBEN, D. L. and PARRISH, E. R.; "Micellar Flooding -Fluid Propagation, Interaction and Mobility", (1974) SPEJ 14, 633-644.

15. HEALY, R. N., REED, R. L. and CARPENTER, C. W.; "A Laboratory Study of Microemulsion Flooding", (1975) SPEJ 15, 87-100.

THE EACN OF A CRUDE OIL: VARIATIONS WITH COSURFACTANT AND WATER OIL RATIO

MIN KWAN THAM and PHILIP BOALT LORENZ

U.S. Department of Energy
Bartlesville Energy Technology Center

ABSTRACT

The EACN concept, which allows the substitution of a crude oil by an alkane or an alkane mixture for phase volume or interfacial tension studies, has been generally accepted. In this paper, it was shown that such parameters as alcohol type, crude oil composition, and water-oil-ratio could have an effect on the EACN of a crude oil. The partition behavior of the alcohol was traced as one of the causes for this aberration. Interaction of surfactant with heavy crude oil components was thought to be another. Experiments testing the later hypothesis is in progress.

INTRODUCTION

The term Equivalent Alkane Carbon Number (EACN), was coined by the research group from University of Texas[1-3]. This concept arises from the observation that the interfacial properties of any oil with a surfactant can be modeled by the behavior of alkanes. Thus, heptane, heptylbenzene, and butyl cyclohexane all exhibit "optimum" conditions, i.e., minimum interfacial tension (IFT) for the same combinations of surfactant, cosurfactant, and salt concentration. In general, the benzene ring appeared to have EACN = 0, and the cyclohexane ring EACN = 3. In addition, the EACN of a mixture of hydrocarbons follows the simple mixing rule[1-3],

$$(EACN)_{mixture} = \sum_i X_i \ EACN_i, \quad ---(1)$$

where X_i is the mole fraction of component i.

This concept was later found to be applicable to crude oils and pseudo crudes[4], whereby an alkane or alkane mixture can be found to model the IFT behavior of a crude oil. An important finding of theirs is that the EACN of an oil (crude, pseudocrude, or hydrocarbon) is independent of the surfactant formulation, and that this equivalence always holds. Crude oil, being dark in color and usually quite viscous, can make equilibrium attainment very slow and phase volume observation difficult. Replacing the crude with hydrocarbon will facilitate screening of surfactant formulation, and therefore, the EACN concept is a very valuable one.

Recently, the Texas group and Glinsmann[5], extended the concept of equivalent optimal salinity to high concentration surfactant systems (> 2%). Here, also, the EACN of a crude oil is independent of the alcohols and surfactants in the formulations.

As part of our supporting research program for the DOE micellar-polymer pilot test in Nowata County, Oklahoma[6], we determined the EACN of the Delaware-Childers (D.C.) oil from that field, using several surfactant systems, and water-oil-ratios (WOR). It was found that the EACN was not a constant value[7]. This paper reports the results in our investigation on the probable causes for this variation.

Glinsmann's method[5] of measuring the EACN of an oil was followed, in which the optimal salinities of a surfactant system with a series of alkanes were determined. By comparing the optimal salinity of the crude oil with the same surfactant system, the EACN was determined. Of the different criteria of defining optimal salinities[8,9], the one used here was the equal solubilization[8] from phase volume measurements.

Various surfactant systems were studied first, with special emphasis on the effect of alcohol type, because studies have shown the strong influence of alcohols on phase behavior and IFT[10-15]. The effect of crude oil components was then studied. Finally, the effect of WOR was also studied.

EXPERIMENTAL

<u>Materials</u>

The surfactants used (and their properties) are listed in Table I. They were used without purification.

The alkanes were pure-grade hydrocarbon from Phillips Chemical Company. The phenyl dodecane was from Eastman Kodak Company.

<u>Procedure</u>

For phase-volume studies, surfactant solutions were mixed with oil in glass tubes (precision bore to 0.474 ± 0.001 cm i.d.), and shaken for one minute in a mechanical shaker (40 Hz). Except where noted otherwise, the WOR was set at unity. The tubes were kept in an air bath at 30° for equilibration. Usually, one week to six months were required for complete equilibration. Some of the solutions--especially those with high viscosity--were shaken a second time to ensure thorough mixing.

Table I. Properties of Surfactants

	Floodaid 141	TRS 10-410[a]	Suntech I[b]
Type	Blend of petroleum sulfonates mixed with Amoco Cosurfactant 122[c]	Petroleum sulfonate	Sulfonates of mixed xylenes and propylene tetramer
% Active	45	62	65
Equivalent Weight	450	418	372
Equivalent Weight Distribution	wide	400-450 (80%)	344-390 (92%)

(a) Witco Chemical Company.
(b) An experimental sulfonate (Sample No. I, Suntech Lot 768511) prepared by Suntech Tech, Inc.[16]
(c) Amoco Cosurfactant 122 is a mixture of ethoxylated alcohols.

Phase volumes were measured with a cathetometer. Standard correction for the round-bottom end of the glass tubes, and for the oil and water menisci, were obtained by weight measurements. Solubilization calculations were fashioned after the work of Glinsmann. The following assumptions were made in the calculations: (a) all the surfactant and cosurfactant is in the surfactant phase (this is an incorrect assumption as can be seen later, but the effect on the phase volumes is negligible); (b) the volumes are additive. In the present work, surfactant and electrolyte concentrations refer to the concentration in the aqueous phase.

Some experiments were performed with crude oil components. Distillation of crude oil into distillates and heavy ends were done at 400°F and 10 mm Hg. Vacuum. Analysis for acids and bases was by column liquid chromatography[16]. Asphatene determination was by pentane precipitation. Alcohol concentrations were measured with a gas chromatograph.

RESULTS AND DISCUSSION

Optimal Salinities

The optimal salinities for a number of systems with normal alkanes are plotted in Figure 1. The observed behavior is the same as that reported in the literature[5, 11-13], that is, the optimal salinity increases with increases in (1) hydrocarbon chain length, (2) water solubility of the cosurfactant (the solubilities are in the order IBA < TAA < Amoco 122), and (3) concentration of the water soluble cosurfactant.

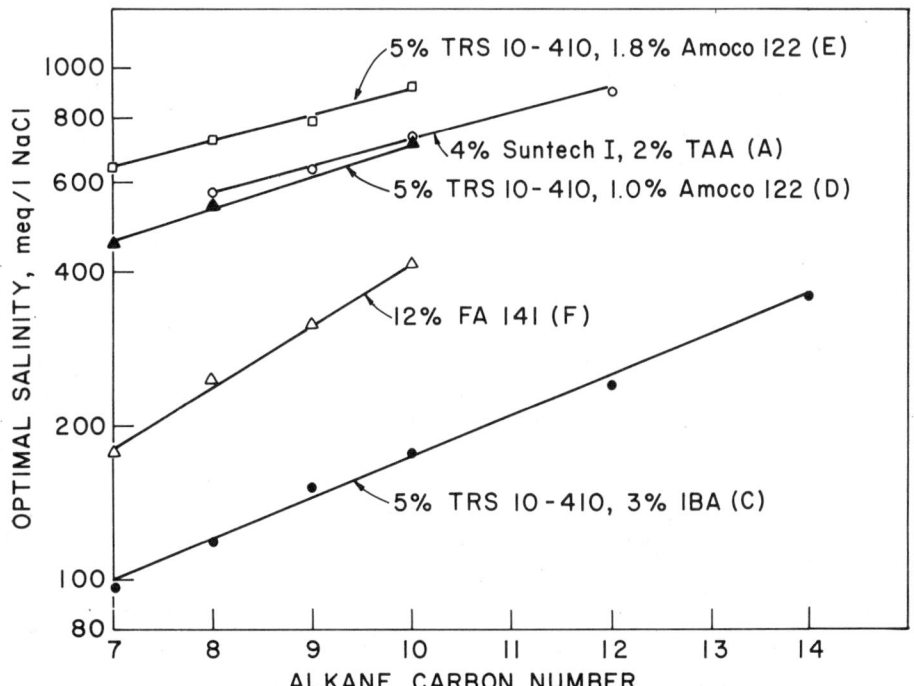

Figure 1. Optimal salinity of surfactant systems with normal alkanes. TAA = tertiary amyl alcohol; IBA = isobutyl alcohol: Amoco 122 = Amoco Cosurfactant 122.

The observed linear relationship of ln (S_ϕ^*) versus alkane carbon number (ACN) was reported by Salager[9,18], who found that the slopes for all the sulfonate systems were 0.16 ± 0.01. The value for the 5 percent TRS 10-410 - 3 percent isobutyl alcohol (IBA) system obtained by least square fit is 0.17, which is in good agreement with his values. However, the slopes were 0.11 for tertiary amyl alcohol and 1.8 percent for cosurfactant 122, and 0.14 with 1.0 percent cosurfactant 122. This is in contradiction to the prediction of Salager's equation[9], which predicts slope independent of the alcohol. The slopes for the Suntech and Floodaid surfactant were 0.12 and 0.28, respectively.

Effect of Surfactant Formulations on the EACN of Delaware-Childers Oil

The EACN of D. C. oil was determined by comparing its optimal salinity with that of the alkanes for a given surfactant formulation. The several surfactant systems were used to determine the constancy of its EACN. Table II shows the results.

Table II. The optimal salinity and EACN of D. C. oil with different surfactant formulations

Surfactant system	Optimal salinity meq/1 NaCl	EACN
A 4% Suntech I 2% Tertiary amyl alcohol (TAA)	680	9.5
B 5% TRS 10-410 3% TAA	197	9.3
C 5% TRS 10-410 3% Isobutyl alcohol	193	10.9
D 5% TRS 10-410 1% Amoco 122	410	6.15
E 5% TRS 10-410 1.8% Amoco 122	590	6.2
F 12% FA 141*	222	7.7

* Sulfonate content equivalent to 7.6% TRS 10-410 or Suntech I

The spread of 4.7 units in the values indicates that EACN as usually determined is not a constant quantity. From Figure 1 and Table II, it is necessary to conclude that the currently accepted concept apply only over a narrow range of conditions.

Thus, Systems A and B, with fairly similar surfactants, give nearly identical EACN values. Also, in Systems D and E, a twofold variation in alcohol concentration has no influence on EACN. But the transition from C to B to D (with a significant increase in water solubility of the cosurfactant at each step) shows that the cosurfactant species has a major influence on the results. There are two properties of System F that could contribute to its different EACN value: a wide distribution of equivalent weight and a different type of alcohol.

The EACN of an oil was determined by comparing the optimal salinity of the oil with those of alkanes. The variation in EACN observed above, necessarily reflects differences in properties between the oil and alkanes. It is therefore of interest to study the effect of surfactant formulation on the EACN of a number of oils.

EACN of Several oils With Systems C and F

Surfactants Systems C and F were used to compare the behavior of different oils. System C was chosen because it has been widely studied[5,12,19], and because the optimal salinities, phase behavior and EACN of most of the oils studied with this system followed a "regular" pattern. On the other hand, System F was chosen for its "irregularities". Table III lists the optimal salinities and EACN of these systems with a number of oils. The El Dorado oil results show differences in the EACN between the two systems, even though the deviation is not as large as the D. C. oil case. Bradford oil shows an even smaller difference. These variations among the various crude oils may be compared with the differences in the crude oil composition (Table IV). Bradford oil is high in paraffin, and D. C. oil contains a larger quantity of heavy bases and acids. These heavy compounds are known to complex with the sulfonates[20].

Table III.   Optimal salinities and EACN of crude oils and crude oil fractions

	System C		System F	
	S*	EACN	S*	EACN
El Dorado oil	169	10.0	261	8.3
Bradford oil	196	11.0	425	10.1
Bradford Distillates	130	8.2	197	7.3
Bradford heavy-ends + decane[a]	238	12.4 (20)	615	11.4 (15.8)
D. C. Oil	193	10.9	222	7.7
D. C. distillate	103	6.7	132.5	5.8
D. C. heavy-ends + decane[a]	185	10.6 (12.5)	295	8.7 (4.6)

(a) Equal weight ratio of heavy ends and decane

The behavior of the components of these crude oils is quite revealing. The distillates show a downward shift in EACN as compared with that of the whole crudes, as expected. Interestingly, the large differences between Bradford and D. C. oils with respect to the surfactant Systems C and F disappeared. Both show a difference of 0.9 units with the two systems, as compared to 0.9 and 3.2 for the whole Bradford and D. C. oils, respectively. Yet, the fact that the distillates having different EACN with different surfactant systems indicates that there are certain components in the distillates behaving differently from the alkanes, which are the standards. Actually, it has been recognized that the equivalence between alkanes and other series of compounds is not exact[4]. There are deviations in the alkyl benzene and alkyl cyclohexane series that are greater, the farther one moves away from EACN of 8. Table V presents some further data on this, showing that the deviation can be quite large when less conventional materials are used. The mixing of benzene with phenyl dodecane (to give an EACN of 8) shows normal EACN with System C. A downward shift in EACN with System E, similar to that

Table IV. Crude oil properties

	Delaware–Childers oil	Bradford oil	El Dorado oil
Gravity°API	31.9	44.3	36.0
Nitrogen, percent	0.07	0.01	0.07
% Aromatic through fraction 12[a]	4.04	3.82	5.50
% Acids[17]	2.17	0.13	–
% Bases[17]	1.58	0.3	–
% Total Asphaltenes	1.46	0.02	–
% Paraffin through cut 7[b]	34.6	64.0	55.3

(a) Cut temperature 437°F at 40 mm Hg. (corresponding to molecular weight of 280).
(b) Cut temperature 392°F (corresponding to molecular weight of 150).

Table V. Optimal salinities and EACN of akyl benzenes

	System C S_ϕ^*	System C EACN	System E S_ϕ^*	System E EACN
Benzene – phenyl dodecane mixture, 1:2 molar ratio	110	7.9	360	1.9
Phenyl dodecane	145	9.0	557	5.9

with crude oils was observed. On the other hand, phenyl dodecane does not observe the simple scaling law[3,4] for both surfactant systems. Under this law, phenyl dodecane should have an EACN of 12. The observed EACN differs greatly from this value and cannot be explained totally by the smaller deviation previously reported for the case without alcohol[4].

The data on the heavy ends in Table III are a lot more "irregular". For practical experimental purposes, it was necessary to cut the viscosity by mixing with equal weights of decane. To get the EACN of the heavy ends by themselves from Equation (1), an assumption was necessary on the molecular weight (MW). No value of MW could be found that was consistent with the EACN values, even for "regular" System C. The distillation temperature suggested a MW of 450, which corresponds to an alkane of carbon No. 32, but gave EACN 20 for the Bradford heavy ends and 12.5 for those from D. C. The weight fractions of distillates and heavy ends from D. C. oil (which lost only 4 wt-% in distillation) required MW 254 (EACN 18) for consistency with the whole-oil EACN. The EACN of the decane mixture obeyed equation (1) only with MW = 160 (EACN = 11.3) for heavy ends. It is obvious that heavy ends are not equivalent at all to alkanes even with System C; and the discrepancy between Systems C and E are very large.

Alcohol Partitioning and Its Effect on EACN

We have shown earlier that replacing isobutyl alcohol with Amoco 122 in a surfactant formulation causes a downward shift in EACN. It is therefore of interest to study the partitioning behavior of these alcohols, because alcohol partitioning is known to be the prime determinant of the phase behavior, interfacial tension, and optimal salinity of a surfactant-oil system[10-12,21].

Since determination of alcohol concentrations in crude oil poses considerable problem due to its wide boiling range—choosing the right column is difficult—only the partition coefficients in hydrocarbons were measured. It was suspected that the large differences in behavior of alkanes and alkyl benzenes would be reflected in the alcohol partitioning and suggest one cause for the difference between alkanes and crude oils. The results of the partitioning experiments are listed in Table VI. The numbers are relevant only at optimal salinity, but data under other conditions are given for illustration. Partition coefficients are not very sensitive to salinity up to 3 percent; the table shows that the same value was obtained for Co-surfactant 122 in pure water and in System E at optimal salinity of 3.4 percent. The differences between the partition coefficients of Amoco 122 in octane and phenyl dodecane is striking. In addition, there is a strong preferential partitioning of the heavier alcohol compounds (components 2 and 3) into the oleic phase of the phenyl dodecane system (Table VII). Thus, in comparison with octane, the aqueous alcohol concentration in phenyl dodecane is lowered. It is not known what will be the effect of this change of alcohol composition and concentration on the optimal salinity and EACN. It is certain, however, such changes will make the effort to estimate a "true EACN" impossible. That is, it is not possible to modify the definition of EACN to account for this change in alcohol concentration. In agreement with the findings of Tosh, et al[22], the presence of surfactant did not affect the alcohol partitioning behavior for the systems studied.

Table VI. Partition coefficients of alcohols

	Octane	Phenyl dodecane
3% IBA	0.39^a	$0.32^{b,d}$
1.8% Amoco 122	$0.68^{a,d}$	$5.5^{c,d}$

a = alcohol originally in deionized water.
b = alcohol originally in 0.9% NaCl.
c = alcohol originally in 3.3% NaCl.
d = in the presence of 5% TRS 10-410 at optimal salinity.

Table VII. Distribution of alcohol components, System E

	Phase	Alcohol concentration, %		
		Component 1	Component 2	Component 3
Octane	Upper	0.26	0.25	0.15
	Middle	0.5	0.84	0.53
	Lower	0.3	0.24	0.25

Partition coefficient = 0.8

	Phase	Alcohol concentration, %		
		Component 1	Component 2	Component 3
Phenyl	Upper	0.26	0.44	0.4
dodecane	Middle	0.6	0.69	0.69
	Lower	0.19	<0.01	<0.01

Partition coefficient = 5.5

Solubilization at Optimal Salinity

The solubilization at optimal salinity, being related to the interfacial tension[23-25], is an interesting property to examine further. Under optimal condition $(V_o/V_s) = (V_w/V_s) = (V/V_s)_{S_\phi}^*$. The value decreases with increases in alkane carbon number[5,12]. Hsieh and Shah[12] found a correlation between $(V/V_s)_{S_\phi}^*$ and alkane density. Puerto and Gale[21] related $(V/V_s)_{S_\phi}^*$ and the side chain length of an alkyl orthoxylene sulfonate. Figure 2 is a plot of $(V/V_s)_{S_\phi}^*$ versus alkane carbon number for System C. Within experimental error, the plot is linear for alkanes. Such linearity was observed also in the solubilization of alkanes in micelles, according to Klevens[26], who studied solubilization as a function of molar volumes, structure, and other characteristics of the solubizates. Reed and Healy[8] had also noted some parallel developments between oil and water solubility and solubilization in micelles. In this work, we found that the data for crude oils and crude oil distillates also fall on the alkane line. This shows that with System C, oils modeled by optimal salinity are also modeled by the degree of solubilization.

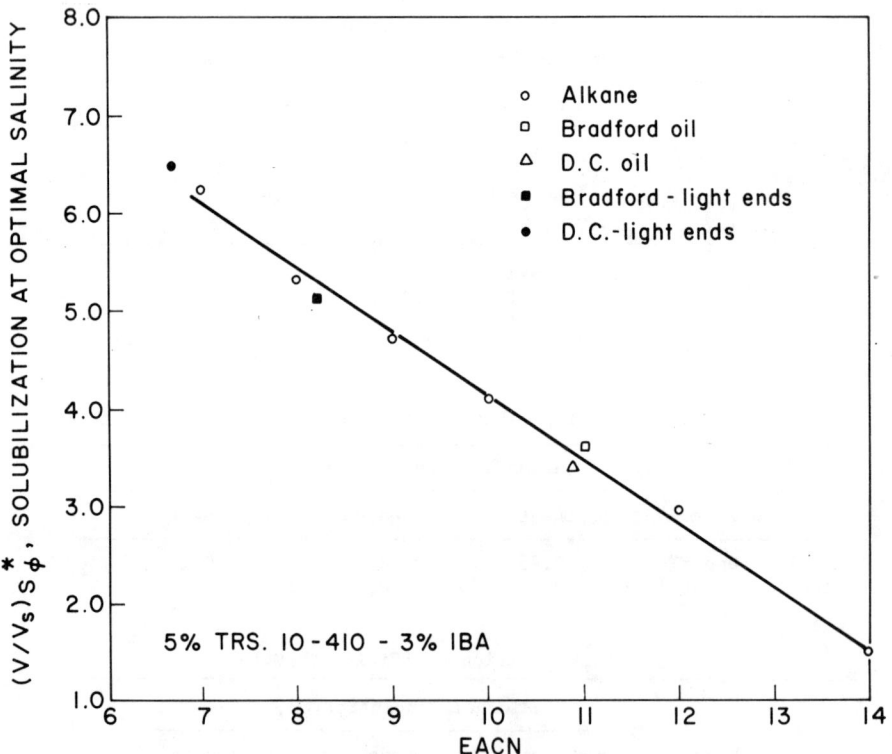

Figure 2. $(V/V_s)_{S_\phi}^*$ versus equivalent alkane carbon number

Figure 3 is a similar plot for System F. Again a linear relationship for alkanes was observed. In fact, Systems D and E also give this linear relation with alkanes (not shown), which shows that this relationship is quite general. In this case, the crude oils and distillates do not fall on the line. If the EACN values determined with System C are used, the fit is much better. This suggests that degree of solubilization might give more consistent values of EACN than optimal salinity. Even so, D. C. oil does not fit the correlation very well, perhaps due to its high content of acids, bases, and asphaltenes.

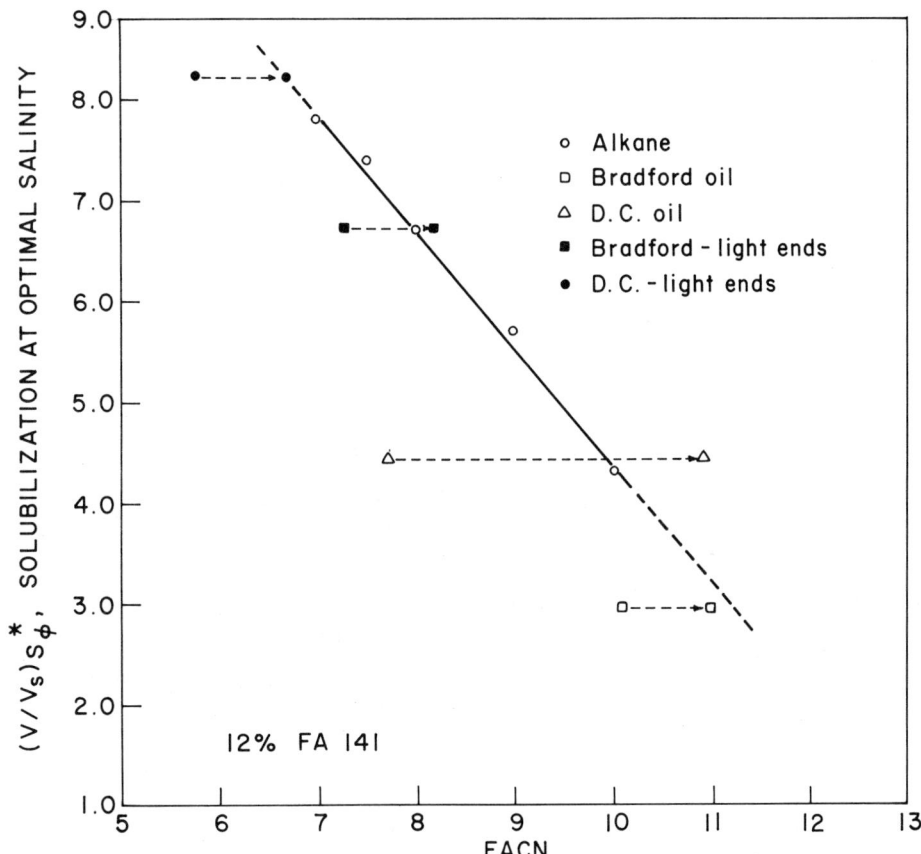

Figure 3. $(V/V_s)_{S_\phi}^*$ versus equivalent alkane carbon number

Effect of Water-Oil-Ratio on EACN

The effect of water-oil-ratio (WOR) can be seen from Figure 4. It is noted that by increasing the WOR from 1 to 2, the position of the octane and D. C. oil lines are interchanged. That is the EACN of crude oil changed from 7.7 to higher than 8, by simply increasing the WOR. It is plausible that the WOR effect is related to alcohol partitioning. Consider the case of phenyl dodecane, with the data of Table VII. Increase of WOR reduces the proportion of the oil phase, which would mean that less of component 1 would be extracted from the aqueous phase. The proportion of water-soluble component in the aqueous phase would increase. According to Figure 1, this should lead to an increase in S_ϕ^*. Since there is no such fractionation with octane, and presumably with other alkanes, the increase in EACN is as expected.

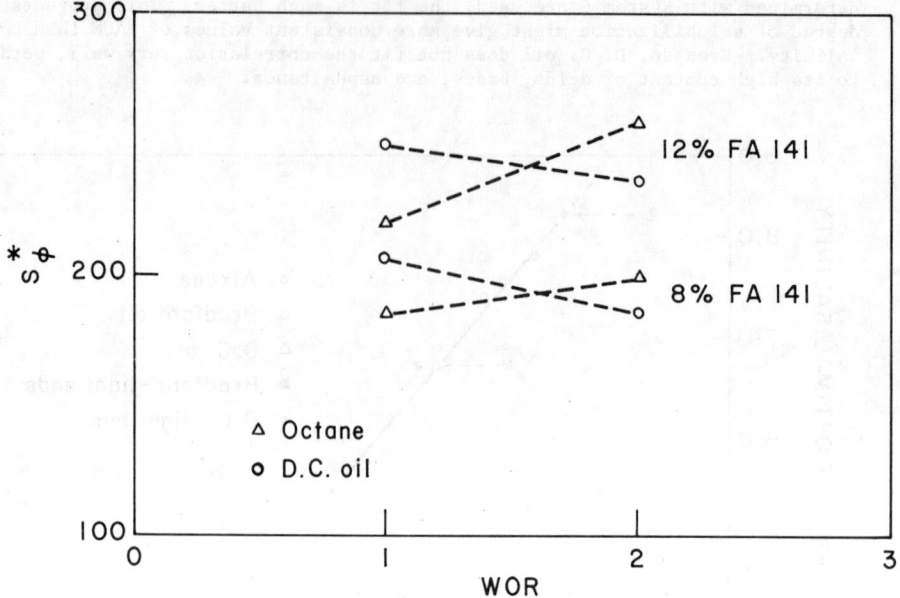

Figure 4. Variation of optimal salinity with water-oil-ratio

CONCLUSIONS

The EACN concept was found to be in error when systems involving ethoxylated alcohols and/or aromatics were used. Alcohol partitioning is found to be an important factor causing this deviation. The higher boiling, non-hydrocarbon components of the crude oil might have contributed partially to this "abnormal" behavior. This is under investigation. We would like to advise caution when applying the EACN concept.

ACKNOWLEDGMENT

The authors wish to acknowlege the help of J. B. Green and J. Lacina for the analyses on crude oils components.

NOMENCLATURE

ACN   Alkane carbon number.

EACN   Equivalent alkane carbon number.

S_ϕ^* Optimal salinity for phase behavior, meq/1 NaCl.

V_o Volume of oil solubilized in the surfactant phase.

V_s Volume of surfactant in the surfactant phase.

V_w Volume of water solubilized in the surfactant phase.

$(V/V_s)_{S_\phi^*}$ Volume of water or oil solubilized per unit volume of surfactant at optimal salinity.

REFERENCES

(1) Cash, R. L., Cayias, J. L., Fournier, R. G., Jacobson, J. K., Schares, T., Schechter, R. S., and Wade, W. H. "Modeling Crude Oils for Low Interfacial Tension," Paper 5813 presented at the SPE Symposium on Improved Oil Recovery, held in Tulsa, Okla., March 22-24, 1976.

(2) Cayias, J. L., Schechter, R. S., and Wade, W. H.: "The Utilization of Petroleum Sulfonates for Producing Low Interfacial Tensions between Hydrocarbon and Water," J. Coll. Int. Sci., 59, 31-38 (1977).

(3) Cash, L., Cayias, J. L., Fournier, G., MaCallister, D., Schares, T., Schechter, R. S., and Wade, W. H.: "The Application of Low Interfacial Scaling Rules to Binary Hydrocarbon Mixtures," J. Coll. Int. Sci., 59, 39-44 (1977).

(4) Cayias, J. L., Schechter, R. S., and Wade, W. H.: "Modeling Crude Oils for Low Interfacial Tension," SPE J., 16, 351-357 (1976).

(5) Glinsmann, G. R.: "Surfactantflooding with Microemulsions Formed In situ-Effect of Oil Characteristics." Paper SPE 8326 presented at the 54th Annual Fall Technical Conference and Exhibition of the SPE-AIME, held in Las Vegas, Nevada, September 23-26, 1979.

(6) Walker, C. J., Burtch, F. W., Thomas, R. D., and Lorenz, P. B.: "ERDA's Micellar Polymer Flood Project in Nowata County." Oil and Gas J., 74, 60-68 (1976).

(7) Lorenz, P. B., and Tham, M. K.: "Calcium Effect in the DOE Surfactant-Polymer Pilot Test." Oil and Gas J. To be published.

(8) Reed, R. L. and Healy, R. N.: "Physicochemical Aspects of Microemulsion Flooding - A Review," in Improved Oil Recovery by Surfactant and Polymer Flooding. Shah, D. O. and Schechter, R. S., eds., Academic Press (1977).

(9) Salager, J. L., Morgan, J. C., Schechter, R. S., and Wade, W. H.: "Optimum Formulation of Surfactant/Water/Oil Systems for Minimum Interfacial Tension or Phase Behavior." SPE J., 19, 107-115 (1979).

(10) Jones, S. C. and Dreher, K. D.: "Cosurfactants in Micellar System Used for Tertiary Oil Recovery." SPE J., 16, 161-167 (1976).

(11) Salter, S. J.: "The Influence of Type and Amount of Alcohol on Surfactant-Oil-Brine Phase Behavior and Properties." Paper 6843, presented at the 52nd Annual Fall Technical Conference and Exhibit of the SPE-AIME, in Denver, Colorado, October 9-12, 1977.

(12) Hsieh, W. C. and Shah, D. O.: "The Effect of Chain Length of Oil and Alcohol As Well As Surfactant to Alcohol Ratio on the Solubilization, Phase Behavior, and Interfacial Tension of Oil/Brine/Surfactant/Alcohol Systems." Paper SPE 6594, presented at the SPE-AIME International Symposium on Oilfield and Geothermal Chemistry, La Jolla, California, June 27-28, 1977.

(13) Wade, W. H., Morgan, J. C., Jacobson, J. K., Salager, J. L., and Schechter, R. S.: "Interfacial Tension and Phase Behavior of Surfactant Systems." SPE J., 18, 242-252 (1978).

(14) Baviere, M., Schechter, R., and Wade, W. H.: "The Influence of Alcohols on Microemulsion Composition." J. Coll. Int. Sci., 81, 266-279 (1981).

(15) Dominguez, J. G., Willhite, G. P., and Green, D. W.: "Phase Behavior of Microemulsion Systems with Emphasis on Effects of Paraffinic Hydrocarbon and Alcohols," in Solution Chemistry of Surfactants. Vol. 2, Mittal, K. L., ed., Plenum, New York, pp. 673-697 (1979).

(16) Malmberg, E. W.: "Large-Scale Samples of Sulfonates for Laboratory Studies in Tertiary Oil Recovery, Preparation and Related Studies," Report No. FE-2605-20, National Technical Information Service, U. S. Department of Commerce, Springfield, Virginia (1979).

(17) Green, J. B. and Hoff, R. J.: "Liquid Chrometography on Silica Using Mobile Phases Containing Aliphatic Carboxylic Acids II - Applications in Fossil Fuel Characterization," J. Chrom. 209, 231-250 (1981).

(18) Salager, J. L.: "Physico-Chemical Properties of Surfactant-Water-Oil Mixtures---Phase Behavior, Microemulsion Formation and Interfacial Tension." Ph.D. Dissertation, The University of Texas at Austin, 1977.

(19) Miller, C. A. and Fort, T. Jr.: "Low Interfacial Tension and Miscibility Studies for Surfactant Tertiary Oil Recovery Processes." Report No. DOE/BC/10007-4, National Technical Information Service, U.S. Department of Commerce, Springfield, Virginia 22161 (1979).

(20) Clementz, D. M. and Gerbacia, W. E.: "Deactivation of Petroleum Sulfonates by Crude Oils." J. Pet. Tech., pp. 1091-1093, September 1977.

(21) Puerto, M. C. and Gale, W. W.: "Estimation of Optimal Salinity and Solubilization Parameters for Alkylorthoxylene Sulfonate Mixtures." SPE J., 17, 193-200 (1977).

(22) Tosch, W. C., Jones, S. C., and Adamson, A. W.: "Distribution Equilibria in a Micellar Solution System." J. Coll. Int. Sci., 31, 297-306 (1969).

(23) Healy, R. N., Reed, R. L., and Stenmark, D. G.: "Multiphase Microemulsion Systems." SPE J., 16, 147-160 (1976).

(24) Huh, C.: "Interfacial Tensions and Solubilizing Ability of a Microemulsion Phase That Co-exists With Oil and Brine." J. Coll. Int. Sci., 71, 408-426 (1979).

(25) Fleming, P. D., III, Vinatieri, J. E., and Glinsmann, G. R.: "Theory of Interfacial Tension in Multicomponent Systems." J. Phys. Chem., 84, 1526-1531 (1980).

(26) Klevens, H. B.: "Solubilization." Chem. Rev., 1-74, 1950.

DYNAMIC INTERFACIAL PHENOMENA RELATED TO EOR

J. H. CLINT, E. L. NEUSTADTER and T. J. JONES

*The British Petroleum Company Limited, BP Research Centre,
Chertsey Road, Sunbury-on-Thames, Middlesex, TW16 7 LN*

ABSTRACT

The relevance of dynamic interfacial tension and interfacial rheology to EOR
is discussed. A technique developed by BP, the "Drop Volume Dynamic Tensiometer"
allows dynamic interfacial tension to be determined over a wide range of rate
of fractional area change. The behaviour of aqueous surfactant systems against
crude oil is very different for fresh systems compared with systems where the
phases have been pre-equilibrated. The application of these measurements
to EOR systems is illustrated with examples of surfactants which give widely
different oil displacement profiles.

A new method for the measurement of interfacial dilatational rheological parameters
of oil/water interfaces is described. This is the pulsed drop experiment which
has experimental advantages over the interfacial trough method and allows
parameters to be determined over a wider range of frequencies. The effect of
interfacial dilatational rheology on coalescence phenomena is illustrated with
data for water-in-oil demulsifiers.

The ease of oil bank formation is influenced by the kinetics of coalescence, which
in turn is controlled by film drainage from between colliding droplets. For
crude oil films in water, increasing interfacial shear viscosity greatly reduces
the rate of thinning. For the reverse system, increasing interfacial shear
viscosity can reduce coalescence rates for oil drops in water almost to zero.
This would have a very adverse effect on oil bank formation.

INTRODUCTION

In an enhanced oil recovery process, oil ganglia which have been trapped at
small pore throats are released by lowering the interfacial tension, prevented
from being retrapped by maintaining a low tension (dynamic) and encouraged to
coalesce to form an oil bank. In all except the initial release it could be
argued that it is the dynamic properties of the interface such as the dynamic
interfacial tension and the interfacial rheology which will govern each
individual and hence the overall process.

This paper reports some novel methods for measuring dynamic interfacial tension
and interfacial dilatational rheology which work very well for crude oil-water
systems. Techniques will be illustrated with results for pure oils as well as
crude oils, and the significance of these data for EOR processes will be
discussed.

DYNAMIC INTERFACIAL TENSION

This technique is essentially an extension of the drop volume method for interfacial tension and is illustrated in Figure 1.

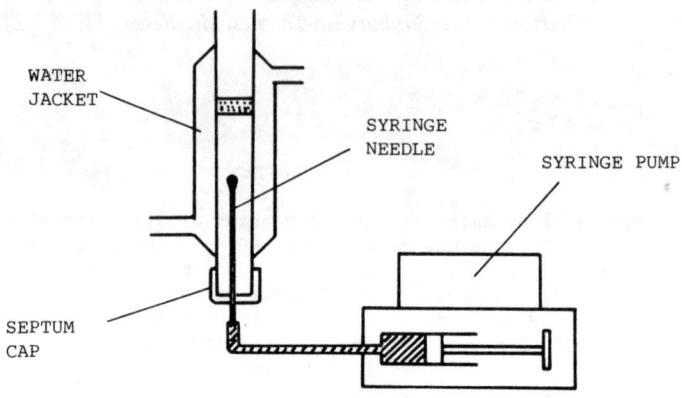

FIGURE 1 - DROP VOLUME DYNAMIC TENSIOMETER

Oil from a syringe pump is pumped at an accurately known volume flow rate to a syringe needle inserted through a septum cap into a small glass cell surrounded by a water jacket. The tip of the syringe needle is ground flat and the inside and outside diameters determined accurately. For convenience of observation an image of the tip and drops formed is obtained using a microscope and TV camera and displayed on a monitor screen.

The experiment consists very simply of measuring the number of drops formed in a fixed period of time and repeating at a whole range of volume flow rates Q. If n is the number of drops per unit time then the volume of each drop

$$V = \frac{Q}{n} \qquad \ldots (1)$$

The interfacial tension γ can then be calculated using the usual formula

$$\gamma = \frac{g(\rho - \rho')V}{2\pi R} \qquad \ldots (2)$$

where $\rho - \rho'$ is the density difference between the oil and water phases, and R is the radius of the tip to which the drop is attached. The latter may be the inside or outside tip radius depending on the wetting conditions.

If we make the assumption that the drops are spherical then the rate of fractional area change at the time when the drop detaches can be shown to be

$$\frac{dA/A}{dt} = \frac{2Q}{3V} = \frac{2n}{3} \qquad \ldots (3)$$

Hence we are able to estimate both the interfacial tension and the rate of fractional area change simply by measuring the rate of formation of drops at a known volume flow rate.

Figures 2 and 3 illustrate the type of results obtained using Forties crude oil against two different surfactant systems. The crude oil used was a well head sample free of any additives such as demulsifiers or corrosion inhibitors. All aqueous solutions were made up in filtered sea water. There were large differences in the results depending on whether the oil/water systems were preequilibrated or whether they were fresh. Figure 2 shows the dependence of dynamic interfacial tension on rate of fractional area change for a surfactant system "A" at 70°C.

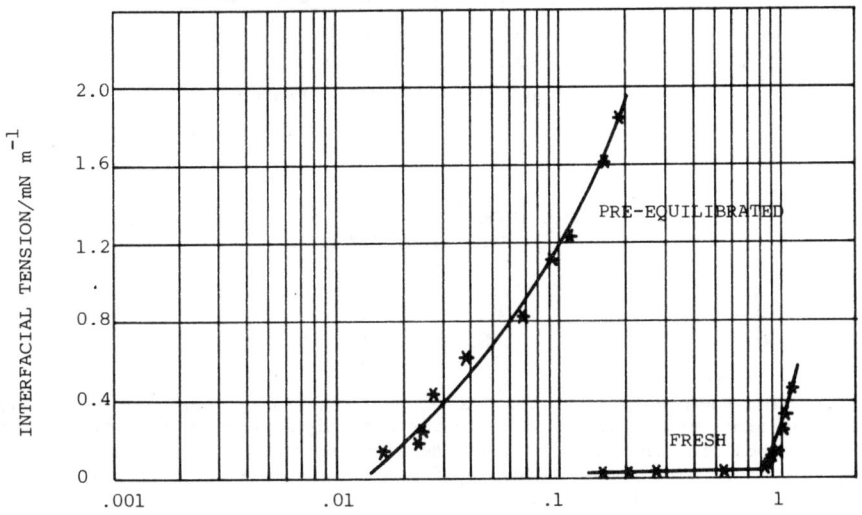

FIGURE 2 - DYNAMIC INTERFACIAL TENSION - FORTIES CRUDE/5000 PPM
SURFACTANT "A" AT 70°C

138

The difference between fresh and pre-equilibrated systems is immediately
apparent. The preequilibrated tension rises rapidly at moderate rates of
area increase whereas the tension of the fresh system stays remarkably low
until very high rates of area change are reached where the area is roughly
doubling every second. In contrast to this is the behaviour of the surfactant
system "B" shown in Figure 3.

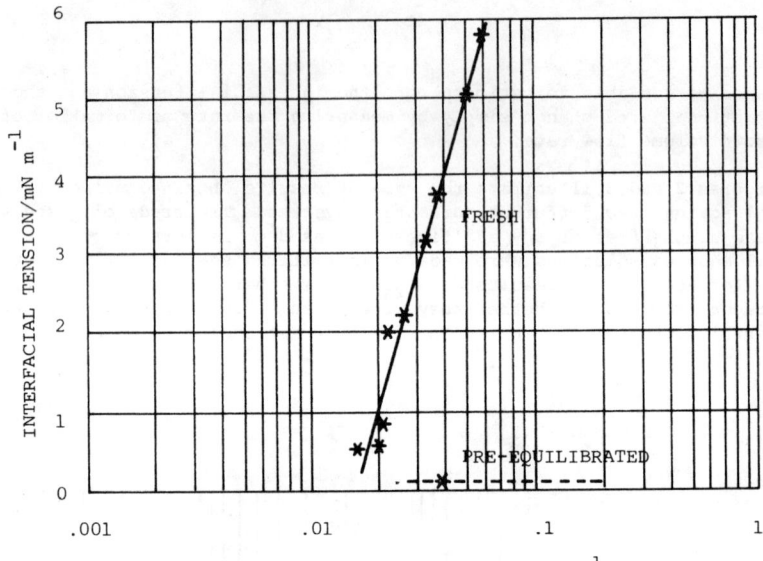

FIGURE 3 - DYNAMIC INTERFACIAL TENSION ÷ FORTIES CRUDE/5000 PPM
SURFACTANT "B" AT 70°C

This time the pre-equilibrated system gave interfacial tensions which were very small
and at times unmeasurably so (only the one which could be measured is shown).
The dashed line indicates that the tension remains low even at high rates of
area change. The tensions for the fresh system showed the normal dynamic effect
rising rapidly with modest rates of area increase.

The interesting point about these two systems is that they give totally different
oil removal profiles when tested in a model sand column test. For surfactant "A"
which gave low fresh tension but high equilibrium tensions, removal of oil was
rapid but incomplete. About 35 per cent of residual crude oil was removed in
less than 2 pore volumes (PV). For surfactant "B", which gave high fresh
tensions but very low equilibrium tensions, removal of residual oil was complete
but required a very large number (15) of PV.

Admittedly the shape and duration of the oil displacement curve will be dependent on more than just the dynamic tension behaviour. Surface wettability and the degree of adsorption will also be important factors. However, the distinction between the two systems above is clear and the oil displacement behaviour is logically related to the dynamic tension properties.

INTERFACIAL DILATATIONAL RHEOLOGY

For the measurement of interfacial dilatational rheology the method employed in the past has been that of dilatational modulus measurements at various frequencies using an interfacial film balance (1). The method involves propagation of longitudinal waves of the frequency of interest and measuring changes of interfacial tension with a Wilhelmy plate. These changes, together with the phase differences between them and the area changes, allow calculation of ε_d, the dilatational elasticity and η_d, and dilatational viscosity, at each frequency. This technique suffers from a number of disadvantages including

(a) Measurements are reliable only at fairly low frequencies where the wavelength of longitudinal waves is long compared with the distance between oscillating barrier and Wilhelmy Plate.

(b) Good results depend on the rapid response of the Wilhelmy plate and the maintenance of a well defined contact angle.

(c) The method uses large quantities of oil with a large area exposed to air allowing loss of light ends. Also the apparatus is not easily used at temperatures much above ambient.

We have developed a new technique which uses a small drop of oil pulsed in water. Area changes are calculated from drop diameters and the tip diameter, and tension is calculated by measuring the excess pressure inside the drop with a sensitive pressure transducer. The experimental arrangement is shown in Figure 4.

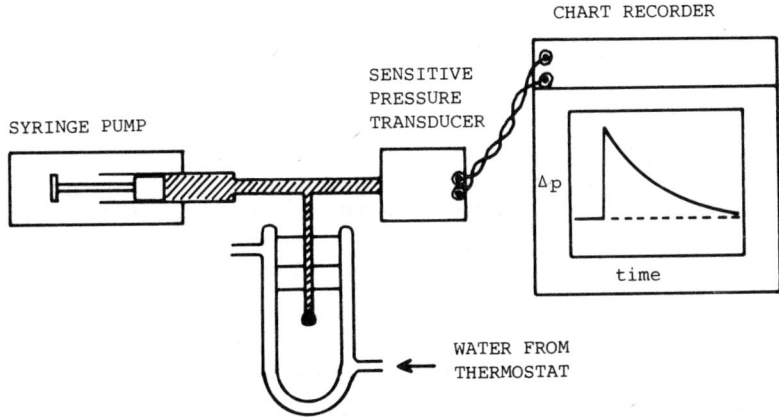

FIGURE 4 - PULSED DROP METHOD FOR INTERFACIAL DILATATIONAL RHEOLOGY

The oil drop is formed at a ground glass or stainless steel tip. The radius of tip needed depends on the region of interfacial tension being investigated. the excess pressure inside the drop was measured using a transducer from SE Labs (EMI) Ltd, type SE 1150/WG. Output from the transducer is displayed on a chart recorder. Instead of the conventional oscillatory method for dilatational modulus measurements, the single pulse Fourier transform method was used (2). When the cell containing the aqueous solution of interest is sufficiently well thermostatted the drop radius (r_1) is measured, a fixed volume pulse is injected from the syringe pump over a short period of time which increases the radius to r_2 and then the variation of pressure with time is followed on the chart recorder. The shape of a typical pressure trace is shown in Figure 5.

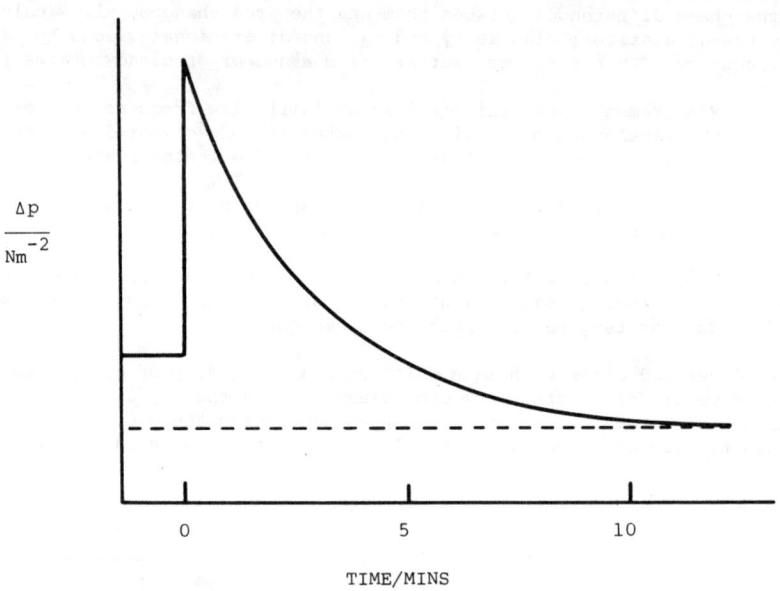

FIGURE 5 - TRANSIENT PRESSURE INSIDE DROP FOLLOWING SUDDEN EXPANSION

The equilibrium pressure after the experiment is lower than that at the beginning because the drop radius is larger. All of the pressure trace after the rapid rise is assumed to take place at a constant drop radius, the final radius r_2. Then the interfacial tension at any time $\gamma(t)$ is given by

$$\Delta p(t) = \frac{2\gamma(t)}{r_2} \qquad \dots (4)$$

The interfacial modulus is usually written:-

$$\epsilon^* = d\gamma/d\ln A = \epsilon' + i\epsilon'' \qquad \dots (5)$$

Taking Fourier transforms of the numerator and denominator coverts the perturbation time function $\Delta A(t)/A$ and the response time function $\gamma(t)$, to the frequency function. Thus:-

$$\epsilon^*(\omega) = \frac{\int_{-\infty}^{\infty} \Delta\gamma(t) \, e^{-i\omega t} \, dt}{\int_{-\infty}^{\infty} \frac{\Delta A}{A}(t) \, e^{-i\omega t} \, dt} \qquad \dots (6)$$

For a perfect step function (instantaneous area change):-

$$\int_{-\infty}^{\infty} \frac{\Delta A}{A}(t) \, e^{-i\omega t} \, dt = \frac{\Delta A/A}{i\omega} \qquad \dots (7)$$

Therefore:-

$$\epsilon^*(\omega) = \frac{i\omega}{\Delta A/A} \int_{0}^{\infty} \Delta\gamma(t) \, [\cos \omega t - i \sin \omega t] \, dt \qquad \dots (8)$$

The real part gives us the dilatational elasticity:-

$$\epsilon' = \epsilon_d(\omega) = \frac{\omega}{\Delta A/A} \int_{0}^{\infty} \Delta\gamma(t) \, \sin \omega t \, dt \qquad \dots (9)$$

The imaginary part gives the dilatational viscosity:-

$$\epsilon'' = \omega\eta_d(\omega) = \frac{\omega}{\Delta A/A} \int_{0}^{\infty} \Delta\gamma(t) \, \cos \omega t \, dt \qquad \dots (10)$$

where ω = angular frequency (radians per second).

Equations 9 and 10 can be used to calculate ϵ_d and η_d at any frequency from the decay curve. A desk top microcomputer is adequate although a little slow. It is convenient to take approximately 100 readings from the decay curve for use in these computations.

The method was evaluated using a model system of 10 ppm stearic acid dissolved in n-decane against distilled water adjusted to pH 2.5 to prevent ionisation of the acid. Results are shown in Figure 6 for the real (elastic) component of the modulus and in Figure 7 for the imaginary (frequency x viscosity) component.

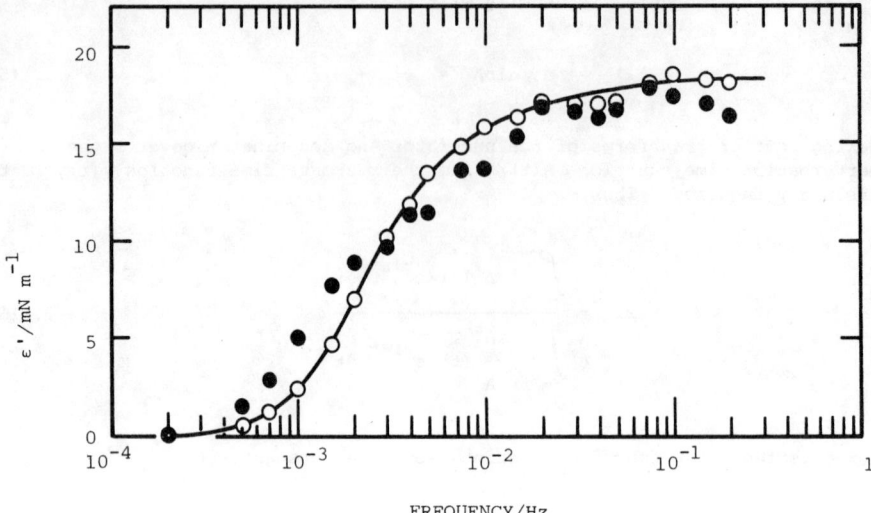

FIGURE 6 - REAL PART OF INTERFACIAL DILATATIONAL MODULUS FOR 10 PPM STEARIC ACID IN n-DECANE/DISTILLED WATER pH 2.5 AT 25°C. OPEN CIRCLES - TROUGH METHOD. FILLED CIRCLES - DROP METHOD

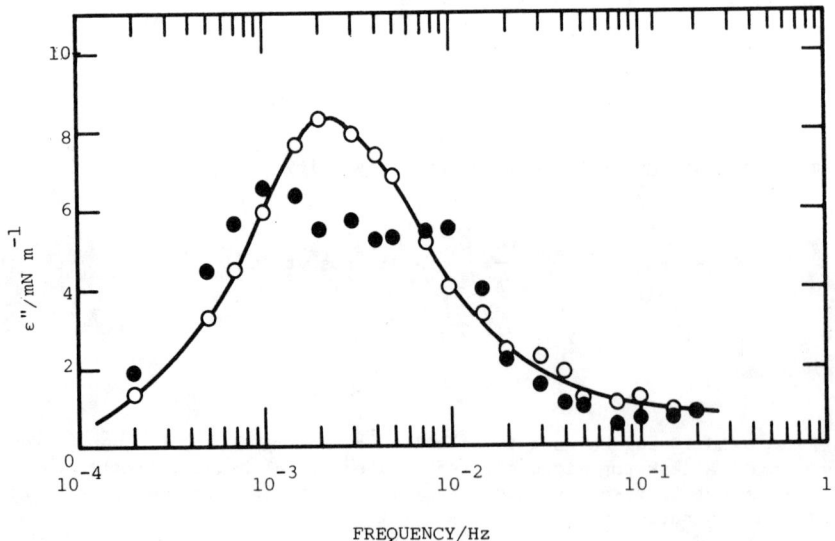

FIGURE 7 - IMAGINARY PART OF INTERFACIAL DILATATIONAL MODULUS. SYSTEM AND SYMBOLS AS FOR FIGURE 6

In each case the results are shown in comparison with data obtained previously
using the interfacial trough technique, also using the Fourier transform
method. Each set of data is the average of three separate runs. Agreement
between the drop and trough methods is very good over most of the frequency
range except possibly for the values of ε'' at intermediate frequencies.

The shapes of the curves of ε' and ε'' are very close to those expected for a
single relaxation mechanism. This is illustrated more strikingly in Figure 8
where a Cole-Cole plot (ε'' against ε') is shown. A single relaxation mechanism
has a semi-circular Cole-Cole plot and the data from interfacial trough
experiments clearly follow a semi-circle quite closely. Again agreement with
pulsed drop data is encouragingly good considering the great difference
between the two techniques. The implication is that the techniques measure
real dilatational parameters and not artefacts.

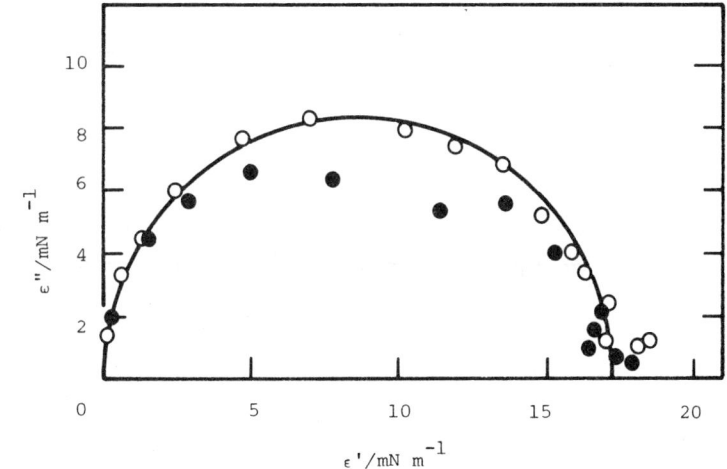

FIGURE 8 - COLE-COLE PLOT FOR INTERFACIAL DILATATIONAL MODULUS.
10 PPM STEARIC ACID IN n-DECANE/DISTILLED WATER pH 2.5.
OPEN CIRCLES - TROUGH METHOD. FILLED CIRCLES - DROP METHOD

The single relaxation mechanism implied by Figures 6, 7 and 8 is presumably
diffusion of the stearic acid from the interface into the bulk decane phase.
The maximum in ε'' which corresponds to the inflection point in ε' occurs at
$\upsilon = 0.0025$ Hz which is an angular frequency $\omega = 2\pi\upsilon = 0.0157$ s^{-1}. This is the
characteristic frequency of the relaxation process. The relaxation time
$\tau = 1/\omega = 64$ sec. This would seem to be a very reasonable relaxation time
for a diffusion controlled mechanism in a dilute system [c = 10 ppm =
3.5×10^{-5} mol dm^{-3}].

144

The main advantages of the drop method over the trough method are

(a) The system can be enclosed so that loss of light ends
     from crude oils is avoided.

(b) The system can easily be thermostatted at high temperatures.

(c) The system is compact and very small quantities of materials are used.

EFFECT OF INTERFACIAL RHEOLOGY ON COALESCENCE PHENOMENA

The pulsed drop method has not yet been used to investigate coalescence phenomena.
However, as an illustration of how interfacial dilatational rheology is involved
in coalescence processes which are essential to oil bank formation, dilatational
parameters for the Forties crude oil/formation water interface can be quoted
which were determined by the trough method. The influence of various water-in
oil demulsifiers was investigated. Results are shown in Figure 9 for ε'' as a
function of frequency and as a Cole-Cole plot in Figure 10.

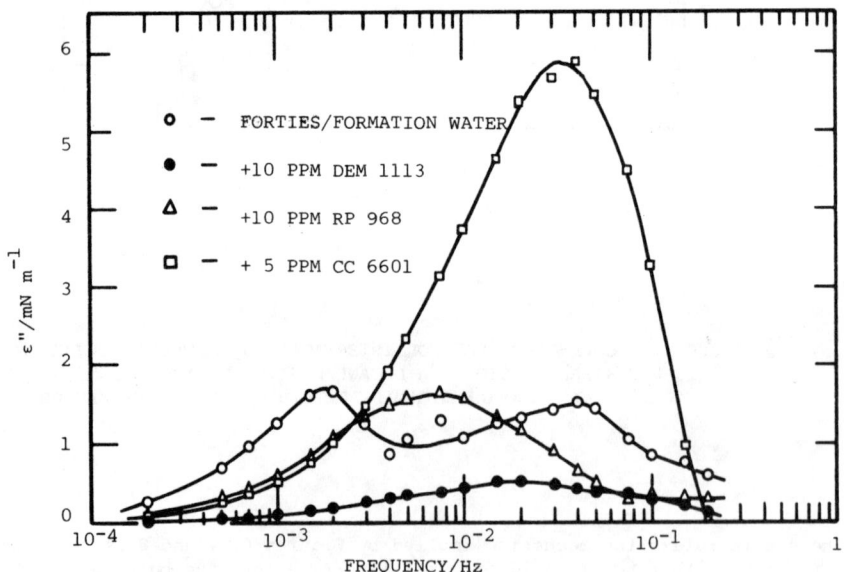

FIGURE 9 - EFFECT OF VARIOUS DEMULSIFIERS ON IMAGINARY (VISCOUS)
COMPONENT OF INTERFACIAL DILATATIONAL MODULUS
FORTIES CRUDE/FORMATION WATER AT 25°C

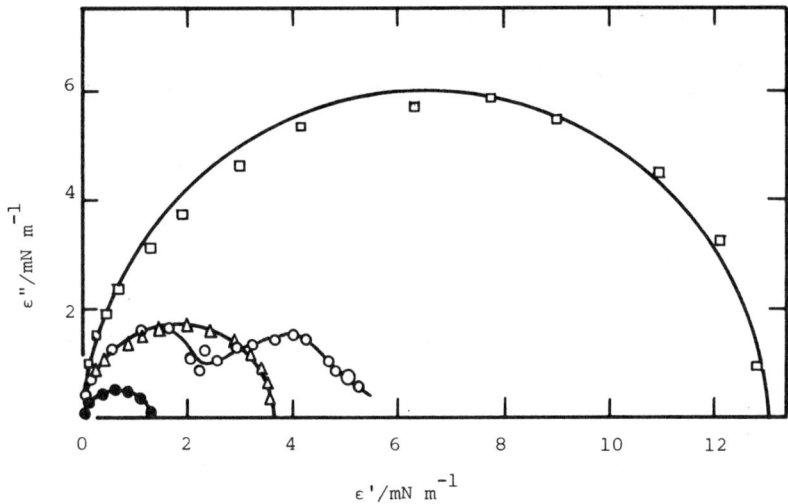

FIGURE 10 - COLE-COLE PLOT FOR SYSTEMS IN FIGURE 9.
SYMBOLS AS IN FIGURE 9.

The interface without additives gives two separate peaks indicating two different relaxation mechanisms are involved. From the positions of the peak maxima we can calculate relaxation times for the two processes of 87 sec and 4 sec. These are compared with relaxation times for systems with low concentrations of three water-in-oil demulsifiers in the table below.

	Relaxation Time (Seconds)
Forties crude/formation water	87
	4
+ 5 ppm CC 6601	4.5
+10 ppm RP 968	22
+10 ppm DEM 1113	9

The major effect of the demulsifiers is to remove the relaxation process characterised by a long relaxation time. Shorter relaxation times are expected to mean more rapid film drainage (3) and therefore more rapid coalescence.

These demulsifiers are also found to reduce the interfacial shear viscosity of the crude oil/water interface. However, from Figure 9 it can be seen that at some frequencies the dilatational viscosity is reduced whereas at other, normally higher, frequencies the dilatational viscosity can be greatly increased. At this stage the mechanistic implications of these observations are not fully understood. Further work on this topic is planned.

MEASUREMENT OF DRAINAGE RATES FOR SINGLE OIL FILMS IN WATER

Direct evidence for the influence of interfacial shear rheology on the kinetics
of drainage of thin films has been obtained by measuring the thickness of crude
oil films in distilled water. The technique was the same as that used to
measure thickness of oil films in air (3), but having the whole cell filled
with water. Measurements of the intensity of light reflected from the single
oil film were used to calculate film thickness as a function of time. Results
for Iranian Heavy crude and for Forties crude in distilled water are shown in
Figure 11.

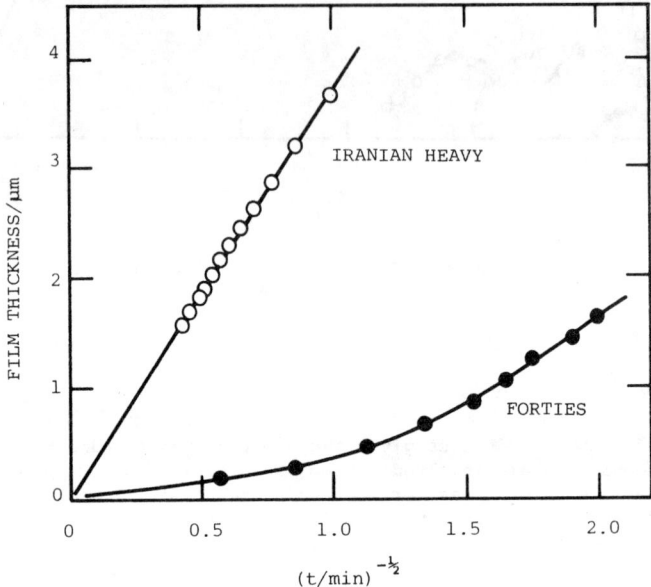

FIGURE 11 - FILM DRAINAGE - CRUDE OIL FILMS IN DISTILLED WATER AT 25°C

For the Iranian Heavy case the thickness is proportional to $t^{-\frac{1}{2}}$ in accordance
with the Stephan-Reynolds equation indicating that drainage is essentially from
between two rigid interfaces. In contrast the Forties crude in water film
drainage curve is not a straight line and indicates much more rapid drainage
of the film than can be accounted for by the lower bulk viscosity of Forties
oil. This implies that the Forties crude/distilled water interface is much
more fluid compared with the Iranian Heavy case. These implications are borne
out by measurements of interfacial shear viscosity at the crude oil/water
interface. Using the biconical bob shear rheometer the results shown in
Figure 12 were obtained. Over a period of hours the shear viscosity of the
Iranian Heavy/distilled water interface builds up to quite high values whereas
that for the Forties/distilled water interface remains low.

The reverse system, drainage of water films from between colliding oil droplets,
is relevant to oil bank formation. Because crude oil is opaque it is not
possible to perform experiments analogous to the single oil film drainage
measurements outlined above. However, there is clear evidence in the literature
for the reduction of coalescence rates for crude oil drops in water when
interfacial shear viscosity is increased (4).

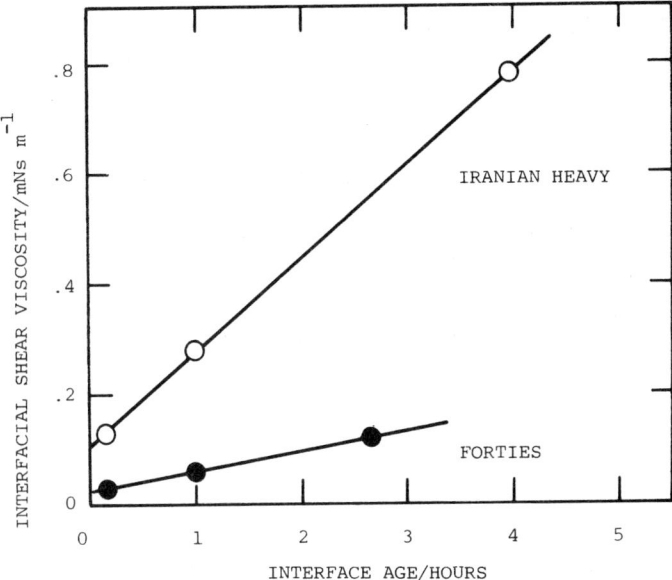

FIGURE 12 - CRUDE OIL/WATER INTERFACIAL SHEAR VISCOSITIES AT 25°C

Clearly an important quality of an EOR surfactant will be the maintenance of low interfacial shear viscosity as an aid to oil bank formation.

CONCLUSIONS

1. A dynamic drop volume technique can be used to determine dynamic interfacial tension in crude oil/water systems as a function of rate of fractional area change.

2. For different surfactant systems which have markedly different oil removal profiles from sand columns, dynamic interfacial tension behaviour can be completely different.

3. A pulsing drop method has been devised which can measure the interfacial dilatational rheological parameters for oil/water systems. The results agree well with those determined using an interfacial trough. Both systems can be used with the single step pulse Fourier transform method.

4. For a pure system of stearic acid in n-decane against distilled water at pH 2.5, the complex dilatational modulus gives a semi-circular Cole-Cole plot indicating that relaxation at the interface is due to a single mechanism, presumably diffusion to and from the interface.

5. For a Forties crude/oil formation water interface, two separate relaxation processes are detected, presumably diffusion and molecular rearrangement. Water in crude oil demulsifiers remove the mechanism with the longer relaxation time.

6. Drainage of crude oils films in water can be followed by reflectance measurements of thickness. Drainage rate depends critically on interfacial shear viscosity.

148

NOMENCLATURE

A   Area of interface (m^2)
Q   Volumetric flow rate ($m^3 \ s^{-1}$)
R   Tip radius (m)
V   Volume of drop (m^3)

g   Acceleration due to gravity ($m \ s^{-2}$)
n   Number of drops per unit time (s^{-1})
Δp Excess pressure inside drop (Nm^{-2})
t   Time (s)

γ Interfacial tension (Nm^{-1})
ϵ^* Complex interfacial dilatational modulus (Nm^{-1})
ϵ' Real part of dilatational modulus (Nm^{-1})
ϵ'' Imaginary part of dilatational modulus (Nm^{-1})
ϵ_d Interfacial dilatational elasticity (Nm^{-1})
η_d Interfacial dilatational viscosity ($Ns \ m^{-1}$)
υ Frequency (cyclic) (Hz)
ρ Density ($kg \ m^{-3}$)
τ Relaxation time (s)
ω Angular frequency (s^{-1})

ACKNOWLEDGEMENT

Permission to publish this paper has been given by The British Petroleum Company
Limited.

REFERENCES

(1)   GRAHAM, D.E., JONES, T.J., NEUSTADTER, E.L. AND WHITTINGHAM, K.P.
         "Interfacial Rheological Properties of Crude Oil Water Systems",
         3rd International Conference on Surface and Colloid Science, Stockholm,
         1979, Plenum Press, in the press.

(2)   LOGLIO, G., TESEI, U. AND CINI, R
         "Spectral Data of Surface Viscoelastic Modulus Acquired Via Digital
         Fourier Transformation"
         J. Colloid Interface Sci, (1979), 71, 316.

(3)   CALLAGHAN, I.C. AND NEUSTADTER, E.L.
         "Foaming of Crude Oils: A Study of Non-Aqueous Foam Stability"
         Chemistry and Industry, 17.1.81, p 53.

(4)   WASAN, D.T., McNAMARA, J.J., SHAH, S.M., SAMPATH, K. AND
         ADERANGI, N.
         "The Role of Coalescence Phenomena and Interfacial Rheological
         Properties in Enhanced Oil Recovery: An Overview"
         J. Rheology, (1979), 23, 181.

BEHAVIOR OF SURFACTANTS IN EOR APPLICATIONS
AT HIGH TEMPERATURES

LYMAN L. HANDY

Department of Petroleum Engineering
University of Southern California

ABSTRACT

Temperature sensitive properties of some anionic and nonionic surfactants
used in EOR operations have been measured. Of particular interest is the
thermal stability. Those surfactants we investigated decomposed by first order
kinetics. The stability can, therefore, be quantitatively expressed in terms
of the half-life of the surfactant. At 180°C half-lifes for petroleum sulfonates varied from 1 to 11 days. Activation energies were measured and these
data can be used to predict half-lifes at other temperatures. Solubility of
nonionics is known to be affected by temperature. At the cloud point they
dehydrate and become less soluble. Anionics appear to form precipitates with
rock minerals. This problem increases with increasing temperature. Adsorption
is temperature dependent although the experimental results for the anionics
were obscured by precipitation. Adsorption of nonionics were observed to
decrease with increasing temperature at low concentrations but to increase with
temperature at high concentrations. Interfacial tensions have also been measured as a function of temperature. The results vary with the surfactant.
Mixtures of sulfonates, however, have all shown an order of magnitude reduction
in interfacial tension at temperatures in excess of 120°C.

INTRODUCTION

Much of the unrecovered oil in the United States occurs in heavy oil
deposits, mostly in California. Large accumulations of heavy oil are also
known to occur in Venezuela, Mexico, Canada and elsewhere. To recover this oil
the viscosity must be reduced by orders of magnitude. The only feasible way to
accomplish this objective is to heat the oil in-place. This can be done by
either steamflooding or in situ combustion. Steam injection is the most frequently used process. This has given rise to the investigation of various
chemical additives which will improve the process. One of the problems with
steam is that it tends to finger through the formation and to override the oil.
Various organic chemicals have been investigated for use with steam as flow
diverters to minimize gravity override. Surfactants are being evaluated as
possible additives which will reduce the residual oil saturation in that portion
of the reservoir which is flooded only with hot water during steam drive.
Although the temperature requirements for chemicals to be used at steam temperatures are much more rigorous, high temperatures are also encountered in the
deeper reservoirs which are currently being considered for enhanced oil recovery.
This has introduced additional requirements with respect to the temperature
compatibility of chemicals used in these reservoirs.

In the present paper we are concerned, primarily, with surfactants, but problems are also encountered with polymers at high reservoir temperatures. Four aspects of the effect of temperature are considered: the effect on the stability of the surfactants, the effect on solubility, the effect on water-oil interfacial tensions and, finally, the effect on adsorption onto the solid matrix.

THERMAL STABILITY

A limited number of studies have been reported in the literature on the stability of surfactants suitable for oilfield operations at temperatures in excess of 100°C. The most extensive of these is that of Handy et al.[1] Data have also been reported by others for the petroleum sulfonate, TRS 10-80, but no temperatures were stated for those experiments.[2] In our earlier report results were presented for anionic and nonionic surfactants. The anionics included sodium dodecylbenzene sulfonate, an acidic Dowfax sulfonate and several petroleum sulfonates. The petroleum sulfonates included TRS 10-80 manufactured by Witco and Petrostep 465 manufactured by Stepan Chemical Corporation. Dowfax 240 was from Dow Chemical Company. The nonionic was an alkylphenoxypolyethanol manufactured under the trademark of Igepal CO-850 by GAF.

The surfactants were mixed at various concentrations without salt and aged at elevated temperatures in Teflon containers in Parr Acid Digestion bombs. Particular care was taken to eliminate air from the bombs. Long term aging tests were conducted in sealed borosilicate glass vials. In comparing our work with that of others, a major factor is the method used for chemically analyzing for the active surfactant. The most common procedure is the Epton titration, which involves a dye transfer between two phases. We found the end points difficult to detect in this procedure. We used instead UV spectrophotometry. The bond which ruptures during high temperature aging is the sulfur-aromatic ring bond. Disubstituted aromatic rings have a characteristic absorption wave lengths at 220-240 nm and 260-280 nm. When the sulfur-aromatic ring bond ruptures, the absorption at these characteristic wavelengths is decreased. The decrease in the concentration of the active surfactant can be measured quantitatively from the change in the peak heights. Concentrations were determined from a comparison of peak heights with those observed for solutions of known concentration. The alkylphenoxypolyethanols could also be analyzed by UV absorption because these compounds also have a disubstituted aromatic ring. A modification of the Epton titration has been proposed by Mukerjee which is reported to be more quantitative than the original method. We have not tested that procedure.[3]

The decomposition reaction for the petroleum sulfonates is the following:

$$ArSO_3^- + 2H_2O \qquad ArH + SO_4^= + H_3O^=$$

It would be possible, therefore, to monitor the reaction from a measurement of the pH.

Representative data from reference 1 are given on Figures 1 and 2. The plot of the logarithm of concentration versus time was linear. pH versus time was also observed to be linear. The other anionic surfactants gave similar behavior. These results indicate that the decomposition reaction for the anionics is first order. The decomposition rate for a reaction following first order kinetics is

$$- dC/dt = kt$$

$$C = C_0 e^{-kt} \quad \text{or} \quad \log C = \frac{-kt}{2.303} + \log C_0$$

In these equations C is concentration in moles per liter; C_0 is the initial concentration; t is time in days and k is the rate constant in days^{-1}. The rate constant is determined from the slope of the semilog plot. One can also show that when $C/C_0 = \frac{1}{2}$, the elapsed time is equal to the half-life of the surfactant.

$$t_{\frac{1}{2}} = \frac{.693}{k}$$

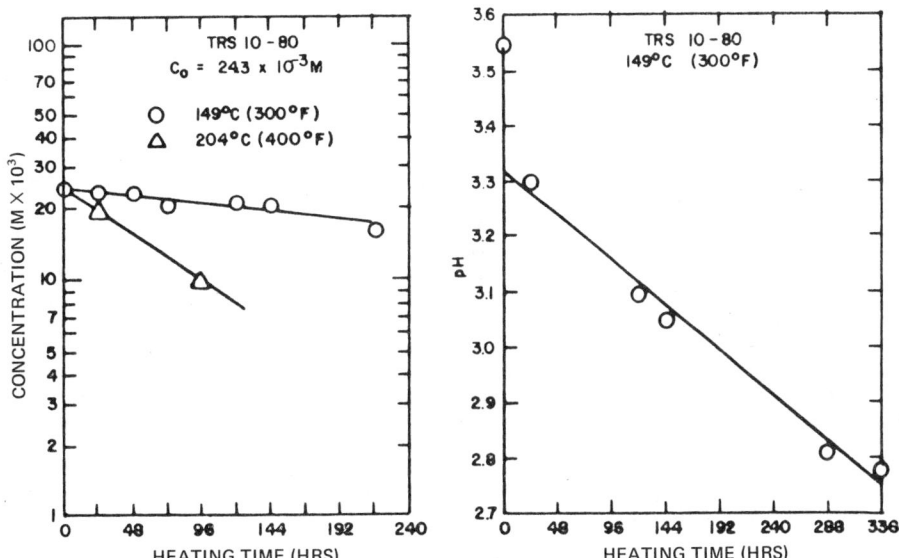

Fig. 1 - Concentration of TRS 10-80 as function of heating time at 149°C and 204°C

Fig. 2 - pH of TRS 10-80 as function of heating time at 140°C

If one has rate constants at several different temperatures one can determine the activation energy for the reaction. With the activation energy one can determine rate constants and half-lifes at other temperatures. This is particularly useful in estimating the stability of surfactants at lower temperatures for which the decomposition rates are low and long times would be required to measure the half-lifes. Figure 3 is a plot of the log of the rate constant versus the reciprocal of the absolute temperature for TRS 10-80. This plot is typical of those obtained for the surfactants which were tested. In the equation

$$\log k = \frac{-E_a}{2.303 \, RT} + B$$

E_a is the activation energy in cals/mole; R is 1.987 cals and T is the absolute temperature in °K. From the slope of the plot one can determine the activation energy.

A summary of decomposition data for several surfactants is given in Table 1. At 180°C Petrostep 465 is the most stable of the surfactants we investigated. Because of its high activation energy relative to the other surfactants, this surfactant would have a half-life of about 16 years at 100°C. None of the surfactants have adequate stability for use at normal steam temperatures. These results would be expected to be representative for aryl sulfonates, but better stabilities have been informally reported for alkyl sulfonates.

SOLUBILITY

Quantitative data on the effect of temperature on the solubility of petroleum sulfonates have not been reported, but evidence has been cited by several authors that precipitation of the sulfonates occurs at the higher temperatures in natural sandstones.[2,4,5] This occurs not as a result of a direct temperature effect on the solubility of the surfactants but, apparently, as a result of an interaction with minerals in the porous media. Reed has measured a significant increase in the solubility of rock minerals at steam temperatures.[6] The petroleum sulfonate ions form precipitates with divalent cations. These precipitates are likely to decrease in solubility with increasing temperature. In general, the presence of salt in the solutions decreases the solubility of the sulfonates.

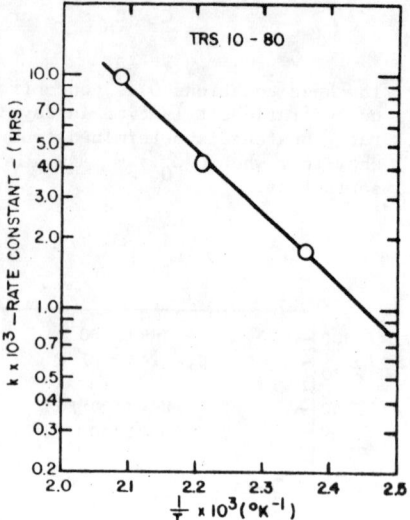

Fig. 3 - The rate constant (k) as function of $\frac{1}{T}(°K^{-1})$ for TRS 10-80

TABLE 1

SUMMARY OF DECOMPOSITION DATA FOR SURFACTANTS

Surfactant	Mol. Wt.	Temp. °C	$t_{\frac{1}{2}}$ (days)	E_a (kcals)
NaDDBS	348.5	130	6.13	24.0
		180	.22	24.0
		150	13.6	26.0
		180	1.75	26.0
Dowfax 2AO	500	177	5.6 (UV)	NA
			6.9 (pH)	
TRS 10-80	415	149	17.4	12.4
		204.5	3.0	12.4
		180	7.0	12.4
Petrostep 465	465	130	444	25.2
		157	108	25.2
		180	11	25.2
Igepal CO-850	1100	130	.75	8.84
		180	.22	8.84

Ziegler observed turbidity in the produced fluid from a Berea sand pack when sodium dodecylbenzene sulfonate solutions were injected at a concentration of 1400 μmols/liter. However, data in Figure 4 show that surfactant precipitated out of a 0.2 molar salt solution could be redissolved when distilled water was injected and when the temperature was increased. In this experiment the sand pack was flushed with 1374 μmols/liter surfactant in 0.2 M NaCl. Then the pack was flushed with salt solution only, with distilled water and, finally, with distilled water at 180°C. Distilled water redissolved sulfonate precipitated out of,

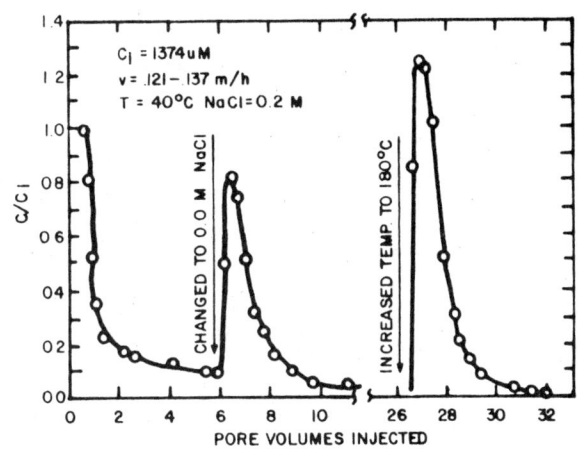

Fig. 4 - Desorption curve for NaDDBS

brine and an increase in temperature to 180°C did redissolve sulfonate still precipitated at 40°C after the distilled waterflood.

The solubility of nonionic surfactants is not as sensitive to salt concentration as that of the anionic surfactants. On the other hand, the solubility of the alkylphenoxypolyethanols shows a marked sensitivity to temperatures. At very specific temperatures called the cloud points, the ethoxy groups in these compounds lose associated water and the solubility decreases abruptly to form precipitates. The cloud point is a function of the molecular weight of the surfactant, the electrolyte composition and the concentration of the surfactant. Cloud points as a function of concentration for Igepal CO-850 are shown in Table 2.

TABLE 2

SUMMARY OF PHYSICAL AND SORPTION PROPERTIES

FOR IGEPAL CO-850

Molecular Weight = 1,100   CMC = 100 μmol/L

Cloud Points

C_o (μmol/L)	Cloud Point (°C)
73	>180
366	113
640	106

Sorption Properties

Temperature (°C)	K_{eq} (dm^3/μmol)	A (μmol/m^2)	k_1 (dm^3/μmol·h)	k_2 (hours^{-1})	$\Delta H°$ (kJ)
45	5.78×10^{-2}	0.524	1.2×10^{-2}	0.21	−40.2
70	2.09×10^{-2}	0.705	1.5×10^{-2}	0.72	
95	7.34×10^{-3}	0.831	2.5×10^{-2}	3.41	

EFFECT OF TEMPERATURE ON SURFACTANT ADSORPTION

If low concentration surfactants are to be used in combination with steam-flooding or hot waterflooding in a reservoir, the effect of temperature on adsorption becomes a matter of considerable importance. Surfactant transport could be combined with heat transport through the reservoir. The surfactant concentration shock could either lead or trail the temperature shock. Data will be presented later which shows that interfacial tensions are reduced at higher temperatures. If this is the case, one would prefer to have the surfactant front remain in the heated portion of the reservoir. In steamflooding, however, it is well-established that the steam overrides the oil. The water transporting the surfactant is likely to be moving primarily in a heated region immediately below the steam zone. In that case the surfactant will be moving in a hot portion of the reservoir under isothermal conditions. Whichever mechanism prevails in the reservoir, adsorption isotherms will be required for the prevailing temperature at which the surfactant is being transported. Consequently, we have made an initial effort to determine adsorption isotherms as a function of temperature for an anionic and a nonionic surfactant.

An abundance of data exists in the literature for adsorption of various surfactants onto various substrates at room temperature. These data were normally obtained by equilibrating the surfactant solutions with the solid surfaces. Measuring adsorption isotherms at steam temperatures is a much more difficult problem.

Ziegler et al. obtained data using a dynamic, chromatographic transport procedure.[4] The porous medium was a disaggregated, fired Berea sandstone, packed in a core holder. The core was saturated with brine or distilled water and placed in an oven to maintain the temperature at the desired value. Surfactant solution was injected, starting at low concentrations. The pore volumes of solution required to move the surfactant through the core were measured. From chromatographic transport theory the quantity of surfactant adsorbed at this concentration could be calculated.

The surfactant concentration in the injected solution was increased step-wise and the volumes required to move each concentration step through the core was measured. The surface area of the sand had been measured by a variation of the BET method. From these data the adsorption isotherm can be calculated. Adsorption isotherms were also measured by the conventional static method at 25°C and 95°C. Dynamic and static adsorption data were obtained for sodium dodecylbenzene sulfonate (NaDDBS) and Igepal CO-850.

As discussed earlier, the NaDDBS has a low solubility in 0.2 molar NaCl and also tended to precipitate at the higher temperatures when in contact with the Berea sandstone. Consequently, only adsorption isotherms obtained by the static method are reported for NaDDBS. These data are shown at 25°C and 95°C for concentrations up to 70 μmols/L on Figure 5. The results show

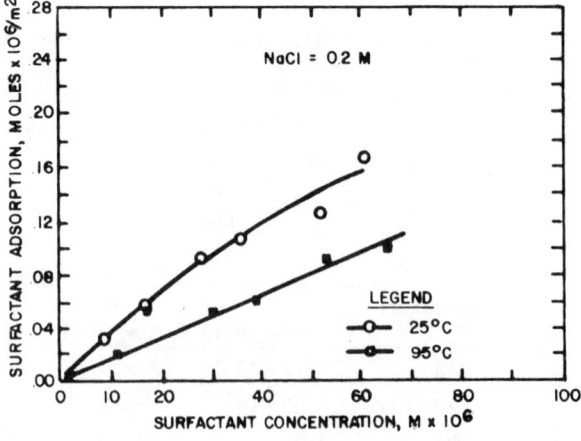

Fig. 5 - Static adsorption isotherms for NaDDBS

that adsorption decreases with increasing temperature as one would expect. Data obtained in the absence of salt show less temperature dependence. Because of the precipitation problem, no dynamic data are reported for NaDDBS. The results of desorption experiments are shown in Figure 4, but the slugs of surfactant being produced after reducing the salt concentration or after increasing the temperature had been explained earlier as being more the result of dissolving precipitated surfactant than desorption of adsorbed surfactant. The slug produced after increasing the temperature, however, may have resulted in part from decreased adsorption at elevated temperatures. This would be consistent with the limited static data showing a decrease in adsorption with temperature.

The experiments with Igepal CO-850 were complicated by the cloud point, which is characteristic of this class of surfactants, and by the instability of this surfactant at high temperatures. Static results are given in Figure 6. Equilibration time for the 95°C curve was limited to three hours. Degradation was a serious problem if significantly longer times were used. The results show a slight temperature dependence. Figure 7 is an example of results obtained by the dynamic method for Igepal CO-850. Surfactant was injected at an initial concentration of 67 μmols/L and at two incremental concentration higher than the initial. Consistent with a Langmuir-type isotherm, the pore volumes of injected surfactant required to produce the incremental step in concentration decreased with increasing concentration. Dynamic data were obtained at 45°C, 70°C and 95°C. Data were not obtained at higher temperatures because of the limit established by the cloud points. Degradation of Igepal is not a problem in the dynamic procedure because the surfactant is at an elevated temperature only while moving through the core.

The dynamic adsorption isotherms for Igepal are given on Figure 8. At low concentrations adsorption decreases with temperature, but adsorption increases with temperature for concentrations in excess of about 200 μmols/L. This effect is also associated

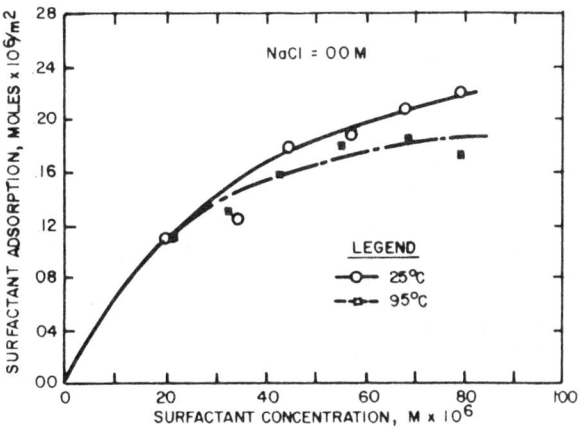

Fig. 6 – Static adsorption isotherms for Igepal CO-850

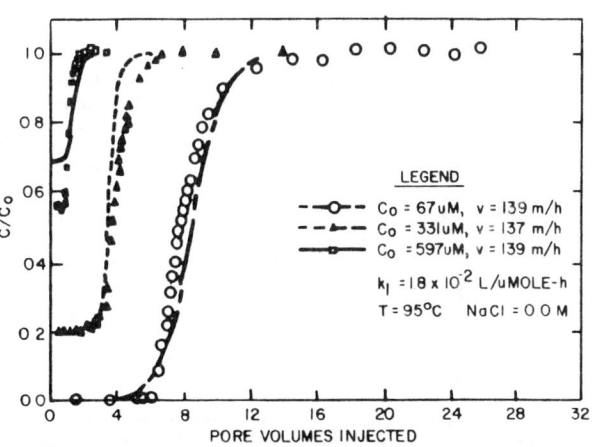

Fig. 7 – Breakthrough curves for Igepal CO-850

with the cloud point. As the ethoxide groups lose their associated water, the surfactant becomes less soluble and would be expected to separate out onto the solid phase more readily. The Langmuir constants for Igepal are given in Table 2.

The results of the dynamic method with Igepal indicate that the method is suitable for determining adsorption isotherms at elevated temperatures, but the surfactants, Igepal and NaDDBS, were not suitable for testing the procedure at temperatures in excess of 100°C because of solubility problems.

EFFECT OF TEMPERATURE ON INTERFACIAL TENSIONS

Few results have been reported giving interfacial tensions of oil-surfactant solutions as functions of temperature.[7,8,9] These data are required for any process using surfactants in reservoirs but, particularly, for the high temperatures associated with steamflooding. We have used two methods for measuring interfacial tensions as functions of temperature

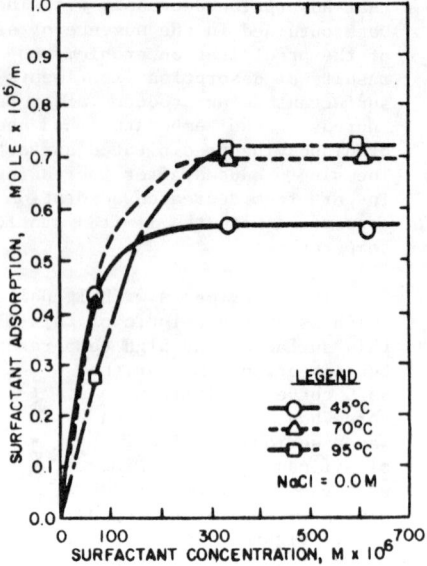

Fig. 8 - Dynamic adsorption isotherms for Igepal CO-850

and pressure. These are the pendent drop and the spinning drop methods. The minimum interfacial tension that can be measured on the pendent drop equipment is about 0.1 mN/m and that is with low precision. The spinning drop reportedly gives data below 0.001 mN/m. If time is an important factor in establishing equilibrium between the surfactant solution and the oil, the pendent drop procedure is also less suitable than the spinning drop. In the pendent drop method a drop can be suspended at the most for one-half hour. On the other hand, a drop can be maintained indefinitely in a spinning drop apparatus. Normally, equilibrium times for surfactant solutions and refined oils are small. The problem arises with caustic solutions and crude oils. Some reports have indicated that equilibrium for these systems has not been established even in matter of days. Although the spinning drop would appear to be the preferred method, we have obtained data by both procedures.

We have modified the spinning drop equipment of Gash and Parrish for use at temperatures to 200°C and pressures to 30 bars.[10,11] The design of the spinning drop equipment is such that it is a simple matter to construct an air bath around the capillary tubes which contain the spinning drop. No bearings need to operate at thermostat temperatures. An epoxy was found that was effective in sealing the capillary tubes at the above temperatures and pressures. The epoxy can be easily drilled out of the tubes to permit using them again. Our equipment is easy and inexpensive to build, but it does not have the versatility of that developed at the Technical University of Clausthal.[11] Their apparatus operates at higher pressures and temperatures and permits the exchange of fluids in the rotating capillary during the experiments.

An important factor in measuring interfacial tensions by either the pendent drop or the spinning drop method is the density difference between the water and the oil. This becomes particularly critical when measuring interfacial tensions at elevated temperatures. The density of water decreases more rapidly with

temperature than that for oil. Consequently, the density difference between
water and oil can become quite small at higher temperatures. A small error in
estimating these densities can have a significant effect on the calculated
interfacial tensions. This problem becomes particularly acute when the oil
phase is a crude oil. For some crudes the density of the oil may, in fact,
become greater than that of the water. The spinning drop equipment cannot be
used under those circumstances. Density data are readily available for water
and can be generated easily for the surfactant and brine solutions. Densities
for refined oils and pure hydrocarbons were determined using data from
"Petroleum Refinery Engineering."[12] To correct crude oil densities for temperature, the volume correction factors from ASTM D-206-36, Group 1, were used.
Density data for water and three crude oils taken from El-Gassier et al. are
shown in Figure 9.

Representative data obtained by the spinning drop method are shown on
Figure 10.[13] Interfacial tensions were measured between mineral oil No. 9 and
TRS 10-80 in various concentrations of salt. The concentration of the TRS
10-80 was kept constant at 0.5 g/L. The interfacial tensions showed little
dependence on temperature up to 180°C, but they are affected substantially by
the salt concentration. The lowest interfacial tensions were observed at salt
concentrations of 5.0 g/L. The lower curves on this figure are duplicate runs
and show reasonable agreement. Interfacial tensions for TRS 10-80 against a
representative crude oil also showed little effect of temperature.

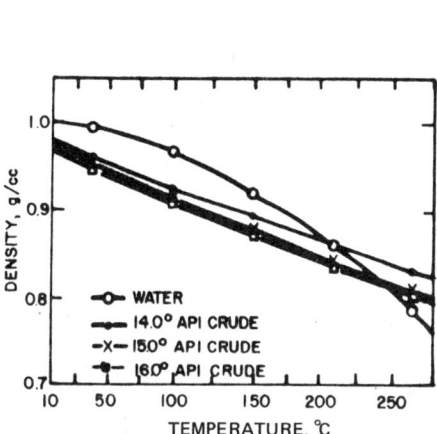

Fig. 9 - Effect of temperature on
density of water and crude oils

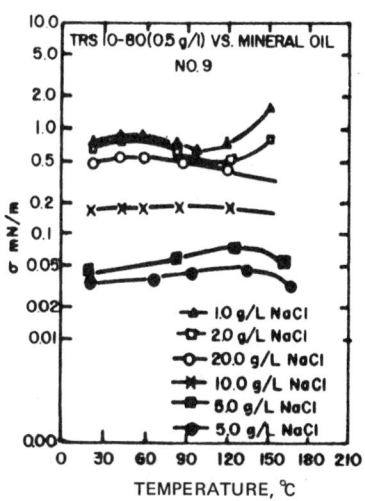

Fig. 10 - Effect of salt concentration and temperature on
interfacial tension of 0.5
g/L TRS 10-80 versus mineral
oil No. 9

The interfacial tension between the nonionic surfactant, Igepal DM-730 and
a 15.9°API California crude oil showed a marked minimum when plotted versus
temperature as shown on Figure 11.[13] No salt was present in this example but
similar data were obtained with the surfactant in presence of salt. The interfacial tension minima for the nonionics coincided with the cloud point for the
particular surfactant concentration. Since the cloud point indicates a decrease
in the surfactant solubility, it is not surprising that the interfacial tension

decreases at this temperature. The decrease in transparency of the aqueous phase at the cloud point was a limiting factor in measuring interfacial tensions of nonionics as a function of temperature by either the spinning drop or the pendent drop method.

A more detailed study of the effects of surfactant concentration, salt concentration and temperature on interfacial tension against a crude oil was made with the pendent drop equipment. Although this equipment is not capable of measuring the ultra low tensions it can show, at least qualitatively, the trend of the effect of these variables. Representative data are shown on Figure 12. The surfactant in this case was TRS 10-80 and the oil was California Wheeler Ridge crude with an API gravity of 15.9°. The temperature was 177°C. The results are interesting in that they indicate an optimum surfactant and salt concentration at 177°C to obtain a minimum interfacial tension. Similar minima were observed for lower temperatures but the minimum interfacial tension increased with decreasing temperature. At 93°C the

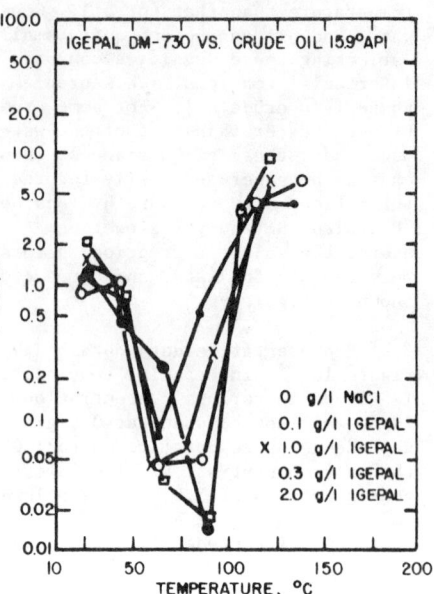

Fig. 11 - Effect of surfactant concentration and temperature on interfacial tension between TRS 10-80 and crude oil. NaCl = 0.0 g/L

minimum interfacial tension was 0.1 mN/m as compared to 0.005 mN/m at 177°C.

Additional data have been obtained using a mixture of surfactants against pure hydrocarbons and mineral oils.[14] The equivalent alkane carbon number, EACN, for the surfactant mixtures was calculated as recommended by Jacobson et al.[15] As shown in Figure 13, these mixtures show an abrupt decrease in interfacial tension at temperatures in excess of 120°C. The experiments are being extended to obtain data for several hydrocarbons and, thereby, evaluate the relation between the change in interfacial tensions with temperature and the EACN concept.

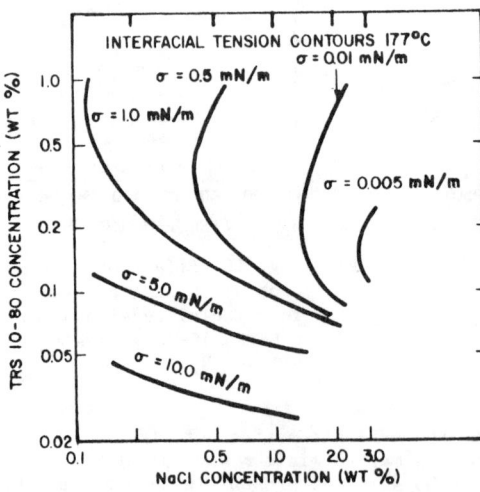

Fig. 12 - Interfacial tensions as functions of NaCl and TRS 10-80 concentrations at 177°C

CONCLUSIONS

The results of our experimental work and data reported by others suggests conclusions about the behavior of surfactants at elevated temperature. Some of these conclusions are quite specific and dependable for the systems to which they apply. Many are tentative. Certainly more work is required to extend the number of surfactants which have been evaluated at high temperatures.

1. The surfactants investigated were observed to decompose by first order kinetics. Therefore, a quantitative measure of the stability of a surfactant at a given temperature is its half-life. Activation energies were determined for several surfactants. Stabilities can be estimated from these energies at higher or lower temperatures than those used in the experiments.

2. The anionic petroleum sulfonates were observed to be more stable than the nonionics. The stability of the best sulfonate would be only marginally accept-able at temperatures to 180°C but other surfactants need to be evaluated. All of the surfactants tested would be adequately stable at normal reservoir temperatures.

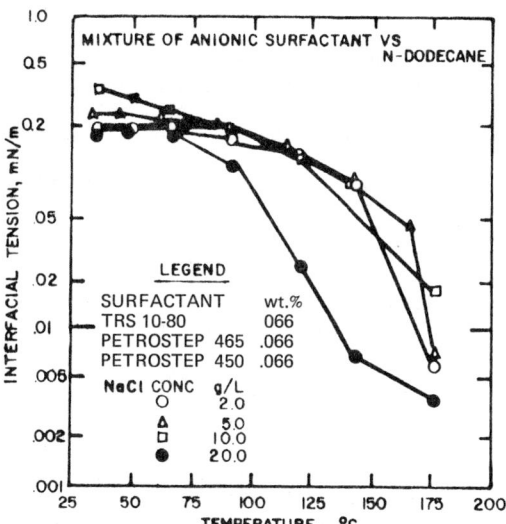

Fig. 13 - Interfacial tensions as functions of temperature and salt concentration for surfactant mixtures against n-dodecane

3. Evidence suggests that the sulfonates may be precipitated at steam temperatures as a result of an interaction with solubilized rock minerals which show limited solubility at elevated temperatures. The solubility of the nonionics decreases abruptly at the characteristic cloud point. This limits the concentration at which these surfactants can be used at higher temperatures.

4. Dynamic and static methods were used for evaluating the temperature effect on adsorption. The data suggest that adsorption decreases for both sodium dodecylbenzene sulfonate and for Igepal CO-850, but the effect is not as substantial as one might have expected. Additional data are required with other surfactants in consolidated sandstones.

5. A substantial amount of data is being accumulated relating interfacial tension and temperature. For specific types of petroleum sulfonates some data indicate little effect of temperature on interfacial tensions. On the other hand, pendent drop data do suggest a significant decrease in interfacial tension with temperature for optimum salt and surfactant concentrations. Other results show a decrease in interfacial tension with temperature for mixtures of sulfonates against pure hydrocarbon or mineral oil. The nonionic, Igepal DM-730, showed a sharp minimum in the interfacial tension at a specific temperature. That temperature appears to be related to the cloud point.

REFERENCES

1. HANDY, L. L., AMAEFULE, J. O., ZIEGLER, V. M., and ERSHAGHI, I.; "Thermal Stability of Surfactants for Reservoir Application", paper SPE 7867 presented at SPE Fourth Intl. Symposium on Oilfield and Geothermal Chemistry, Houston, Jan. 22-24, 1979.

2. ISAACS, E. E., PROWSE, D. R., and RANKINE, J. P.; "The Role of Surfactant Additives in the In Situ Recovery of Bitumen from Oil Sands", Paper No. 81-32-13, presented at the 32nd Annual Technical Meeting of the Petroleum Society of CIM, Calgary, May 3-6, 1981.

3. MUKERJEE, P.; "Use of Ionic Dyes for the Analysis of Ionic Surfactants and Other Ionic Organic Compounds", Analytical Chemistry (May 1956) 28 (5) 870.

4. ZIEGLER, V. M. and HANDY, L. L.; "Effect of Temperature on Surfactant Adsorption in Porous Media", Soc. Pet. Engr. Jour. (April 1981) 21 (2) 218-226.

5. CELIK, M., GOYAL, A., MANEV, E., and SOMASUNDARAN, P.; "The Role of Surfactant Precipitation and Redissolution in the Adsorption of Sulfonate on Minerals", paper SPE 8263 presented at the SPE 54th Annual Technical Conference and Exhibition, Las Vegas, Sept. 23-26, 1979.

6. REED, M. G.; "Gravel Pack and Formation Sandstone Dissolution During Steam Injection," J. Pet. Tech. (June 1980) 941-949.

7. GOPALAKRISHNAN, P., BOREIS, S. A., and CAMBARNOUS, M.; "An Enhanced Oil Recovery Method -- Injection of Steam with Surfactant Solutions", Report of Group d'Etude IFP-IMF Sur les Milieux Poreux Toulouse (1977).

8. SANDVIK, E. I., GALE, W. W., and DENEKAS, M. O.; "Characterization of Petroleum Sulfonates", Soc. Pet. Engr. Jour. (June 1977) 184-192.

9. McCAFFERY, F. G.; "Measurement of Interfacial Tensions and Contact Angles at High Temperature and Pressure", J. of Canadian Petroleum Technology (July 1972).

10. GASH, B., and PARRISH, D. R.; "A Simple Spinning-Drop Interfacial Tensiometer", J. Pet. Technology (January 1977) 30-31.

11. BURKOUSKY, M. and MAX, C.; "Applications for the Spinning Drop Technique for Determining Low Interfacial Tension", Tenside Detergents (1978) 15 (5) 247-251.

12. NELSON, W. L.; "Petroleum Refinery Engineering", (1958) 157-161.

13. HANDY, L. L., EL-GASSIER, M. and ERSHAGHI, I.; "Interfacial Tension Properties of Surfactant-Oil Systems Measured by a Modified Spinning Drop Method at High Temperatures", paper SPE 9003 presented SPE Fifth Intl. Symposium on Oilfield and Geothermal Chemistry, Stanford University, May 28-30, 1980.

14. ZEKRI, A.; Personal Communication.

15. JACOBSON, J. K., MORGEN, J. C., SCHECHTER, R. S., and WADE, W. H.; "Low Interfacial Tensions Involving Mixtures of Surfactants", Soc. Pet. Engr. Jour. (1976) 122-128.

SURFACTANT SLUG DISPLACEMENT EFFICIENCY IN RESERVOIRS; TRACER STUDIES IN 2-D LAYERED MODELS

ROBERT J. WRIGHT, RICHARD A. DAWE and COLIN G. WALL

Petroleum Engineering Section, Imperial College, London SW7 2AZ

ABSTRACT

The effects of layering within porous material with regard to basic flow mechanisms and chemical dispersion have been investigated. Experiments have been performed within unconsolidated glass bead packs. The variables controlled were layer permeability and dimensions, fluid viscosity and flow rate; gravity and capillary pressure influences were eliminated by using model fluids of matched density and complete miscibility. The importance of channeling and crossflow effects are emphasized by the results, and the behaviour of non-unit mobility ratio displacements is predictable using relatively simple conceptual/mathematical models. The dispersion of chemical tracers between layers has also been modelled mathematically and the results have been applied to laboratory tests on heterogeneous cores.

INTRODUCTION

It is well known that the natural heterogeneity of petroleum reservoir
material is one of the major problems in chemical E.O.R. processes. Of
particular consequence are the non-random variations in permeability to
be found within porous rocks. Layering structures are a common feature
of sandstones and their effects have been reviewed in recent literature
with reference to fluid flow (1) and dispersion mechanism (2). The
efficiency of surfactant slugs is probably the most likely application of
these considerations; however the fundamental problems are common to all
E.O.R. processes. We have investigated layered models, both conceptual/
mathematical and physical (visual). Experimentally, flow mechanisms
and dispersion effects have been monitored using dye tracers. Displacements have been of an ideal miscible type and therefore represent perfect
microscopic displacement efficiency. The properties peculiar to surfactants such as adsorption, phase equilibrium and emulsification characteristics have been excluded in the present work. We are taking the
approach that the gross fluid flow and dispersion effects within heterogeneous media should be better understood before laboratory core-flood
results and data from linear homogeneous packs can be applied to the
reservoir system. We have attempted to view miscible and immiscible
displacement mechanisms on a common basis since the two concepts merge
in ultra-low-tension systems.

The experimental work discussed here involved idealized layered models
of packed Ballotini. The flow mechanics of displacements at various
(favourable and unfavourable) mobility ratios were recorded by photo-
graphing dye tracer boundaries under conditions of flow rate for which
diffusion/dispersion effects were small. To quantify dispersion phenomena we have considered equiviscous miscible displacements, and we
describe here numerical predictions with one example application.
Conceptual models were developed, based on simple two layer-channel
interactions. This approach follows contributions within the literature
on dispersion (2) & (3) and crossflow (4) & (5) in such model systems.

FLOW PATTERNS IN LAYERED MEDIA

Model.

It has been found useful to consider simple two-channel conceptual models
in order to account for crossflow behaviour in multilayered and striated
media. Crossflow directions and approximate magnitudes can be demonstrated mathematically by considering the variation of flow potential
along the axes of the channels. Figure 1(i) illustrates two parallel
channels composed of homogeneous and continuous porous media; a high
permeability channel (a) and a less permeable channel (b). The displacement of fluid (1) by fluid (2) within this model (in the x direction) has
resulted in two displacement boundaries (at x_a and x_b). The instantaneous
pressure profiles are plotted for two different viscosity ratios; displacing fluid the more viscous in Fig.1(ii) and the less viscous in Fig.1(iii).

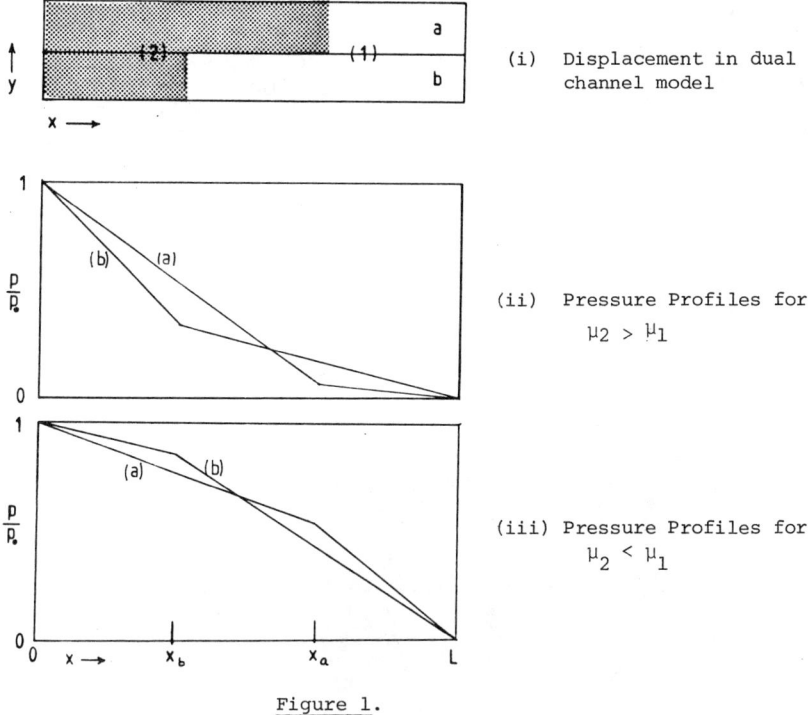

(i) Displacement in dual channel model

(ii) Pressure Profiles for $\mu_2 > \mu_1$

(iii) Pressure Profiles for $\mu_2 < \mu_1$

Figure 1.

This assumes no capillary pressure, dispersion, gravity or compressibility effects; also for the moment, no crossflow between the channels (as if separated by an impermeable barrier). It is, however, a useful method for representing local crossflow tendencies as indicated by pressure drops (at fixed x) between the channel axes. Crossflow would therefore be strongest around the displacement fronts and occurs in the directions indicated in Table 1.

Table 1.

Fig.	$\dfrac{\mu_2}{\mu_1}$	Location.	Fluids Crossflowing	From Channel:-	Into Channel:-
1(ii)	>1	x_b	(2)	a	b
		x_a	(1)	b	a
1(iii)	<1	x_b	(1) & (2)	b	a
		x_a	(1) & (2)	a	b

Experimental.

Displacements were performed in a visual model composed of glass beads
packed to form a central (high permeability) layer surrounded by less
permeable packing. Channel widths were around 1 cm within a total model
width of 10 cm., the length of the flow model was 20 cm, the permeability
ratios between layers was 2.8, and porosity was approximately 38%.
Constant flow rates were used and matched density aqueous solutions were
employed. Glycerol/water mixtures and sodium chloride or sodium sulphate
solutions were used. The photographs (Plates 1 -3) illustrate displacement patterns with fluorescent tracers under three distinct conditions
of viscosity ratio; $\mu_2/\mu_1 = 0.22$ (Plate 1), 1.0 (Plate 2), and 3.0
(Plate 3). Roughly 0.2 pore volumes of displacing fluid has been injected, and flow is in all cases from left to right.

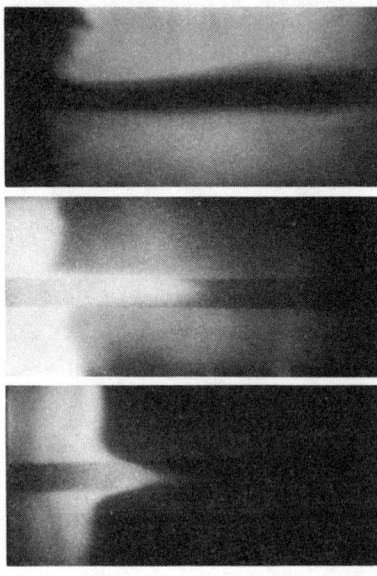

Plate 1: unfavourable
mobilities

Plate 2: equal mobilities

Plate 3: favourable mobilities

The different displacement boundary patterns and the consequent differences
in displacement efficiency can be explained to some extent by the axial
pressure gradients of Figs. 1(ii) and (iii), but the dominant influence is
crossflow as detailed in Table 1. Hence, when displacing fluid is less
viscous (more mobile) the leading "finger" within the high permeability
channel advances rapidly due to crossflow into the finger at its base and
out of the channel around its leading tip. Crossflow is seen to alter
the shape of the finger by swelling its front and squeezing its base.
For "favourable" viscosity ratios these conditions are reversed; penetration into the high permeability channel is retarded, the advancing cusp
being squeezed in at the front (with some tracer dispersed ahead as a
thin plume) and widened at its base where it joins the main displacement
front. Smaller scale differences are apparent around the main displacement
boundary; local fingering in the former case and a sharp stabilized
boundary in the latter favourable mobility case.

Quantitative Results: Unfavourable Mobility Ratio Continuous Displacement

Crossflow can be quantified in the two channel conceptual model described above by means of a numerical crossflow index:-

$$\alpha = A . \frac{k_t}{k_a} . \frac{(\Delta x)^2}{d} \quad ,$$

where A is the crossflow boundary surface area per unit volume of channel (a), k_t is the effective cross permeability between the channel axes, k_a is the permeability of channel (a), Δx is the numerical inter-node fractional distance and d is the separation of the channel axes.

For equal width straight layers of isotropic media, d is equal to the layer width and

$$A = \frac{2}{d} . \frac{k_t}{k_a} = \frac{2}{1 + k_a/k_b} \quad .$$

Fig. 2 illustrates calculated instantaneous channel (a) pressure profiles for $x_a = 0.5$, $x_b \simeq 0$ when $\mu_2/\mu_1 = 0.1$, for the crossflow indexes and layer aspect ratios given below:-

Curve	α ($\Delta x = 0.02$)	layer $\dfrac{d}{L}\left(\dfrac{k_a}{k_b} = 4\right)$
1	0.0001	2.0
2	0.01	0.2
3	0.1	0.06
4	1.0	0.02

These span the extremes of practically no crossflow (curve 1) to near maximum crossflow (almost zero resistance to flow between channel centres, curve 4).

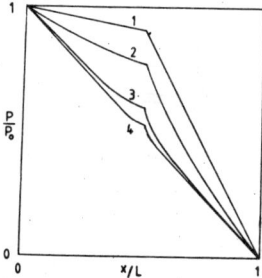

Figure 2.
Possible pressure distributions along channel (a).

Numerical calculations of distance/time tracks have been performed based on incremental advances of a displacement front ($\Delta x = 0.02$) and estimation of pressure gradients, hence displacement velocities (as a function of frontal position). For various values of "α" displacement tracks were calculated for viscosity ratios of 0.5 (Fig.3), 0.1 (Fig. 4) and 0.01 (Fig.5).

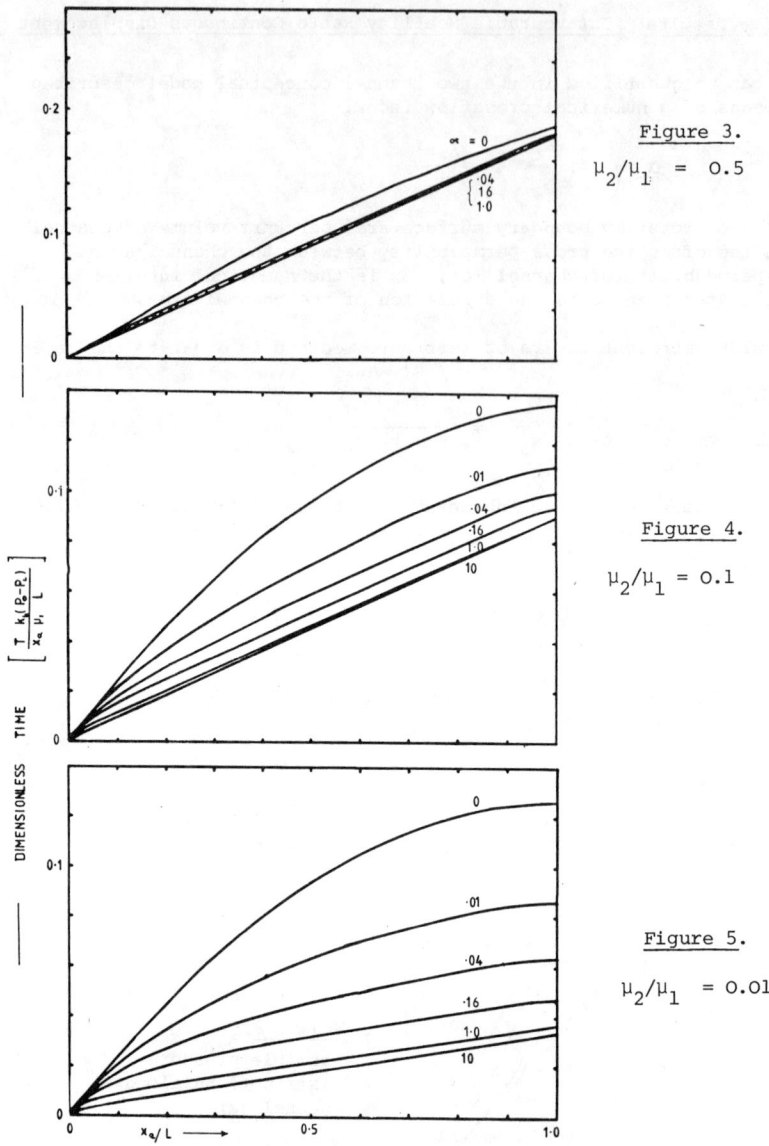

Figure 3.

$\mu_2/\mu_1 = 0.5$

Figure 4.

$\mu_2/\mu_1 = 0.1$

Figure 5.

$\mu_2/\mu_1 = 0.01$

Figs. 3 - 5: Time/Distance tracks for displacement front within high
           permeability channel; numerical calculations.

Clearly, for mobility ratios near unity crossflow is unimportant and
displacement velocities are constant with time. The limiting linear
tracks for high α are approached quite closely for moderate values of
viscosity ratios and layer aspect-ratio. The behaviour is only sensitive
to crossflow index when $d/L > 0.1$ and $\mu_2/\mu_1 < 0.1$, as a general guide.

Experimental Results.

Flow visualization experiments were conducted with matched density fluid
pairs having "adverse" viscosity ratios. The packed bead models were as
described above. Experiments were distinguished by the parameters given
below:-

Table 2.

Experiment	k_a/k_b	d/L	μ_2/μ_1
◊ 1	⎫	⎫ 0.05	⎫
+ 2	⎬ 2.8	⎭	⎬ 0.33
× 3	⎬	⎫ 0.07	⎭
o 4	⎭	⎭	0.22

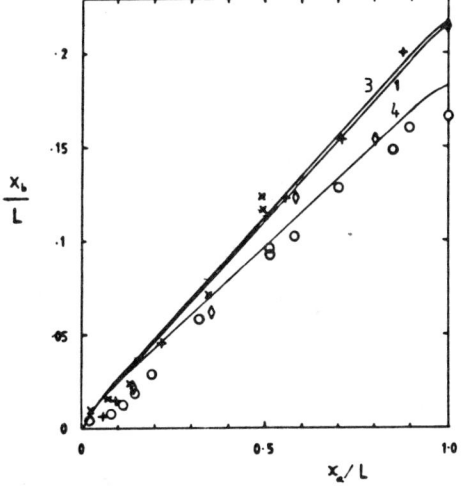

Figure 6.

Relative Front
Positions.

"Mean front" positions were estimated from colour photographs taking into
account dispersion and local fingering. When plotted versus time, approximately straight line tracks were obtained; data scatter being not too
serious. The results in terms of the leading front displacement (x_a) and
the main front (x_b) are plotted on Fig. 6, along with the numerically
predicted curves using the parameters given in Table 2. The correlation
of experiment and calculations is encouraging. However, these predictions
are based on equating x_b/L to the dimensionless time (of Figs.3 - 5)
which is not expected to be a good approximation in all cases. It is
noticeable that there is a significant dependence on viscosity ratio. An
interesting feature of most experiments is the relatively fast initial
penetration into the high permeability layer, a detail contradicted by the
numerical results. Similar findings are described by Peaceman and Rachford
(6) for viscous fingering in randomly variable porous media.

Discussion of Analytical Methods.

It is useful to consider at this point the effectiveness of an analytical
solution method based on 1-dimensional flow theory and "pseudo" relative
permeability functions (7). These are better described as synthetic
functions since they are derived by adding together the effects of the
individual layer properties. The relative permeability to displacing (2)
and displaced (1) phases are plotted versus saturation of phase (2) on
Fig. 7 for the model parameters of experiments 3 and 4. Use of these
functions is ideally restricted to immiscible (no diffusion) processes;
however they can be applied to miscible processes when the effect of
dispersion is negligible. A useful feature of the present displacements
is that they should give results which are similar to perfect ultra-low-
tension displacements (having negligible capillary pressures and 100%
microscopic displacement efficiency).

Predicted saturation/distance profiles based on the above functions using
the viscosity ratios of interest are given on Fig. 8. These extended
distributions are not found in practice even when local fingering is
taken into account; however it is only the averaged displacements within
the fast ($S_2 = 0 - 0.14$) and slow ($S_2 = 0.14 - 1.0$) flowing regimes
which will be considered (dotted lines). The ratio of displacement rates

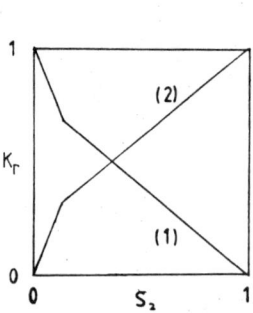

Fig. 7. Relative permeabilities.

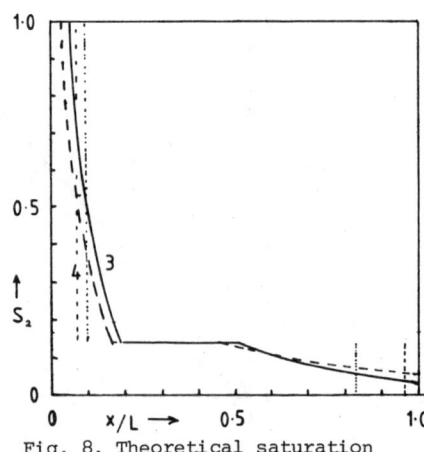

Fig. 8. Theoretical saturation
distributions.

are predicted to be 8.5 for a viscosity ratio of 0.33 and 12.9 for a
viscosity ratio of 0.22. These relative rates are about a factor of two
greater than those indicated in Fig. 6. It is thought therefore that 1-
dimensional flow theory exaggerates the effect of mobility ratio for
reasons concerning crossflow mechanism. It may therefore be possible,
using convenient approximations, to obtain predictions for miscible and
low tension displacements within layered media which are significantly
better than those provided by analytical 1-dimensional methods.

Quantitative Results : Favourable Mobility Ratio Continuous Displacement.

Crossflow is the principle mechanism by which a displacement front may be
stabilized against the influence of local permeability variations. The
dual-channel pressure profiles discussed above can be used to explain
this flow mechanism and the "shock front" concept of 1-dimensional displacement theory (5).

In some preliminary work we used a packed bead model containing four
fast flow channels (permeability ratio 13:1) of different width. The
results reflect a considerable influence of gravity since the displacing
fluid was more dense and was flowed vertically upward. Fig. 9 illustrates
traced displacement fronts (full lines) in relation to the layer boundaries (dashed) for three stages (fractional pore volumes injected
indicated). Here $\mu_2/\mu_1 = 5$, $\Delta\rho = 0.113$ g/cm^3; while on Fig. 10 are
the observations for $\mu_2/\mu_1 = 10$, $\Delta\rho = 0.149$ g/cm^3. Predictions based on
synthetic relative permeabilities for this model lead to the single shock
fronts shown (dotted). The superficial flow rate was greater in the
latter case (1.8 x 10^{-3}cm/sec, as compared with 0.91 x 10^{-3}cm/sec) and
the effect of this is to compensate to some extent for the effect of a
higher viscosity ratio.

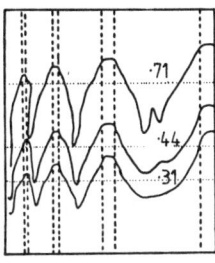

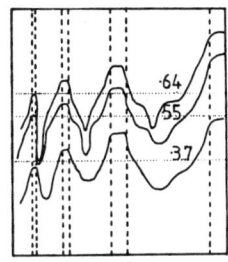

Fig. 9.   Fig. 10

Shock front formation is clearly not observed. The oscillations of frontal
boundary appear to increase in amplitude with increase in channel diameter
(the far right channel is really a half-channel since there is a no-flow
boundary at its side). In the case of the higher viscosity ratio displacement there is little change in the frontal shape with time.

It has been found that the basic characteristics of such stabilized
displacement patterns can be approximated by considering dual-channel
pressure profiles. Figure 11 illustrates the form of such profiles when
viscous crossflow (but not gravity) is allowed for. The stabilization

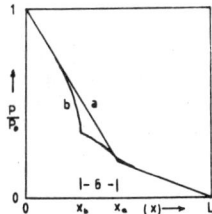

Figure 11.

Pressure profiles for
favourable mobilities with
crossflow.

phenomenon, which tends to discourage channeling into the high permeability
zone, depends upon the crossflow which itself is governed by the region
between the two profiles. The geometry of this region can be approximated
by a triangle enabling an expression to be derived for a stabilized
separation "δ" ($= x_a - x_b$), assuming the velocities of the two fronts are

equal and the ratio of pressure gradients either side of a front to be given by the viscosity ratio. Hence, it has been derived for

$$\mu_2 > \mu_1 \quad \text{and} \quad \frac{\mu_1}{\mu_2} \cdot \frac{k_a}{k_b} < 1;$$

that,

$$\delta^2 \;\simeq\; \frac{\dfrac{2d\, k_a}{A\, k_t}\left(1 - k_b/k_a\right)^2}{\left(1 - \dfrac{\mu_1}{\mu_2}\dfrac{k_a}{k_b}\right)\left(1 - \dfrac{\mu_1}{\mu_2}\right)}.$$

For straight isotropic layers we can say:-

$$\frac{2\, d\, k_a}{A\, k_t} \;=\; \frac{d^2.\,(1 + k_a/k_b)}{2}.$$

Therefore δ is proportional to d, and increases with the ratios μ_1/μ_2 and k_a/k_b. It is however only intended to give an order of magnitude representation of flow separation effects due to the grossly simplifying assumptions used.

Quantitative tests were performed in the packed models described in the previous section, involving single high permeability layers and matched density fluids. The series of experiments is defined below:-

Table 3.

Experiment	k_a/k_b	d/L	μ_2/μ_1
5		0.05	3.0
6	2.8		
7		0.07	4.57
8			

Plate 3 illustrates the general form of the observed displacements (from experiment 6). Of main interest was the extra penetration of tracer into the high permeability channel as related to the displacement within the rest of the packing. This was difficult to quantify because no piston-like front was ever observed. The point of 50% occupancy across the channel by full intensity displacing tracer was therefore chosen as a reference point. This point was not far ahead of the position where the channel was completely filled with displacing fluid. As a mean displacement estimation, this measurement must be regarded as a conservative estimate of penetration. The separation of this point from the main displacement front is plotted as a function of x_b/L on Figure 12. The correspondence of data from experiments 5 and 6 having a 40% difference in d supports the theoretical prediction that δ should be proportional to d. Frontal separations therefore grow towards definite maxima, the value of which is sensitive to mobility ratio. Bars representing these maxima covering the scatter of data are plotted on Figure 13, along with crosses indicating the separations based on the position at which the channel is completely occupied by displacing fluid. Theoretical curves applying to our model, and other permeability ratios (indicated on the curves) are included. These give the equilibrium stabilized frontal separations predicted using the above equations. Assuming the bars to be acceptable as an experimental estimate of this parameter (remembering that no piston-like front is observed in the high permeability channel) then the usefulness of the mathematical approximation is supported. This should be viewed in relation to the predictions of 1-dimensional flow theory based on the synthetic relative permeability functions for these models. Fig. 7 shows these for experiments 7 and 8. The analytical solution indicates a single shock front through the whole system for viscosity ratios greater than 2.82, i.e. $\delta = 0$. Our experimental results clearly demonstrate that this is not the case and will be of more serious consequence to displacement efficiency as layer (or other channel) diameters increase.

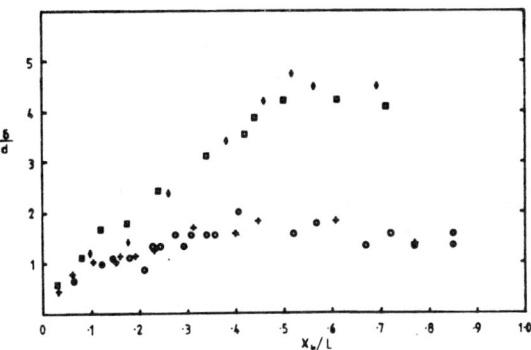

Figure 12. Frontal Separations, Experiment Nos. ◊ 5, □ 6, + 7, ○ 8.

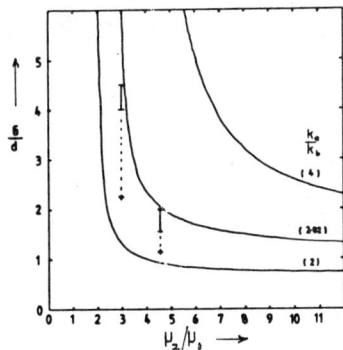

Figure 13. Viscosity Ratio Effect, experimental (bars) and theoretical (lines).

SLUG DISPLACEMENTS

It has been found that continuous injection tests model well the development of displacement boundaries at front and rear of a "slug" up to the time when overtaking occurs. Plate 4 shows a low viscosity slug fingering and channelling ahead in a similar way to the displacements discussed above involving continuous injection. Behind the slug we have a favourable mobility ratio displacement of typical pattern. The high viscosity slug of Plate 5 shows a stabilized form at its front. It is pushed by a similar liquid, without dye, exhibiting a typical equiviscous displacement.

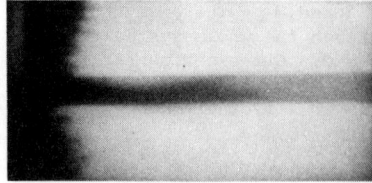

Plate 4

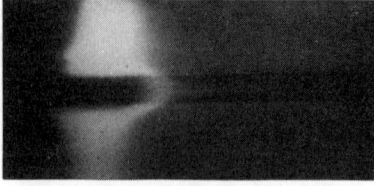

Plate 5

The permeability ratio was as before (2.8) and the viscosity ratios involved in the displacement of fluid (1) by fluid (2) by fluid (3) were for Plate 4, $\mu_3:\mu_2:\mu_1 = 3:1:3$, and for Plate 5, 4.6:4.6:1.

Although the volumes of these slugs are about 20% of the pore volume, loss of slug integrity occurs. The low viscosity slug (Plate 6) is continuing to be squeezed from the low permeability medium into the fast flow channel; however the slug is near being divided into three portions. The high viscosity slug (Plate 7) has been split by the chase fluid which has channelled through and is crossflowing out of the high permeability layer, particularly near the front of the slug.

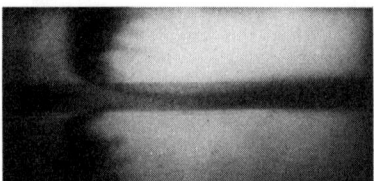

Plate 6.

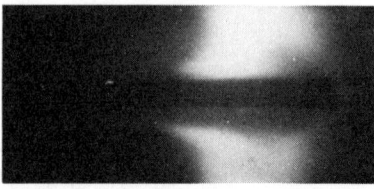

Plate 7.

The breakdown of slug integrity could possibly be resisted by chemicals, added to the chase fluid, designed specifically to resist certain crossflow processes and the mixing of out-of-sequence fluids. An example could be the in-situ gelling polymers which are sensitive to salinity environment (8). This is a possibility which will be investigated in future modelling work.

For surfactant slugs the fluid redistributions discussed above will be combined with considerable adsorption, dispersion, mass-transfer and gravity effects. Capillary pressure effects could also be important even though interfacial tensions may be low, since mobilized oil banks will be partly or wholly composed of discontinuous oil whose flow will be highly non-Newtonian (9).

DISPERSION IN LAYERED MEDIA

The stability of chemical slugs within channelled porous media can be
strongly affected by diffusion/dispersion processes. Here we consider
a two-layer model,following the approach of Lake and Hirasaki (2) and
Koonce and Blackwell (3) for chemical dispersion and Satman and Zolotukhin
(10) for the analogous problem in heat transfer.

To scale these effects it is useful to define a transverse dispersion
number (2):

$$N_{TD} = \frac{14\,L}{d^2} \cdot \frac{K_t}{V} \, ,$$

where L and d are the length and width of the system, K_t is the transverse
dispersion coefficient, V is the superficial flow rate in the high
permeability layer.

Lateral dispersion is insignificant when $N_{TD} < 0.2$, while when $N_{TD} > 5$
composition is practically constant over any cross-section through the
system and the behaviour can be represented by a single effective longitudinal dispersion coefficient (2). We examine here the intermediate
range of N_{TD}, between 0.2 and 5, which could apply to common field conditions if d is of the order of 1m and to laboratory core tests if layers of
a few mm width are present within the porous medium.

We consider flow parallel to the layers and tracer dispersion normal to
this direction (longitudinal dispersion coefficient is zero). The lateral
dispersion coefficient has been taken to be constant,independent of concentration, position and flow rate. For reservoir rates it is generally
found to be of the order of the molecular diffusion coefficient (11).

Figure 14 shows computed isoconcentration contours (at 0.1 intervals)
within a two layer system, the upper one (between Y values 0.5 and 1.0)
flowing from left to right, the lower is stagnant but receives injected
tracer by lateral dispersion from the permeable layer. Tracer injected
at unit concentration is dispersed as shown at three values of the
dimensionless time:-

$$t = \frac{T\,K_t}{d^2} = \frac{N_{TD}}{14} \, ,$$

where T is the absolute time from the start of the displacement.

It is of interest to obtain convenient analytical approximations to the
mean tracer concentration within a given cross-section of the flow
channel (and of the non-flowing matrix). Figure 15 shows numerical points
and analytical curves representing the distribution of average concentration with distance in the flow direction (normalized for t = 0.2)
The analytical approximations were derived using solutions to the zero-
concentration-boundary-condition case (12), evaluated for short dimensionless times. Expressions of similar form are applicable to other channel
geometries (e.g. cylindrical) provided times are short. The approximations
derived for the heat transfer problem (10) involve also a square root of
time dependence; however these integral solutions are very complex because
they are intended to cope with a large time range. Our approximation is:-

174

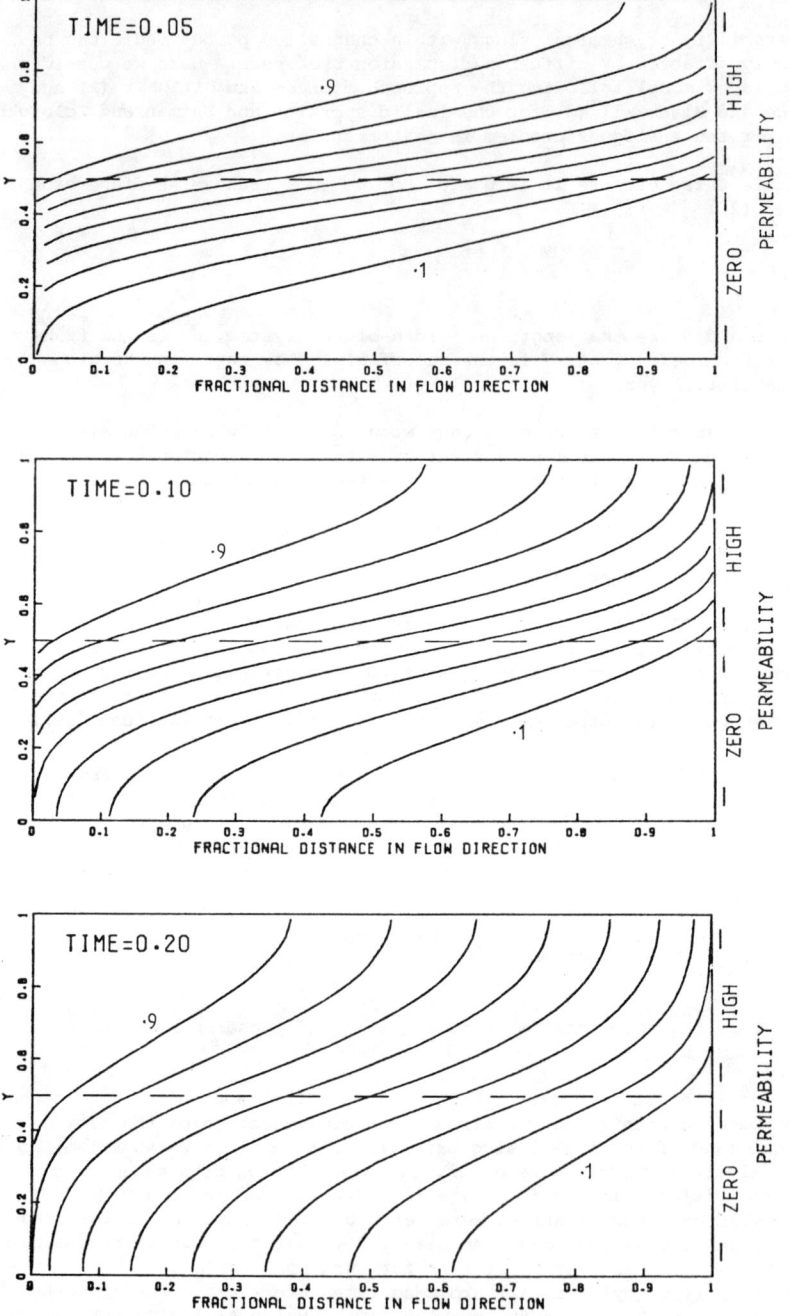

Figure 14. Isoconcentration contours in two layer system.

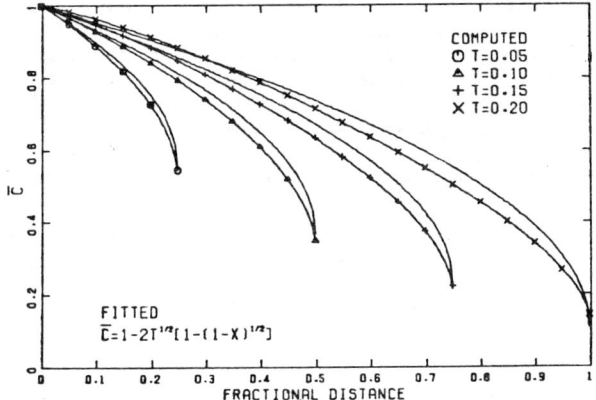

Figure 15.   Cross-sectional averaged concentrations in flow channel.

$$\bar{C} = 1 - 2.0 \, t^{\frac{1}{2}} (1 - (1 - X)^{\frac{1}{2}})$$

where $\bar{C}$ is the average injected tracer concentration within the cross section (at X) of the flowing channel; X is the fractional distance equal to $x/V.\tau$.

A similar method can be used to approximate the averaged concentrations within the non-flowing matrix ($\bar{C}_m$) for the two equal-capacity layers here considered:-

$$\bar{C}_m = 2.0 \, t^{\frac{1}{2}} (1 - X^{\frac{1}{2}}) .$$

Applications to a Multichannel Problem.

One approach to a multichannel problem is to consider each individual channel as interacting with a surrounding matrix which possesses the suitably averaged properties of the rest of the porous body. Generally a non-zero flow rate will apply to the external matrix in contrast to the stagnant case   as   above. This necessitates consideration of the problem as one of relative flow rates using moving co-ordinate methods.

Tracer effluent profiles have been analysed in terms of various models intended to account for heterogeneity (13),(14), (15). Laboratory tracer tests on layered reservoir materials are of interest for two reasons; first, conventional methods for characterizing dispersion co-efficients for miscible displacement and relative permeability functions for low tension immiscible displacement may be unreliable; second, such laboratory systems can model similar problems on the reservoir scale.

To estimate mass transfer rates for the channels (e.g. layers) within a heterogeneous core sample displacement flow tests of different rates have to be compared. Unfortunately, very little raw data of this kind is to be found within the petroleum literature. Our main source is the high quality recent work of Spence and Watkins (16). Handy (17) has used dual tracers to evaluate diffusion effects and we have begun tests on layered sand-stones using ultra-violet absorption monitoring techniques.

Using the above approximations, and the assumption that flow within the
matrix surrounding any given channel "i" is approximated by the mean
velocity of the whole displacement ($\bar{V}$), a method of effluent curve analysis has been derived (18). Thus if the fractions of displacing fluid
at its moment of breakthrough at the effluent end of the single channel
"j", ($f_{j(1)}$ and $f_{j(2)}$), are known from two experiments (1 and 2) performed
at different rates, we can estimate the fractional cross section of the
channel ($\delta_j S$) and its effective mass transfer coefficient (M_j) thus:-

$$\delta_j S = \frac{\bar{V}_2}{V_{j(2)}} \left(f_{j(2)} + \frac{f_{j(1)} - f_{j(2)}}{1 - \left[\frac{T_{j(1)}}{T_{j(2)}} \right]^{\frac{1}{2}}} \right)$$

and,

$$(M_j)^{\frac{1}{2}} = \frac{f_{j(1)} - f_{j(2)}}{\delta_j S \left[(T_{j(2)})^{\frac{1}{2}} - (T_{j(1)})^{\frac{1}{2}} \right]}$$

where, $\bar{V}_{(1)} > \bar{V}_{(2)}$

$\bar{V}_{(1)} T_{j(1)} = \bar{V}_{(2)} T_{j(2)}$

$V_j = L/T_j$, L being the length of the test core.

$V_{j(1)} T_{j(1)} = V_{j(2)} T_{j(2)}$

$M_t = 4K_t/d^2$ for a layer .

These expressions can be applied to effluent composition values measured
shortly after the first detected breakthrough of displacing phase from the
multichannel system. This characterizes the fastest flow channel(s) of
the sample. Subsequent tracer measurements have to be processed to
allow for the (time dependent) contributions from all the faster-flowing
channels. The gross composition measured in one experiment is F (a function
of time) and the individual ("breakthrough") channel "j" contribution can
be obtained using the following algorithm:-

$$f_j = F - \sum_{i=1}^{i=(j-1)} \delta_i S \left[u_i + (M_i \bar{T})^{\frac{1}{2}} \left\{ t_j^{\frac{1}{2}} - \frac{(1 - t_j)^{\frac{1}{2}}}{u_i - 1} \right. \right.$$

$$\left. \left. - u_i \left(t_j^{\frac{1}{2}} - \frac{(u_i t_j - 1)^{\frac{1}{2}}}{u_i - 1} \right) \right\} \right]$$

where: $\bar{T} = L/\bar{V}$
$t_j = T_j/\bar{T}$
$u_i = V_i/\bar{V}$, V_i being the mean flow rate within channel i.

The effluent profile is analyzed forward in time as presented above for
$T < L/\bar{V}$; while for $T > L/\bar{V}$ steps are taken backward in time redefining F as
the concentration of displaced fluid. A satisfactory analysis can be performed with a programmable calculator (a program suitable for an "HP 41C"
is available from the authors). Small time steps should be avoided since
errors due mainly to intralayer longitudinal dispersion, ignored in the
present analysis will become important; ten to twenty steps for each
effluent curve have been found to be satisfactory.

Example results based on some of the tracer composition profiles of Spence and Watkins are indicated on Figures 16 - 18. Mass transfer coefficient distribution over the cross section of the samples is given on Figure 16 for a sandstone and a carbonate. M_t values of 10^{-5} and 10^{-4} could be interpreted in terms of layers of about 2.0 cm and 0.5 cm width respectively. Figure 17 gives the "no dispersion" velocity profiles of the porous media. The latter can be represented as relative permeability functions (Fig. 18) applying to the ideal "no dispersion" case or to the ideal near-zero-interfacial tension immiscible displacement case. Predictions of mobility ratio effects could therefore be made using conventional 1-dimensional displacement theory. However, for highly heterogeneous media allowance for crossflow effects, as discussed above, should be included.

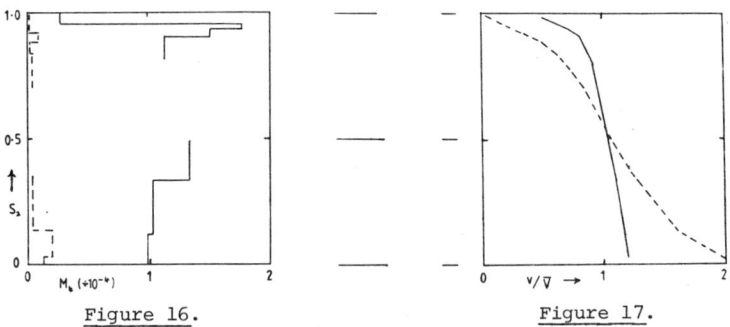

Figure 16.   **Figure 17.**

Mass transfer coefficient distributions.   Velocity distributions.
Full lines:"Sandstone SS2"
Dashed lines: "Carbonate B17"

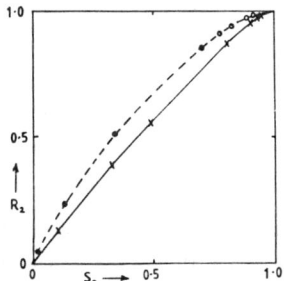

Figure 18.

Miscible type relative
permeabilities.

CONCLUSIONS

Surfactant E.O.R. slugs will be susceptible to layer and streak permeability heterogeneities found within reservoirs due to disturbance of flow patterns and increased dispersion. Mathematical approximations have been found which are capable of modelling the channelling and crossflow effects present in non-unit mobility ratio displacements. Experimentally, loss of integrity due to flow mechanism has been observed in slugs of around 20% pore volume. Diffusion/dispersion effects can be large, depending on the width of layers. For short dimensionless times it is possible to model these phenomena analytically to match numerical simulations and to analyze tracer test data.

ACKNOWLEDGEMENTS

Dr. M. Allmen is thanked for performing the dispersion computations and
Mr. M. Hughes for technical help. We are grateful to the Department of
Energy for financial support.

REFERENCES

1. WEBBER, K., Influence On Fluid Flow of Common Sedimentary Structures
   In Sand Bodies., S.P.E. Paper 9247

2. LAKE, L. & HIRASAKI,G., Taylor's Dispersion In Stratified Porous Media.,
   S.P.E. Paper 8436.

3. KOONCE, T. & BLACKWELL, R., Idealized Behaviour of Solvent Banks in
   Stratified Reservoirs., Soc.Pet.Eng.J.(Dec. 1965) 5,(6), 318 - 328.

4. HAWTHORNE, R., The Effect of Capillary Pressure In a Multilayer Model
   of Porous Media. Soc.Pet.Eng.J. (Dec. 1975) 15, 467 - 476.

5. WRIGHT, R. & DAWE,R., An Examination Of The Multiphase Darcy Model Of
   Fluid Displacement In Porous Media. Rev.Inst.Fr.du Petrole (Nov-Dec 1980)
   35, (No.6) 1011 - 1024.

6. PEACEMAN, D. & RACHFORD, H., Numerical Calculation of Multidimensional
   Miscible Displacement. Soc.Pet.Eng.J. (Dec 1962) 2, 327 - 340.

7. HEARN, C., Simulation Of Stratified Waterflooding By Pseudo Relative
   Permeability Curves, (July 1971), 23, 805 - 813.

8. MACK, J., Process Technology Improves Oil Recovery, Oil & Gas J.(Oct.1979)
   77, No. 40, 67 - 71.

9. EGBOGAH, E., WRIGHT, R. & DAWE, R., A Model of Oil Ganglion Movement In
   Porous Media, S.P.E. Paper 10115.

10. SATMAN, A. & ZOLOTUKHIN, A., Application of the Time-Dependent Overall
    Heat Transfer Coefficient Concept to Heat Transfer Problems In Porous
    Media, S.P.E. Paper 8909.

11. PERKINS, T. & JOHNSON, O., A Review of Diffusion and Dispersion in Porous
    Media, Soc.Pet.Eng.J. (March 1963) 3, 70 - 84.

12. CRANK, J., The Mathematics of Diffusion, Oxford Univ. Press. 1975,Sec.4.3.

13. KOVAL, E., A Method For Predicting The Performance Of Unstable Miscible
    Displacements In Heterogeneous Media, Soc.Pet.Eng.J. (June 1963)3,145-154.

14. JOHNSON, C. & SWEENEY, S., Quantitative Measurements Of Flow Heterogen-
    eity In Laboratory Core Samples And Its Effect On Fluid Flow Characteri-
    stics, S.P.E. Paper 3610.

15. ROSMAN, A. & SIMON, R., Flow Heterogeneity In Reservoir Rocks, S.P.E.
    Paper 5631.

16. SPENCE, A. & WATKINS, R., The Effect Of Miscroscopic Core Heterogeneity
    On Miscible Flood Residual Oil Saturation, S.P.E. Paper 9229.

17. HANDY, L., An Evaluation Of Diffusion Effects In Miscible Displacement,
    Trans. AIME (1959) 216, 61 - 65.

18. WRIGHT, R. et. al., Heterogeneous Porous Media; A Miscible Displacement
    Model;- to be submitted for publication.

SOME ASPECTS OF THE INJECTIVITY OF NON-NEWTONIAN FLUIDS IN POROUS MEDIA

PETER VOGEL and GÜNTER PUSCH

Institut für Tiefbohrkunde und Erdölgewinnung, Technical University Clausthal, West Germany

ABSTRACT

In existing numerical models, the rheological behaviour of polymer solutions is commonly described by the power law, which is not satisfactory at very low shear rates and at relatively high shear rates. An improvement of the mathematical description was achieved by using the Carreau viscosity equation and deriving a filter law for porous media. The validity over a wide range of shear rates was proven by experimental results obtained from flood tests in sand packs with one typical product each of the three polymer classes (PAA, HEC, BPS) used in enhanced oil recovery.

On the basis of typical reservoir data, the behaviour of an injection well during polymer injection is investigated by calculating the pressure profile around a wellbore. From these data, conclusions are drawn for the selection of polymers according to their rheological properties.

INTRODUCTION

Flooding with viscous media has aroused increasing interest in the field of enhanced oil recovery. Numerous pilot projects are currently in progress or have already been terminated /1,2/. The importance which is at present attached to this field of research is thus evident.

Chiefly aqueous polymer solutions are employed as viscous flooding media. A characteristic feature of these polymer solutions is that the decisive parameter for the description of their flow properties, the viscosity, varies as a function of the shear rate. In general, the solutions exhibit pseudoplastic behaviour, that is, a decrease of the viscosity with augmenting shear stress.

In the field of enhanced oil recovery, the viscous behaviour of polymer solutions in porous media has become of vital importance as far as their injectivity is concerned. The investigations were initiated by the following two questions:
- How can the viscosity values indicated in a rheogramme be applied to flow processes in porous media?
- Can these polymer solutions be injected into the reservoir without exceeding the fracturing pressure of the rock?

In the following, a method which allows a calculation of the injectivity of polymer solutions on the basis of the rheogrammes and of the knowledge of the characteristic reservoir data is presented.

CHARACTERIZATION OF THE POLYMERS EMPLOYED

Information about the flow behaviour of non-NEWTONian fluids is provided by their rheogramme, that is, the plot of the viscosity as a function of the shear rate; this is both important and experimentally easy to obtain. All of the considerations discussed in the following are based exclusively on the information gained therefrom.

To begin, the rheogrammes of the polymer solutions used here are presented. The liquids employed are aqueous solutions

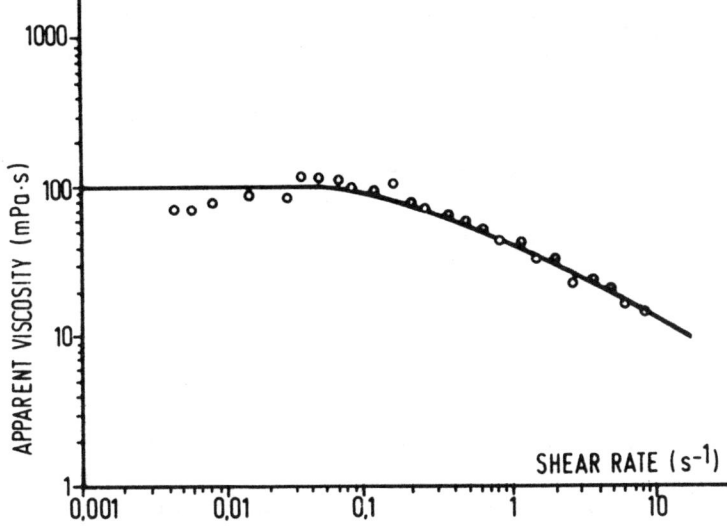

Figure 1:   Viscosity behaviour of a polysaccharide solution

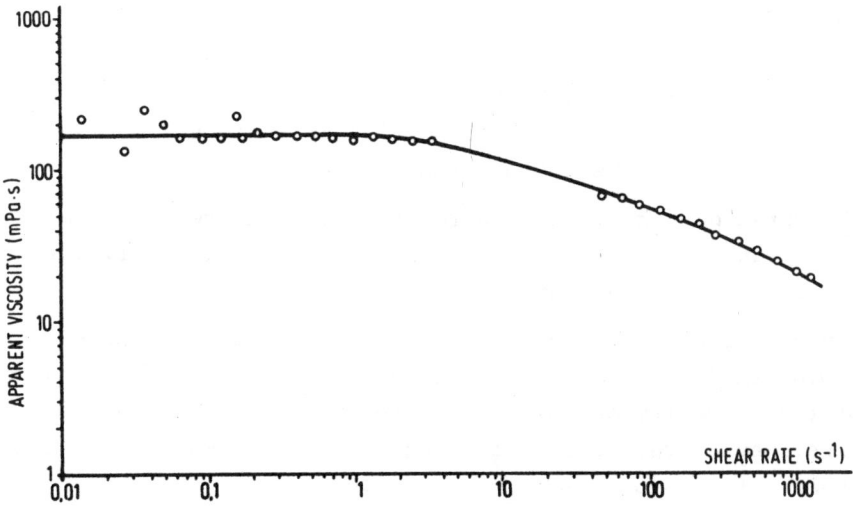

Figure 2:   Viscosity behaviour of a hydroxyethylcellulose
            solution

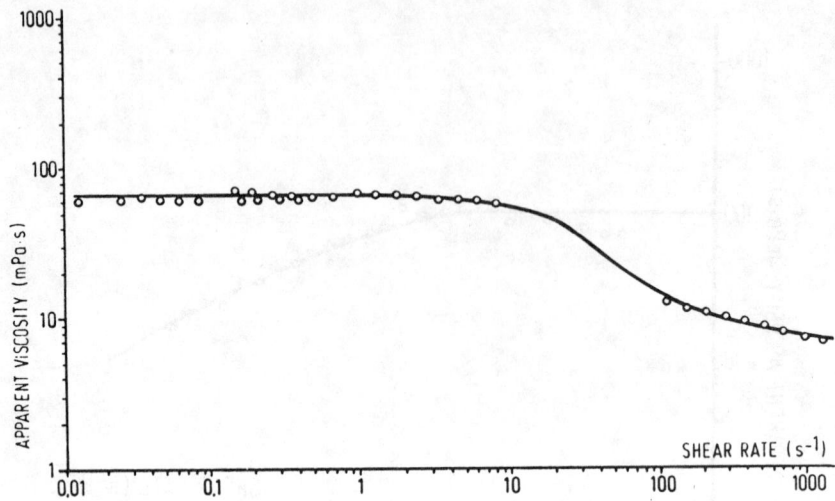

Figure 3: Viscosity behaviour of a polyacrylamide solution

(original brine with a salt concentration of 100 g/l;
reservoir temperature of 50°C) of a typical, representative
product in each of the three classes of polymers used in
enhanced oil recovery. Polymer solutions which yield a mutually
comparable additional oil recovery (m³ of additional oil per
m³ of polymer solution consumed) in flooding tests were
thereby selected.

Figure 1 shows the rheogramme for a polysaccharide, figure 2
that for a hydroxyethylcellulose, and figure 3 that for a
polyacrylamide solution. A double logarithmic scale has
been chosen for the graphic representation.

The three curves display characteristic features in common:
A plateau occurs in the range of low shear rate; a linear
decrease is observed at higher values.

For the calculation of the flow behaviour of these non-
NEWTONian fluids, an analytical expression for the dependence
of the viscosity on the shear rate, which represents the
experimental values of the rheogramme over a wide range of
shear rate, is of special importance.

The preceding figures show that the four-parameter equation
found by CARREAU /3/

$$(1) \qquad \frac{\eta - \eta_\infty}{\eta_0 - \eta_\infty} = \left[1 + (\lambda \dot{\gamma})^2\right]^{\frac{n-1}{2}}$$

provides a good fit to the experimentally determined rheogramme for the polymer solutions under investigation here.

The significance of the parameters in the CARREAU equation, as well as a simple method for determining them, are briefly explained. η_0 denotes the viscosity at the shear rate $\dot{\gamma} = 0$, and can be determined directly from the horizontal portion of the curve in the range of very low shear rates. By means of supplementary measurements performed in the range of high shear rates, values indicative of η_∞ are obtained. n-1 is the slope of the linearly decreasing part of the curve. The plateau for the range of low shear rate and the linearly decreasing part of the curve intersect at a point whose abscissa is approximately equal to $1/\lambda$.

In the following, the essential steps in the development of a filter law for CARREAU fluids are described. The power law frequently employed in previous publications is considerably simpler to handle analytically, and is therefore preferred for the treatment of concrete problems. For the polymer solutions investigated in this work, however, a power-law dependence of the viscosity on the shear rate does not describe the experimentally observed behaviour with sufficient accuracy. Consequently, sizable errors can result in the description of the flow processes in porous media, as will be shown by means of an example. For a wide range of shear rates, an extension, as described in this work with respect to the viscosity model, is indispensable.

A FILTER LAW FOR CARREAU FLUIDS

Filter laws for non-NEWTONian fluids are known only for a few special cases /4, 5, 6, 7/. The procedure common to their derivations is as follows: First the capillary flow is treated analytically for the liquid in question, in order to obtain a filter law with the use of an appropriate capillary bundle model.

This procedure is adopted in the following as well; a filter

184

law is thereby derived for CARREAU fluids, and the porous medium is replaced by a capillary bundle which is hydrodynamically equivalent with respect to porosity and permeability.

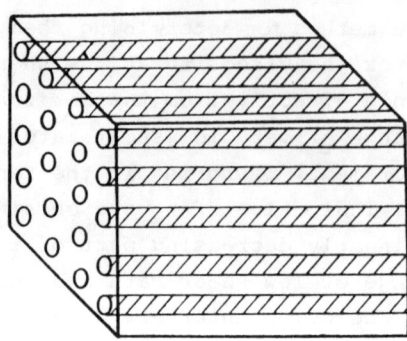

Figure 4:   Straight capillaric model of a porous medium

The simplest capillary model of a porous medium /8, 9/ consists of a bundle of circular cylindrical capillaries of equal radius R. Figure 4 illustrates this concept.

A comparison of the DARCY filter law with the law of HAGEN-POISEUILLE yields the "hydraulic equivalence radius" for this simple model:

(2) $$R = \sqrt{\frac{8K}{\phi}}$$

By means of this concept, the flow through a porous medium is related to the capillary flow of the liquid in question and can be treated accordingly.

On the basis of this theory, a filter law for CARREAU fluids can be derived. The procedure is justified by experimental results. In the following considerations, it is remarkable that no empirical corrections are required.

It is necessary first to calculate the flow behaviour of CARREAU fluids in capillaries; for this purpose the velocity profile and the average velocity of the capillary flow must be known. For the derivation, a circular cylindrical capillary

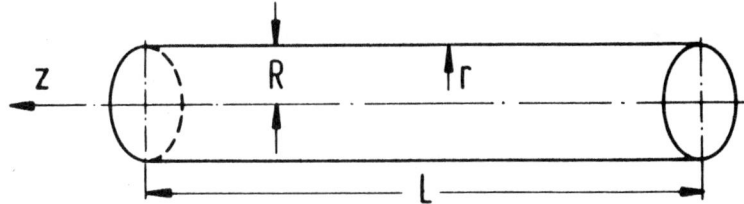

Figure 5: Flow through a circular tube

of radius R and length L is considered - Figure 5 - and a cylindrical coordinate system is introduced. The z-axis and the capillary axis are identical; the direction of flow is taken to be that of the positive z-axis.

The differential equation for the radial velocity distribution $v(r)$ is

$$(3) \qquad r = \frac{-2L}{P_0 - P_L} \left[\eta_\infty + (\eta_0 - \eta_\infty)\left(1 + (\lambda\frac{dv}{dr})^2\right)^{\frac{n-1}{2}} \right] \cdot \frac{dv}{dr}$$

whereby $p_0 - p_L$ denotes the applied pressure difference. This differential equation is transcendental in the derivative of the function being sought, $v(r)$; this fact proved to be a considerable problem in the further course of the calculations.

The introduction of the wall shear rate $\dot{\gamma}_R$ as a parameter is decisive for the solution of this problem. The calculation /10/ finally yields an analytical expression for the average velocity during capillary flow. By means of the hydraulic equivalence radius, this expression can be easily transformed to a filter law. In the case of the capillary bundle model used, the one-dimensional filter law takes the following form:

$$(4) \qquad v_f = \frac{K}{\eta_\infty} \cdot \frac{P_0 - P_L}{L} \left| \frac{4}{3} \cdot \frac{1}{g(\lambda \dot{\gamma}_R)} - \frac{1}{3} \cdot \frac{1}{g^4(\lambda \dot{\gamma}_R)} \right.$$

$$- 4 \frac{\eta_0 - \eta_\infty}{\eta_\infty} \cdot \frac{1}{g^4(\lambda \dot{\gamma}_R)} \cdot \frac{T_1(\lambda \dot{\gamma}_R)}{(\lambda \dot{\gamma}_R)^4}$$

$$- 4 \frac{(\eta_0 - \eta_\infty)^2}{\eta_\infty^2} \cdot \frac{1}{g^4(\lambda \dot{\gamma}_R)} \cdot \frac{T_2(\lambda \dot{\gamma}_R)}{(\lambda \dot{\gamma}_R)^4}$$

$$\left. - \frac{4}{3} \frac{(\eta_0 - \eta_\infty)^3}{\eta_\infty^3} \cdot \frac{1}{g^4(\lambda \dot{\gamma}_R)} \cdot \frac{T_3(\lambda \dot{\gamma}_R)}{(\lambda \dot{\gamma}_R)^4} \right|$$

In order to save space, the following substitutions have been made:

$$(5.1) \qquad g(\lambda \dot{\gamma}_R) := 1 + \frac{\eta_0 - \eta_\infty}{\eta_\infty} \left(1 + (\lambda \dot{\gamma}_R)^2 \right)^{\frac{n-1}{2}}$$

$$(5.2) \qquad T_1(\lambda \dot{\gamma}_R) := \frac{1}{n+1} \left[1 + (\lambda \dot{\gamma}_R)^2 \right]^{\frac{n+1}{2}} (\lambda \dot{\gamma}_R)^2$$

$$- \frac{2}{(n+1)(n+3)} \left[\left(1 + (\lambda \dot{\gamma}_R)^2 \right)^{\frac{n+3}{2}} - 1 \right]$$

$$(5.3) \qquad T_2(\lambda \dot{\gamma}_R) := \frac{1}{2n} \left[1 + (\lambda \dot{\gamma}_R)^2 \right]^{n} (\lambda \dot{\gamma}_R)^2$$

$$- \frac{1}{2n(n+1)} \left[\left(1 + (\lambda \dot{\gamma}_R)^2 \right)^{n+1} - 1 \right]$$

$$(5.4) \qquad T_3(\lambda \dot{\gamma}_R) := \frac{1}{3n-1} \left[1 + (\lambda \dot{\gamma}_R)^2 \right]^{\frac{1}{2}(3n-1)} (\lambda \dot{\gamma}_R)^2$$

$$- \frac{2}{(3n-1)(3n+1)} \left[\left(1 + (\lambda \dot{\gamma}_R)^2 \right)^{\frac{1}{2}(3n+1)} - 1 \right]$$

With the exception of a correction factor, the external form
of this filter law is identical to that of the DARCY law.
This factor depends on the parameters of the CARREAU equation
and on the maximal shear rate $\dot{\gamma}_R$ occurring in the capillary
bundle model. The maximal shear rate is obtained from the
transcendental equation

(6)
$$\lambda \dot{\gamma}_R = \frac{\lambda \frac{P_0 - P_L}{2L} \sqrt{\frac{8K}{\Phi}} \cdot \frac{1}{\eta_0 - \eta_\infty}}{\frac{\eta_\infty}{\eta_0 - \eta_\infty} + \left[1 + (\lambda \dot{\gamma}_R)^2 \right]^{\frac{n-1}{2}}}$$

which admits an iterative solution according to the BANACH
fixed-point theorem.

The algorithm necessary for the numerical solution of
equations (4) and (6) requires the following steps:
After the parameters of the CARREAU equation, as well as the
permeability and porosity of the porous medium have been
determined, $\dot{\gamma}_R$ is calculated from (6) for predetermined
values of the pressure gradient, and the corresponding filter
velocity is determined from (4).

COMPARISON OF THEORETICAL AND EXPERIMENTAL RESULTS

The theoretical results are verified by experiment; no
empirical correction factors are thereby required.

In order to carry out the required flood experiments, an
apparatus similar to that already used by DARCY was
employed. Sand packs of 50 percent porosity and 5 D
permeability, compacted by vibration, served as porous media.

If the DARCY equation is solved for the viscosity, the
result is

(7)
$$\eta = \frac{K}{V_f} \frac{P_0 - P_L}{L}$$

With the use of the present results, the effective viscosity
in the porous medium was determined directly from the
measured data according to (7) on the one hand, and by means
of the previously derived filter law, on the other hand.

For comparison, polysaccharide and hydroxyethylcellulose, which exhibit a dominantly linear, decreasing range in their rheogrammes, were treated as power-law fluids.

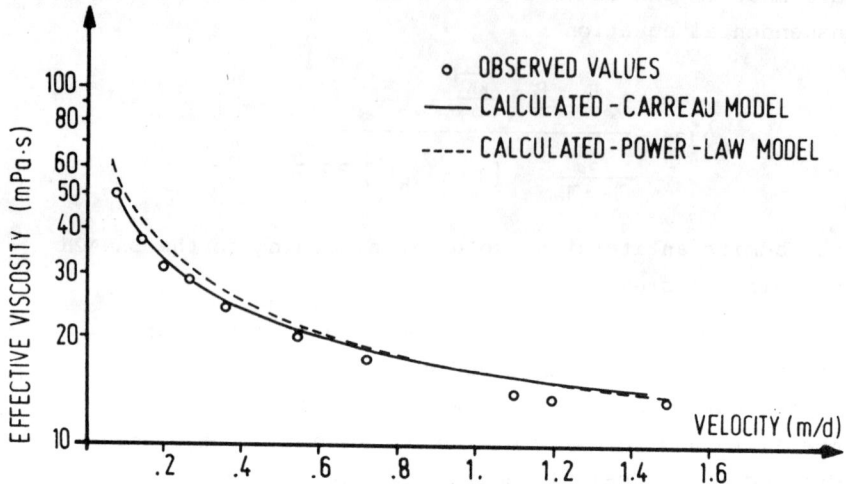

Figure 6: Effective viscosity for flow of polysaccharide solution in porous media

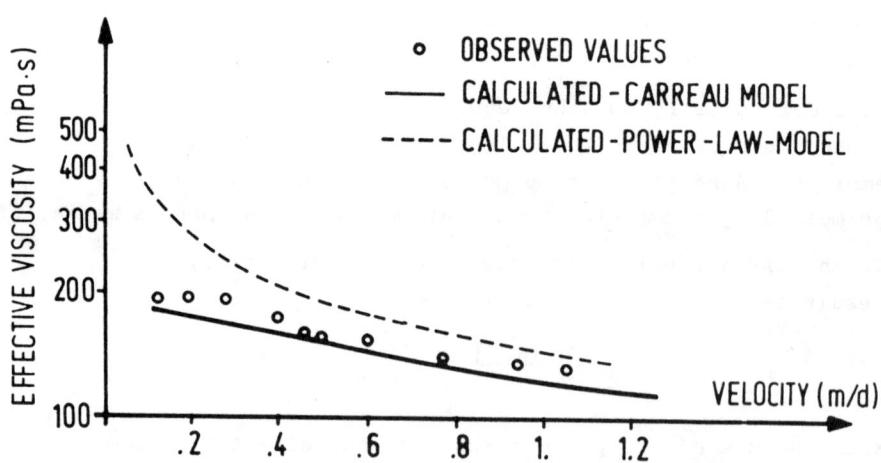

Figure 7: Effective viscosity for flow of hydroxyethyl-cellulose solution in porous media

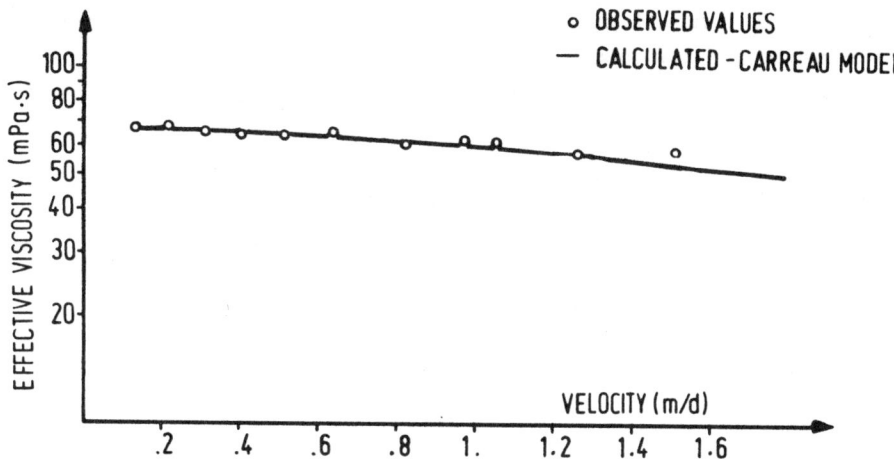

Figure 8:   Effective viscosity for flow of polyacrylamide
solution in porous media

From the filter law for power-law fluids, the effective
viscosity in a porous medium was likewise calculated.

Figures 6, 7, and 8 show the dependence of the viscosity on
the filter velocity and compare the experimental and
theoretical results.

For the CARREAU model, the deviation between the experimental
and theoretical results is less than 10 percent for the
polysaccharide and polyacrylamide solutions, and less than
15 percent for the hydroxyethylcellulose solution. Hence
the agreement between theory and experiment can be regarded
as good.

The power-law model describes the dependence of the viscosity
on the filter velocity with sufficient accuracy in the case
of polysaccharide, whereas considerable deviation occurs for
hydroxyethylcellulose.   These examples demonstrate the
advantages of the new filter law for the questions under
investigation.

CALCULATION OF THE INJECTIVITY BEHAVIOUR

During enhanced oil recovery, the pseudoplastic behaviour of
the polymer solutions used exerts a pronounced influence on
their injectivity. Once the questions concerning filtration

adsorption, stability, etc. have been clarified for a given
reservoir in the course of the product selection procedure, the
question of the injectivity of the polymer solution involved
remains to be answered by the reservoir engineer. At this
juncture, an important decision of whether or not a selected
product is suitable for field application must be made; this
is a vital cirterion because of the high financial risk
involved. A method must be provided for predicting the behaviour
in the field on the basis of laboratory data; thus a criterion
for decision must be established.

In the following, the flowing pressure and radial
distribution of pressure around the injection well are
calculated for an injector in a radially symmetric reservoir
and for a predetermined injection rate, with the use of the
filter law just presented.

The multitude of influential parameters necessitates a
restriction to a typical case encountered in practice. The
following, realistic, geometrical and physical reservoir
data are employed for the model calculations:

Reservoir:

Permeability	$K = 1000$	mD
Porosity	$\phi = 0.24$	
Effective reservoir thickness	$h = 4$	m

Well:

Cased with 7" diameter and ideally perforated in the
reservoir zone

Wellbore radius	$r_w = 0.089$	m
Injection rate	$q = 100$	m^3/d
Depth	$= 1000$	m

FORMULATION OF THE SELECTION CRITERION

From the standpoint of reservoir engineering, the essential
criterion for the injectivity of a polymer solution is that
the fracturing pressure of the rock must not be exceeded during
the injection. The predetermined injection rate and the
average reservoir pressure also affect the decision. For a
depth of 1000 m and under the assumption that the average
reservoir pressure $\bar{p}$ corresponds to the hydrostatic pressure,

a value of $\bar{p}$ = 100 bar results. The order of magnitude of
the fracturing gradient typical for sedimentary rocks lies
between 0.18 and 0.24 bar per metre of depth. For the
injector under consideration here, this results in a maximal
bottom-hole flowing pressure of 180 to 240 bar; hence the bottom
hole flowing pressure may exceed the average reservoir pressure
by a maximum of 80 to 140 bar during polymer injection.
Furthermore, a radially symmetric reservoir is thereby
assumed. The range of influence of the injector is selected
at r_e = 200 m; the reservoir pressure of 100 bar is assumed
to prevail at the outer boundary. Thus, the following criterion
for decision is obtained: The polymer solution is injectable
provided the pressure drop over a distance of 200 m from the
bore hole does not exceed 80 to 140 bar.

CALCULATIONAL PROCEDURE

The object of the calculation is to determine the relationship
between the pressure gradient and the distance from the well.
This function is subsequently integrated.

Because of the complicated structure of the filter
law previously derived, the entire calculation is performed
numerically.

As a result of its structure, the filter law just developed
allows only the determination of the corresponding filter
velocity for given values of the pressure gradient. With
reference to /11/, the following procedure is adopted for
determining the locally prevailing pressure gradient. From
the equation of continuity the following expression is
obtained for the radial velocity distribution:

$$(8) \qquad\qquad v_f = \frac{q}{2\pi h} \cdot \frac{1}{r}$$

whereby r denotes the distance from the wellbore axis. This
provides a possibility of determining the distance from the
well corresponding to given values of the pressure gradient
by means of the filter law and equation (8).

The calculation starts with the determination of the pressure

192

gradient at the bore hole. For this purpose, two values of
the pressure gradient, of which one is smaller and one larger
than that prevailing at the well, are initially assumed. By
nesting of intervals a sequence of pressure gradient
values is constructed in such a way that the values
of the radius determined from the filter law and equation (8)
converge toward the wellbore radius. The procedure is
truncated as soon as the wellbore radius has been approached
with the required accuracy. The value of the pressure
gradient corresponding to the radius thus determined is then
taken as the pressure gradient at the well.

Subsequently, this value is decreased stepwise, and the
corresponding values of the radius are determined from the
filter law and equation (8). Thus, a tabular representation
of the pressure gradient as a function of the distance from
the well is obtained. The calculation of the total pressure
drop is subsequently performed by means of numerical
integration.

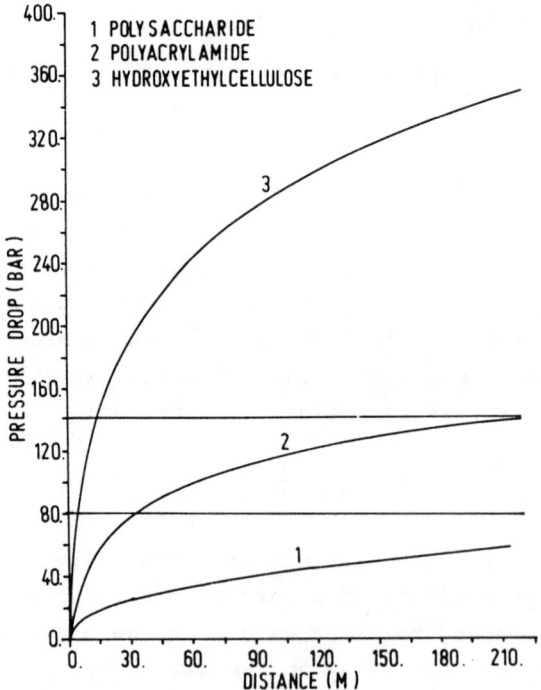

Figure 9: Calculated pressure profile during polymer injection

RESULTS OF THE MODEL CALCULATION

In figure 9, the pressure difference occurring during injection, as referred to the pressure at the injection well, is plotted as a function of the distance from the well for the three polymer solutions under investigation. Moreover, the maximal values of 80 and 140 bar for the injection over-pressure are indicated. According to the criterion formulated here, the polymer solutions are suitable for injection provided the pressure difference remains less than 80 to 140 bar over a distance up to 200 m from the injection well. This condition is fulfilled for the polysaccharide, and partially fulfilled for the polyacrylamide in this case. In contrast, the hydroxyethylcellulose exhibits a decidedly deviating behaviour. The pressure difference, as referred to the well, already amounts to 140 bar at a distance of about 20 m, and increases to more than 350 bar over a distance of 200 m. It must be emphasized that this is a model calculation, whereby the effects described are attributed solely to the dependence of the viscosity on the filtration velocity.

If, in a practically relevant case, the model calculations indicate that the maximal permissible injection pressure will be exceeded, the concentration of the polymer solution to be used must be reduced; the viscosity is thus decreased. The parameters of the CARREAU equation are then determined from the rheogramme, and the calculation is repeated with the use of these values.

A further possibility is the purely theoretical plotting of rheogrammes for injectable fluids by the variation of parameters in the CARREAU equation.

CONCLUSIONS

The rheological behaviour of aqueous polymer solutions is well described by the CARREAU model. A filter law derived for such fluids is described and experimentally verified. With the use of the new filter law, the radial pressure distribution around the injection well during the injection

of polymer solution is calculated. A polymer solution is
judged as suitable for injection as far as the bottom hole
flowing pressure does not exceed the fracturing pressure of
the rock at the bottom of the hole. Among the products
investigated here, the polysaccharide solution fully, and
the polyacrylamide solution conditionally satisfies this
criterion under the given conditions.

Nomenclature

h	Formation thickness
K	Permeability
L	Length
n	Power-law index
$\bar{p}$	Average pressure
$p_O - p_L$	Pressure drop
q	Injection rate
r	Radial coordinate
r_e	External boundary radius
r_w	Wellbore radius
R	Radius of the tube
v	Velocity
v_f	Filtration velocity
$\dot{\gamma}$	Shear rate
$\dot{\gamma}_R$	Shear rate at the tube wall
η	Viscosity
η_o	Zero-shear-rate viscosity
η_∞	Infinite-shear-rate viscosity
λ	Time constant
ϕ	Porosity

References

1   CHANG, H. L.;
    Polymer Flooding Technology - Yesterday and Tomorrow
    J. Pet. Tech. (Aug. 1978), 1113 - 1128

2. GRODDE, K.H., SCHAEFER, W.;
    "Experience with the Application of Polymer to Improve
    Water Flood Efficiency in Dogger Reservoirs of the
    Gifhorn Trough, Germany"
    Erdoel-Erdgas-Zeitschrift 94 (July 1978) 7, 252 - 259

3. CARREAU, J.P.;
    "Rheological Equations from Molecular Network Theories"
    Ph.D. Thesis, Univ. of Wisconsin, Madison 1968

4. BIRD, R.B., STEWART, W.E., LIGHTFOOT, E.N.;
    "Transport Phenomena"
    J. Wiley a. Sons, New York (1960), 206 - 207

5. SADOWSKI, T.J.;
    "Non-Newtonian Flow Through Porous Media"
    Trans. Soc. Rheol. 9 (1965) 2, 251 - 271

6. SADOWSKI, T.J., BIRD, R.B.;
    "Non-Newtonian Flow Through Porous Media"
    Trans. Soc. Rheol. 9 (1965) 2, 243 - 250

7. PARK, H.C., HAWLEY, M.C., BLANKS, R.F.;
    "The Flow of Non-Newtonian Solutions Through Packed Beds"
    Polym. Eng. Scie. (1975) 15, 761 - 773

8. SCHEIDEGGER, A.E.;
    "Theoretical Models of Porous Matter"
    Producers Monthly 17 (Aug. 1953) 10, 17 - 23

9. SCHEIDEGGER, A.E.;
    "The Physics of Flow Through Porous Media"
    University of Toronto Press (1963), 115 - 117

10. VOGEL, P.;
    "Untersuchungen zur Berechnung des Fließverhaltens wäßriger
    Polymerlösungen in Sandpackungen"
    Ph.D. Thesis, TU Clausthal 1980, West Germany

11. BONDOR, P.L., HIRASAKI, G.J., THAM, M.J.;
    "Mathematical Simulation of Polymer Flooding in Complex
    Reservoirs"
    Soc. Pet. Eng. J. (Oct. 1972), 369 - 382

BASIC RHEOLOGICAL BEHAVIOR OF XANTHAN POLYSACCHARIDE SOLUTIONS IN POROUS MEDIA: EFFECTS OF PORE SIZE AND POLYMER CONCENTRATION

G. CHAUVETEAU and A. ZAITOUN

Institut Français du Pétrole,
B.P. 311, 92500 Rueil Malmaison - France

ABSTRACT

The basic rheological behavior of xanthan polysaccharide solutions has been extensively investigated by varying polymer concentration, pore size and the chemical nature of porous media. The rheological characterization of solutions has shown that xanthan macromolecules behave like rigid rods in the salinity conditions selected. All microgels were carefully removed from solutions in order to study the behavior far away from injection wells.

In fine cylindrical pores, mobility reduction at low shear rates was found to be constant and lower than the Newtonian viscosity at low shear rates, except for pore diameters smaller than macromolecule length. Water permeability was not reduced after polymer flow, showing that the rheological behavior was not influenced by retention or adsorption phenomena. The ratio between mobility reduction and relative viscosity decreases as pore size decreases and polymer concentration increases. This is explained by the existence near the pore wall of a depleted layer in which polymer concentration and thus viscosity is smaller than in the bulk. This depletion is due to steric effects and does not depend on chemical nature and pore shape. A model based on this physical hypothesis is proposed for calculating mobility reduction as a function of pore size and polymer solution properties. The model's predictions are in agreement with experimental results.

In various unconsolidated porous media, such as packs of glass beads, carborundum particles and sand grains, the same behavior is observed. The mobility reduction is less than in large capillaries and decreases with pore size. Moreover, the depleted layer effect decreases with shear rate until it vanishes at high flow rates. A comparison between flow curves and rheograms gives an estimation of effective shear rates in pore throats of porous media as a function of average velocity.

The experiments carried out in Fontainebleau sandstones having different permeabilities confirm this observation and show that pore throat diameters in consolidated porous media are larger than predicted by the usual capillary models.

In all types of porous media, no dilatant behavior was detected even at the highest flow rates.

The practical applications of this study for EOR are 1) xanthan solutions are better sweeping fluids in heterogeneous reservoirs than conventional fluids having the same average viscosity; 2) they can be used in less permeable formations than previously claimed; 3) very good injectability is expected for microgel-free solutions.

INTRODUCTION

Both hydrolyzed polyacrylamide and xanthan polysaccharide solutions are candidates for enhancing oil recovery.

Up to now, hydrolyzed polyacrylamides have undoubtedly been more extensively studied in the laboratory and used in field applications. However, the macromolecular flexibility of this type of polymer causes several detrimental effects (1): 1) The viscosity decreases sharply as salinity increases, due to the screening of charged groups, particularly in the presence of bivalent ions. 2) The dilatant behavior at high flow rates which decreases injectability. This behavior is due to the coil-stretch transition of macromolecules in converging zones of porous media (2). 3) The mechanical degradation which occurs when hydrodynamic forces on the stretched molecules overcome the strength of carbon-carbon bonds (3). Moreover, the hydrolysis of acrylamide groups at high temperatures, observed even in neutral conditions (4), can lead to precipitation in the presence of calcium ions. So their use is limited to low salinity and temperature reservoirs.

The rigid rodlike conformation of xanthan polysaccharide molecules in most reservoir conditions enables the problems mentioned above to be avoided. The viscosity is almost insensitive to salinity, except in a very low salinity range, and neither dilatant behavior nor mechanical degradation has been observed in oil recovery conditions. So this polymer is potentially very attractive, particularly for high salinity reservoirs. But, up to now, the poor quality of most industrial products available on the market has excluded xanthan polysaccharides from many field applications. The poor solubility of some products and the existence of both microgels and cellular debris, particularly in powders, is well documented (5). The influence of these microgels on their flow behavior has been extensively studied in well-defined porous media (6). However, recent improvements in manufacturing processes, particularly for fermentation broths, reduce to a great extent the risks of well plugging, so that xanthan solutions could be widely used in the near future.

These newly manufactured polymers contain so few microgels that they will be adsorbed or retained at a short distance from the injection well. In these conditions, most of the oil to be recovered which is located far from the injection well will be swept by a polymer solution without microgels. Thus knowing the basic rheological properties of such a solution in porous media is very important from a practical point of view.

The first experiments carried out in porous media with a microgel-free solution (7) showed that the apparent viscosity or mobility reduction is less than the viscosity determined in a viscometer, mainly at the lowest shear rates in the Newtonian regime. Further experiments, performed with a well-characterized polymer solution and well-defined porous medium, showed that this phenomenon was related to the existence of a depleted layer near the wall, due to steric effects (8). The present investigation aims to study the influence of polymer concentration and rock permeability in order to estimate the effects of this depleted-layer phenomenon on the sweeping properties of xanthan solutions.

POLYMER SOLUTIONS

The xanthan polymer used is a sample manufactured in a fermentation-broth form by Rhône-Poulenc laboratories with a fermentation process specially designed to avoid microgel formation. Its molecular weight should be close to 0.8×10^6. All solutions were obtained by dilution with salted water, clarified and filtered at very low shear rate to remove any possible remaining microgel - with a method previously described (6). The addition of 400 ppm NaN_3 protected solutions against bacterial attack. In the conditions chosen (salinity = 5 g/l NaCl, pH = 7,

$\theta = 30°C$), the polymer molecule was shown to behave like a rigid rod having a 0.62 µm length and 16.5 Å diameter (8).

BULK RHEOLOGICAL PROPERTIES

<u>Shear flow</u>

Viscosity measurements were performed with a series of glass capillary viscometers, previously described, over a wide shear rate range (0.1 to 3000 s^{-1}) for various polymer concentrations (25 to 2400 ppm) using Rabinowitch-Mooney correction for power-law fluids. The plots of shear viscosity versus shear rate in log-log coordinates (Fig.1) show how solutions behave in pure shear flow. The following can be observed:

1) A Newtonian regime, at very low shear rates, in which relative viscosity η_r which is the ratio between polymer solution and brine viscosities is independent of shear rate and equal to η_{ro}.

2) A transition zone, characterized by a critical shear rate, equal to the inverse of a rotational relaxation time τ_r.

3) A shear-thinning regime, in which relative viscosity decreases with shear rate according to a power law whose exponent is 2 m.

Over the shear rate range tested, the experimental data fit very well with the Carreau model A (9).

$$\eta_r = \frac{\eta_{ro}}{\left[1 + (\tau_r \times \dot\gamma)^2\right]^m} \tag{1}$$

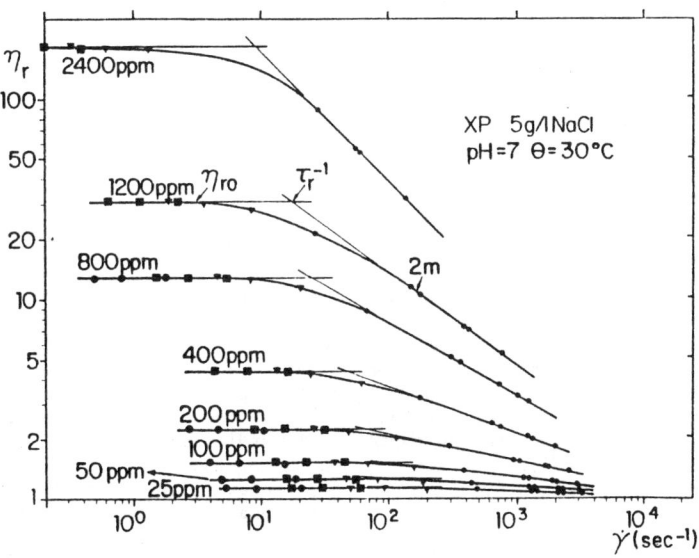

Figure 1.

Viscosity-shear rate curves for various polymer concentration

Converging flow

An estimate of viscous friction in converging flows can be made by measuring the apparent relative viscosity in a model consisting of successive short capillaries separated by cylindrical expansions for which the geometry is shown in Figure 2.

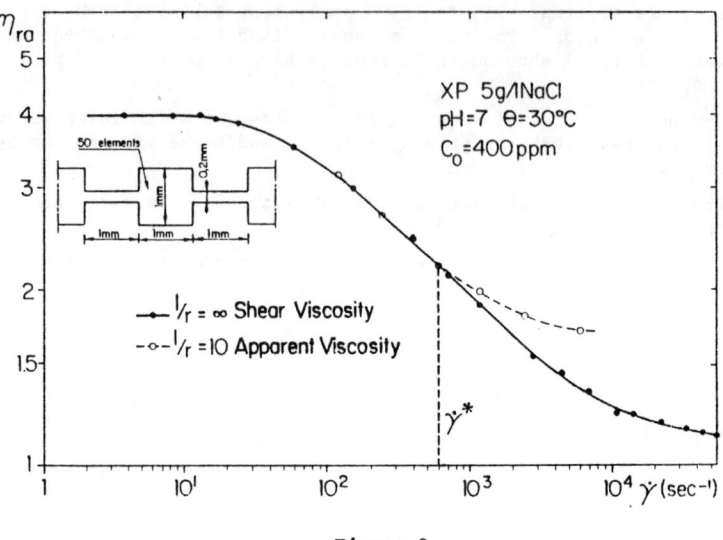

Figure 2.

Influence of converging flow zones on apparent viscosity

The capillary radius was chosen sufficiently small so as to avoid any inertia effect in our experimental conditions. For shear rates less than a critical value $\dot{\gamma}^* \approx 600$ s^{-1}, the apparent viscosity in the model was found to be equal to the shear viscosity, meaning that relative viscosities are equal in both converging and shear flow. For $\dot{\gamma} > \dot{\gamma}^*$, the apparent viscosity becomes greater than shear viscosity. This increased viscous friction occurring in converging flow near the entrance to the capillary is explained by the strong orientation of the rods in the flow direction when the product of relaxation time by elongation rate is sufficiently high (2) (10). However, this increase in apparent viscosity is very small, compared to that obtained with polyacrylamide solution with the same flow conditions (2). Indeed, the polyacrylamide molecule is both stretched and orientated in the flow direction by the converging flow. The high stretching degree (the stretched length may be 50 times the initial coil diameter) explains the magnitude of the viscous friction increase with polyacrylamide, thus involving dilatant behavior.

WALL EFFECT IN FLOW THROUGH FINE CAPILLARIES

The effects of pore size on apparent viscosity were first investigated in a very simple system, namely with a well-characterized rodlike polymer solution flowing through fine cylindrical capillaries, in order to make the interpretation easier.

Experimental facility

Nuclepore membranes were selected for these experiments because their pores have a well-defined cylindrical shape. The average diameters and areal pore densities corresponding to nominal diameters (ranging between 0.4 and 12 μm) were determined by electron microscopy (8), and the average diameters are given in Figure 3.

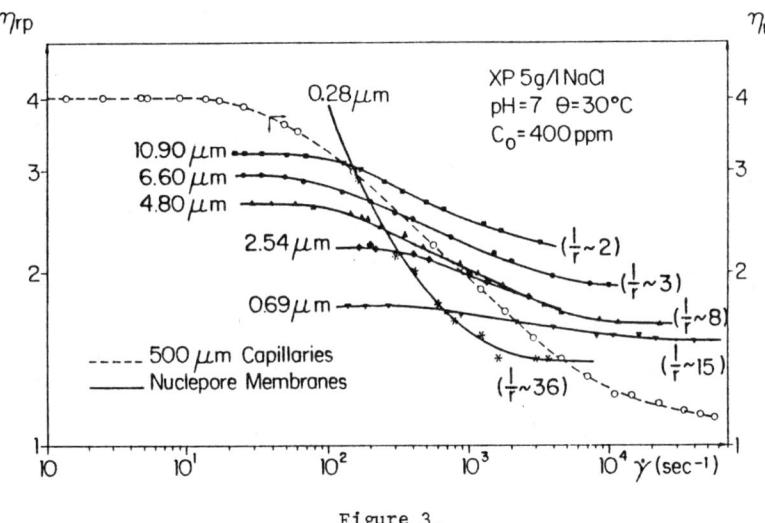

Figure 3.

Influence of pore diameter on apparent viscosity

A series of six filter holders, each one containing five membranes separated by nylon grids, was used to obtain sufficient pressure drops measured by oil-water manometers. The thickness of the Nuclepore membranes is constant and approximately equal to 10 μm, so that the capillary length to radius ratio l/r given in Figure 3 depends on pore diameter.

Results and discussion

The results of flow experiments are shown in Figure 3. Bulk relative viscosity versus shear rate is plotted as a dashed line. The solid-line curves show the variations of relative apparent viscosity measured during flow through membranes having different pore diameters.

The most important result is that in the Newtonian regime the apparent viscosity in fine pores is found to be lower than in bulk solutions and decreases with pore diameter, except for the smallest one whose diameter (0.28 μm) is less than molecule length (0.62 μm). In this last case the macromolecules are retained on the upstream side of the membrane, causing an extra-pressure drop and thus a curve upturn in low shear range. At the highest shear rates, the macromolecules are oriented by hydrodynamic forces and can easily pass through the membranes. In all experiments, the water permeability was unchanged after polymer flow, showing that flow properties were not disturbed by adsorption or retention phenomena. It must be noted that a comparison between apparent viscosities is valid in the Newtonian regime even for cylindrical pores having different length to-radius ratios. At higher shear rates, the entrance effects can increase apparent viscosity in relatively short capillaries (11), and the shear rate dependence must be studied with models having similar geometric shapes such as glass-bead packs (see below).

This decrease in apparent viscosity as pore diameter decreases has been interpreted (8) by the existence of a depleted layer near the pore wall. This depletion is due to the steric hindrances which reduce the probability that the macromolecular center of mass may be at a distance less than one macromolecular half-length from the wall as shown in Figure 4. Thus, the polymer concentration will increase from zero at wall contact up to bulk concentration at a distance close to half the length of a macromolecule. Such a depleted-layer has been theoretically predicted for both coil polymers (12) and rodlike particles (13), and it physically explains the apparent slip at the wall predicted for concentrated solutions (14). As a consequence of this depleted layer, the increase in viscosity due to the polymer is less near the wall than in the bulk, causing a lower overall apparent viscosity in fine pores than in the bulk. This effect increases as pore diameter decreases.

A coaxial two-fluid flow model has been proposed to schematize polymer solution flow (Fig. 4). The bulk solution with a relative viscosity η_{rb} flows in the center of the capillary inside a radius equal to (r-δ). A depleted solution having a relative viscosity η_{rw} flows in an annulus having thickness δ surrounding the bulk solution. The velocity is zero at the wall and equal in both

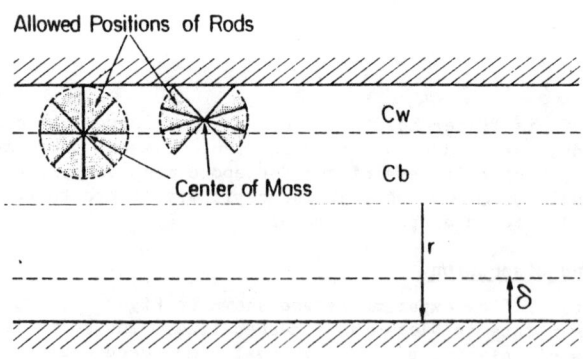

Figure 4

Schematic view of polymer solution flow through fine
pores with a depleted layer effect

the bulk solution and the depleted solution at a distance $r - \delta$ from the axis. From this model, an analytical equation has been derived to calculate apparent relative viscosity η_{rp} as a function of pore diameter $2\ r$:

$$\eta_{rp} = \frac{\eta_{rw}}{1 - (1 - 1/\rho)(1 - \delta/r)^4} \tag{2}$$

where $\rho = \eta_{rb}/\eta_{rw}$

Very good agreement between the experimental apparent relative viscosity in the Newtonian zone (Fig. 3) and the predictions of this model was found in choosing the following values for depleted layer characteristics:

$$\delta = 0.3\ \mu m \qquad \eta_{rw} = 1.77$$

The value of δ is close to half the length of a macromolecule ($L/2 = 0.31\ \mu m$), and the value of η_{rw} is consistent with the physical hypothesis proposed. Moreover AUBERT and TIRRELL (15) have recently proposed a calculation based on the finitely extendable nonlinear elastic dumbbell as a molecular model and the exclusion of all molecule configurations intersecting the walls. Good agreement is found between their calculations and our experimental findings.

Thus, the relation between the diameter dependence of apparent viscosity and the depleted- layer phenomenon seems to be very well established. Moreover, the same behavior was recently observed with polyacrylamide solutions when there are no effects of adsorption on flow properties (1).

FLOW THROUGH UNCONSOLIDATED POROUS MEDIA

Pore size dependence

Calibrated glass beads having different diameters (see Table 1) were packed to obtain porous media having similar pore shapes but different pore sizes. The flow experiments were performed with a 400 ppm xanthan solution, and the absence

TABLE I

Bead diameter D_b (μm)	Perme-ability k (μm^2)	Porosity $\emptyset$	Apparent viscosity η_{rp}	Shear-thinning index $2\ m_p$	Shear-rate constant α	Pore-throat diameter $2\ r_p$ (μm)
400-500	137	0.40	3.90	0.185	1.7	117
200-250	36	0.40	3.75	0.180	1.7	43
80-100	8.4	0.40	3.60	0.175	1.4	26.5
40-50	2.4	0.41	3.37	0.160	1.05	15.5
20-30	0.66	0.41	2.97	0.130	1.10	7.6
10-20	0.21	0.41	2.72	0.110	1.25	5.3
8-15	0.11	0.41	2.37	0.080	1.75	3.35

of permeability reduction after polymer flow was checked for every bead pack to ascertain the absence of any adsorbed layer effect.

The flow-experiment results are quite similar to those observed in flow through cylindrical pores (Fig. 5). The apparent viscosity in the Newtonian zone is

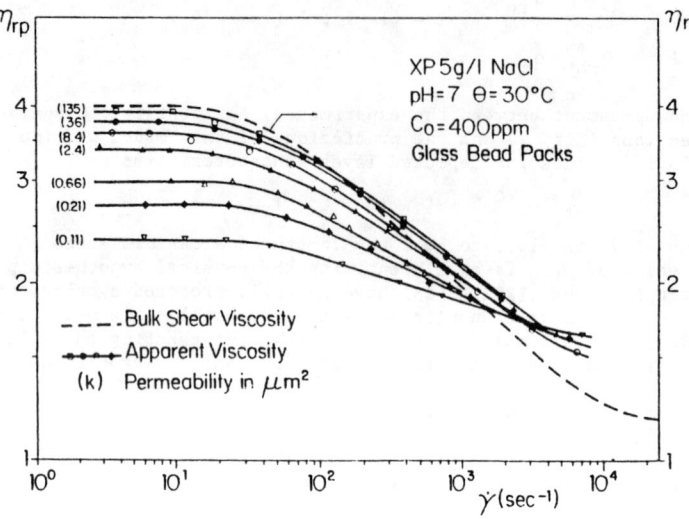

Figure 5.

Pore size dependence of apparent viscosity
in flow through glass-bead packs

found to be lower than the bulk viscosity and decreases with average pore size evaluated by pack permeability as shown in Figure 5. The maximum wall shear rate in the average pore throat diameter was calculated by:

$$\dot{Y} = \alpha \times 4 \ v \ (8 \ k \ \phi^{-1})^{-0.5} \tag{3}$$

Where α is a shape parameter characteristic of the pore structure. The value of α should be one for a bundle of capillaries having the same diameters. For porous media, the value of α is experimentally determined as being that which gives the same critical Y_c corresponding to the onset of shear-thinning behavior for both the shear viscosity-shear rate curve and the apparent viscosity-shear rate curve in the porous medium under consideration. The α value was found to be equal to 1.7 for packs of large spheres having the same diameter (8). It decreases with the pore size-molecule length ratio and increases as pore structure heterogeneity increases. This is the case when the bead-diameter distribution becomes wider or when the consolidation degree of sands giving sandstones increases (8).

Due to the statistical homothety of bead packs, the shear rate dependence of the depleted-layer effect can be deduced from flow experiments in this type of porous media. As expected, the rod orientation with shear decreases the depleted layer effect as the flow rate increases, and apparent viscosity becomes independent of pore size at high shear rates ($\dot{Y} > 3000 \ s^{-1}$). At the highest flow rates, the apparent viscosity overcomes the shear viscosity. This can be explained by the increase in viscous friction in converging zones of porous media where the macro-molecules are orientated in the direction of flow (Fig. 2).

As shown by results obtained with polymer flow through Nuclepore membranes, Equation (2) gives the relation between apparent viscosity and pore diameter. Thus an effective diameter can be calculated for each glass bead pack from the apparent viscosity measured. This effective diameter corresponds to an average hydrodynamic diameter of pore throats where polymer flows. On the other hand the mean pore size is proportional to the square root of the permeability for homothetic porous media. In Figure 9, the effective pore-throat diameter deduced from the polymer apparent viscosity is plotted versus pore diameter calculated from the simplest capillary model, $2 \; r = 2 \; (8 \; k \; \emptyset^{-1})^{0.5}$. All the points corresponding to experiments performed with glass-bead packs are lined-up on the first bissectrix. So the average hydrodynamic diameter of pore throats is approximately equal to $2 \; (8 \; k \; \emptyset^{-1})^{0.5}$ for homogeneous bead packs.

Additional points deduced from experiments carried out in sand packs (8) are also lined-up on the same curve.

Polymer concentration effects

The influence of polymer concentration was systematically studied by using a Carborundum pack having a permeability equal to $0.1 \; \mu m^2$, a porosity equal to 0.48 and an effective pore diameter of $2.6 \, \mu m$. The polymer concentration was varied from 200 ppm to 1600 ppm, and the absence of permeability reduction was checked after every polymer flow experiment.

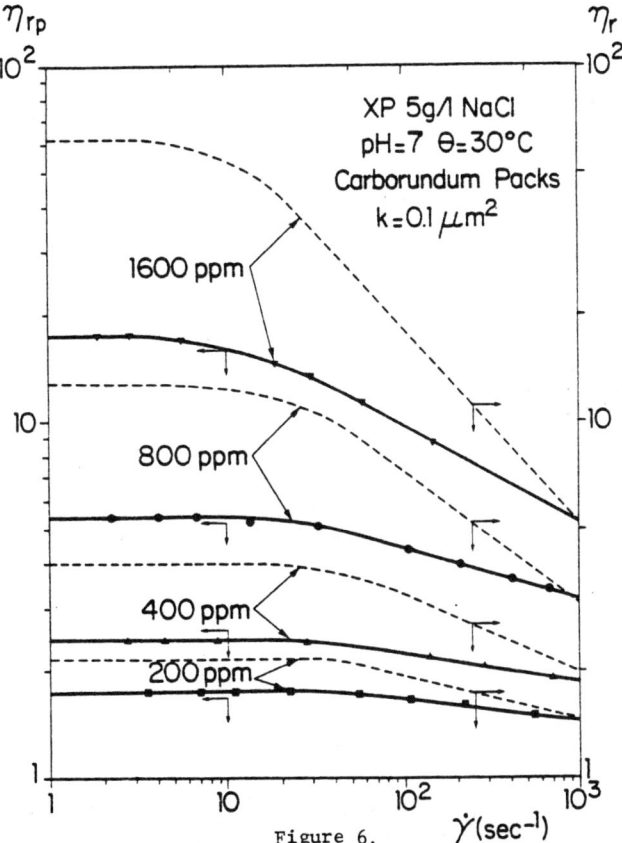

Figure 6.

The depleted–layer effect as a function of shear rate
at different polymer concentrations in Carborundum packs

Both shear viscosities in dashed lines and apparent viscosities in solid lines are plotted in Figure 6.

The first observation is that the general behavior is quite similar to that observed in glass-bead packs. The depleted-layer effect appears to be insensitive to the pore shape and chemical nature of porous media. This result is consistent with the steric origin of the phenomenon.

Moreover, the magnitude of the effect, i.e. the ratio between apparent viscosity and shear viscosity, increases sharply with the polymer concentration, at low shear rates (Fig.7).

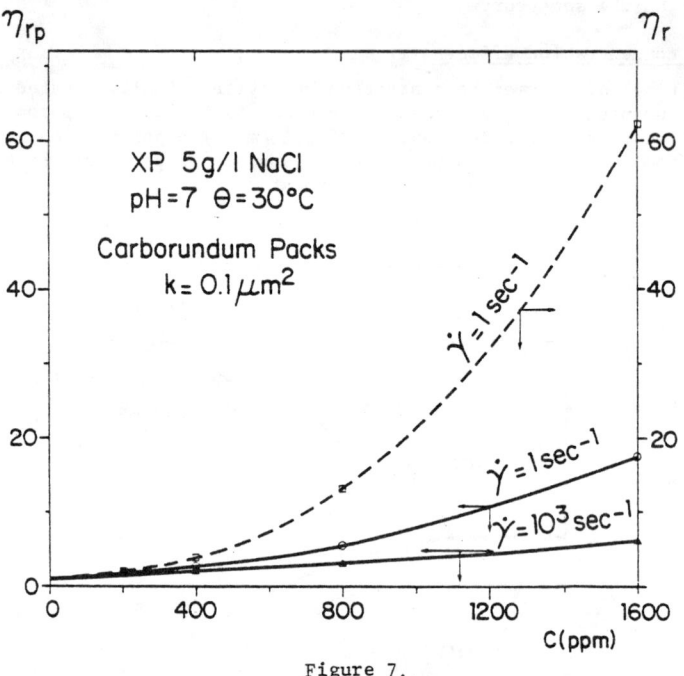

Figure 7.

The influence of polymer concentration on the magnitude
of depleted layer effect

At the highest concentration tested (c = 1600 ppm), the apparent viscosity (η_{rp} = 17.5) is less than one third of bulk shear viscosity (η_{rb} = 62).

This polymer concentration effect could also be predicted. Indeed, after both division by η_{rb} and inversion, Equation (2) can be written:

$$\frac{\eta_{rb}}{\eta_{rp}} = \rho \left[1 - (1 - \frac{1}{\rho}) \ (1 - \frac{\delta}{r})^4 \right] \qquad (4)$$

For dilute solutions, the thickness of the depleted layer δ is expected to be constant so that:

$$\eta_{rb}/\eta_{rp} = k + (1 - k)\rho \tag{5}$$

where $k = (1 - \dfrac{\delta}{r})^4$ is a positive constant, always less than 1 for a given porous medium. As a consequence, the depleted-layer effect increases linearly with $\rho = \dfrac{\eta_{rb}}{\eta_{rw}}$.

In the concentration range tested, the C_b/C_w ratio is expected to be constant (13). Since the viscosity of these polymer solutions is roughly an exponential function of polymer concentration (8), the η_{rb}/η_{rw} ratio increases very sharply with polymer concentration, thus explaining the concentration dependence observed for the depleted-layer effect.

FLOW THROUGH SANDSTONES

Permeability effects

As shown above, the depleted-layer effect depends only on pore size for a given polymer solution. However, the well-known relation between pore size and permeability deduced from the simplest capillary model

$$2r_c = 2 (8 \, k \, \emptyset^{-1})^{0.5} \tag{6}$$

is valid only for homothetic unconsolidated packs.

For natural porous media such as sandstones, this relation is no longer valid, and electron microscopy observations (16) have shown that pore throat diameters are generally larger than those calculated by Equation (6). As a consequence, the influence of permeability cannot be predicted by a simple model.

TABLE II

Flow through sand packs and sandstones
(XP solution, $\eta_r = 4.0$, 2 m = 0.22)

Grain diameter $D_g (\mu m)$	Permeability k (μm^2)	Porosity $\emptyset$	Apparent viscosity η_{rp}	Shear-thinning index 2 m_p	Shear-rate constant α	Pore-throat diameter 2 r_p (μm)
Sand 1 80-120	5.0	0.38	3.5	0.165	2.5	21
Sand 2 50-100	3.4	0.43	3.4	0.165	1.4	15.7
Sandstone 1	0.256	0.119	3.32	0.087	4.2	13.6
Sandstone 2	0.0373	0.084	2.95	0.062	5.6	8.3
Sandstone 3	0.0206	0.075	2.83	0.060	5.6	7.0
Sandstone 4	0.0096	0.056	2.69	0.056	9.1	6.0
Sandstone 5	0.0033	0.056	2.49	-	14.3	4.4

Some cores of quartzitic Fontainebleau sandstones having permeabilities ranging from 3×10^{-3} to 0.4μ m^2 (Table II) were selected to obtain a quantitative evaluation of the depleted-layer effects. All the cores were preflushed by a hydrochloric acid solution to remove the slight quantity of iron contained in the sample in order to avoid possible interactions with the polymer. After polymer flow experiments, the initial permeability of each core was exactly restored, even for the less permeable sample ($3.3 \times 10^{-3} \mu$ m^2).

The experimental results shown in Figure 8 are similar to those observed in unconsolidated porous media. The apparent viscosity is less than in the bulk in the Newtonian zone, which indicates a depleted-layer effect that increases as permeability decreases.

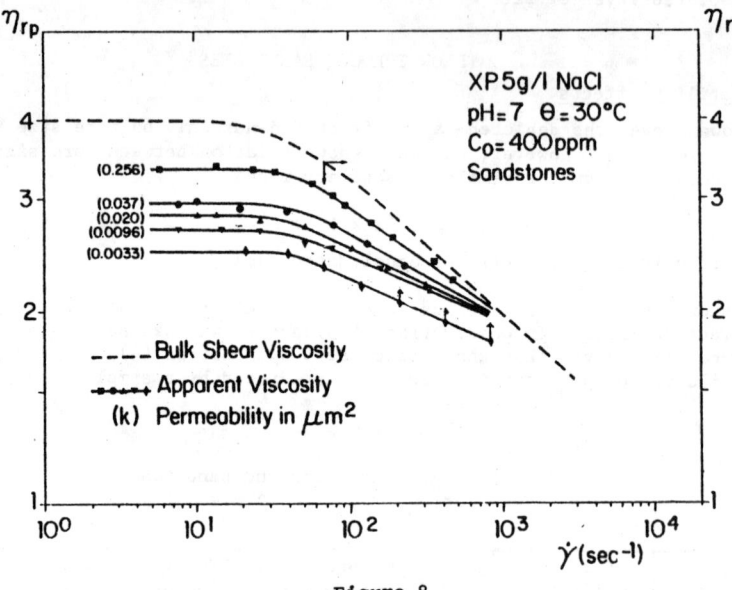

Figure 8.

The depleted layer effects in flow through
Fontainebleau Sandstones

Using polymer-solution characteristics in the bulk ($\eta_{rb} = 4$) and near the wall ($\delta = 0.3 \mu$m and $\eta_{rw} = 1.77$), deduced from experiments with Nuclepore membranes, Equation (2) gives the effective diameter of pore throats $2 r_p$ as a function of permeability for Fontainebleau sandstones (Table II). As expected, this effective diameter $2 r_p$ is always larger than $2 r_c$; and the ratio r_p/r_c increases as permeability decreases, as shown in Figure 9, in which experimental points corresponding to sandstones are plotted as solid circles. This trend is consistent with the secondary crystallization process which explains the decrease in permeability for Fontainebleau sandstones. The ratio r_p/r_c should be nearly one for Fontainebleau sand packs having the same grain diameter ($10 < k < 20 \mu$ m^2), but reaches values as high as 3 in very low permeability range ($10^{-3} < k < 5.10^{-3} \mu$ m^2). These results are consistent with Dullien's observations, but are more accurate because of the method and the use of a homogeneous series of sandstones. Moreover, an increase in α values is observed as permeability decreases (Table II), as could be expected from the pore diameter heterogeneity increase with the consolidation process.

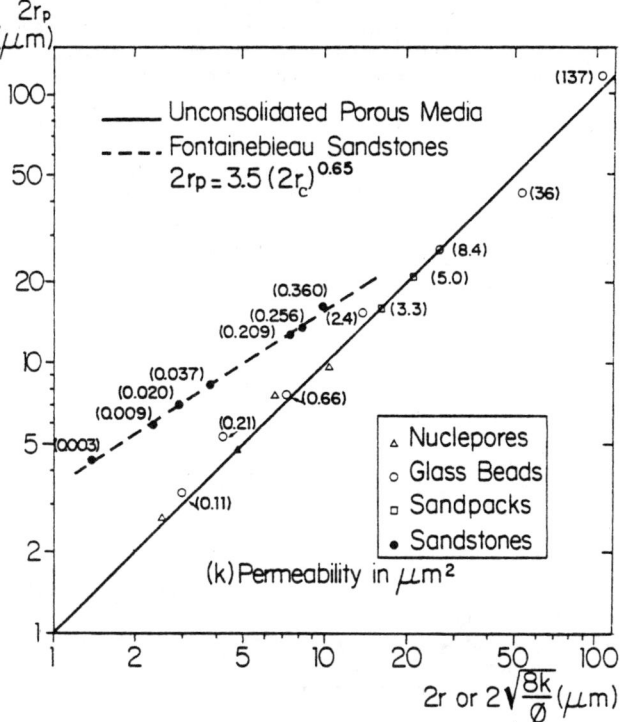

Figure 9.

Comparison of pore throat diameters determined by polymer
injection method with measured or calculated pore diameter
in various porous media

From a practical point of view, these results show that xanthan solutions can
pass very easily through even the low permeability zones of reservoirs. The
lowest limit for use of such xanthan solutions should correspond to pore throat
diameters equal to macromolecular length ($1 \simeq 0.6 \ \mu m$), i.e. to permeability
much lower than $10^{-3} \ \mu m^2$ for sandstones having a structure similar to Fontainebleau sandstones. Practically, the use of such polymers is never limited by
polymer dimensions.

Hydrodynamic retention

The first type of hydrodynamic retention, which is related to thermodynamic effects
(17) and thus does not depend on pore-molecule relative dimensions, was found to
be almost negligible for these xanthan solutions having rodlike molecules. As
theoretically expected, the entropy differences due to molecular alignment are
too small to induce large concentration differences between the different zones

of the porous medium. In high permeability sandstones, the concentration differences observed after sudden flow-rate changes (from Newtonian to shear-thinning
regimes, namely from 6 to 700 sec^{-1}) were very small ($\Delta\eta/\eta < 1\%$), in comparison
with those observed in dilatant regime with coiled polymers (18) (19) (20) (1).

The second type of hydrodynamic retention, which depends on pore-molecule relative dimensions and which can be explained by the slow accumulation of polymer
molecules in the zones of porous media where pore-throat diameters can be smaller
than molecular size, was observed in very - low - permeability sandstones
(k = 3 and 9 x 10^{-3} μm^2). Indeed, when the flow rate has been suddenly increased
and kept constant in the shear-thinning regime, large concentration decreases in
the effluent have been noted and the apparent viscosity was found to increase
slowly as indicated by the arrows in Figure 8.

CONCLUSIONS

The basic rheological behavior of xanthan solutions in porous media has been
studied with solutions without microgels, i.e. as they are in reservoirs far
away from the injection wells. Indeed, the microgels contained in injected
solutions are retained in a zone located around the injection well.

The main conclusions of these investigations are the following:

1) The apparent viscosity of polymer solutions flowing through fine pores is
always less than bulk shear viscosity at low shear rates in the Newtonian regime
and decreases as pore size decreases. This phenomenon is interpreted by the
existence near the pore wall of a depleted layer where average polymer concentration and viscosity are lower than in the bulk.

2) An analytical equation derived from a schematization of polymer solution flow
as a two-fluid concentric flow is proposed to predict apparent viscosity as a
function of pore size and polymer solution characteristics. Flow experiments
performed in well-calibrated cylindrical pores established the validity of this
equation and provided the characteristics of the depleted layer. Particularly,
its thickness close to the half-length of the macromolecule is consistent with
our interpretation of the origin of the depleted layer.

3) The depleted-layer effect decreases as shear rate increases so that, at the
highest shear rates, the apparent viscosity becomes independent of pore size.
Moreover, the rodlike conformation of xanthan molecules minimizes viscous friction
in zones of converging flow inside the porous structure, so that no dilatant
behavior is observed even at the highest flow rates such as those existing around
the injection well. The injectability of microgel-free xanthan solutions should
be excellent.

4) The depleted-layer effect is also observed in Nuclepore membranes, glass bead
packs, Carborundum and sand packs, and sandstones. Thus, this effect seems to be
independent of the pore shape and chemical nature of porous media. This is
consistent with the steric origin of this phenomenon.

5) The magnitude of the depleted-layer effect increases sharply with the polymer
concentration, as predicted by our model, so that this effect becomes very significant from a practical point of view.

6) The magnitude of the depleted-layer effect increases as sandstone permeability
decreases. Microgel-free xanthan solutions can pass easily, with a small
apparent viscosity and without any permeability reduction, through sandstones
having very low permeabilities.

7) The average hydrodynamic diameter of pore throats in a given sandstone can be deduced by measuring the apparent viscosity of a well-known polymer solution in the Newtonian regime. Thus polymer injection is a new method for investigating pore structure.

8) The effect of the depleted layer, which decreases apparent viscosity mainly in low permeability zones, enables xanthan solutions to sweep oil better in heterogeneous formations than conventional fluids having a viscosity that is independent of pore size.

Acknowledgments

This research was supported by the Association de Recherches sur les Techniques d'Exploitation du Petrole (ARTEP) , and Rhône-Poulenc Industries provided the polymer sample. The authors wish to acknowledge the contribution of Ph. Delaplace and R. Tabary who performed laboratory experiments.

REFERENCES

1. CHAUVETEAU, G.; "Molecular Interpretation of the Different Properties of Coil Polymer Solution Flow Through Porous Media in Oil Recovery Conditions", paper SPE 10 060 presented at the Annual Technical Conference and Exhibition, San Antonio, Oct. 4-7, 1981.

2. CHAUVETEAU, G. and MOAN, M.; "The Onset of Dilatant Behavior in Non-Inertial Flow of Dilute Polymer Solutions through Channels with Varying Cross Sections", Journal de Physique-Lettres, 42 (1981) L-201 - L-204.

3. GHONIEM, S., MOAN, M., CHAUVETEAU, G. and WOLFF, C.; "Mechanical Degradation of Semi-Dilute Polymer Solutions in Laminar Flows", accepted for Publication in Journal of Canadian Chemical Engineering.

4. MULLER, G., FENYO, J.C., and SELEGNY, E.; "High Molecular Weight Hydrolyzed Polyacrylamides. III Effects of Temperature on Chemical Stability", J. Appl. Pol. Sci., (1980), 25, 627-633.

    MULLER, G.; "Thermal Stability of High Molecular Weight Polyacrylamide Aqueous Solutions", Polymer Bulletin (to be published).

5. KOHLER, N., and CHAUVETEAU, G.; "Xanthan Polysaccharide Plugging Behavior in Porous Media: Preferential Use of Fermentation Broth", J. Pet. Techn. (Feb. 1981) 23, 349-358.

6. CHAUVETEAU, G., and KOHLER, N.; "Influence of Microgels in Xanthan Polysaccharide Solutions on Their Flow Behavior Through Various Porous Media", Paper SPE 9295 presented at the 55th Annual Technical Conference and Exhibition, Dallas, Sept. 21-24, 1980.

7. CHAUVETEAU, G.; "Ecoulement laminaire en milieu poreux de solutions de macromolécules de taille non négligeable devant les dimensions des pores", C.R. Acad. Sci. Paris, (Feb. 1979), 288, 107-110.

8. CHAUVETEAU, G.; "Rodlike Polymer Solution Flow through Fine Pores: Influence of Pore Size on Rheological Behavior", Submitted for publication in Journal of Rheology, 1981.

9. CARREAU, P.J.; "Rheological Equations from Molecular Network Theories", Trans. Soc. Rheol., (1972), 16, 99-127.

10. HOA, N.T., CHAUVETEAU, G., GAUDU, R., and ANNE-ARCHARD, D.; "Relation entre le champ de vitesse d'élongation et l'apparition d'un comportement dilatant d'une solution de polymère diluée dans un écoulement convergent non-inertiel", Submitted for publication in C.R. Acad. Sci. (1981).

11. MOAN, M., CHAUVETEAU, G., and GHONIEM, S.; "Entrance Effects in Capillary Flow of Dilute and Semi-Dilute Polymer Solutions", J. Non.Newt. Fluid. Mech., (1979), 5, 463-474.

12. JOANNY, J.F., LEIBLER, L., and DE GENNES, P.G.; "Effects of Polymer Solutions on Colloid Stability", J. of Pol. Sc. (1979), 17, 1073-1084.

13. AUVRAY, L.; "Solutions de macromolécules rigides : Effets de paroi, de confinement et d'orientation par un écoulement", Journal de Physique (Janv. 1981), Vol. 42, 79-95.

14. DE GENNES, P.G.; "Ecoulements viscosimétriques de polymères enchevêtrés". C.R. Acad. Sci. Paris, (April 9, 1979), 288, B, 219-220.

15. AUBERT, J.H., and TIRRELL, M. ; "Effective Viscosity of Dilute Polymer Solutions near Interfaces", ACS Polymer Preprint (1981) 22, 1, 82-83.

16. BATRA, V.K., and DULLIEN, F.A.L. ; "Correlation between Pore Structure of Sandstones and Tertiary Oil Recovery", Soc. Pet. Eng. J. (Oct. 1973), 13, 256-258.

17. METZNER, A.B. ; "Flow of Polymeric Solutions and Emulsions through Porous Media", in "Improved Oil Recovery by Surfactant and Polymer Flooding", Acad. Press. Inc., New-York (1977), 439-451.

18. CHAUVETEAU, G., and KOHLER, N.; "Polymer Flooding: the Essential Elements for Laboratory Evaluation", Paper SPE 4745, presented at the Improved Oil Recovery Meeting, April 22-24 (1974).

19. WILLHITE, G.P., and DOMINGUEZ, J.C.; "Mechanisms of Polymer Retention in Porous Media", in Improved Oil Recovery by Surfactant and Polymer Flooding, Acad. Press. Inc. New-York (1977),511-553.

20. CHAUVETEAU, G.; "The Effects of Rheological Properties and Polymer-Rock Dimension-Sensitive Interactions on Polyacrylamide Solution Flow through Porous Media", Soc. Rheol. Meeting, Houston, (1978) Oct. 22-26.

THE CHATEAURENARD (FRANCE) POLYMER FLOOD
FIELD TEST

A. LABASTIE and L. VIO

Elf Aquitaine (Production)

Abstract

A polymer flood is operated by Elf Aquitaine in the Chateaurenard (France) field, located in the Paris Basin.

The pilot is developped with one injector and seven producers, in a layer of unconsolidated sand, 5 meters thick, at a depth of 600 m ; the 44 ha pattern encloses a very important pore volume (700 000 m3). The oil is paraffinic and has a viscosity of 40 cPo at reservoir temperature (30°C).

The water being almost fresh (0,4 g/l TDS) it has been decided to use hydrolyzed polyacrylamides. Several commercial products have been tested, mainly for viscosity and injectivity ; a liquid polymer, dissolved in produced water, is presently being used for the pilot. The water must be carefully treated before dissolution to avoid polymer degradation and formation plugging.

The injection has been started in 1977 and on account of the quantities injected so far, we have not yet seen any response in the six main producers (which are at a distance of 400 to 500 m from the injector). But the seventh intermadiate producer, drilled at a shorter distance (280 m) from the injector, has shown very interesting results with a sharp decrease of the WOR (9000 tons of tertiary oil have been produced) ; this response is due to the effect of mobility control, maybe amplified by a local reservoir heterogeneity.

Field description

The Chateaurenard field, outlined in Fig. 1, is part of the Neocomian (Lower Cretaceous) oil reservoirs, found in the southern part of the Paris Basin ; it is located 100 km SSE of Paris, and the oil zone extends over an area of 20 km^2.

214

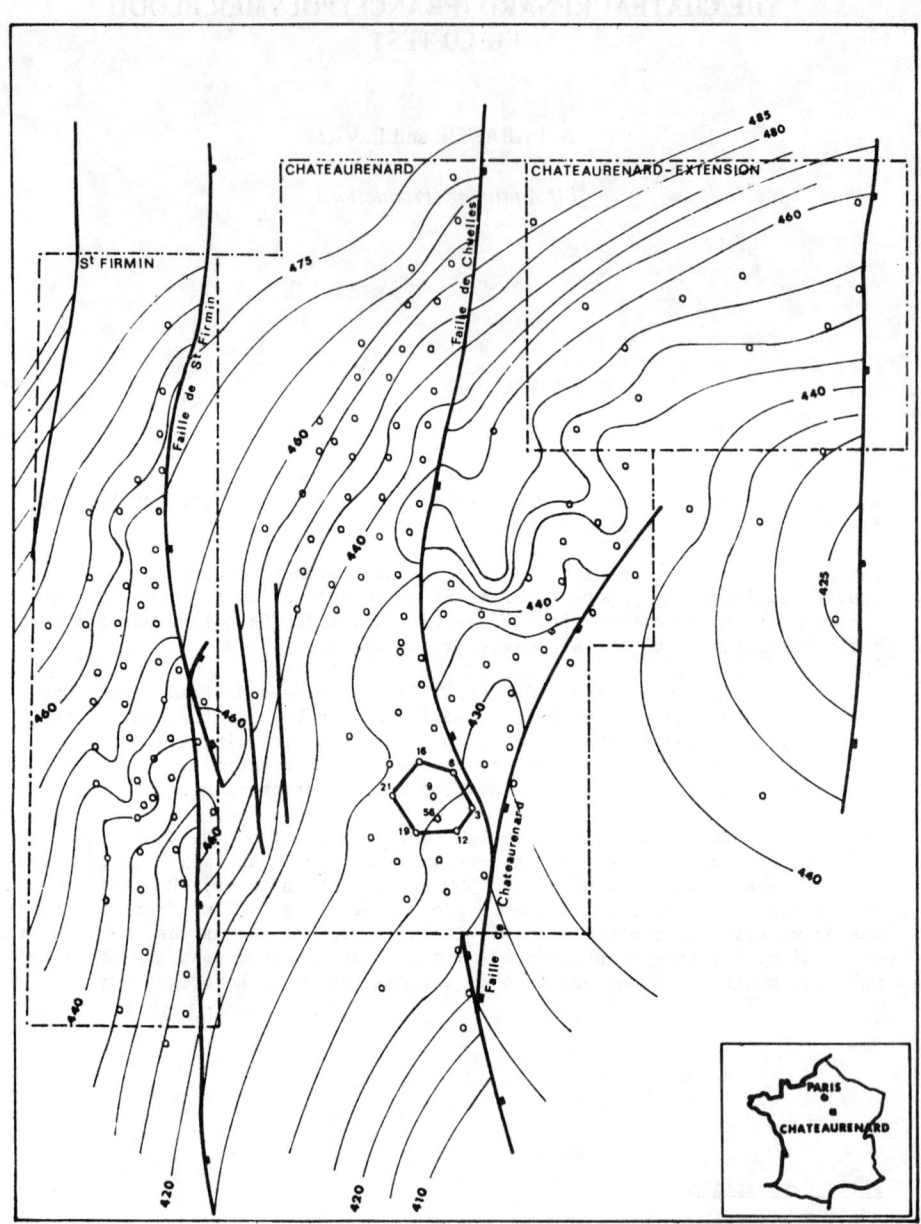

Fig. 1 - Map of Chateaurenard field

Geology

There are three distinct structures, separated by North-South faults with
throws of 15 to 20 m. Situated at a depth of about 600m, the reservoir is
formed of three layers of unconsolidated sands separated by shale as depicted
on the type - log of Fig. 2 ; the dip of these layers is very slight, about 1°.
The reservoir concerned by the polymer injection is formed by the two upper
levels (R1 and R2, belonging to the Hauterivian stage of the Neocomian)
of the central structure ; these two levels can be considered as a single
reservoir, because the clay layer between them is discontinuous and does not
form a tight barrier. The reservoir forms a roughly triangular monocline whose
closures are a fault in the east and the wedgeout of the sands in the south.

The deposit of these sands is in the form of submarine channels ; this sedimentation type gives massifs with sharp lateral variations of facies.

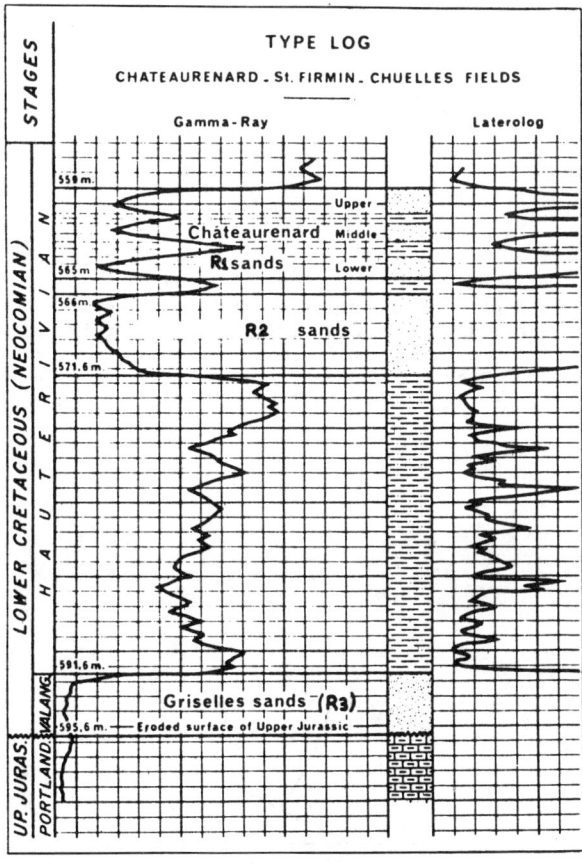

Fig. 2 - Type log

Characteristics of R1/R2 reservoir ; fluid properties

The reservoir is formed by unconsolidated sand, with some amount of clay (2 to 15 %). The average total thickness in the pilot area is 5 metres, with a porosity of 30 %. This sandstone is relatively fine grained (average 150 μ), but with a wide grain size distribution (80 to 350 μ). The average permeability is 1 Darcy (1 μm^2), but with rather large and unforeseeable variations on account of the channel type sedimentation system.

The fluids are a relatively viscous oil (40 cPo at 30°C, reservoir temperature) of paraffinic type and without dissolved gas, and an almost fresh water ; the relevant characteristics are indicated in the following table :

Depth, m (ft)	600 (1970)
Porosity, %	30
Permeability, mD	1000
Clay content, %	2 to 15
Initial water saturation, %	30
Residual oil saturation, %	30
Current field average oil saturation, %	55
Temperature, °C (°F)	30 (86)
Oil gravity, g/cm3 (°API)	0,89 (27)
Oil viscosity, mPa.s (cPo)	40 (40)
Water salinity (TDS), ppm	400
Water hardness (Ca + Mg), ppm	70

Production history

The field was discovered in 1958, initial oil in place was estimated at 11 Mm3 (69 millions bbl), with half for R1/R2 reservoir. In 1980, cumulative production was 26 % of OOIP, mainly through the action of an edge water drive. Because of very small dip angle and adverse mobility ratio (see in Fig. 3 the relative permeability curves) water appeared early in the production and water cut increased quickly ; in 1980, its average value for the field was 89 %. Peak oil production reached 267 000 m3 in 1964 ; 1980 production was 95 000 m3.

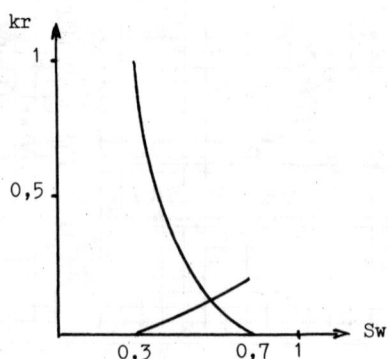

Fig. 3 - Relative permeabilities

POLYMER PILOT DESIGN

Pattern selection

The Chateaurenard field has been developped with an average well spacing of
400 m, and for this polymer pilot it has been decided to use this spacing.
The R1/R2 was interesting for this test, because of the two layers with a
discontinuous separation, maybe responsible of poor sweeping efficiency ; but
it was necessary that the two layers were not separated at the polymer injector
well.

After several interference tests, a seven spots pattern has been selected,
with one injector (CR 9 bis) and six main producers (CR 3 - CR 6 - CR 12 -
CR 16 - CR 19 Bis - CR 21 bis) at distances of 400 to 500 meters from the
injector ; a seventh intermediate producer (CR 56) has been drilled at a
shorter distance from the injector, in order to get an earlier response. This
pattern is outlined in Fig. 4.

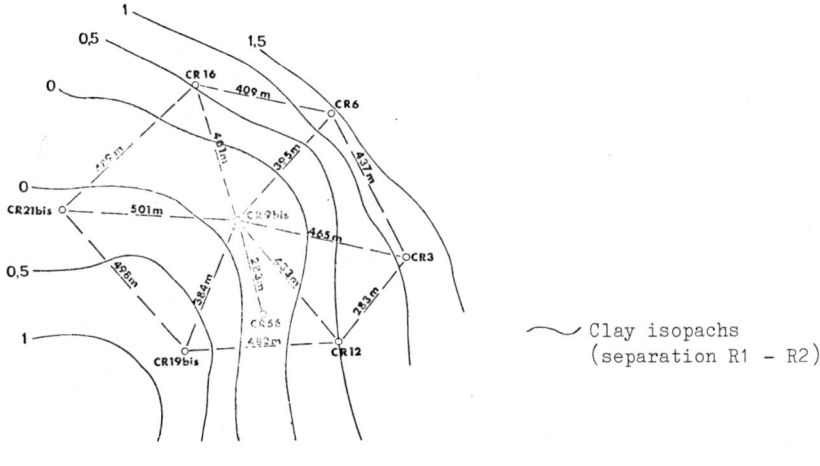

Clay isopachs
(separation R1 - R2)

Fig. 4 - Polymer pilot pattern

The surface area of this pattern is 44 ha (110 acres), and it encloses a pore
volume of 700 000 m3. The mobility ratio is very adverse and before polymer
flood, after many years of waterflooding, the oil saturation was still 55 %.

Polymer choice and slug design

The water being almost fresh and the temperature low, hydrolyzed polyacrylamides were selected. To dissolve polymer in produced water was the easiest
for field operations, but it can be detrimental for polyacrylamides stability
(1). So we have studied the degradation of polymer in presence of oxygen and
iron (little amounts are present in produced water). It has shown that we must
avoid the presence of both iron and oxygen, but that little amounts of one of
them is not detrimental (see Fig. 5 - 6).

218

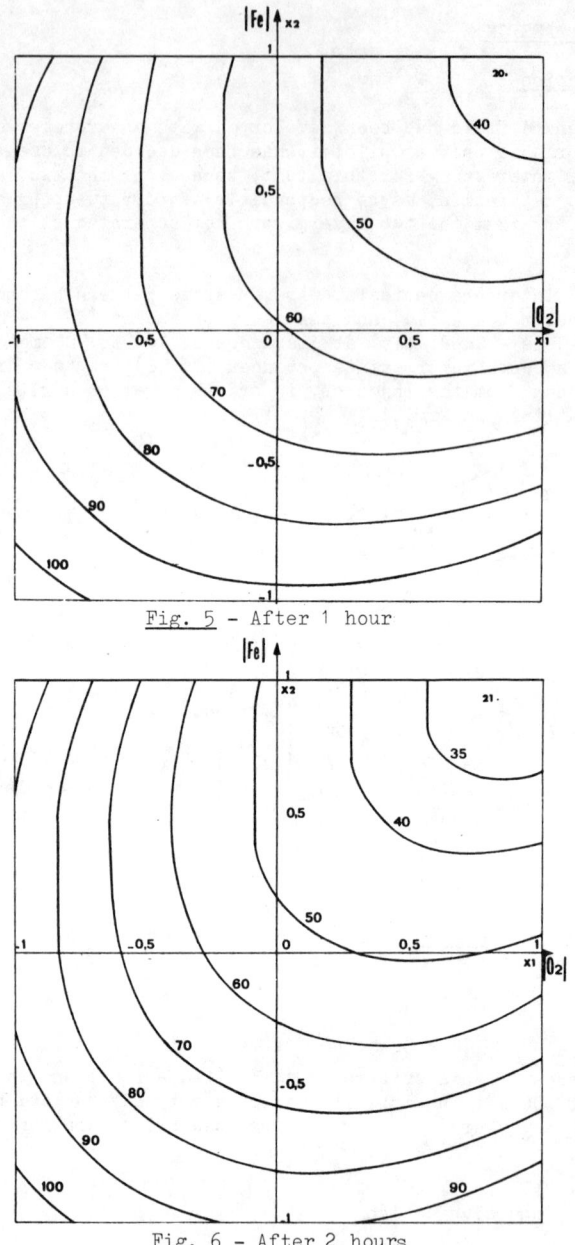

Fig. 5 - After 1 hour

Fig. 6 - After 2 hours

Chemical degraduation of polyacrylamides : viscosity (% of initial viscosity) function of Fe and O_2 content

	− 1	0	1
O_2	0,1 ppm	3,73ppm	7,47 ppm
Fe	0	5 ppm	10 ppm

Commercial products were tested mainly for viscosity and injectivity. The injectivity test is a constant flow rate test through a 5 μ millipore[R] filter , with pressure drop measurement ; the pore size (5 μ) is similar to the one of Chateaurenard reservoir, and the flow rate is chosen to give same shear rates as in the field flood. This is important because a plugging behaviour can be hidden if the flow rate is to high, due to microgel deformations in important pressure gradients (2).

This test seems a good screening procedure for comparison of products, even if sandpacks floods may be necessary for further investigation. Some results are given in Fig. 7 as an example.

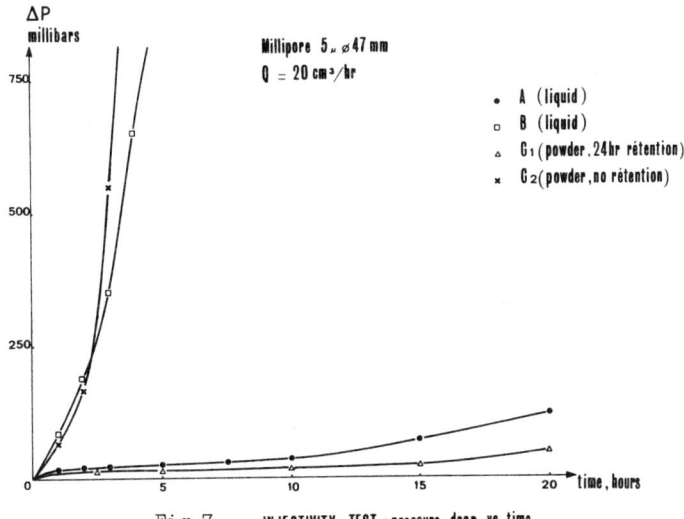

Fig.7 — INJECTIVITY TEST : pressure drop vs time

It has been found that dry polymers (powder) need a retention of several hours after dissolution to become satisfactory on injectivity test. In this example, products A and C1 are good, products B and C2 are not satisfactory (plugging behaviour) ; product A is used for this operation.

Polymer concentration and slug size have been determined by performance predictions with a reservoir model ; a slug of 0,33 PV with 700 ppm polymer has been chosen.

The polymer slug will be followed by water injection, with viscosity decrease designed to prevent deleterious effects of viscous fingering (3).

Surface installations

The polymer solution is prepared with produced water, available in great quantities but not clean : it is contaminated by iron (Fe^{++} and Fe^{+++}), residual oil (1000 ppm), oxygen (< 1 ppm) and clay particles. Removal of oil, clay and insoluble iron is carried out in a flotator using nitrogen, which also strips the water of oxygen traces ; a nitrogen blanket prevents anay oxygen entry. After treatment, the water is good for polymer dissolution, with low amounts of oil (< 3 ppm), oxygen (< 0,01 ppm) and iron (< 5 ppm). A bacteria killing agent is added to the water, that is filtered before polymer dissolution.

At the beginning of the operation, dry powder polymer was used, with residence
time between dissolution and injection to get good injectivity characteristics.
It has been changed for liquid polymer (emulsion), easier to handle and good
for injection immediately after dissolution. The fluid flow diagram of Fig. 8
outlines the equipment, that is designed for first dissolution of polymer
in a concentrated master solution, and then final dilution ; polymer solution
is 5 μ filtered at the wellhead before injection, without any problem.

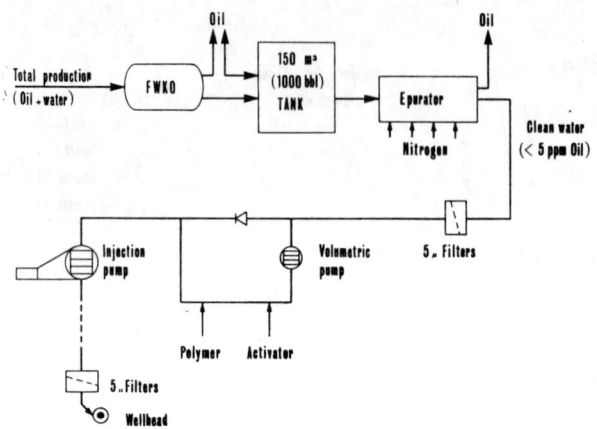

Fig.8 - Fluid flow diagram

PILOT PERFORMANCE : FIRST RESULTS

The polymer injection has begun in 1977. Until 1979, the flow rate was
135 m3/day (850 bbl/day) ; since 1980, it has been increased to 250 m3/day
(1600 bbl/day) ; no injectivity problems were encountered with polymer
solution. A slug of 235 000 m3 is to be injected, with decreasing concentration
at the rear front to prevent fingering, followed by water.

Up to june 15, 172 700 m3 of polymer solution have been injected (24,7 % pore
volume).

The six main producers have not yet shown any response, which is normal on
account of quantities injected so far.

However, the seventh intermediate producer (CR 56), has given very interesting
results (see fig. 9). This well, drilled at a distance of 280 m from the
injector CR9 bis, has been put into production in 1978. Untill mid 1979, the
water cut was about 90 %, as in the other well of this area, then we observed
a sharp decrease to 20 % of the water cut, which is only 55 % now (increasing).

9000 tons of tertiary oil have been produced.

The time of oil bank breakthrough is in accordance with our predictions, but
it is not possible to explain such an oil cut assuming an homogeneous repar-
tition of permeabilities and saturations before polymer flodd (see predicted
and observed production in Fig. 9).

A very poorly waterflooded zone (of lower permeability) has been reached by polymer flood, due to mobility control effect. This very good result would have not been possible without mobility control, so it must be attributed to the effect of polymer injection. However, this effect has been very important because of a local reservoir heterogeneity, so such a result cannot be generalized.

But with this channel type sedimentation system, some other poorly waterflooded zones may exist, which can be swept by polymer flood in good conditions.

We have now to wait for the response of other producers to have a good idea of polymer flodding performance in this field.

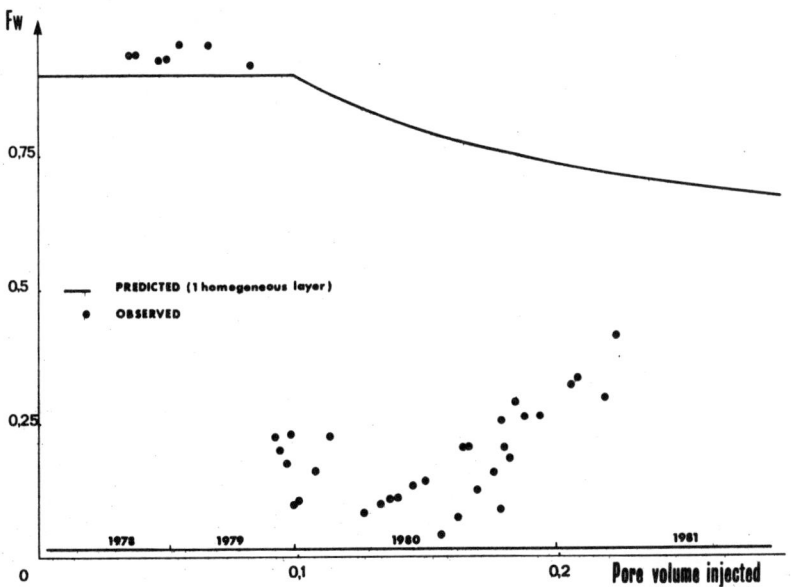

Fig.9 - OBSERVED AND PREDICTED WATER CUT PERFORMANCE OF WELL CR 56

CONCLUSIONS

1 - The polymer solution is prepared without problem using produced water, which is carefully treated.

2 - The solution is easily injected and does not show any plugging behaviour.

3 - A significant decrease of WOR has been observed in the closest producer well ; a poorly waterflooded zone has been swept by polymers, due to mobility control effect. 9000 tons of tertiary oil have been produced.

REFERENCES

1 - G. Chauveteau et N. Kohler - Conditions de stabilité des solutions de polymères lors d'une injection sur champs.

International Symposium on hydrocarbon exploration, drilling and production technics (Paris, 10 - 12 dec. 1975).

2 - G. Chauveteau - The effect of rheological properties and polymer rock dimension sensitive interactions on polyacrylamide solution flow through porous media - 49th Annual Meeting of the Society of Rheology (1978).

3 - E. L. Claridge - A method of design of graded viscosity banks - Paper SPE 6848 presented at 52nd SPE Annual Fall Meeting (Oct. 1977).

CAUSTIC FLOODING IN THE WILMINGTON FIELD, CALIFORNIA LABORATORY, MODELING, AND FIELD RESULTS

VERNON S. BREIT

Scientific Software Corporation

EDWARD H. MAYER

THUMS Long Beach Company

JOHN D. CARMICHAEL

City of Long Beach Department of Oil Properties[1]

ABSTRACT

A caustic enhanced waterflood test is being conducted in the Ranger Reservoir of the Long Beach Unit, Wilmington Field, California by the Department of Oil Properties of the City of Long Beach and its field contractor, THUMS Long Beach Company, in association with the United States Department of Energy. The purpose of the pilot demonstration is to evaluate the efficiency of the caustic displacement mechanism in the environment of a stratified, heterogeneous, high oil viscosity reservoir where primary waterflood recovery is relatively poor. The test area is located in the Ranger Zone of Fault Block VII within the Wilmington Field. The pilot test involves the injection of caustic solution into a modified staggered line drive well pattern consisting of eight injection wells which surround eleven active producers in an area of approximately ninety-three acres.

Laboratory investigations conducted jointly by THUMS and the Department of Oil Properties indicated that Ranger Zone crude could be readily emulsified in the presence of water containing as low as 0.1% by weight sodium hydroxide. Additional oil was recovered in core floods when 1.0 weight percent sodium chloride was added to the alkaline solution.

The results of the laboratory core test work and tests of the reaction between alkaline solutions and reservoir sands used in reservoir simulations indicated oil rate response and total incremental oil recovery are very dependent upon the caustic concentration and caustic slug size. Alkaline consumption calculated to be very large.

---

1. Now with Xtra Energy Corporation, Signal Hill, California.

This paper summarizes the results of the caustic core floods which were performed to evaluate the entrainment mechanism of oil displacement and laboratory tests to evaluate the long term consumption of hydroxide ions by the reservoir sands. The past performance of the field and the reservoir simulation history-match of that past performance are discussed. The predicted future performance of the field for both continued waterflooding and a caustic flood is summarized.

Alkaline facilities were completed and placed in operation on March 27, 1980. Pre-flush injection consisted of 11.5 million barrels of softened fresh water with an average of 0.96 weight percent salt. The pre-flush amounted to approximately 10 pore volume percent. Alkaline solution containing 0.4 weight percent sodium orthosilicate and 1.0 weight percent salt in softened water is being injected.

INTRODUCTION

The Wilmington Field is the largest field in California, Fig. 1. It has seven basic reservoir zones with crudes that generally have a relatively low gravity, high viscosity and high organic acid content. The recovery efficiency for the waterflood in the Ranger Zone of the Wilmington Field has been low due primarily to a highly unfavorable mobility ratio between water and oil and significant reservoir stratification.

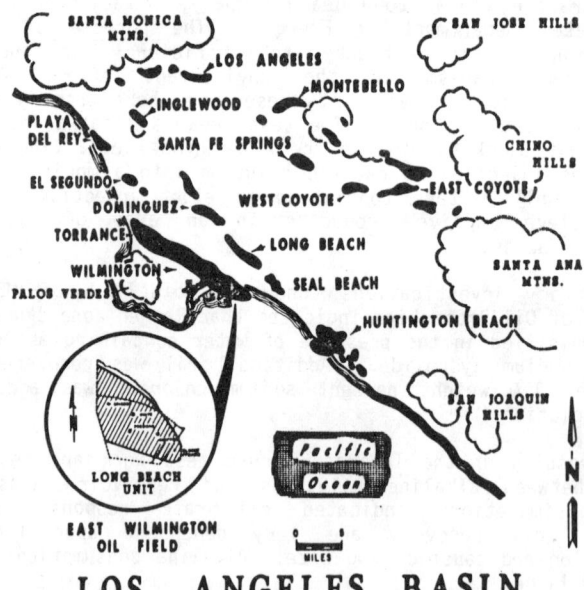

LOS ANGELES BASIN

FIGURE 1 - FIELD LOCATION MAP

The concept of activating the natural surfactants present in the crude oil by contact with alkaline water, although limited to reservoirs with suitable crude oils, has potential economic advantages over commercial surfactant flooding owing to the high cost of the surfactants and the low cost of alkaline materials. Several mechanisms have been postulated for the improved oil recovery resulting from alkaline waterflooding. Included among these are emulsification and entrainment, wettability reversal, and emulsification and entrapment[1] The relationship between these possible mechanisms is necessarily more complicated in caustic waterflooding than surfactant injection due to the complexity of the alkali-crude oil reaction which would take place in the reservoir.

LABORATORY STUDIES

Laboratory investigations have been performed for this alkaline pilot project to provide comparison core flood tests between waterflood and alkaline flood recovery and define the extent of caustic consumption by the reservoir rock. The comparative core floods were performed with preserved core material which was cut parallel to the core axis. The plugs measured approximately two inches in diameter by five inches long. The long term alkaline comsumption tests were performed with sand packs which were prepared in Lucite columns and varied in length from six to twelve inches with a diameter of approximately one and a half inches.

Comparative Core Flood Studies

Frozen preserved core samples were jacketed on coring in plastic tubing. In the laboratory the plugs were placed in a modified Hassler sleeve apparatus, thawed and confined at 1600 psi overburden pressure. The cores were heated to reservoir temperature of 125°F and dynamically driven to minimum water saturation using Ranger Zone crude oil. The samples were then water driven at a rate varying from six feet per day prior to water breakthrough to one foot per day after water breakthrough (the reduction was done to avoid excessive pressure gradients within the core). The cores were waterflooded to residual oil saturation (see Fig. 2). Then the cores were again dynamically driven to minimum water saturation using crude oil. The enhanced or alkaline water drive tests were then performed. These tests consisted of the following steps:

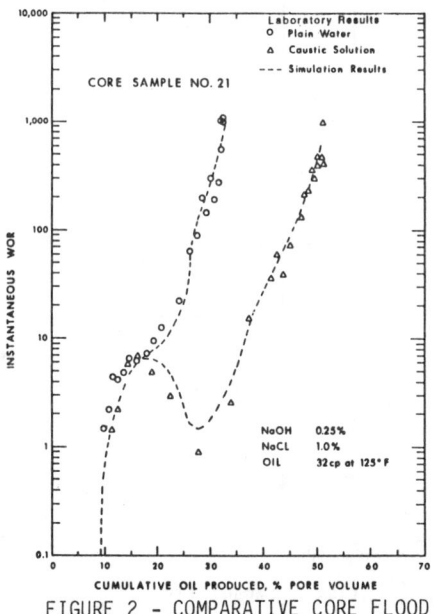

FIGURE 2 - COMPARATIVE CORE FLOOD

1. Inject water to breakthrough or a pre-determined water/oil ratio;
2. Inject a pre-flush containing 1% sodium chloride brine in softened water; and
3. Follow with an alkaline-soft water solution containing sodium chloride at a concentration of 1%. (Different alkaline concentrations were used in the various tests performed.)

Figure 2 shows a typical response to this type entrainment mechanism alkaline waterflooding. (Sixty-two comparative core flood tests were performed.) On the average, improvement in oil recovery was approximately 10 pore volume percent. No strong correlation was found between improvement in oil recovery and the concentration of alkali injected. Therefore, all of the core flood tests were combined and analyzed to obtain a more statistically meaningful average core response to alkaline flooding. These tests were used to obtain relative permeability to oil and water for both the waterflood and the alkaline flood performance (Fig. 3). These relative permeability phenomena were used in reservoir simulation matches of individual core

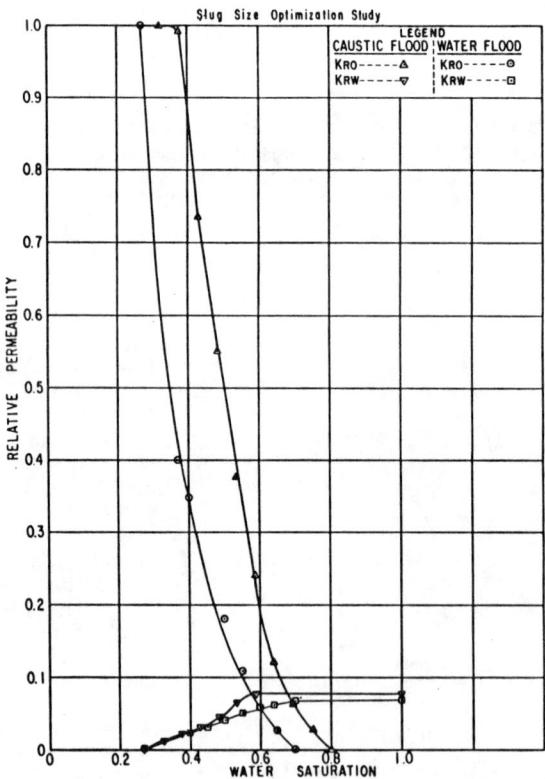

FIGURE 3 - OIL/WATER RELATIVE PERMEABILITY

tests. They then were scaled for the two-dimensional slug size optimization cases and for a three-dimensional model of the pilot area for the caustic flood prediction cases.

Long Term Alkaline Consumption In Reservoir Sands

The effect of alkaline consumption is a critical economic consideration. As a result, studies were undertaken in an attempt to define the magnitude of the caustic consumption which can be expected to occur in an alkaline flood in the Ranger Zone of Fault Block VII. The tests conducted included static equilibrium tests, reversible adsorption chemical consumption tests, sensitivity of caustic consumption to flow rate, and long-term flow tests. In the latter type tests sand packs were prepared, and following waterflooding to breakthrough, soft water-alkali solutions with 1% sodium chloride were injected into the sand packs for periods ranging between 30 and 104 days. Static periods of varying length followed after which the alkaline injection was resumed. The outlet caustic concentrations were measured daily during the entire flow test. It was evident from the tests that the consumption of alkaline material is a long term phenomenon. The number of pore volumes of injection required for concentration of output solution to reach the concentration of the injected solution ranged upward to 38 pore volumes. Reducing the flow rate of the injection increased the number of pore volumes required to reach an effluent concentration nearly equal to the inlet concentration. The upper curve of Fig. 4 shows typical results for long term alkaline consumption results where the amount of consumption, in terms of mass per unit volume is plotted versus the concentration residence time product. The consumption laboratory work is described in more detail in the "Fourth Annual Report" of this project prepared for the U. S. Department of Energy.[2]

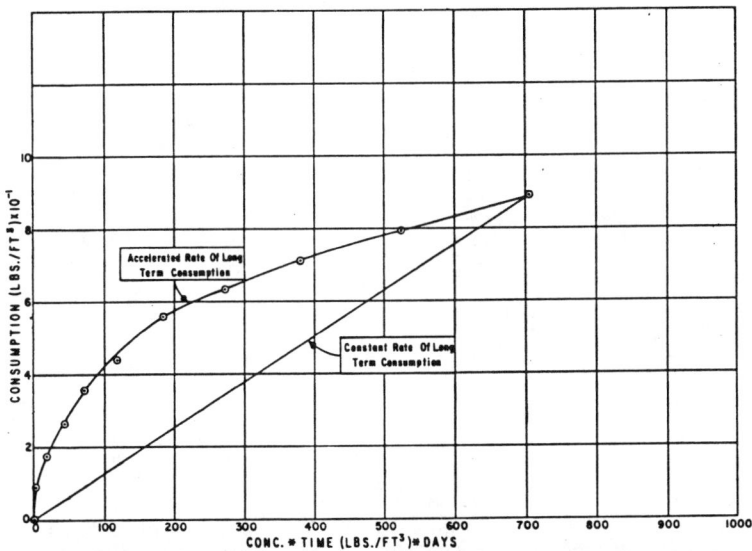

FIGURE 4 - LONG TERM ALKALINE COMSUMPTION RELATIONSHIPS

RESERVOIR DESCRIPTION

The Wilmington field is located in the south-western portion of Los Angeles, California as shown in Fig. 1. It is the largest field in California and one of the major fields in North America. Cumulative oil production to date is in excess of 1 billion barrels.

The field is an asymmetrical anticline with a north-west south-east axis broken by a series of transverse normal faults. The faults divide the reservoir into pools and have proven to be effective barriers to fluid and pressure communication. Dips range from a maximum of 20° on the northern flank to approximately 60° on the southern flank. The entire structure is eleven miles long and three miles wide underlying approximately 13,000 acres. Producing zones in the Field (Tar, Ranger, Upper Terminal, Lower Terminal, Union Pacific, Ford, and 237) lie between the depths of 2,000 and 7,000 feet subsea and range in age from late Miocene to Pliocene. The upper four zones containing low gravity, high viscosity crude are the major oil reservoirs. The reservoir rock in all zones is sandstone with different degrees of consolidation and varied silt and clay content. The pilot pattern area is in the eastern portion of the Long Beach Unit of Wilmington Field between the Junipero and Temple Avenue Faults in the Ranger zone and is shown in Fig. 5.

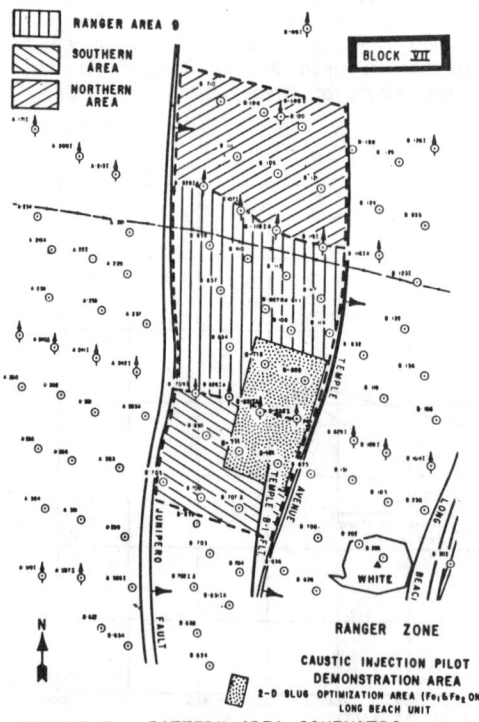

FIGURE 5 - PATTERN AREA SCHEMATIC

The modified line drive configuration of the pattern represents a typical waterflood well pattern for the Ranger Zone of the Long Beach Unit. Ranger is the largest and most prolific of the Unit's reservoirs. It consists of several distinct intervals or subzones separated by impermeable shale sections (see Fig. 6). Each subzone is an integrated sequence of shales and unconsolidated to semi-consolidated, poorly sorted, medium-to-fine grained sands. These six subzones lie at depths of 2,600 to 3,400 feet with a net thickness of 305 feet. The properties of each zone are summarized in Table 1. Productive subzones underlying the pilot area contain crudes with a wide range of

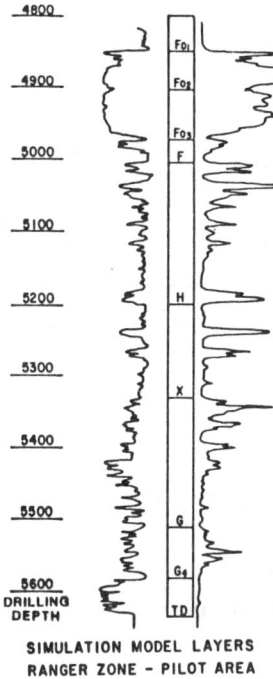

SIMULATION MODEL LAYERS
RANGER ZONE - PILOT AREA

FIGURE 6 - TYPE LOG

TABLE 1 - RESERVOIR CHARACTERISTICS BY ZONE

Ranger Subzone	Porosity* (fraction)	Perm* (md)	Net Pay (feet)	Net Volume (Acre Feet)
F_O	.260	270	109	9347
F	.274	321	52	4578
H	.246	179	41	3647
X	.265	173	48	4339
G	.289	220	29	2655
G4	.270	131	26	2331

* Based on 1600 PSI Confining Pressure Core Analysis Data and Special Logging Programs.

physical properties. The general characteristics of these properties are: an oil gravity range of 14°-27° API; the oil gravity within a subzone depends upon structural position with the higher subzone gravity at the high structure positions and low gravities at the lower structure positions. From subzone to subzone the oil gravity depends upon geological age with the lower (older) subzones containing higher gravity crude and the upper subzones containing the lower gravity crude.

PRODUCTION HISTORY

Initial development in the pilot area began in August 1967. At the time of the development, pressure gradients existed across the pattern with the average reservoir pressure being approximately 85% of hydrostatic. This phenomenon is due to communication between the pilot and older producing areas in the vicinity. Waterflooding operations began concurrently with development. The modified three producing row line drive flood pattern was aided by peripheral aquifer injection. The initial development of the pattern was completed in early 1975. Oil production for the pilot pattern area as of September 30, 1980 was 11,490,000 STB. Cumulative water injection by the eight surrounding injection wells was 55,000,000 STB. Of this amount, the pilot area had produced 38,000,000 STB of water. (Performance of the confined pattern is shown in Fig. 7.)

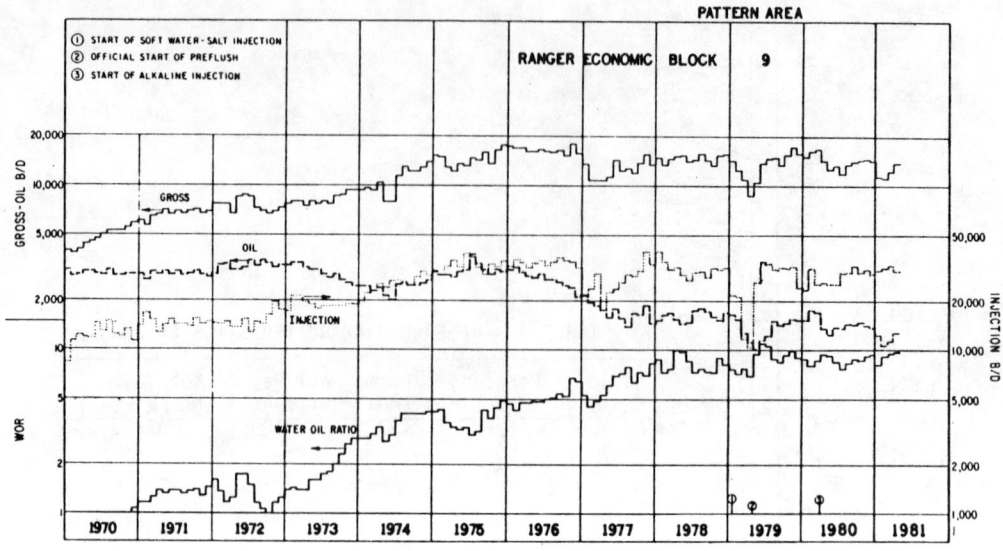

FIGURE 7 - PATTERN AREA OF PERFORMANCE HISTORY

RESERVOIR SIMULATION MODELING

The formulation of the caustic simulation model has been reported in an earlier paper by Breit, et al[3]. For enhanced waterflooding, the simulator accounts for the injection and production of up to six different active agents in an aqueous phase. Any or all of these agents may be caustic or polymer type fluids or a combination of these types of fluids. The primary displacement effects of a caustic fluid are represented by changes in relative permeabilities to oil and water. This simplified approach permits the modeling of enhanced recovery projects without the necessity of determining the exact mechanisms of the displacement in minute detail. The model also accounts for the consumption of active material within a caustic slug by three different mechanisms: The interaction between a caustic slug and formation water to form precipitates of divalent cations, the instantaneous or equilibrium adsorption of caustic solution, and the long term kinetically controlled interaction between caustic and the rock matrix itself. The permeability changes resulting from the divalent ion precipitation were not considered in the simulator work for the Range VII pilot.

The simulation work was based on the use of the relative permeability curves produced in the laboratory experiments shown in Fig. 2, an instantaneous caustic consumption of 0.42 pounds and a maximum long term consumption of 0.84 pounds per cubic foot. The match of the laboratory response for core experiment No. 21 using these parameters is shown in Fig. 2.

Caustic Flood Optimization Study

A small area of the F_0 subzone, Fig. 5, including two injectors was selected for slug optimization studies using the N-HANCE reservoir simulation model. Injection was scaled from the planned field injection rate of 34,000 STB per day. The reservoir and waterflood characteristics of this area are summarized in Table 2.

TABLE 2 - ALKALINE SLUG OPTIMIZATION STUDIES
BASE DATA FOR STUDY AREA

Reservoir Pore Volume of Study Area	1,879,000 RB
Initial Oil in Place (TOIP)	1,231,000 STB
Waterflood Oil Recovery:	
To 5-01-79	562,000 STB
% TOIP	45.65
@ PV	29.91
Waterflood Oil Recovery 5-01-79	
To economic limit of 150 WOR (11-01-85)	84,000 STB
Cumulative Waterflood Oil Recovery	646,800 STB
% TOIP	52.54
% PV	34.42
Injection Rate 5-01-79 on to End	4,806 B/D
(7.5 PV%/Yr.±)	
Preflush Injection - 1.0% Salt in	
Softened Water Solution (390 Days)	1,874,340 STB
% PV	99.75
Start of Alkaline Injection 5-24-90	

Performance of this area of the field and simulator was characterized by rapid water breakthrough followed by a gradual rise in WOR to its current average value of 50. Injection surveys have confirmed that over half the injection water in these two injectors has entered the F_0 subzone leaving it at a much higher average water saturation than lower zones. Results of the optimization study runs are summarized in Table 3. In all but one case, discussed subsequently, the relative permeability adjustment was made linearly between alkaline and waterflood behavior depending on the active alkaline concentration in each cell. The low alkaline concentration of 0.4 weight percent in the largest pore volume slug, 60%, produced the greatest amount of incremental oil. However, this increase in production tended to be at low rates, continuing on to late in the life of the producers being modeled. In contrast, the higher concentration, smaller slug volume cases produced a more rapid oil rate response as can be seen in Fig. 8. This figure also illustrates the effect of long term caustic consumption on the projected results.

As can be seen by the results of the three 0.8 weight percent alkali cases, the incremental oil recovery increases approximately 50% when no long term consumption is assumed to be present. In addition, the oil rate reaches a maximum value approximately 15% higher in the absence of long term consumption.

The relative success of an alkaline flood will be more dependent on the oil recovery at wells far removed from the injection rows than of the wells directly adjacent to the injectors, because those areas

TABLE 3 - ALKALINE SLUG OPTIMIZATION STUDIES OIL RECOVERY DATA

Alkaline Slug Description	Long Term Consumption	Oil Recovery MST Bbl.	Economic Limit (150 WOR) Date	Incremental Oil Recovery Above Water Flood		
				M Bbl.	% TOIP	% PV
Waterflood	-	84.8	11-01-85	-	-	-
Pro-Rated Alkaline Relative Permeability Adjustment						
60 PV % 0.4% Alkali	Accelerated Rate	174.0	7-03-89	89.2	7.25	4.75
32 PV % 0.8% Alkali	Accelerated Rate	137.9	9-01-86	53.1	4.31	2.83
32 PV % 0.8% Alkali	Constant Rate	143.5	6-05-86	58.7	4.77	3.12
32 PV % 0.8% Alkali	None	159.8	7-01-86	75.0	6.09	3.99
24 PV % 1.0% Alkali	Accelerated Rate	122.1	12-06-85	37.3	3.03	1.99
24 PV % 1.0% Alkali	None	142.3	10-21-85	57.5	4.67	3.05
48.6 PV % Variable*	Accelerated Rate	162.0	2-16-88	77.5	6.30	4.12
Minimum Threshold Alkaline Relative Permeability Adjustment						
60PV % 0.4% Alkali	Accelerated Rate	226.0	5-28-90	141.2	11.47	7.51

* 0.4% orthosilicate for first 1.5 years, 1.0% for next 1.0 year, 0.8% for 1.0 year, 0.4% for 1.0 year and 0.2% for 2.0 years.

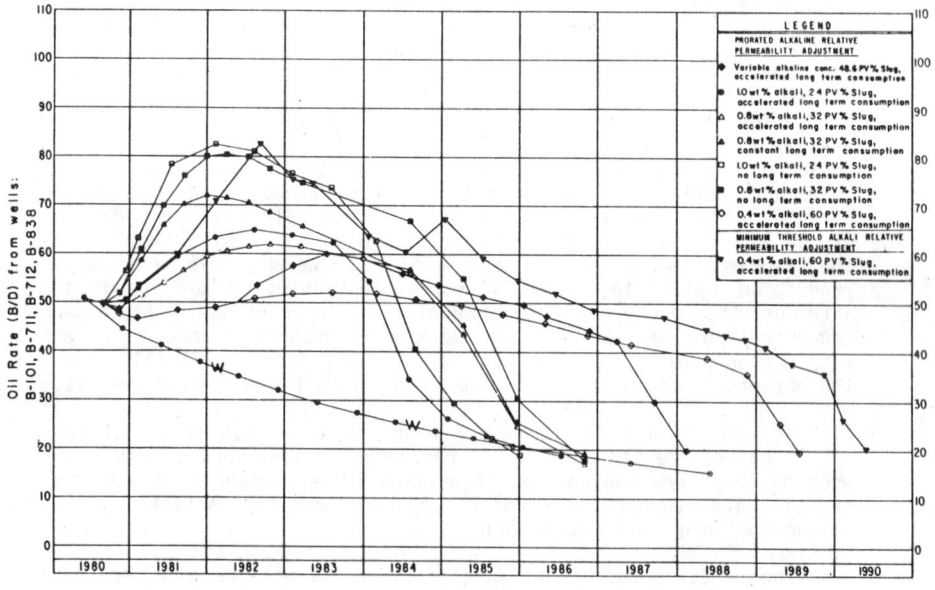

FIGURE 8 - OIL PRODUCTION RATES SLUG SIZE OPTIMIZATION

near the injectors have been more completely swept. The amount of caustic that can be transmitted through these closer areas without being consumed is of considerable interest. The lower the long term consumption the greater is the transmission of the caustic through the area mear the injectors and out into the rest of the reservoir, (Fig. 9). This long term consumption is .dependent upon both the concentration of the caustic in an area and its residence time. Increased transmission rates could also be expected for constant concentration slug injections at accelerated injection rates.

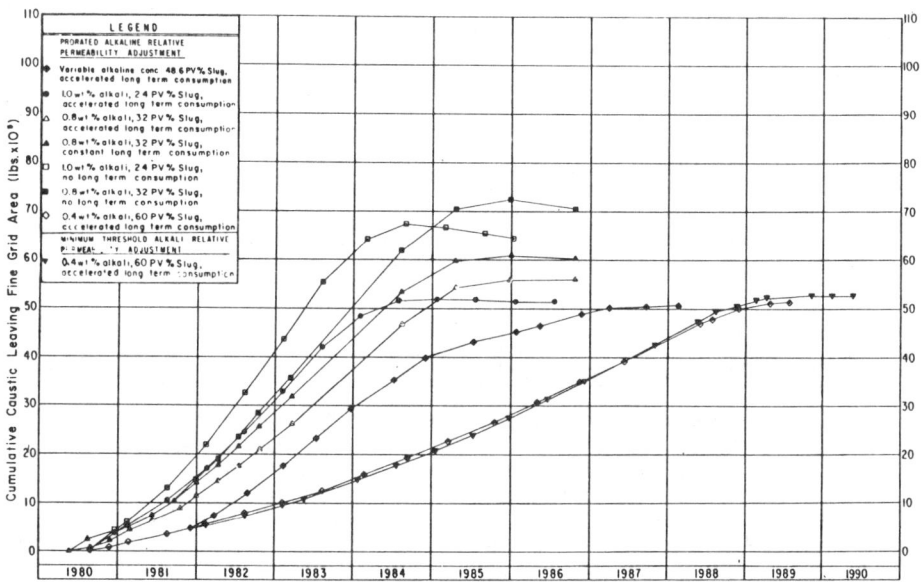

FIGURE 9 - ACTIVE ALKALI TRANSMITTED THROUGH OPTIMIZATION STUDY AREA

Two additional variations from the constant concentration cases already discussed were run. First, a variable concentration case was run in which the alkaline content was varied from the 0.4 weight percent currently being injected in the field to a maximum of 1.0 weight percent and then tapered back to a concentration of 0.2 weight percent. Total alkali injected was the same as in the prior runs however. Although this run did show some minor acceleration in the oil rate response, the cumulative production and the caustic moving outside the area were both disappointing in comparison to the constant concentration injection cases.

Similarly, another run was made in which the change to the enhanced recovery relative permeability curves within the model was made at a minimum threshold alkali concentration which corresponded to the decrease in interfacial tension from the laboratory experiments. This case did show a considerably greater oil production and a signficantly higher oil production rate than the linear shifting from normal water/oil relative permeability curves to caustic relative permeability curves. Currently, we are unable to determine which of

234

these two relative-permeability shift techniques more accurately represents true reservoir phenomena. Owing to the fact that the greatest cumulative oil production was achieved for the continuous injection of 0.4 weight percent alkali and the equipment limitations on injection concentration and its rate in the field, the 0.4 weight percent constant concentration cases were selected for prediction of the performance of the entire pilot area.

Performance Match of the Pilot Area

The historical performance of the pattern area was matched using a black oil reservoir simulation model containing 1770 grid cells in seven layers, two within the F_O subzone, and one layer in each of the remaining subzones. The model included areas to both the north and south of the pattern, Fig.5.

The injection into each subzone was specified on the basis of surveys of the injection wells. The performance shown in Fig. 7 was matched by controlling the amount of fluid which migrated off the pattern area to the north and south, and by minimal changes in the truncation of relative permeability curves between the pattern and the areas to the north and south.

Predicted Performance of the Pilot Area

Oil recovery prediction cases were run for continued waterflood operation and for the two caustic flood cases. One of these used the prorated alkaline flood relative permeability adjustment, Case I, and the other the minimum threshold alkaline relative permeability adjustment, Case II, outlined in an earlier section. The results of these predictions are summarized in Table 4 and Fig. 10. The

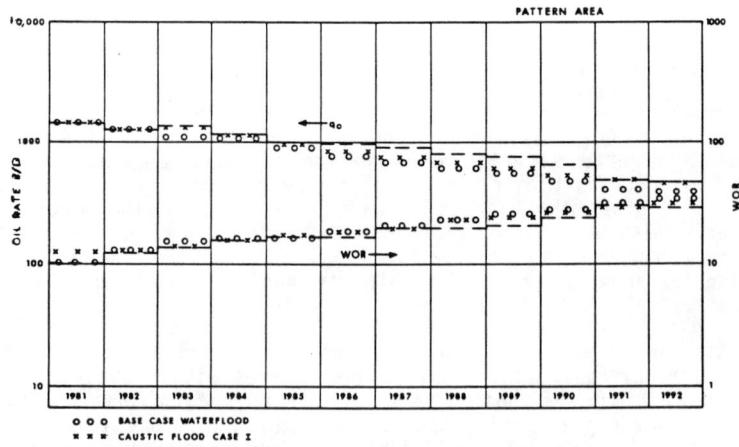

FIGURE 10 — OIL PRODUCTION RATES, WATERFLOOD AND CAUSTIC FLOOD CASE I AND CASE II

TABLE 4 - PATTERN AREA PREDICTED PERFORMANCE THROUGH DECEMBER, 1994

Subzone	Waterflood Cum Oil MSTB	Alkali Case I Cum Oil MSTB	Alkali Case II Cum Oil MSTB
F_0	4390	4497	4758
F	2871	2990	3048
H	1742	1751	1762
X	3560	3580	3622
G	1787	1791	1795
G4	699	706	715
TOTAL	15049	15315	15700

predicted performance of the alkaline flood in comparison to continued waterflood injection is disappointing. The runs indicate that those wells in the first row of producers away from the injection wells do respond to caustic injection. However, these were well swept by the injection water prior to caustic injection leaving less oil to respond to injection of caustic. Additionally, throughout the history of the field, the F_0 subzone has taken over half of the injected water. In the prediction cases half the caustic water continued to flow into this F_0 subzone. Late in the life of the field, significant parts of the oil production from the wells within the pattern is occurring from zones lower than the F_0 subzone. As a consequence, the increase or incremental production in the F_0 subzone is overwhelmed or masked by the production from lower zones.

CONCLUSIONS

The field performance to date and the predicted performance from the simulation studies indicate that there are many complicating factors to the successful application of alkaline flooding in a heterogeneous reservoir. The laboratory core tests have confirmed the applicability of alkaline flooding to Ranger Zone crude and reservoir rock. However, the high degree of consumption indicated by laboratory work and simulation results can be a controlling factor in the success of any caustic injection project. Actual field results are needed to calibrate the consumption parameters to field conditions.

Minimization of this consumption appears possible by injecting the alkaline solution at a higher concentration and/or injecting at a higher rate to minimize the residence time in the reservoir. Furthermore, the simulation experience has indicated the desirability of control of the placement of injection fluid by subzone (vertically) for a more efficient alkaline injection project.

REFERENCES

1. Johnson, C.E.: "Status of Caustic and Emulsion Methods". J. Pet. Tech. (January, 1976) 85-94.

2. "Caustic Waterflooding Demonstration Project, Ranger Zone, Long Beach Unit, Wilmington Field, California. Annual Report for the Period June 1979 - May, 1980." Report SAN/12047-4. Prepared for DOE by the City of Long Beach, Department of Oil Properties and THUMS Long Beach Company under Contract No. DE-AC-03-76ET-12047.

3. Breit, V.S., Mayer, E.H., and Carmichael, J.D.: "An Easily Applied Black Oil Model of Caustic Waterflooding". SPE Paper No. 7999, Presented at the 1979 California Regional Meeting of SPE of AIME, Ventura, California, April 18-20, 1979.

MISCIBLE DISPLACEMENT: ITS STATUS AND POTENTIAL FOR ENHANCED OIL RECOVERY

R. J. BLACKWELL

Exxon Production Research Company

Miscible flooding continues to be one of the most intriguing enhanced oil recovery methods because of its potential for recovering all of the oil flushed by solvent; and one of the most exasperating, because only in rare instances have actual field performances come anywhere close to the high recovery efficiencies potentially possible from this process.

History

The concept of miscible flooding is quite old. Its potential was generally recognized by the petroleum industry well over 50 years ago and several papers were published in the 1920's describing early research in this area. During the 1930's and early 1940's, interest in enhanced recovery techniques was low; however, following the end of World War II, there was a dramatic increase in research directed toward improving our knowledge of what might be called the "physics and chemistry of fluid flow in porous media" and toward the development of the three basic areas of enhanced oil recovery--thermal, chemical, and miscible. Investigations into the use of miscible flooding techniques to improve oil recovery was a significant part of this increased effort.

It was an exciting era. Laboratory tests were conducted to determine which fluids could be used for miscible flooding. Almost every available fluid including alcohols, ketones, propane, butane, LPG, nitrogen, carbon dioxide, methane and mixtures of many of the above were tested. Some of the first research was on completely miscible[1,2] (frequently called first-contact miscible) systems in which all mixtures of the solvent and oil form a single phase fluid. However, two multiple-contact methods of achieving miscible displacements, the high pressure or vaporizing gas method[3] and the enriched gas process,[4] were also developed during the 1950's. Both of the latter involve injection of a fluid which is initially not miscible with the crude, but is able to generate a solvent bank within the porous medium during the displacement process. In the high-pressure gas process, the injected gas is enriched with intermediate and higher molecular weight components vaporized from the first crude contacted. If the phase behavior of the gas oil system is favorable, a self sustaining solvent bank is formed le ing a small volume of denuded crude as residual in the reservoir. In the enriched-gas process, the enriching components in the injected gas transfer to the crude oil and generate a solvent bank consisting of a modified crude oil. In this process, essentially all of the oil (possibly excluding some asphaltenes) is flushed from the region contacted.

Following the discovery of these three basic approaches and the development of methods for determining the conditions that each must meet in order to have a miscible displacement, attention was turned (about 1954) to determining the conditions required for effective use of each method in field applications. Initially, the primary objective was to design miscible floods using the smallest amount of LPG or enriched gas possible. In order to do this, measurements of the amount of mixing that occurs as fluids flow through porous media were needed and several companies including Exxon initiated work in this area. At about the same time, we and other research laboratories began our first experimental and theoretical studies of viscous fingering. It rapidly became apparent that, although the effects of mixing must be considered in any attempt to predict miscible flood performance, it was viscous fingering or solvent channeling that would likely dominate the behavior of a miscible flood.

Viscous fingering was studied in long models, short models, narrow models and wide models. Tests were conducted in Hele-Shaw models, in homogeneous sand or glass bead packs, and in models containing various permeability heterogeneities. In order to establish the physical principles responsible for viscous fingering and to quantify its effects, a number of floods were run in each model using fluid systems with different oil/solvent viscosity ratios, viscosity levels and fluid densities. At the beginning of this period, some people hoped that viscous fingering would turn out to be a "laboratory artifact" in the sense that it would be less of a problem in the field than it was in laboratory floods. However as confirmed later in many field tests, solvent channeling was a serious problem in field applications and at least as detrimental there as it was in laboratory floods.

Nevertheless, there was a general air of optimism during the middle 50's. Many believed that the remaining problems (such as viscous fingering) would soon be solved and that miscible flooding would usher in a new era of high enhanced oil recoveries. Because of this optimism, a number of field tests were initiated. However, incremental oil recoveries of only 5 to 10% OOIP were obtained in many of these tests. These incremental recoveries not only fell far short of original expectations but were far from being economically attractive. The earlier optimism turned suddenly into pessimism.

The principal reason for the poorer than anticipated recovery efficiencies was severe channeling of the solvent banks. In some instances, the dominant cause of this channeling was reservoir heterogeneities such as permeability variations in different strata, fractures, etc. However, viscous fingering and gravity overriding were invariably major factors--either causing solvent channeling or aggravating the channeling associated with reservoir heterogeneity.

It is perhaps worthwhile to point out that viscous fingering and the closely related gravity override phenomenum had been recognized sometime earlier. Hill[5] in 1952 and Dietz[6] in 1953 discussed gravity stabilization of displacement fronts and established the critical rate concept for the control of viscous fingering by gravity segregation. Unfortunately, practical production rates can be achieved in only a limited number of reservoirs without exceeding the critical rate; hence the incentive for developing other methods of controlling solvent channeling was and remains quite high.

Several papers and patents were published during the late 50's and early 60's describing methods that might increase the reservoir volume swept by the solvent bank. One of the methods was gas-water injection which later became known as the WAG process. Another was the use of foaming agents. Several other methods, including the use of polymers, were also investigated. Unfortunately, none appeared particularly attractive at that time; and with the growing disenchantment with miscible flooding, research in mobility control methods dropped to a low level. By default, the WAG approach became the generally "accepted" method for mobility control, but one with obvious deficiencies. In the middle 1970's, several laboratories renewed their research activity in this area and several new patents and papers describing the addition of various surfactants and other modifications to the WAG process have appeared recently. I regret to say however, that in my opinion, no dramatic breakthroughs have occurred to date and our ability to control solvent channeling has not changed much since 1960.

Nevertheless, miscible flooding technology has advanced significantly during the past two decades. Results from both field applications and laboratory studies have provided additional insight into the dynamics of miscible displacement processes, particularly in the area of CO_2 - miscible flooding. Numerical methods for simulating process performance have been developed along with better techniques for arriving at the critically important reservoir descriptions used for analysis of field results and in making predictions of the performance and economic viability of a miscible flood in a specific field.

Field Applications

Miscible gas projects have provided the industry with valuable field data and operating experience during the past 15 years, particularly in the use of CO_2. Although detailed flood performance information is available from only a few of these projects, the results generally lead one to the same conclusions reached in 1950's--that is, high displacement efficiencies can be achieved in the regions flushed by solvent, but high volumetric sweep efficiencies are possible only if solvent channeling can be controlled effectively.

The early breakthrough of CO_2 in the largest miscible-CO_2 flood in the United States, the 30,000 acre Sacroc project in the Kelly-Snyder field of West Texas, provides a dramatic illustration of the problem. CO_2 injection was initiated in this project in January 1972. Breakthrough occurred in June of the same year after injection of less than 2% HCPV of CO_2 and increased rapidly. In November 1972, it became necessary to curtail CO_2 injection when CO_2 production exceeded the capacity of the existing gas plants to extract the CO_2 from produced gas. A paper by Kane[8] describes in detail efforts to maintain control of CO_2 production. Two important steps were taken. First, the WAG ratio was increased from its initial value of about 0.5:1 to 3:1, and then second, a zonal injection program was initiated to provide an improved distribution of the gas and water into all zones and thereby improve the overall sweep efficiency. CO_2 channeling and production continued to be an exasperating but manageable problem throughout the flood. Nevertheless, the extra investment and operating costs involved in recovery, purifying, and reinjecting the produced CO_2 were significant factors in the 1977 decision to reduce the volume of CO_2 injected from 20% HCPV as planned originally to about 12% HCPV. The corresponding reduction in the estimated incremental oil recovery was from 107 million STB (8.1%OOIP) to 88 million STB (or 6.7%OOIP).

In general, the performance of Sacroc and other field experience obtained
to date suggests that incremental oil recoveries (over that possible by water
flooding) will most often fall in the range of only 6 to 10 percent of the
original-oil-place (OOIP), less frequently in the 10 to 15% range, and will
rarely exceed 15% OOIP.

Current Laboratory Research

At the present time, there are a number of industrial and university
laboratories actively engaged in miscible displacement research. In recent
years these research efforts have emphasized work on CO_2 flooding, although
there has been a limited amount of work on the use of other gases such as
nitrogen, flue gas and CO_2 enriched with intermediate hydrocarbons such as
propane, butane, etc. As mentioned earlier, several laboratories are investigating different methods for improving volumetric sweep.

Several laboratories are engaged in fundamental studies of phase behavior and
CO_2 flood performance often including measurements of the composition,
density and viscosity of individual equilibrium phases of fluids produced
during laboratory floods in slim tubes, or through reservoir cores. Equilibrated samples taken from PVT cells are also being analyzed and compared
with the slim tube results. A recent U. S. Department of Energy report by Orr
et al is an excellent example of this type of study. The report describes
their comprehensive study of the complex phase behavior of a particular CO_2-
crude oil system in some detail and carefully delineates the reservoir conditions under which liquid-liquid, liquid-liquid-vapor and liquid-vapor equilibrium mixtures were observed. It also describes their chromatographic
analysis procedure which permits the characterization of the hydrocarbons
present in the various fluid combinations described above throughout the C_{11} -
C_{36} range (as well as the usual C_1 - C_{10} range).

Similar studies are being carried out in other laboratories using other CO_2 -
crude oil systems. Nevertheless, there is a need for additional carefully
scaled experimental and theoretical studies of the interactions of phase
behavior, viscous fingering, gravity segregation, rock lithology and heterogeneity, and the relationship between oil remobilization and rock wettability.

Research efforts to develop better mobility control techniques were mentioned
earlier. The use of foams is again being investigated and some encouraging
laboratory and field test results have recently been reported. Even though
I remain skeptical that foam injection, as such, is the solution to the problem
of viscous fingering, I feel additional research is merited. In recent years,
the greatest strides in micellar-polymer technology have been made as the
result of fundamental studies of the basic mechanisms involved. Similar
comprehensive studies are needed using several CO_2, crude oil, brine and
classes of surfactants at pressures above and below the minimum miscibility
pressure and temperatures spanning the range from 20° to $100^{\circ}C$.

Mathematical Simulation

Although worthwhile improvements in our numerical technique for simulating
miscible floods have occurred during the past 15 years, further improvements
are greatly needed. The ideal computer program for modelling miscible CO_2
displacements, for example, must simulate the generation of the miscible
solvent bank, the potential precipitation of a solid (asphaltene) phase, and
predict the amount and compositions of the various phases present in every

grid block each time step. Any grid block may contain a number of mobile and immobile immiscible hydrocarbon liquid phases (each containing over 30 components), and carbonated water. The simulator must be able to model block-to-block flow of both miscible and immiscible phases, and correctly model the dispersion or mixing of the miscible components transfer of components between immiscible phases.

Compositional simulators with sophisticated phase behavior packages and other simulators with specialized capabilities have been developed to model miscible gas processes. Current compositional simulators can model most of the physical and chemical phenomena involved in a miscible flood; unfortunately, none provide all of the features that one might desire. Numerical dispersion remains a serious problem for the grid block sizes typically required in most reservoir studies; viscous fingering is difficult to model and is usually approximated empirically using a mixing parameter model; and computing costs are normally high because of the overall complexity of the simulator. Consequently, fully compositional models are frequently not as practical for field wide reservoir engineering studies as they are for special studies such as the simulation of the flood performance of a cross section or a small reservoir pattern area. A particularly important application is their use in conjunction with laboratory tests. This type of application is not only a good way to test the capability of the computer program, but it is also a good way to test our understanding of the chemical and physical processes involved in a miscible gas flood.

Greatly simplified compositional simulators are frequently used for reservoir performance predictions, comparison of different gas injection programs, etc. These simplified simulators normally use a limited number of gas and pseudo components to represent the injected gas, and the natural gas and crude oil (including the asphaltenes).

The number and composition of the required pseudo-oil components can be determined by comparing reservoir model results obtained by use of the simplified computer program with those obtained using a fully compositional simulator. Typical applications of the resulting simplified model include sensitivity studies of flood performance for various geological models of the reservoir, optimization of the WAG ratios, and large scale or field wide studies.

Other types of simulators, such as modified black oil simulators, are also frequently useful for specialized applications. Effective use of these specialized simulators requires that the user understand the limitations of the various simulators since interpretation of results is often complicated by the simplying assumptions used. A recent paper by Todd[18] includes comparisons of the advantages and disadvantages of the principal types of "miscible" simulators.

RESERVOIR DESCRIPTION

The need for a reliable description of reservoir geology and other reservoir engineering data can hardly be overemphasized. No matter how well we know the chemistry and physics involved in a miscible displacement, nor how precisely we are able to model these phenomena mathematically, it is not possible to make useful reservoir performance predictions of miscible processes without having a reliable reservoir description. It must be recognized that a much better reservoir description is required for predicting miscible flood performance than is normally required for a comparable study of a water flood in the same field. Surprisingly small changes in the reservoir description can lead to significant differences in prediction of miscible flood performance and project economics; whereas, these same changes in reservoir description may be unimportant when predicting the performance and economics of a waterflood.

In the past, the amount, type, and quality of routinely available reservoir description data have been dictated primarily by reservoir engineering needs for conventional primary and secondary recovery processes. Many of the same types of data are needed for predicting miscible flood performance.

Useful geological input includes the depositional environment of the reservoir. Depositional environment data and information on subsequent diagenetic changes of rock matrix can be particularly valuable in predicting continuity of permeable zones, shale deposits or tight streaks, and the frequency and distribution of the openings (or windows) through these impermeable layers. Acquisition of this additional reservoir description data can be both difficult and costly, but its acquisition and careful interpretation is absolutely necessary.

EOR Potential For Miscible Processes

United States: During the past decade, there have been numerous studies of the future potential of miscible gas processes and other enhanced recovery processes in the United States. Since the basic displacement mechanisms and phase behavior concepts used to predict miscible flood performance are well known, one might assume that the incremental oil recovery from miscible flooding could be easily estimated and that the incremental oil volumes predicted by the various studies would be similar in magnitude with perhaps some differences in timing. However, this is not what one finds primarily because of the uncertainties in volumetric sweep caused by inadequate reservoir description data, in the estimates of the incremental oil recovery possible over waterflooding and in the economic assumptions used. Estimates of the incremental recovery over that possible from waterflooding, range from an "almost assured" 2 billion barrels* to "possibly optimistic" estimates of over 30 billion barrels. Our own estimates for the incremental reserves that can reasonably be added by the year 2000 fall into the 3 to 5 billion range. Hopefully, these estimates will turn out to be far too conservative.

Despite this apparent conversatism, I believe that the United States and possibly Canada will begin to see significant production from miscible gas processes during this decade. Miscible processes have the most potential of the various enhanced oil recovery processes for near-term production of light oil and could begin to make its contribution felt by the mid-1980's. But timing for this increased production will be critically dependent on near term investments and development of CO_2 supplies.

A recent study by Frost and Sullivan[11] includes a breakdown of their projections of expenditures for enhanced oil recovery in the United States during the 10-year period 1979-88. F&S predicts that expenditures for miscible gas processes will grow at a rate of about 25% per year from a level of about $0.7 billion per year in 1980 to $1.4 billion per year in 1984 and should reach a level of about $2.5 billion per year (of which $2.1 billion is for injected gases) in 1988. The total expenditure allotted to miscible gas processes during the

*2 x 10^9 barrels

10-year period was $13.75 billion or 36% of the $38 billion projected
for all EOR processes. These estimates include projections for oil field
equipment and services as well as the cost of the injected fluids.

Plans are nearing completion for three new pipelines which will bring over one
billion(10^9) scf/day of CO_2 to the Permian basin in West Texas from formations
in Colorado and New Mexico. Current plans call for completion of the first
two pipelines in early 1983.

If we assume that injection of 10 k scf of CO_2 will provide approximately one
barrel of enhanced oil recovery production, then CO_2 from these pipelines would
result in an oil production rate of 100 k B/D. This would more than double
current U. S. production (currently about 70 k B/D) from all miscible gas
projects.

Most of the near term activity will continue to be concentrated in West Texas,
but new miscible gas projects are also being considered for several other
regions of the U. S., including Louisiana and the Mid-continent area. Most
will employ CO_2 although some projects will use nitrogen or methane enriched
with LPG. The use of nitrogen will usually be limited to deep high temperature
reservoirs containing high gravity crudes because of miscibility pressure
restrictions. For example, nitrogen will be used by Exxon in the 15400 foot,
$285°F$ Jay and Black Jack Creek Fields in Florida. However, in some areas of
the U. S. (e.g. in offshore reservoirs) or in Canada, acquisition of adequate
supplies of CO_2 at a reasonable cost may not be possible and miscible hydrocarbon gases may be used despite their high cost. Production of 100 k B/D
is not anticipated outside the Permian Basin of West Texas, until the late
1980's or early 1990's.

Canada: In a recent (March 1980) study[12] of the potential of enhanced oil
recovery in Canada, it was estimated that miscible gas processes could increase
oil recovery by 1.885 billion barrels, 1.352 from hydrocarbon miscible and
0.533 from CO_2. This base case estimate was made for an assumed oil price of
$20 per barrel although higher prices ($25 and $100) were used in sensitivity
studies. The study utilized the tax and royalty regulations of the federal
government and the province of Alberta which were in place or announced in
1978. Consequently, the study will need to be updated when current negotiations
between the federal and provincial governments have been completed.

The study found that the base case estimate of 1.885 billion barrels is
extremely sensitive to small changes in the values assumed for recovery
efficiency, operating costs, etc. For example, a reduction of only 15% in
the assumed recovery efficiency reduced the estimated recovery to 0.476 billion
barrels. This reduction of almost 75% indicates that a significant volume of
marginally profitable (high risk) oil is included in the base case estimate.
Past experience dictates that without significant increases in oil price, very
few projects with marginal screening study economics remain as economically
attractive prospects after more detailed studies have been completed. One
reason is that early recovery estimates almost invariably drop as reservoir
geology becomes better defined.

Thus the potential for miscible flooding in Canada remains highly uncertain
but it appears likely that an incremental production of about 1 billion
barrels of oil could be achieved if current technical, economic, and political
problems can be resolved.

North Sea: In recent years, consideration of potential applications of EOR processes frequently starts when a field is still in the early stages of its productive life. Thus it is not surprising that the evaluations of the potential of various EOR methods in the North Sea have already progressed past the screening stage and more detailed engineering studies are currently in progress for a number of fields.

Although several reservoirs should be good miscible CO_2 flood candidates, the volume of CO_2 available is limited. Hence, the first miscible gas projects in the North Sea may, in fact, use hydrocarbon gases rather than CO_2.

Until the CO_2 supply problem is resolved, it is premature to estimate probable incremental recoveries that can be attributed to future use of miscible gas processes.

Other Areas: Although North America may have the largest number of existing or planned miscible-gas projects, the two largest miscible gas projects are in Libya and Algeria.[13] Both are hydrocarbon miscible. The Intisar D project in Libya was started in 1969 and has been producing at approximately 100 k B/D. The Hassi Messaoud project in Algeria was started in 1964 and has been producing at approximately 60 k B/D. There are several other small project in various parts of the world with a total production rate of perhaps 500-1000 B/D.

The ultimate potential for the use of EOR processes in the North Sea remains to be determined. However, I anticipate that several EOR projects will be initiated in North Sea fields before the end of the decade. Plans for a miscible CO_2 project onshore in a depleted East Midland oil field and a miscible hydrocarbon project offshore have already been announced. Similarly I understand that consideration is being given to a surfactant flooding pilot offshore. Undoubtedly, other projects will follow but the bulk of EOR activity in the North Sea will probably not occur until the next decade.

I am optimistic about the future potential of EOR. Currently there is a shortage of trained scientists and engineers in the area. However, the number is increasing rapidly and the outlook for solving the remaining technical problems and designing economically attractive projects is promising. I feel that the thousands of man-years of research that the industry has devoted to the development of enhanced recovery technology are finally beginning to bear fruit. Interest and activity in applying miscible gas and other EOR processes are expanding rapidly throughout the world. The industry is once again becoming increasingly optimistic about EOR potential. But there is a difference between the optimism of the 50's and that of the 80's; EOR technology of the 80's is more mature than it was then. We have a much better understanding of the capability and limitations of the various methods and the role that EOR can realistically be expected to play in our efforts to meet the world's energy needs.

I recognize that significant and challenging problems remain to be solved, but I am confident that the solution to many of these problems can be found and that substantial volumes of EOR production will become economically feasible in the future. Miscible gas processes should make a significant contribution to this objective.

REFERENCES

1. Everett, J. P. et al: "Liquid-Liquid Displacement in Porous Media as Affected by the Liquid-Liquid Viscosity Ratio and Liquid-Liquid Miscibility," Trans., AIME 198 (1950) 215.

2. Henderson, J. H.: "A Laboratory Investigation of Displacement From Porous Media by A Liquified Petroleum Gas" Trans., AIME 198 (1953) 33.

3. Whorton, L. P., and Kieschnick, W. F.: "A Preliminary Report on Oil Recovery by High-Pressure Gas Injection," Drilling and Production Pract. API, 1950, 247.

4. Stone, H. L. and Crump, J. S.: "The Effect of Gas Composition Upon Oil Recovery by Gas Drive," Trans., AIME (1956) 207, 105.

5. Hill, S.: "Genie Chemique," Chem. Eng. Sci. (1952) I. NO. 6, p. 246.

6. Dietz, D. N.: Proc., Acad. Scie. Amst. B. (1953) 56, 83

7. Caudle, B. H., and Dyes, A. B.: "Improving Miscible Displacement by Gas-Water Injection," Trans., AIME, 213, (1958), 281.

8. Kane, A. V., "Performance Review of a Large Seale CO_2-WAG Project Sacroc Unit-Kelly Snyder Field," SPE 7091, Presented at Fifth Symposium on Improved Methods for Oil Recovery in Tulsa, Okla., April 1978.

9. Orr, F. M., Taber, J. J. et al, "Displacement of Oil by Carbon Dioxide," U. S. DOE/ET/12082-9, May 1981.

10. Todd, M. R.: "Modeling Requirements for Numerical Simulation of CO_2 Recovery Processes," SPE 7998, presented at the 1979 Regional Meeting of the SPE (AIME) in Ventura, Calif., April 1979.

11. ___, "#38 Billion Projected for Enhanced Recovery in '80s": Petroleum Engineer International pages 98-100, Feb. 1981.

12. Prince, J. Philip, "Enhanced Oil Recovery Potential in Canada", Canadian Energy Research Institute ISBNO-0-920522-09-2, March 1980.

13. Chierici, G. L.: "Enhanced Oil Recovery Techniques: State of the Art and Potential" presented at Seminar on Improved Techniques For the Extraction of Primary Forms of Energy sponsored by United Nations Economic Commission for Europe, Vienna, Austria, Nov. 1980.

THEORETICAL ASPECTS OF CALCULATING THE PERFORMANCE OF CO_2 AS AN EOR PROCESS IN NORTH SEA RESERVOIRS

DAVID S. HUGES, JOHN D. MATTHEWS, ROBERT E. MOTT

AEE Winfrith, Dorchester, Dorset, DT2 8DH

ABSTRACT

This paper examines some aspects of calculating the performance of CO_2 as a prospective enhanced oil recovery agent in North Sea reservoirs. The paper falls into two areas.

First the problems of predicting the phase behaviour of CO_2 with reservoir oils are examined. Although experimental PVT data are available for CO_2-hydrocarbon systems, these are at lower pressures than prevail in the North Sea. The Peng-Robinson and Generalised Redlich-Kwong equations of state are compared for existing experimental data, and their predictions for miscibility are reviewed for North Sea reservoir conditions. Some of the problems of pseudo-components and interaction coefficients are discussed in this context.

Second, results are presented of 3-D compositional simulations for a simplified reservoir model based on the Forties Field. The few component equilibrium factors in this model are adjusted to match the equation of state implications discussed above. Current reservoir conditions are found to give an immiscible CO_2-displacement. Good sweep efficiencies are obtained in the watered-out reservoir from the immiscible CO_2-displacement calculations. This occurs because: (i) local displacement efficiency is good as a result of oil swelling and transfer of hydrocarbons into the gas stream and (ii) volumetric sweep is good with component exchange between gas and oil reducing viscosity and density differences. The reservoir pressure is then increased to achieve an MCM displacement. Three-dimensional results are obtained which compare the performance of the miscible and immiscible displacement processes. The immiscible results are slightly more attractive, but modelling approximations in both cases may be giving a false impression of the real comparability.

INTRODUCTION

This paper examines some of the problems of predicting oil displacement behaviour by CO_2 in the context of typical North Sea field characteristics. For this purpose we have considered both immiscible and miscible CO_2 drive in a conceptual reservoir simulation with properties akin to the Forties field.

The first part of the paper is concerned with problems of predicting phase behaviour using equations of state based on the Peng-Robinson and Generalised Redlich-Kwong formulations. Some preference for the latter is given because of its superior prediction of fluid densities. The use of the Redlich-Kwong equation in a few pseudo-component formulation is illustrated, which then dictates the choice of an equilibrium K-factor correlation for use in a compositional reservoir simulation code. In the case of the Forties field the minimum miscibility pressure is predicted to be just above the operating pressure, which implies opportunity to consider both immiscible or miscible CO_2-flooding of the reservoir, following the present conventional water flood.

The second part of the paper examines both immiscible and miscible oil displacement by CO_2 in a conceptual 5-spot pattern with properties akin to the Forties field. Various alternative CO_2/water injection strategies are compared for immiscible displacement based on a vertical two-dimensional stream-tube section. A near optimum process is then evaluated in a full three-dimensional model. The predicted immiscible CO_2-drive is found to be more attractive than expected due to its oil swelling and mass transfer behaviour. Miscible calculations for this same three-dimensional model have also been undertaken using the Todd and Longstaff mixing approximation in the simulation code. The differences in recovery efficiency between these two types of displacement are believed to be within the uncertainty of the methods used. The factors influencing the relative sweep efficiencies under immiscible and miscible drive are discussed.

PVT PROPERTIES OF CO_2 AND RESERVOIR OILS

Choice of Equation of State

When predicting PVT properties of reservoir fluids from a thermodynamic equation of state, a typical approach is to use a two constant cubic equation based on the Redlich-Kwong equation (Ref 1) which gives a satisfactory compromise between simplicity and accuracy. The two equations which are most commonly used, especially in applications to CO_2 systems, are the Peng-Robinson (PR) equation (Ref 2) and the Generalised Redlich-Kwong (GRK) equation in a form first proposed by Zudkevitch and Joffe (Ref 3).

The Peng-Robinson equation takes the form

$$P = \frac{RT}{v-b} - \frac{a(T)}{v(v+b) + b(v-b)} , \qquad (1)$$

where

$$a = \sum_{i,j} (1 - \delta_{ij}) (a_i a_j)^{\frac{1}{2}} x_i x_j , \quad b = \sum_i b_i x_i , \qquad (2)$$

$$a_i = 0.45724 \, \alpha_i^2 \, R^2 \, T_{ci}^2/P_{ci} , \quad b_i = 0.07780 \, R \, T_{ci}/P_{ci} , \qquad (3)$$

$$\alpha_i = 1 + m (1 - T_r^{\frac{1}{2}}). \qquad (4)$$

$$m = 0.37464 + 1.54226\omega - 0.26992\omega^2. \qquad (5)$$

The other symbols have their conventional definitions which are given at the end of the paper.

In the GRK equation

$$P = \frac{RT}{v-b} - \frac{a}{T^{\frac{1}{2}}v\,(v+b)} \qquad , \qquad (6)$$

where

$$a_i = \Omega_{ai}\, R^2\, T_{ci}^{\,2.5}/P_{ci} \quad , \quad b_i = \Omega_{bi}\, R\, T_{ci}/P_{ci} \quad , \qquad (7)$$

and the mixing rules (2) are used to calculate a and b. Ω_{ai} and Ω_{bi} are temperature dependent functions which are calculated for each component by fitting to the vapour pressure and saturated liquid density of the component at the given temperature. For supercritical temperatures Ω_a and Ω_b are assumed to take the same values as at the critical temperature.

The vapour pressure and saturated liquid density are normally derived from correlations in terms of critical properties, normal boiling point and acentric factor, along with a liquid density at a single reference temperature. The alternative procedures suggested by Yarborough (Ref 4) and Coats (Ref 5) give broadly similar results.

The introduction of the Ω-parameters is the most significant difference between the two equations. In the PR equation the critical Z-factor is necessarily 0.307 for all components, whereas for liquid hydrocarbons the critical Z-factor is known to be between 0.20 and 0.26. Thus the PR equation nearly always underpredicts the density of hydrocarbon liquids. For example, at 100°C the density of decane is underpredicted by 6% and the density of pentadecane by 12%.

The use of the Ω-parameters in the GRK equation overcomes this problem (at the cost of a loss of simplicity) and this equation generally gives good predictions of liquid densities if the Ω-parameters are chosen appropriately.

Interaction Coefficients

In both equations the mixing rule for parameter 'a' employs binary interaction coefficients δ_{ij} which must be determined empirically. Interaction coefficients for pairs of hydrocarbon components are generally zero or very small (except for methane-heavy hydrocarbon pairs) but non-zero coefficients for hydrocarbon-CO_2 binaries are essential if accurate predictions are to be obtained, and the choice of interaction coefficients is a major problem when applying an equation of state to CO_2/hydrocarbon mixtures. The conventional approach to this problem is to derive the interaction coefficients from data on binary mixtures, but the pressures in these binary systems are usually much lower than found in reservoirs, and there is some evidence that the resulting values are not optimal for multi-component systems at higher pressures. We have found that an interaction coefficient of 0.10 for all CO_2-hydrocarbon binaries gives reasonable results in the PR equation for ternary and multi-component systems, while binary data suggest rather larger coefficients (eg. 0.13 for butane).

A more systematic approach for CO_2-hydrocarbon mixtures has been proposed by Turek et al (Ref 6) for the GRK equation. A second interaction coefficient for CO_2-hydrocarbon binaries was introduced by modifying the mixing rule for the parameter 'b' in equation (2), to read

$$b = \tfrac{1}{2} \sum_{i,j} (1 + D_{ij})\,(b_i + b_j)\, x_i\, x_j \quad . \qquad (8)$$

This reduces to equation (2) when $D_{ij} = 0$. These interaction coefficients were assumed to be respectively quadratic and cubic functions of hydrocarbon acentric factor, and also Ω_a and Ω_b for supercritical CO_2 were assumed to be quadratic functions of temperature. The polynomial parameters were then determined by a regression analysis using phase equilibrium data on fifteen CO_2-hydrocarbon binaries.

These developments to the GRK equation have emphasised accurate predictions of phase behaviour rather than densities; no density data were used in the regression analysis. A consequence of the changes to the CO_2-parameters from fitting to binary data is the overprediction of CO_2 densities at high pressures; for example, at 100°C the density of CO_2 is overpredicted by 10% at 200 bars and by 19% at 300 bars. CO_2-densities predicted by the PR equation are accurate to within 5%, even at high pressures. Thus, if this form of the GRK equation is to be used to calculate densities in a compositional simulator, some compensating adjustment of the Ω's is needed.

Comparison of Predictions for CO_2/Synthetic Oil Systems

To illustrate some of these points, the PR and GRK equations have been used to predict saturation pressures and densities of mixtures of CO_2 with the 10 component synthetic oil whose composition is given in Table 1. Experimental data on this system at 48.9°C and 65.6°C is given by Turek et al. The calculations were carried out using the VOLE phase equilibrium code (Ref 7) developed at AEE Winfrith. In the calculations with the GRK equation the CO_2-interaction coefficients and Ω-parameters were taken from Reference 6, and interaction coefficients for all hydrocarbon pairs were set to zero except for C_1-C_{10} and C_1-C_{14}, where a value of 0.01 was used in order to match the observed bubble points of the original oil. The PR calculations used an interaction coefficient of 0.10 for all CO_2-hydrocarbon pairs.

Interaction coefficients between C_1, C_2 and C_3 with C_6^+ hydrocarbons were taken from Katz and Firozabadi (Ref 8), except that the C_1-C_{10} and C_1-C_{14} coefficients had to be increased by 0.01 to fit the observed bubble point of the original oil at 65.6°C. However, the predicted bubble point was then in error by 2% at 48.9°C.

TABLE 1

COMPOSITION OF SYNTHETIC OIL (Ref 6)

Component	Mole per cent	Component	Mole per cent
C_1	34.67	n C_6	3.06
C_2	3.13	n C_7	4.95
C_3	3.96	n C_8	4.97
n C_4	5.95	n C_{10}	30.21
n C_5	4.06	n C_{14}	5.04

Predicted saturation pressures are shown in Figure 1. The GRK results agree quite well with the measurements except that the predicted critical points occur at a rather higher CO_2 concentration. The PR predictions are rather less accurate, especially at the lower temperature, although some improvement could probably be made by adjusting individual interaction coefficients.

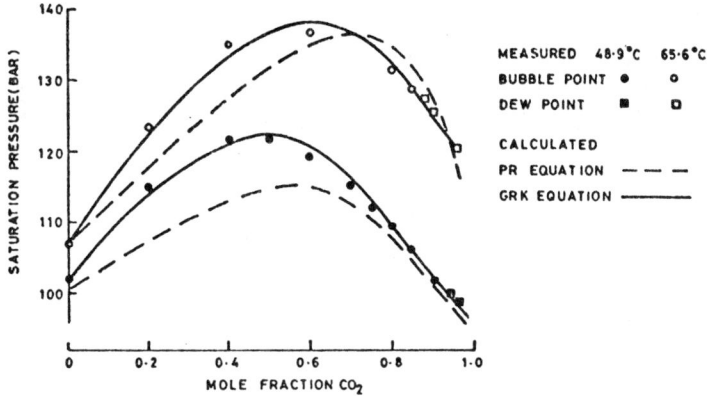

FIG. 1. SATURATION PRESSURES IN THE CO_2- SYNTHETIC OIL SYSTEM.

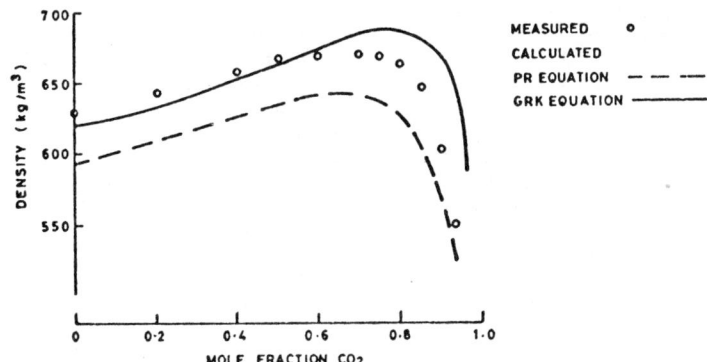

FIG. 2. DENSITIES OF SATURATED FLUID IN THE CO_2 - SYNTHETIC OIL SYSTEM (48·9 °C)

Density predictions at 48.9°C are shown in Figure 2. The GRK equation is accurate at low CO_2 concentrations, but at high CO_2-concentration the density is overpredicted by 10%, since the density of pure CO_2 is overestimated. The PR equation underpredicts densities by 5 or 6% at 48.9°C, and by 3 or 4% at 65.6°C. This is consistent with the work of Sigmund et al (Ref 9), who found that while liquid densities are underpredicted by the PR equation, the fractional changes in volume due to the addition of CO_2 to oil are represented quite well. From these results, and other calculations on synthetic oil mixtures, it appears that neither equation is capable of accurate predictions of oil/CO_2 mixtures across the entire composition range. This implies that some fitting of parameters to experimental data is needed if accurate predictions of density and phase behaviour are to be obtained simultaneously.

Pseudo Component Representation of Reservoir Oils

The application of an equation of state to synthetic oils is relatively
straightforward as individual components can be identified and represented
as such in the theoretical model. However, the heavy fractions of reservoir
fluids contain so many different isomers that it is impossible to identify
them individually, and it is necessary to divide the heavy fractions
(normally C_6 and above) into pseudo components, each pseudo component
representing a group of components having similar properties. The selection
of parameters for these pseudo components is a severe problem when applying
an equation of state to reservoir oils, and is particularly significant in
CO_2/oil systems, where the heavy fractions have a strong influence on the
phase behaviour at high pressure.

A common approach is to divide the heavy fraction into groups, each of which
has individual components whose boiling point lies within a certain range.
This is particularly convenient if a distillation analysis has been carried
out on the oil, as one pseudo component can be assigned to each 'cut' in
the distillation. Specific gravity, average boiling point and molecular
weight are normally determined for each 'cut', but the equation of state
model requires 'pseudo' critical properties as input. Various correlations
have been proposed for determining these properties; for example those of
Cavett (Ref 10) and Whitson (Ref 11) which are most convenient as they use
specific gravity and average boiling point as correlating parameters.

A further problem is the selection of interaction coefficients for these
pseudo components. The systematic approach to CO_2-hydrocarbon interaction
coefficients described earlier has the advantage that it is possible to use
the functional dependence on acentric factor to extrapolate the pseudo
components. This route has been followed in the present work for the GRK
equation, while for the PR equation a value of 0.1 was used for all
CO_2-hydrocarbon pairs. In both cases the interaction coefficient between
methane and the heaviest fractions (C_{16+}) was adjusted to match the observed
bubble point of the original oil without CO_2. This coefficient was always
small (around 0.05) for the GRK equation, but when applying the PR equation
to North Sea oils, it has always been found necessary to use large methane-
heavy hydrocarbon interaction coefficients (up to 0.4) to fit the observed
bubble points; these coefficients varied considerably between different oils
and did not follow any obvious systematic trend.

Calculations on North Sea oils have shown that the GRK equation predicts the
density of oil without CO_2 to within a few per cent; while the PR equation
underestimates oil density by between 10 and 20%, depending on the correlation
used to calculate pseudo component properties. However, to obtain accurate
densities with the GRK equation it was necessary to use the measured density
of each 'cut' when calculating the Ω-parameters.

Assessing the various methods for calculating PVT properties for CO_2/reservoir
oil systems is difficult because of the lack of experimental data on oils
for which a comprehensive analysis of the heavier fractions is also
available. Further work is needed to develop the equation of state method to
the stage where it can give reliable a priori predictions (ie. without
fitting to experimental data) of these properties. In any case it may prove
simpler to make a few PVT measurements on CO_2/oil mixtures, than to carry out
the detailed compositional analysis required for input to a predictive
equation of state model.

Prediction of PVT Properties of CO_2/Forties Oil Mixtures

At the present time, no experimental phase behaviour data are available for CO_2 and North Sea oils. In the absence of such data, detailed equation of state calculations have been performed to generate PVT information for CO_2/Forties oil mixtures, and to derive data for a model with a small number of pseudo components with approximately the same essential properties. Figures 3 and 4 show calculated saturation pressures and swelling factors for CO_2/ Forties oil, using the alternative PR and GRK equations in an 18-component model in the VOLE code (swelling factor is defined as the volume of oil plus CO_2 at saturation pressure relative to volume of original oil at its saturation pressure). Both the Whitson and Cavett correlations were used for estimating the properties of the twelve pseudo components to represent the $C_{6}+$ fraction. The two GRK calculations predict similar saturation pressures, while the two PR calculations give significantly different saturation pressures, depending on which correlation is used for the pseudo component properties. It is to be expected that the GRK predictions are less sensitive to the pseudo component correlation because of the subsequent matching to the measured density using the Ω-parameters. The various methods give similar predictions for the oil swelling factors as a function of CO_2 concentration. However, the swelling factors for saturated oil at a typical pressure of 200 bars vary considerably, with values of 1.50 and 1.72 for the PR calculations, and 1.58 and 1.64 from the GRK equation.

These calculations have shown some significant differences between different methods for predicting the PVT properties of reservoir oil/CO_2 mixtures. The choice of correlation for pseudo component properties appears to be at least as important as the choice of equation of state. Experimental data are needed to resolve the position.

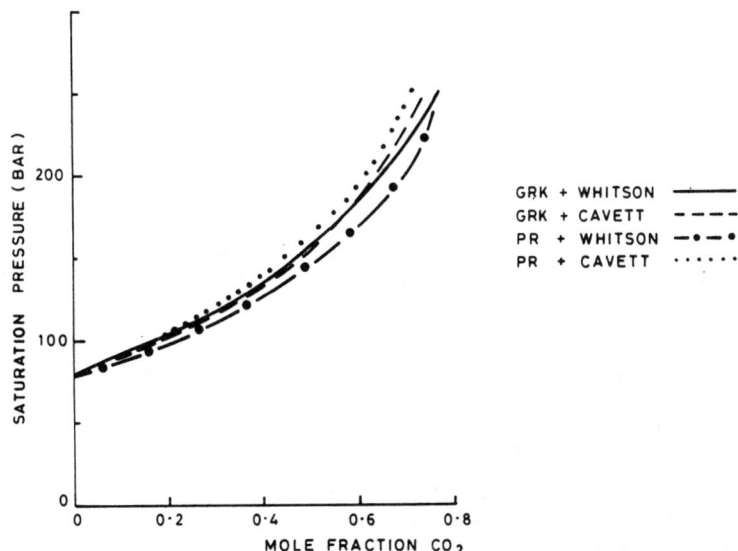

FIG. 3. CALCULATED SATURATION PRESSURES
OF FORTIES OIL - CO_2 MIXTURES

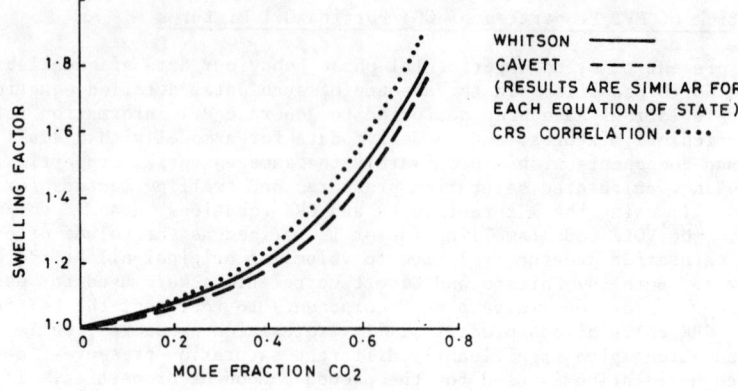

FIG.4. CALCULATED SWELLING OF SATURATED
FORTIES OIL - CO_2 MIXTURES

The equations of state were used to estimate minimum miscibility pressure (MMP), using a simplified model of the multi-step process leading to multiple contact miscibility. In the numerical procedure adopted in the VOLE code at Winfrith, the fluid composition at each stage is determined by mixing original oil with the gas phase from the previous state in a multi-step process. The MMP for Forties was predicted by both PR and GRK models to be between 210 and 220 bars, compared with a figure of 240 bars from the latest correlation of Holm and Josendal (Ref 12). In the Forties field the reservoir pressure was intially 220 bars, but has fallen to around 180 bars. Thus the calculations suggest that CO_2-injection in Forties would give rise to an immiscible displacement, but the pressure is only just below the MMP.

Few Component Representation of Forties Oil

To realistically represent the PVT behaviour of reservoir oils some 15 to 40 different components (or pseudo components) are needed. However, when using a compositional simulator, it is essential to reduce the number of components so that computing costs are acceptable. When mathematically simulating a CO_2 flood, the methane and CO_2 are normally kept as single components, and between two and four pseudo components are used to represent the C_2^+-fraction of the oil. It is unlikely that such a coarse representation will give accurate results unless certain parameters have been adjusted to fit data from experiments, or from more detailed calculations. This can be done by non-linear regression analysis using an equation of state phase equilibrium code in which the Ω-parameters (or critical properties in the PR equation), and interaction coefficients are adjusted until a good match is obtained. When used in a few component model within a compositional simulator, there is little to choose between different equations of state; in all cases it should be possible to obtain an accurate representation by tuning appropriate parameters. In essence, the few component equation of state model becomes a sophisticated correlation with a wider range of validity than the K-value approach.

In an immiscible CO_2 flood two main recovery mechanisms operate. They are the swelling of the oil through solution of CO_2 and evaporation of certain components from the oil into the produced gas stream. One-dimensional calculations using an equation of state compositional simulator which has recently been developed have demonstrated that these phenomena could be accurately represented in a few component model. Two calculations were performed for a CO_2 flood in a one-dimensional geometry. The first used an 18-component representation of the reservoir oil, and the second a 6-component representation in which the parameters were obtained by an averaging procedure chosen to give the same values of the equation of state 'a' and 'b' coefficients. After some minor adjustments of the methane-heavy hydrocarbon interaction coefficients, the 6-component and 18-component predictions of density, viscosity and bubble points agreed everywhere to within 1 or 2%. The oil composition used was similar to that found in the Forties reservoir, and the six components selected were CO_2, C_1, C_2-C_5, C_6-C_{10}, C_{11}-C_{19} and C_{20+}.

Figure 5 shows the oil recovery and GOR as a function of the amount of CO_2 injected, for the two different representations. There is close agreement between the two cases, with oil recoveries differing by at most 0.5%. The recoveries of individual components are also in close agreement. These results suggest that a six component model can give an adequate representation of the PVT properties of the fluids, so long as the parameters have been adjusted to match data from experiments or more detailed calculations.

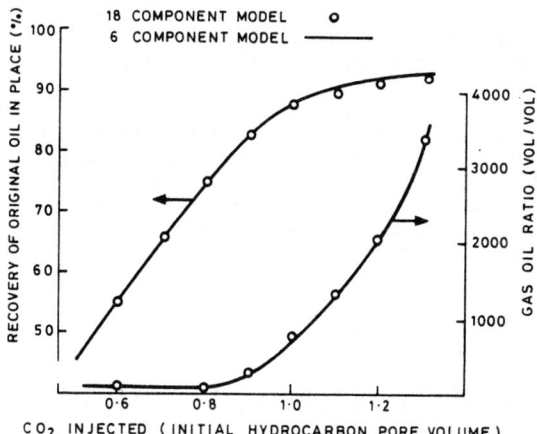

FIG. 5. PREDICTED PERFORMANCE OF ONE DIMENSIONAL
CO_2 DISPLACEMENT OF FORTIES OIL AT 210 BAR

THE DISPLACEMENT OF OIL BY CO_2 IN A CONCEPTUAL RESERVOIR

Calculations of oil displacement by CO_2 have been performed using a simplified three dimensional conceptual reservoir with properties broadly similar to those found in the Forties field (Ref 13). A repeated 5-spot pattern has been assumed in a 100 m thick sandstone with uniform porosity of 27% and permeabilities of 400 mD in the horizontal direction and 40 mD in the vertical direction. The assumed well separation between an injector and a producer was a typical value of 700 metres. The connate water saturation was 23% and the residual oil from waterflooding was 30%. Water viscosity was 0.42 cp.

The assumption of homogenetity in the 5-spot pattern allows eight-fold symmetry to be invoked and a three-dimensional representation of a symmetry sector was used in the calculations. To reduce mesh orientation effects, a curvilinear areal mesh pattern was adopted, based upon unit mobility potential flow. This areal mesh consisted of 4 stream tubes with 12 subdivisions along each tube the subdivisions being selected to give equal areas in the shortest stream tube (see Figure 7); this reduces the time step penalty which occurs in explicit calculations of saturations. In the vertical cross section, 6 vertical layers were assumed, the thickness of the uppermost layers (6.7 and 13.3m) being smaller than that of the lower layers (20 m) to allow gravity override to be followed whilst economising on the number of meshes. The broad adequacy of this basic mesh arrangement for the current comparative exercise was examined by performing mesh refinement calculations in two dimensions.

Before undertaking calculations of oil displacement by CO_2 injection, it was first necessary to define the state of the model reservoir after waterflooding. Calculations of waterflood behaviour were performed assuming the 5-spot pattern to be produced at a constant rate of 5000 m^3/day (31000 BPD) with an equal injection of water into each injector. Relative permeabilities were derived assuming that the sandstone was predominantly water wet, and the Corey approximation (Ref 14) was used to obtain water relative permeabilities. Imbibition oil relative permeabilities were derived by applying the Land concept of reduced free saturations (Ref 15) to the Corey drainage approximation. With the rather coarse level of mesh definition described above, it was not thought necessary to include the marginal effects associated with the use of capillary pressure curve, but the calculated distribution of water saturations arising from the model is expected to be typical of a possible condition for a CO_2 displacement process.

The calculations of waterflood behaviour were performed using the compositional simulator described later. Breakthrough of water occurs at the producer after 1800 days, with a recovery of 51% of the oil in place. The water cut rises to 50% at 2100 days when the recovery is 59% of the oil in place. At this condition, most of the unswept oil was confined to the two outer stream tubes around the production well. Some underriding of oil by water had occurred so that the top two layers contained about half of the unswept oil. This condition when the water cut reached 50% has been used as the starting point for the CO_2 displacement calculations, since this may be a typical limiting situation for off-shore water handling.

Modelling Assumptions Adopted in Immiscible and Miscible Displacement Calculations

Displacement of oil by CO_2 in the model reservoir has been studied under immiscible and miscible conditions by performing calculations for pressures just below and just above the minimum miscibility pressure. The Intercomp Compositional Reservoir Simulator, CRS, (Ref 16) used for the multi-dimensional immiscible displacement calculations employs equilibrium K-value correlations for calculating phase behaviour, with an equation of state for densities. The six component representation of Forties oil described earlier was used in the study, the parameters in the K-value correlation being adjusted to match the saturation pressure and swelling factors of oil/CO_2 mixtures predicted by the detailed equation of state model (see Figure 4), and to match measured oil densities and viscosities.

The miscible calculations were carried out using the Todd and Longstaff mixing model (Ref 17). This is a model for a two component, oil and solvent, system which assumes direct miscibility of the phases. The method relies on the assumption of an effective oil viscosity and density, and an effective gas viscosity and density, using a mixing parameter, ω, which has to be defined by empirical means. The CRS code was modified to provide a suitable vehicle for this mixing model. This involved bypassing the phase equilibrium calculation and incorporating the Todd and Longstaff equations. Unmixed oil and CO_2 properties were entered in tabular form. Using a value of $\omega = 0.67$, Todd and Longstaff calculated the oil displacement for an areal bead pack system which gave good agreement with experimental data obtained by Lacey (Ref 18). Good agreement with Lacey's results was also obtained in the present work using the modified CRS model with a curvilinear grid and $\omega = 0.67$. There is, however, no experimental evidence that the mixing model is correct for a vertical cross section geometry and the effective densities assumed in the model may not be valid.

Whilst Todd and Longstaff recommend a value of $\omega = 0.67$ for areal studies, Warner (Ref 19) has recommended a value of $\omega = 0.8$ for vertical cross sections and for 3-dimensional calculations. In the present studies, a value $\omega = 0.7$ was used for most of the calculations, but some calculations were also performed using $\omega = 0.5$ and $\omega = 1.0$ to assess the sensitivity of the results to changes in this factor.

The relative permeability treatment in the models for miscible and immiscible displacement are very different. For immiscible displacement, three phase relative permeabilities are evaluated using Stone's second method (Ref 20). In this approach, the water relative permeability is a function of the water saturation only and the gas relative permeability is a function of the gas saturation only. The oil relative permeability is a function of both water and gas saturations and is given by

$$k_{ro}(S_g, S_w) = \left[k_{row}(S_w) + k_{rw}(S_w)\right]\left[k_{rog}(S_g) + k_{rg}(S_g)\right]$$
$$- \left[k_{rw}(S_w) + k_{rg}(S_g)\right] \tag{9}$$

Each of the relative permeabilities in immiscible displacement are subject to three-phase hysteresis effects as the flow regimes change (Ref 21). In a gas drive the gas becomes mobile at an initial saturation of about 0.05, whereas when gas is displaced a trapped gas saturation of 0.30 is typical (Ref 22). Failure to account for the hysteresis in gas relative permeabilities results in optimistic recoveries, since little gas is trapped. The CRS code was therefore modified to allow for gas hysteresis effects. Gas relative permeabilities were obtained by applying the Land method (Ref 15) to the drainage curve calculated from the Corey approximation (Ref 14). The "free" gas saturation (S_{gF}) defined by Land is first computed as a function of the current gas saturation (S_g) and the highest value of gas saturation previously reached within each grid block (S_{gmax}). The equations are

$$S^*_{gr} = S^*_{gmax}/(1 + C.\ S^*_{gmax}) \tag{10}$$

$$S^*_{gF} = 0.5\left[S^*_g - S^*_{gr} + \sqrt{(S^*_g - S^*_{gr})^2 + 4(S^*_g - S^*_{gr})/C}\right] \tag{11}$$

where
$$S^*_g = \frac{S_g - S_{gc}}{1 - S_{wc} - S_{gc}} \tag{12}$$

A value of $C = 1.86$ was obtained from Equation 10 by substituting a trapped gas saturation (S_{gr}) of 0.30 when the maximum gas saturation had reached 0.77 (ie 1-S_{wc}). Hysteresis effects in the oil and water relative permeabilities were considered to be small and were ignored.

The relative permeability treatment in the Todd and Longstaff miscible model is based upon the assumption that oil and solvent behave as a single phase. For consistency with the waterflood, this single hydrocarbon/CO_2 phase was given the same relative permeability (k_{rhc}) as that of oil in water. This implies a residual hydrocarbon/CO_2 saturation after water drive of 0.30. The separate oil and gas relative permeabilities were then obtained by making the linear assumption.

$$k_{ro} = \frac{S_o}{S_o + S_g} k_{rhc} \tag{13}$$

$$k_{rg} = \frac{S_g}{S_o + S_g} k_{rhc} \tag{14}$$

Hysteresis effects in gas relative permeability were not included in the miscible calculations, it being argued that oil and gas behave as the same phase under miscible conditions.

Two-Dimensional Studies

A number of comparative immiscible displacement calculations have been performed using a vertical, two dimensional model to examine various CO_2-injection strategies after waterflood. This vertical model consisted of the second longest of the four stream tubes in the mesh scheme identified previously. Continuous injection of CO_2 (Case a) was used as the basis of comparison. The following alternative strategies were then examined for 0.22 PV of injected CO_2 followed by chase water until the produced water reaches 90%.

Case b   Injection of a single slug of CO_2 over the full height of the section

Case c   Alternating 100 day injections of CO_2 and water across the full height of the section

Case d   Simultaneous injection of CO_2 into the lower half of the section and water into the upper half

Case e   Alternating 100 day injections of CO_2 and water, with the CO_2 injection restricted to the lower half of the section and water injection restricted to the upper half.

Cases d and e, if practicable, would both require dual completions with the injected CO_2 flowing down a central pipe and water through a surrounding annulus. The practical problems of dual completions have not been considered in detail. However, separation of CO_2 and water will considerably reduce the corrosion problems associated with alternating CO_2 and water injection.

The results of these calculations are illustrated in Figure 6. With Case a, the asymptotic value of oil recovery is 45% of the target, with only 15% of the oil being recovered when 0.22 PV of CO_2 is injected. Inclusion of chase water after 0.22 PV CO_2 in Case b increases oil production to 34%, for two reasons, namely more of the swollen oil is displaced, and some of the CO_2 is moved to become effective in other parts of the reservoir. Alternating water with the CO_2 in Case c reduces the peak value of CO_2 saturation attained in any grid block, so that the amount of CO_2 trapping is also reduced. As a result, more of the CO_2 can be mobilised during the subsequent chase water flood and the recovery increases to 39%.

In each of the above cases, there is a significant amount of gravity override. By injecting CO_2 into the lower half of the reservoir with water into the upper half, Case d, gas override is reduced and the oil recovery is increased to 59% of the target. Alternating the gas and water injections, whilst restricting the gas to the lower half of the reservoir and water to the upper half, Case e, produces a further slight improvement in oil recovery to 62% of the target.

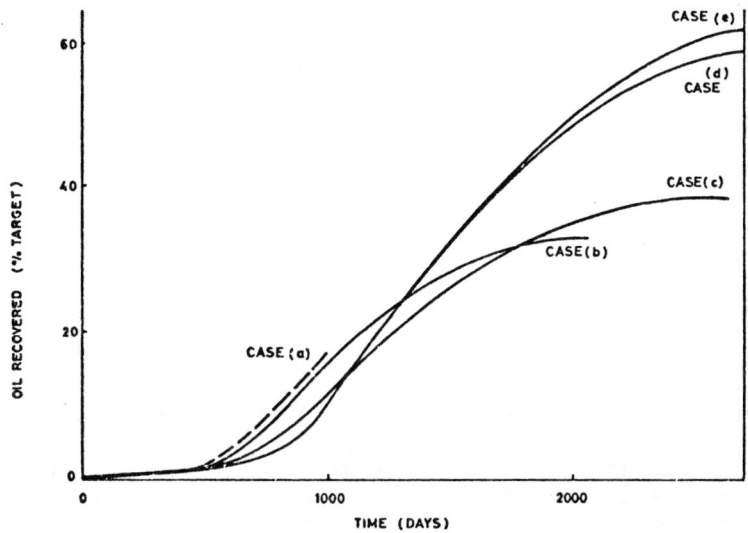

FIG.5 COMPARISON OF OIL RECOVERIES FOR
DIFFERENT INJECTION STRATEGIES

The calculations for CO_2 injection into a vertical two dimensional model reservoir have been repeated for miscible displacement conditions. As shown in Table 2, the results showed the same trends as those for immiscible displacement, with Case e again producing the highest level of oil recovery.

The miscible displacement calculations produced oil recoveries slightly below the immiscible values. This trend in difference between miscible and immiscible displacement was observed in all of the calculations reported in this paper. The calculational models described earlier for the two processes involve very different physical concepts. As is discussed later, current limitations in representing these physical concepts are thought to be the reason for the somewhat lower calculated oil recoveries for the miscible condition.

Areal sweep effects were examined for both miscible and immiscible displacement of oil when a total of 0.22 PV of CO_2 is injected, with alternating 100 day slugs of gas and water, followed by chase water. For these calculations, a two dimensional areal model was used with a single mesh block in the vertical direction. The immiscible and miscible displacements showed oil recoveries of 61% and 54% of the target oil respectively. The resulting areal distributions of oil for the two processes are illustrated in Figure 7. It can be seen that in the immiscible case the CO_2 has been effective for a greater distance from the injector than in the miscible case. This result is a direct consequence of the different treatments for gas and oil relative permeabilities in the two models, which produce different values for the residual gas

TABLE 2

COMPARISON OF IMMISCIBLE AND MISCIBLE DISPLACEMENT IN A VERTICAL, TWO DIMENSIONAL
CROSS SECTION FOR A TOTAL INJECTION OF 0.22 PV OF CO_2

	Strategy	Oil Recovery (% of Target)	
		Immiscible	Miscible
a	Single slug injection of CO_2	15	12
b	Single slug of CO_2 followed by chase water	34	29
c	Alternating 100 day injections of CO_2 and water, followed by chase water	39	39
d	Simultaneous injection of CO_2 into lower half of reservoir and water into upper half followed by chase water	59	51
e	Injection of CO_2 into lower half of reservoir and water into upper half with alternating 100 day cycles between CO_2 and water injection, followed by chase water	62	52

saturations. This highlights an important factor in the mathematical
representation of gas displacement processes which needs further theoretical
development coupled with experimental information.

The high areal sweep efficiency obtained with the immiscible displacement is
caused by two effects:

(i) component exchange between oil and gas reduces the initial viscosity
ratio of 20:1 to 3.2:1

(ii) up to one third of the gas dissolves in the oil keeping gas saturations
low and hence maintaining low gas mobilities.

The effect of varying the Todd and Longstaff mixing parameter was examined in
the two dimensional studies. Results of calculations for ω = 0.5, 0.7 and
1.0 for alternating 100 day injections of CO_2 and water over the full height
of the section, followed by chase water, are shown in Table 3. It might have

TABLE 3

EFFECT OF VARYING TODD AND LONGSTAFF MIXING PARAMETER ON OIL RECOVERY FOR WATER
ALTERNATING WITH GAS IN TWO DIMENSIONAL STUDIES

	Oil Production (% of target)		
	ω = 0.5	0.7	1.0
Vertical Cross Section	32	39	45
Areal Model	44	54	63

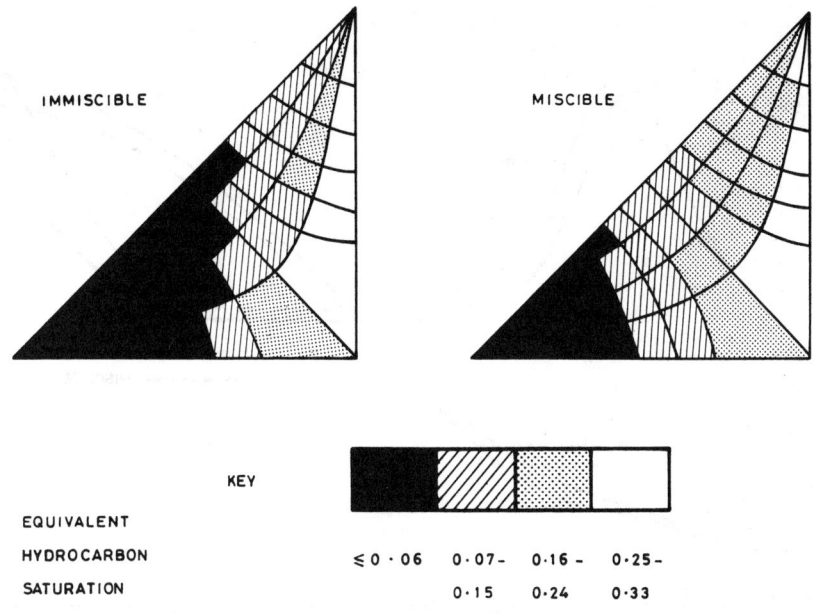

IMMISCIBLE

MISCIBLE

KEY

EQUIVALENT

HYDROCARBON

SATURATION

≤ 0·06   0·07-   0·16 -   0·25-
         0·15   0·24   0·33

FIG 7 RESIDUAL HYDROCARBON DISTRIBUTION RESULTING FROM
2000 DAY WAG IN AREAL MODEL

been expected that the sensitivity to the value of ω would be different in
the horizontal and vertical planes, because in the latter the effective
densities will influence the override behaviour. This has not occurred in an
obvious manner, although the direct link between effective viscosities and
densities in the Todd and Longstaff model may not be realistic.

Three-Dimensional Calculations

Calculations of immiscible and miscible displacement of oil have been performed
using the three-dimensional conceptual reservoir model described earlier.
These calculations were performed assuming the CO_2-injection strategy which
produced the highest oil recovery in two dimensional studies; ie. injection of
CO_2 restricted to the lower half of the reservoir and water injection
restricted to the upper half with alternating 100 day injections of CO_2 and
water, followed finally with chase water across the full height of the column.
The results are presented in Figure 8. The calculated oil production for
immiscible displacement was 56% of the oil remaining after waterflood, compared
with 51% for miscible displacement. In both cases the water cut continued to
rise immediately after initiation of CO_2-injection,but decreased to a level of
50% for about 1000 days before it increased rapidly once more. Thus both
processes imply a need to handle high water-cuts, but not as high as would be
incurred from continued injection of water without CO_2.

The distributions of hydrocarbon in the vertical cross section along the second
longest streamtube are shown in Figure 9. These illustrate the effects of
gravity on the recovery and show the additional penetration made by the CO_2

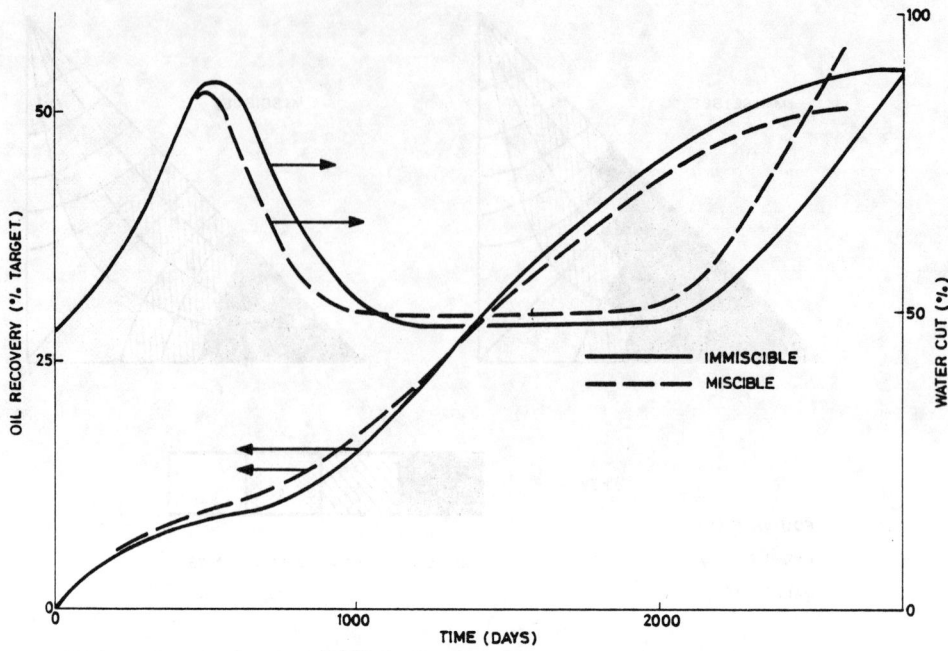

FIG. 8. COMPARISON OF IMMISCIBLE AND MISCIBLE DISPLACEMENTS FOR THE
THREE DIMENSIONAL RESERVOIR MODEL

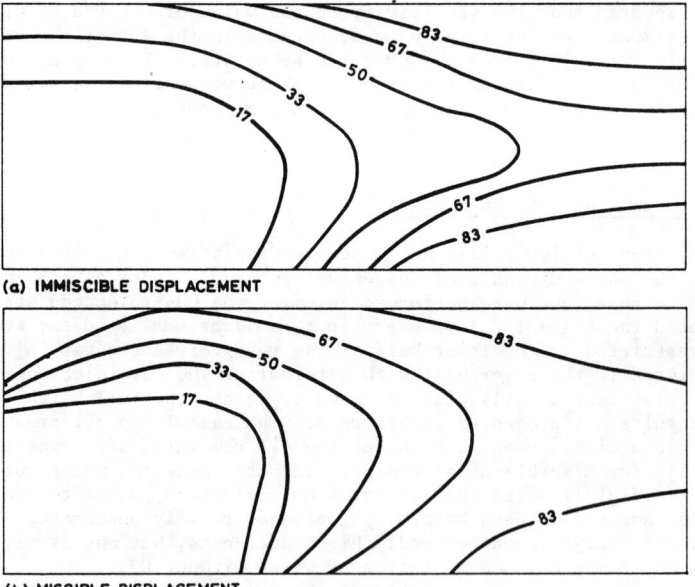

(a) IMMISCIBLE DISPLACEMENT

(b) MISCIBLE DISPLACEMENT

FIG. 9. DISTRIBUTIONS OF HYDROCARBON IN A VERTICAL CROSS
SECTION OF THE THREE DIMENSIONAL MODEL
(HYDROCARBON MASS AS PERCENT WATER FLOOD
RESIDUAL)

during immiscible displacement. These results are broadly similar for both immiscible and miscible cases with the previous two-dimensional cross section calculations, and this indicates that selection of that model was a good basis for comparing general injection strategies.

Modelling Factors Influencing the Calculations

The higher oil recovery calculated for immiscible displacement compared with miscible displacement was unexpected. However, there are a number of features in the simulation models which need further theoretical and experimental investigation.

(i)   Although hysteresis has been introduced into the gas relative permeabilities for the immiscible calculation, the model adopted has certain limitations. The treatment used has considered variations in trapping of the gas phase only, whereas a full treatment should consider all non-wetting phases. However, this introduces new problems in estimating the proportions of oil and gas that are trapped, and there is no experimental evidence to resolve this problem. The relative permeability treatment for the immiscible case, which assumes distinct gas and oil phases, should also change as the displacement process approaches miscibility. In the mixing model for the miscible cases, only a single CO_2/hydrocarbon phase exists, and there is no equivalent to the hysterises in trapping of CO_2 assumed in the immiscible relative permeability model.

(ii) The mixing model used for the miscible calculations is designed to provide effective viscosities in the presence of viscous fingering, whereas the effects of viscous fingering are ignored in the multi-component immiscible model. Immiscible viscous fingering may well be important, particularly when miscibility conditions are approached. Predicted recoveries from the immiscible model are therefore likely to be optimistic.

(iii) The method of calculating effective oil and solvent densities in the Todd and Longstaff model has not been validated. Gravity override is an important characteristic of miscible gas displacement processes and there is a need to develop and validate an independent mixing model for effective densities. Gravity override may be a partially stabilising influence on fingering in the areal plane.

(iv) The assumption of instantaneous equilibration over the whole grid block with the multicomponent model causes immiscible predictions to be optimistic. This phenomenon is particularly emphasised in a coarse mesh arrangement, since it allows substantial amounts of CO_2 to dissolve in the oil ahead of the displacement front, causing the oil to swell artificially and consequently increasing the mobility ratio. Similarly, the formation of free gas behind the front is inhibited by the coarse mesh mixing.

(v)   Oil recoveries with both processes are slightly optimistic because no allowance has been made for the solubility of CO_2 in water. The effect of water blocking, preventing the CO_2 from contacting some of the oil has also been neglected.

All of the above factors require more detailed quantatative analysis supported by experimental measurements to allow more definitive comparisons to be made.

264

Nomenclature

a,b	Parameters in equations of state
C	Imbibition trapping constant
D	Binary interaction coefficient for the parameter 'b'
K	Vapour-liquid equilibrium partition coefficient
K_r	Relative permeability
m	Characterisation constant in PR equation
P	Pressure
R	Universal gas constant
S	Saturation (S* effective saturation)
T	Absolute temperature
v	Molar volume
x	Mole fraction
α	Parameter in PR equation
δ	Binary interaction coefficient for the parameter 'a'
ω	Acentric factor; empirical mixing parameter
Ω	Parameter in GRK equation.

Subscripts

c	Critical property
F	Free
g	gas
gc	critical gas
hc	hydrocarbon
i,j	components
max	maximum
o	oil
r	reduced property
w	water
wc	connate water

Acknowledgement

The work reported in this paper has been funded by the Department of Energy.
The authors acknowledge the advice given by Dr F J Fayers. Dr T P Fishlock
and Mr R I Hawes and the help of Mr I R Hawkyard in undertaking computations.

References

1. REDLICH, O. and KWONG, J.N.S., "On the Thermodynamics of Solutions. V.
    An Equation of State.Fugacities of Gaseous Solutions", Chemical Reviews
    (February 1949), 44, 233-244.

2. PENG, D.Y. and ROBINSON, D.B., "A New Two-Constant Equation of State",
    Ind. Eng. Chem. Fundam. (1974) 15, 59-64.

3. ZUDKEVITCH, D. and JOFFE, J., "Correlation and Production of Phase
    Equilibria with the Redlich-Kwong Equation of State", AIChE Jnl (1973),
    16, 112-119.

4. YARBOROUGH, L.; "Application of a Generalised Equation of State to
    Petroleum Reservoir Fluids", in "Equations of State in Engineering",
    Advances in Chemistry Series, 182, American Chemical Society, Washington
    DC. (1979), 385-435.

5. COATS, K.H.; "An Equation of State Compositional Model", Soc.Pet.Eng.Jnl.
    (October 1980), 363-376.

6. TUREK, E.A. et al; "Phase Equilibria in Carbon Dioxide - Multicomponent
    Hydrocarbon Systems: Experimental Data and an Improved Prediction
    Technique", paper SPE 9231 presented at the SPE Annual Fall Technical
    Conference and Exhibition, Dallas, September 21-24 1980.

7. MOTT, R.E., "Development and Evaluation of a Method for Calculating the
    Phase Behaviour of Multi-Component Hydrocarbon Mixtures from an Equation
    of State", AEEW.- R 1331 (1980).

8. KATZ, D.L. and FIROZABADI, A; "Predicting Phase Behaviour of Condensate/
    Crude-Oil Systems using Methane Interaction Coefficients", J. Pet. Tech.
    (November 1978), 1649-1655.

9. SIGMUND, P.M. et al, "Laboratory CO_2 Floods and their Computer Simulation",
    paper PD10(5) presented at the 10th World Petroleum Congress, Bucharest,
    1979.

10. CAVETT, R.H., "Physical Data for Distillation Calculations - Vapour-Liquid
    Equilibrium", Proc. 27th API Mid-year Meeting, San Francisco, 1962.

11. WHITSON, C.H., "Characterizing Hydrocarbon Plus Fractions", paper EUR 183
    presented at the European Offshore Petroleum Conference and Exhibition,
    London, October 1980.

12. HOLM, L.W. and JOSENDAL, V.A., "Effect of Oil Composition on Miscible
    Type Displacement by Carbon Dioxide", paper SPE 8814, presented at the
    1st SPE/DOE Symposium on Enhanced Oil Recovery, Tulsa, April 1980.

13. HILLIER, G.R.K, COBB R.M., DIMMOCK, P.A., "Reservoir Development Planning
    for the Forties Field", Paper EUR 98 presented at European Offshore
    Petroleum Conference and Exhibition, London, October 1978.

14. COREY, A.T., "The Inter-relation Between Gas and Oil Relative Permeabilities", Producer's Monthly Vol XIX, 1, Nov 1954.

15. LAND, C.S., "Calculation of Imbibition Relative Permeability for Two and Three Phase Flow From Rock Properties", Trans AIME (1968), 243, 149-156.

16. NOLEN, J.S., "Numerical Simulation of Compositional Phenomena in Petroleum Reservoirs", SPE Reprint Series No 11, Numerical Simulation (1973), p 269-284.

17. TODD, M.R. and LONGSTAFF, W.J., "The Development, Testing and Application of a Numerical Simulator for Predicting Miscible Flood Performance", J. Pet. Tech. (July 1972), 874-882.

18. LACEY, J.W., FARIS, J.E., BRINKMAN, F.H., "Effect of Bank Size on Oil Recovery in High Pressure Gas-Driven LPG-Bank Process", J.Pet.Tech (August 1961), 806-816.

19. WARNER, H.R., "An Evaluation of Miscible CO_2 Flooding in a Waterflooded Sandstone Reservoir", J. Pet. Tech (Oct 1977), 1339-1347.

20. STONE, H.L., "Estimation of Three Phase Relative Permeability and Residual Oil Data", J. Can. Pet. Tech. (Oct 1973), 12, 53-61.

21. BREIT, V.S. and GRAUE, D.J., "Scaling of Flow Parameters for Miscible Gas Flood Simulation Studies", Paper SPE/DOE 9804, Presented at the Second Joint Symposium on Enhanced Oil Recovery, Tulsa, April 1981.

22. HOLMGREN, C.R. and MORSE, R.A., "Effect of Free Gas Saturation on Oil Recovery by Waterflooding" Trans AIME (1951), 192, 135-140.

OIL RECOVERY BY CARBON DIOXIDE
THE RESULTS OF SCALED PHYSICAL MODELS AND FIELD PILOTS

TODD M. DOSCHER, MAHGUIB EL ARABI,
SIAVASH GHARIB and RICHARD OYEKAN

*Department of Petroleum Engineering, University of Southern California,
Los Angeles, California 90007*

I. ABSTRACT

Considerable efforts have been expended in the past decade on the potential use of carbon dioxide for the recovery of residual crude oil, however the results of field pilot operations have indicated that very large quantities of carbon dioxide are required to recover the residual oil.

Scaled physical model studies have been undertaken in an attempt to determine whether the large ratios of injected carbon dioxide to produced crude oil, observed in field demonstration projects, are to be expected or are due to some defects in the application of the process in the field. The results of this study have rather unequivocally indicated that the high ratios of injected carbon dioxide to recovered crude oil should in fact be expected.

The experiments have also shown that a very heightened efficiency of the process can be achieved by using a slug of carbon dioxide followed by water. However, the results need to be qualified by noting that the prototype reservoirs used in these studies are very favorable for the use of CO_2.

The recovery mechanism appears to be chiefly the solution of carbon dioxide in the oil, its swelling and consequent increase in mobility, and then the displacement of the swollen, mobile oil by a gas drive (if carbon dioxide is injected continuously) or by water if a slug of carbon dioxide is followed by the latter. The efficiency of the process is thwarted by the high mobility of the carbon dioxide which leads to viscous fingering and its low density, compared to water, which leads to gravity segregation.

Experiments have indicated that nitrous oxide, which also displays a high solubility in organic compounds, is as effective as carbon dioxide in these model studies.

II. INTRODUCTION

Much of the optimism concerning the potential of carbon dioxide in recovering residual crude oil has been based on slim tube experiments. A 50 to 100 foot length of 3/8 inch tubing is packed with sand, filled with crude oil and displaced with carbon dioxide at a high pressure, which changes but little between entrance and outlet because of the high permeability of the system. A typical result of a slim tube experiment is shown in Figure 1. On the same plot is shown the variation in density and viscosity of the carbon dioxide as a function of the pressure.

268

Earlier investigators have defined[1,2,3,4], and then debated [5,6,7], the predictability of a "minimum miscibility pressure"; that pressure at which 95% of the oil contained in the slim tube is recovered before carbon dioxide breakthrough. There appeared to be an implicit suggestion that good recovery would be achieved as long as the displacement pressure was equal to or greater than this value. Gardner, et.al.[8] however showed that the high recoveries obtained in slim tubes is related to the attainment of a low physical dispersion coefficient (D/vL), and it can be readily shown that the dispersion coefficients achieved in slim tubes are very different from the dispersion coefficients obtained in real reservoirs[9]. Earlier work on viscous fingering of course showed that stable piston-like displacement of one fluid by another is effected whether the fluids are miscible or not; even despite an adverse mobility ratio if the diameter of the flow system approaches the thickness of the fingers that can be generated in the system under study[10]. The juxtaposition of the physical properties of carbon dioxide with the slim tube recovery as a function of pressure in Figure 1 certainly suggests that the increasing recovery is due to gradually decreasing adverse mobility and gravity ratios.

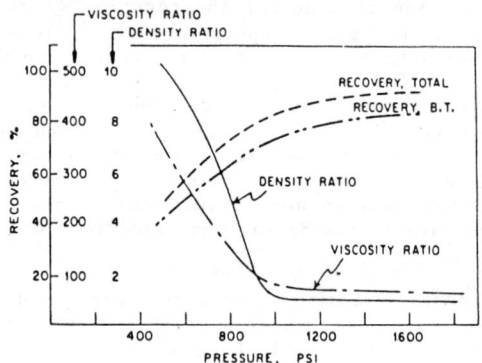

Fig. 1: THE EFFECT OF PRESSURE ON THE
DISPLACEMENT OF A 45° CRUDE IN A SLIM TUBE

Warner[11] studied the numerical simulation of carbon dioxide displacement of residual crude oil with a model that depended on the mixing of the carbon dioxide with the oil for its mobilization and transport. No oil bank was developed in these simulations and the parameter that exercised the chief control over the process was the gravity segregation of the injected carbon dioxide in the water-filled reservoir. Warner's numerical results would have still been poorer had he included the very fine grid that Claridge[12] had earlier shown was necessary for viscous fingering to be properly exhibited in a numerical model.

The phase behavior as described in the current literature, e.g., Reference 8, shows the existence of a miscibility gap[13] in carbon dioxide-crude oil systems. As the carbon dioxide content of the system is increased, the solution of carbon dioxide in the crude fractionates into two or more phases. One is a solution of carbon dioxide in the heavy components of the crude. For virtually all crude oils that have been studied and reported in the literature, the solubility is about 60 mol percent carbon dioxide. The second phase is the mixture of the light components of the crude and the excess carbon dioxide in the system. The latter phase remains a highly mobile fluid and can be expected to finger through the reservoir just as would any low viscosity fluid. Even in the absence of viscous fingering, a Buckley Leverett analysis[14] of the injection of solvent into a watered out reservoir shows that early breakthrough of a mobile solvent should be expected.

The field performance of tertiary pilot operations have indicated that indeed carbon dioxide can mobilize and displace residual crude oil. The Sacroc tertiary pilot recovered 3% of the residual oil at a CO_2/OIL ratio of 36 MSCF/Bbl[15]. At Little Creek, Mississippi, as much as 60% of the oil contained in a pilot pattern may have been recovered at a ratio of 27.6 MSCF/Bbl[16]. At Lick Creek, Arkansas, a pilot operation in the Meakin sand indicates the ratio of recovered oil to injected carbon dioxide will be 28.4 MSCF/Bbl, and the last available figures for the Two Freds field in West Texas indicate a ratio of 18 MSCF/Bbl[16].

If these ratios are projected to be valid for full scale operation, then the economic viability of carbon dioxide injection projects for the recovery of residual oil must be re-examined. Even re-injection of produced carbon dioxide, after required purification and drying, and accounting for interest charges due to the delay between injecting carbon dioxide and recovering the crude oi, the cited ratios of carbon dioxide injected to produced oil would add over $20 to the cost of recovering a barrel of oil by the injection of carbon dioxide.

These studies were carried out in an attempt to determine whether such high carbon dioxide/oil ratios are to be expected or are due to specific peculiarities of the pilot and demonstration tests from which they have emenated.

III. RESULTS OF THE EXPERIMENTAL STUDIES

The experimental work was carried out in physically scaled models of a direct line drive pattern, see Table 1. The details of the scaling procedures that have been used and the construction of the required high pressure (to 5000 psi.), high temperature (to 250°F.) equipment is described elsewhere[9]..

TABLE 1.

PARAMETER	MODEL VALUE	PROTOTYPE VALUE
PERMEABILITY, MDS.	3,000	20.
INJECTION PATTERN	LINE DRIVE	LINE DRIVP
SPACING, INJECTOR TO PRODUCER, FT.	3.08	462.
RESERVOIR THICKNESS, FT.	0.1875	28.2

The reservoir chosen for this particular set of experiments, it can be seen, was one of relatively low permeability. This was done to minimize the effects of gravity segregation and viscous fingering and thus secure a relatively favorable performance.

A. The Production History of Residual Oil Recovery by Carbon Dioxide

Figures 2 through 4 present a typical production history for the displacement of residual crude oil by sub-critical carbon dioxide at 1400 psi and 73°F. It is readily seen that water alone is produced at first; it is the only mobile phase in the reservoir following a water flood. Carbon dioxide appears at the producing end of the system after the injection of about 0.2 of a pore volume, and crude oil production is initiated simultaneously with that of carbon dioxide.

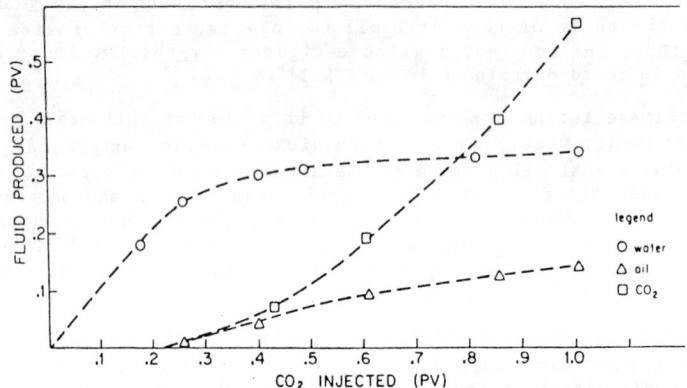

Fig. 2. PRODUCTION HISTORY FOR A TYPICAL CO_2 DISPLACEMENT
45°A.P.I.CRUDE, S_o=0.21, P=1400 PSI, T=73°F.

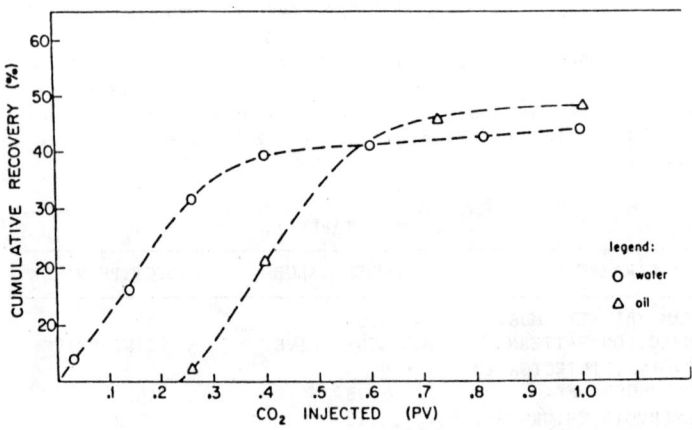

Fig. 3. CUMULATIVE RECOVERY OF OIL AND WATER FOR FIG. 2

The production rate of oil reaches a maximum value within 0.2 to 0.3 of a pore volume following breakthrough, but the ratio of oil to carbon dioxide in the effluent continuously decreases after its first appearance.

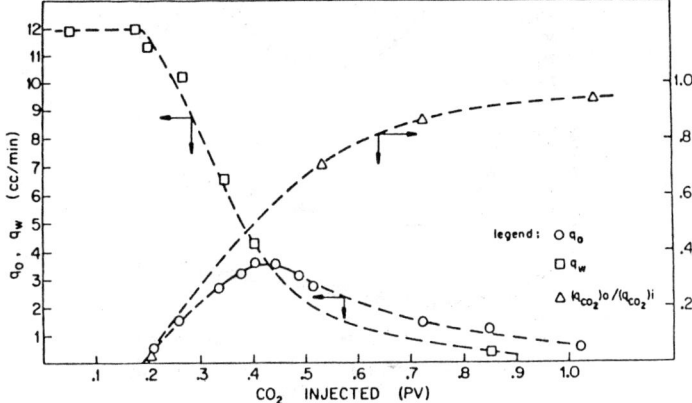

Fig. 4. PRODUCTION RATE HISTORY FOR FIG. 2

The recovery of the residual oil is never complete even with the injection of several pore volumes of carbon dioxide, and this is true even which a completely miscible hydrocarbon (dodecane) is substituted for the crude oil.

It is informative to note that the average molar concentration of carbon dioxide in the oil phase within the model, calculated by material balance, reaches a value of well over 0.6 by the time carbon dioxide breaks through in the effluent, Figure 5. The importance of this observation in connection with the relationship of phase behavior to recovery will be discussed later.

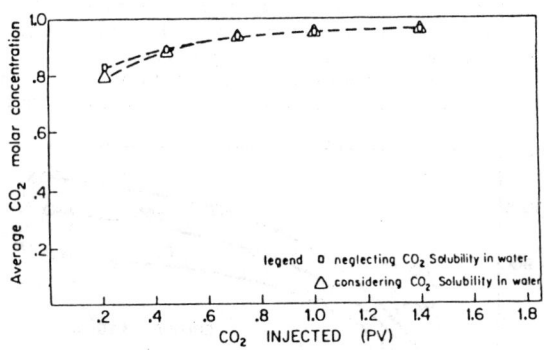

Fig. 5. AVERAGE MOLAR CONCENTRATION OF CARBON DIOXIDE IN OIL PHASE
FOR EXPERIMENT DESCRIBED BY FIGURE 2.

The efficiency of the recovery process, expressed in terms of MSCF/Bbl, is shown in Figure 6 for various initial saturations of crude oil. It is seen that the CO_2/oil ratios are indeed in the range of 20 to 30 MSCF/Bbl when residual saturations of less than 30% are being recovered.

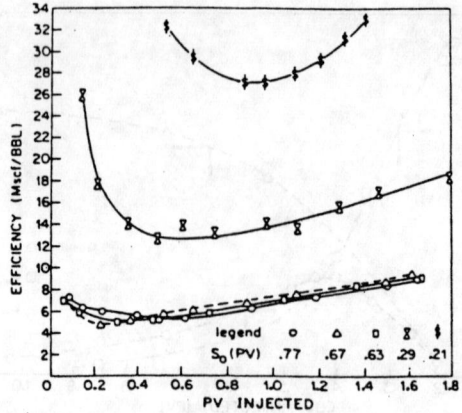

Fig. 6. THE EFFICIENCY OF OIL RECOVERY BY CARBON DIOXIDE
AS A FUNCTION OF INITIAL OIL SATURATION, P=1400 PSI, T = 73°F.

B. Effect of Temperature and Pressure

The temperature and pressure affect the displacement of oil by carbon dioxide in two important aspects. The first is related to the fact that these parameters affect the physical properties, viscosity and density, which, in turn, affect fluid flow including the displacement of one fluid by another. The second is the effect of temperature and pressure on the solubility of the carbon dioxide in the crude oil.

From chemical thermodynamics it is known that for the case where there is no net volume change attendant upon mixing, solubility of one liquid in another is not appreciably affected by pressure, and is increased by temperature. This indeed was found to be the case for the solubility of liquid carbon dioxide in hydrocarbons[17]. The solubility of supercritical carbon dioxide in hydrocarbons increases with increasing pressure and decreases with increasing temperature. The solubility of carbon dioxide in the principal crude oil used in this study is shown in Figures 7 and 8.

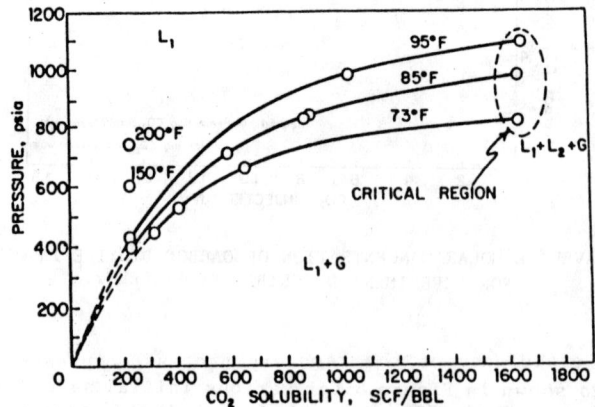

Fig. 7. SOLUBILITY OF CARBON DIOXIDE IN 45° CRUDE, SCF/Bbl.

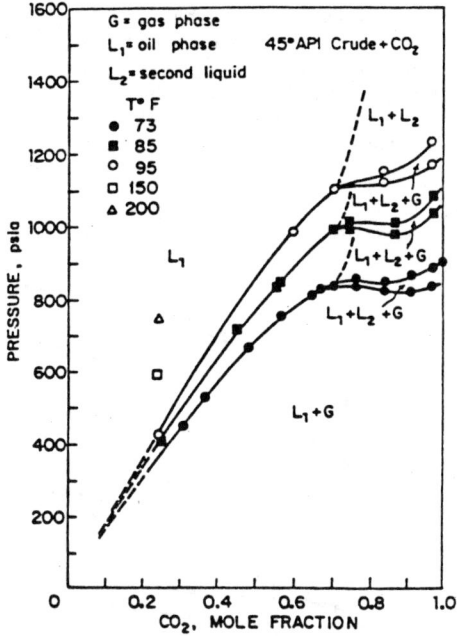

Fig. 8. SOLUBILITY OF CARBON DIOXIDE IN 45⁰ CRUDE, MOL PERCENT

With this background in hand, some twenty runs were conducted to evaluate the effect of temperature and pressure on the displacement of crude oil by carbon dioxide as revealed in this physically scaled model study. Both crude oil and a miscible hydrocarbon (dodecane) were used at residual and high initial saturations, at temperatures to 130°F., and over the pressure range of 650 psi. to 2650 psi.

The effect of pressure on the recovery of the 45°A.P.I. crude oil at sub-critical temperatures is shown in Figure 9 for both residual and high initial saturations. It is apparent that the effect of pressure at a sub-critical temperature is very significant if the pressure is less than the saturation value (about 900 psi at 75°F.), but a further pressure increase above the saturation value has only a minor effect on recovery. For this high initial oil saturation, the recovery increases from a meager 20% to over 60% upon increasing the pressure from 650 psi. to 1,000 psi, but increases from 63% to only 71% when the pressure is increased another 1000 pai to 2000 psi.

The dramatic increase in the recovery as the pressure is increased from 650 psi to 1000 psi at 75°F. parallels a marked increase in the density of carbon density from 0.11 to 0.74 g/cc. The density of the carbon dioxide increases from 0.22 to 0.70 to 0.76 as the pressure increases from 1150 to 2150 psi. and finally to 2650 psi. This translates to a density difference between displacing fluid and displaced fluid (water) of 0.78, 0.30, and 0.24 g/cc. A simple force balance then shows that the same degree of gravity segregation that occurs at a pressure of 2150 psi can be achieved at 1150 psi only by incresing the rate by a factor of three. Hence, the poorer efficiency at 1150 psi as compared to that at 2150 psi is probably due primarily to a greater degree of gravity segregation at the lower pressure. Between 2150 psi and 1650 psi, the small difference in recovery is probably due to the combined effect of small differences in both solubility and gravity segregation.

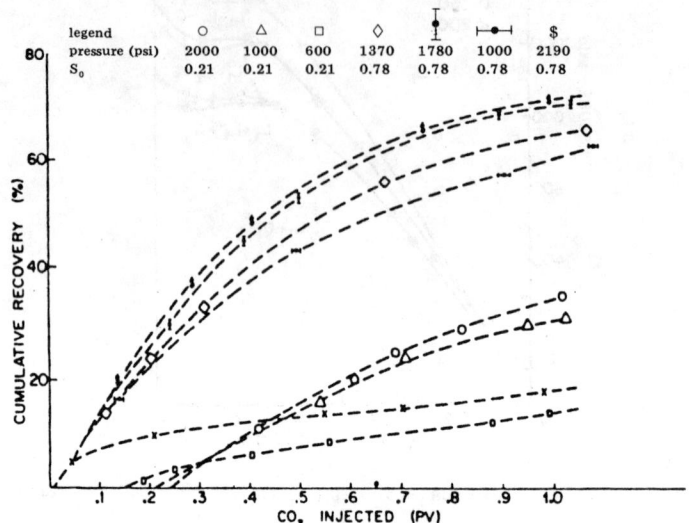

Fig.9. THE EFFECT OF PRESSURE ON RECOVERY
OF CRUDE OIL BY CARBON DIOXIDE AT 73°F.

At a super-critical temperature the effect of pressure on the recovery of residual crude oil is similar to its effect at sub-critical values although the recovery levels are somewhat less than at the lower temperature, Figure 10.

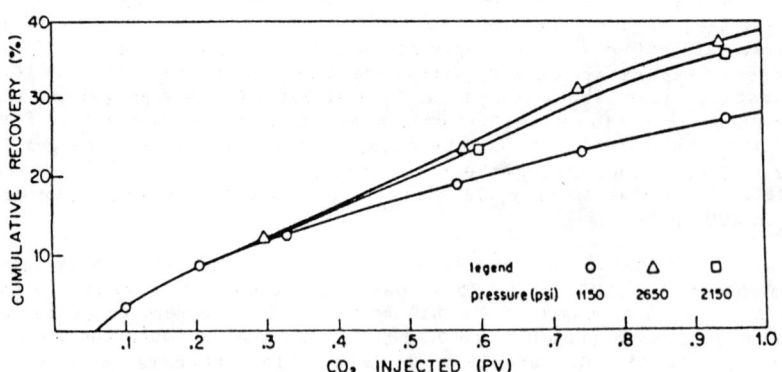

Fig. 10. THE EFFECT OF PRESSURE ON
RECOVERY OF CRUDE OIL AT 130°F.

To further study the role of density on the displacement of carbon dioxide, a set of experiments were performed over a range of temperatures and pressures where the density could be maintained relatively constant by manipulating these two parmeters. At 1400 psi. and 75°F. the density of carbon dioxide is 0.82, as it is at 2700 psi. and 125°F. even though the temperatures are below and above the critical temperature, respectively. The results of two runs, one at each of the foregoing sets of parameters, is shown in Figure 11 and it is readily seen that the results of the two runs can be superimposed on each other.

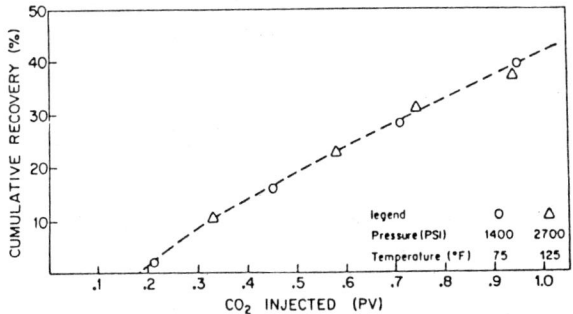

Fig. 11. THE COMPENSATING EFFETS OF TEMPERATURE AND PRESSURE
ON THE DISPLACEMENT OF RESIDUAL CRUDE OIL

To gain still further insight into the displacement of crude oil by carbon dioxide, the effect of pressure on the displacement of a completely miscible hydrocarbon was investigated as a function of pressure, Figure 12. It is apparent that the effect of pressure is the same for the displacement of dodecane as it is for that of crude oil; it is the rate of change of solubility with pressure that affects the rate of change of recovery with pressure whether the displacing fluid is miscible or not.

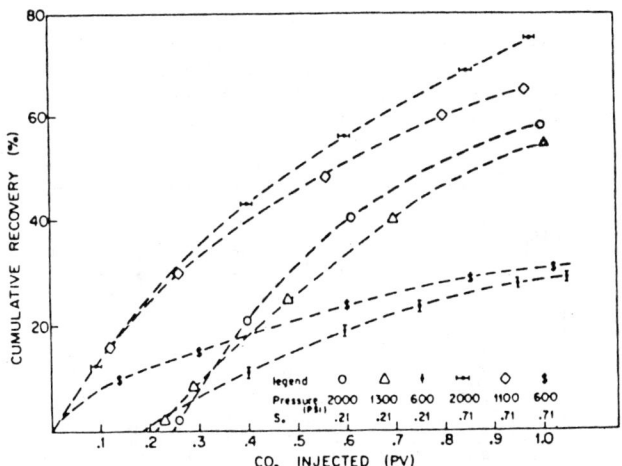

Fig. 12. THE EFFECT OF PRESSURE ON THE DISPLACEMENT OF
DODECANE BY CARBON DIOXIDE

C. Effect of Injection Rate

Experiments were conducted at prototype velocities varying from 0.05 to 0.4 foot per day. Fig. 13 shows that increasing the velocity over the indicated range results in a slight improvement in recovery. However, the effect is sensed only in the later life of the flood.

Some experiments were conducted at still lower rate, but gravity segregation dominated the system and the crude oil recovery decreased rapidly.

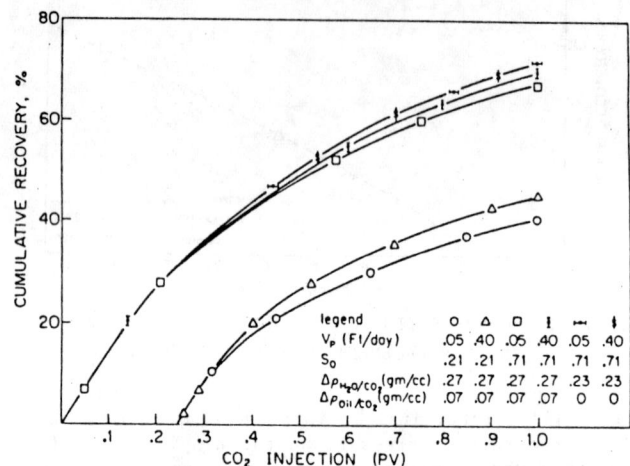

Fig. 13. THE EFFECT OF VELOCITY (OR INJECTION RATE)
ON THE RECOVERY OF CRUDE OIL BY CARBON DIOXIDE

D. The Effect of Initial Oil Saturation.

The initial oil saturation has a very direct effect on both the fractional oil recovery, and the resulting carbon dioxide/oil ratio, see Fig. 14. The oil recovery is only 50% of the residual saturation but climbs to 80% of the initial oil saturation when the latter is 80%.

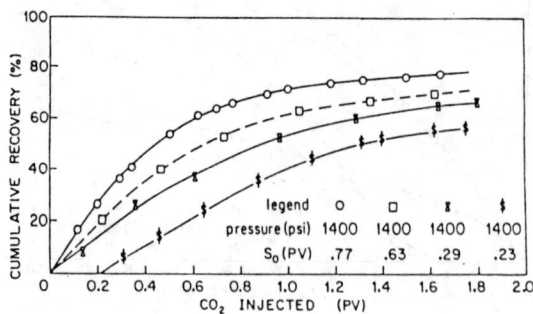

Fig. 14. THE EFFECT OF INITIAL OIL SATURATION ON OIL RECOVERY

The difference in the efficiency of the displacement is directly related to the fact that when a low oil saturation is displaced, much of the injected carbon dioxide is being used merely to displace the water in the reservoir (which is necessary before the carbon dioxide can contact and dissolve in the crude oil). When the initial oil saturation is above the residual value, both oil and water are produced throughout the run. The flow of oil, especially after carbon dioxide breakthrough, is much higher than would be predicted from the relative permeability relationship for the oil-water system; this leads to the conclusion that the mobility of the oil in which carbon dioxide has dissolved has been increased. It is obviously the viscosity reduction and, perhaps, most importantly, the swelling of the crude oil phase which causes this increase in mobility. The swelling of the 45° crude oil used in this study is shown in Figure 15 it is substantial.

The CO_2/OIL ratio is only about 7 MSCF/Bbl for an initial saturation of 0.77, but increases to a value of 15 MSCF/Bbl when the initial saturation is dropped to 0.29.

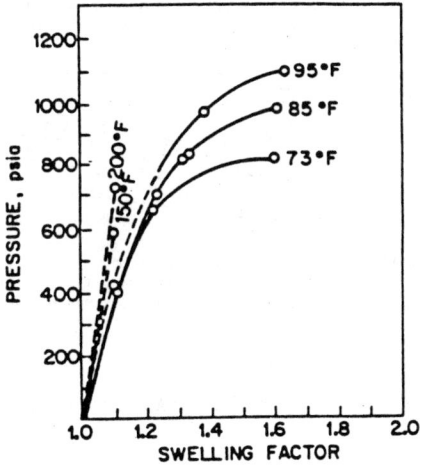

Fig. 15. THE SWELLING FACTOR OF A 45°A.P.I. CRUDE BY CO_2

E. The Effect of Oil Composition

Four different "oils" were used to study the effect of oil composition: dodecane, which is completely miscible, hexadecane which displays a miscibility gap, the 45°A.P.I. crude, and a solution of a 14° A.P.I. crude in the 45° crude. The results are shown in Figure 16.

The overall recovery is highest when carbon dioxide is completely miscible with the oil phase, viz., dodecane. However, only a slightly lower oil recovery is achieved when hexadecane is substituted for the dodecane. There is in fact virtually no difference in the recovery of the oil phase, as long as its viscosity is less than 6 centipoises, miscible or not, during the injection of the first 0.5 pore volume of carbon dioxide. The slight differences in recovery that develop upon the injection of more carbon dioxide are perhaps best understood by referring to the experiment using a mixture of 14° and 45° crude oils.

278

This mixture had a viscosity of 20 centipoises and a gravity of 35°; its recovery by continuous carbon dioxide injection was noticeably less than the recovery with the other oils which had a viscosity of less than 6 centipoises. Exacerbated viscous fingering, lower solubility and swelling probably all contribute to the lower recovery for the more viscous crude. It is informative to note, Fig. 17, that most of the oil that is recovered has the same gravity as that of the mixture. Only at the tail end of the recovery is there some evidence of the lighter fraction being preferentially recovered. Because of the fractionation of the system into two or more phases when the carbon dioxide content of the system exceeds a mol fraction of 0.6 to 0.7, the mobile phase containing a large fraction of carbon dioxide and a small fraction of the light ends of the crude oil is being preferentially produced. Additional work has shown that when a live crude oil is used, the results are not significantly different from those that have been described for the dead crude. Although a good part of the methane appears to be stripped from the crude as the carbon dioxide dissolves in it, it also appears that fractionation occurs at a somewhat lower molar concentration of carbon dioxide and a more volatile oil is produced somewhat earlier. However the overall recovery does not seem to be substantially affected by the presence of moderate amounts of methane in the crude oil.

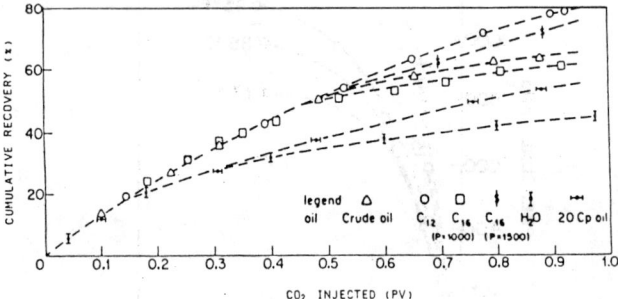

Fig. 16. COMPARISON OF THE DISPLACEMENT OF VARIOUS OILS

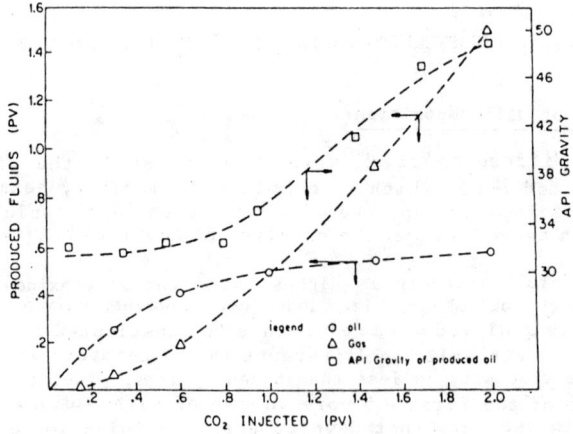

Fig. 17. CHANGES IN COMPOSITION OF A RECOVERED CRUDE OIL

F. The Effect of Slug Size

The effect of using slugs of carbon dioxide on the recovery of residual crude oil was studied, and the results are presented in Figures 18 and 19. The 45° crude was used in all these experiments, and the residual oil saturation was consistently brought down to 0.21 p.v. before initiating the test. It is important to note in the following discussion, that the comparisons that will be made on the efficiency of the various slugs will be for a <u>limited volume of total fluid injected</u>, carbon dioxide or carbon dioxide and water.

For operating conditions of 1000 psi. and 73°F. the oil recovery increases linearly with an increse in slug size from 0.11 to 0.22 pore volume for a total injection of 1.0 to 1.2 pore volumes. However, when the size of the slug is increased above 0.22 pore volume, and the total fluid injected is kept constant at about one pore volume, the recovery does not increase any further. As a matter of fact, as long as the total fluid injected is limited to 1.2 p.v., the recovery actually decreases as the slug size is increased above a value of 0.22.

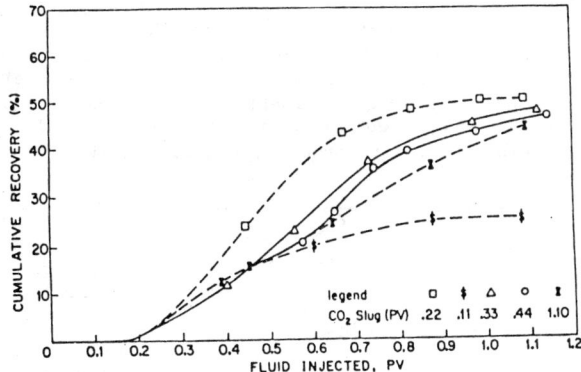

Fig. 18. THE EFFECT OF SLUG SIZE ON THE RECOVERY OF RESIDUAL CRUDE OIL
1400 PSI, 73°F.

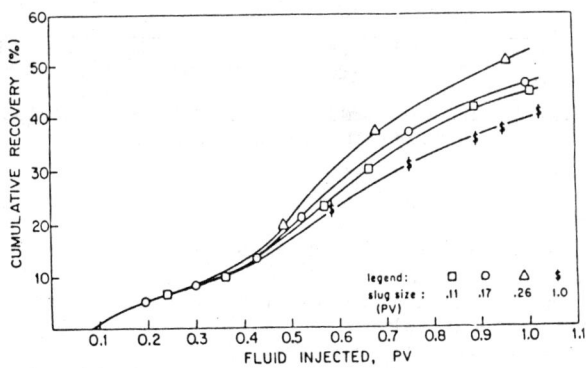

Fig. 19 THE EFFECT OF SLUG SIZE ON THE RECOVERY OF RESIDUAL CRUDE OIL
18 PSI, 130°F.

280

Over the range of temperatures and pressures investigated in the study of the slugs of carbon dioxide, 73° to 130°F., and from 1000 psi to 1800 psi., the optimum slug size showed no consistent change; it ranged from 0.20 to 0.26 pore volume. It is hypothesized that the optimum slug size is that volume of carbon dioxide which can be injected into the system without establishing a free and continuous saturation throughout the entire model. Once such a mobile gas saturation is established, any further injection of carbon dioxide results in the development of a (dense) gas drive, which is relatively inefficient in displacing the swollen crude oil. On the other hand, if carbon dioxide injection is halted before a free gas phase saturation is established throughout the model, then the swollen crude oil phase, rendered mobile by the increase in its pore volume saturation, will be much more efficiently displaced by a relatively viscous fluid, viz, water.

The increased efficiency of slugs of carbon dioxide in recovering residual crude oil is well illustrated by the results of this work which are plotted in Figure 20. Again, it must be noted that the reservoir prototype modelled in this work is one which should show up carbon dioxide at its very best.

The efficiency of the ultimate displacement can be increased slightly if the viscosity of the chase water is increased by the addition of a glycol or a polymer. If, following the injection of an optimum slug of carbon dioxide, nitrogen is injected; then the resulting recovery of oil is markedly reduced. The nitrogen is an inefficient displacing fluid; moreover it strips some of the dissolved carbon dioxide from solution in the crude oil, thereby defeating the entire process, see Figure 21.

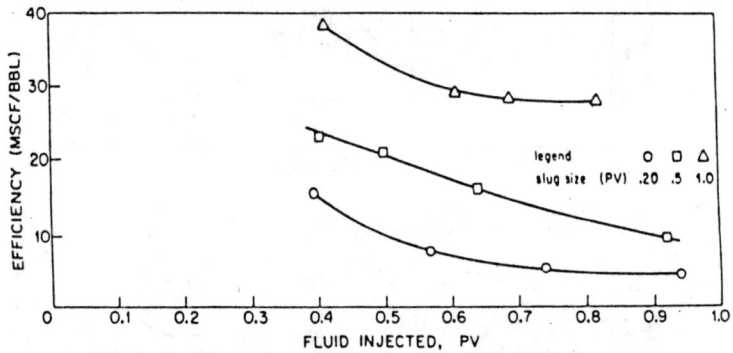

Fig. 20. THE EFFICIENCY OF SLUGS OF CARBON DIOXIDE
IN RECOVERING RESIDUAL CRUDE OIL

G. Nitrous Oxide

In order to gain further corroboration for the hypothesis that it is the swelling of the residual crude oil is the key factor in the recovery of the latter by the injection of carbon dioxide, a search was made for other substances that would dissolve to the same extent and swell the crude oil equivalently: nitrous oxide has been described to be virtually equivalent to carbon dioxide in many physical properties[18]. Experiments proved that nitrous oxide performed in the physical models in a virtually identical manner to carbon dioxide. (It is a far more expensive substance.)

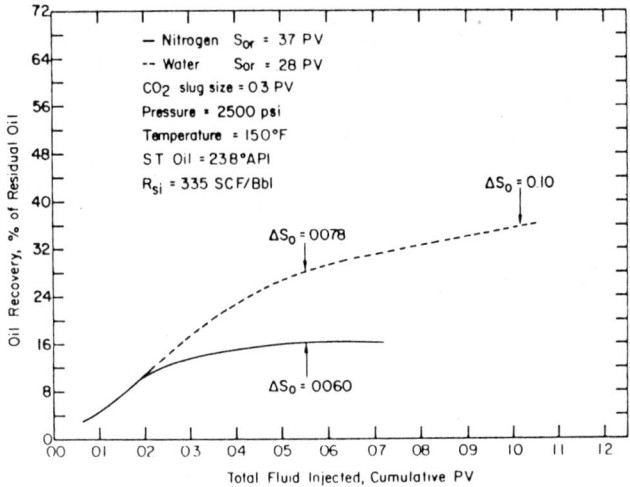

Fig. 21. THE INEFFICIENCY OF NITROGEN AS A CHASE FLUID AFTER A CO_2 SLUG

IV. CONCLUSIONS

Physically scaled model studies of the displacement and recovery of crude oil by carbon dioxide yield results which are consistent with the results of field demonstrtion and pilot projects, and consistent with the principles of fluid flow and phase behavior.

Continuous injection of carbon dioxide will recover a significant fraction of a waterflood residual oil saturation, but the resulting carbon dioxide/oil ratios will be above 20 MSCF/B, and may be as high as 30.

The use of slugs of carbon dioxide followed by water will effectively reduce the resulting carbon dioxide/oil ratio without seriously affecting the amount of oil that can be recovered by the injection of a total of about one pore volume of fluid. Although values approaching 5 MSCF/B have been achieved in these model studies, it is cautioned that the model used was a very favorable one, viz., low permeability, uniform and linear. Even minor heterogeneity in a field operation will encourage channeling, and the decrease in the viscous to gravity forces encountered in radial flow away from the well bores will encourage gravity segregation.

The performance of the displacement experiments leads to the conclusion that the mechanism by which carbon dioxide displaces residual crude oil is comprised of three sequential steps: 1) the immiscible displacement of the oil-occluding, mobile water, 2) the solution of carbon dioxide in the crude oil and its subsequent swelling that develops oil phase mobility, and 3) the immiscible displacement of the mobile solution of carbon dioxide in oil by the continuing flow of carbon dioxide or water.

Although the residual saturation of the oil phase (a solution of carbon dioxide in oil) can be lowered by continuing the flow of carbon dioxide, resulting in some continuing evaporation of crude oil fractions, the resulting incremental carbon dioxide/produced oil ratios will be very high. The more practical limit to the recovery is reached when the residual saturation of the low viscosity oil phase to the subsequent gas or water drive is approached.

Nitrous oxide, which dissolves in and swells crude oils similarly, is as effective as carbon dioxide in recovering crude oil. The substitution of nitrogen for water as a chase fluid injures the recovery because the gas is not as good a displacing agent for the swollen crude.

The complex phase behavior of carbon dioxide with crude oil appears to contribute little to the recovery process; the effect of the fractionation of the crude in the presence of carbon dioxide results in some slight additional recovery at the tail end of the flood.

Slim tube experiments since they do not correctly model the dispersion coefficients and the relations between gravity and viscous forces do not provide adequate insight into a reservoir recovery process. The so-called minimum miscibility pressure as interpreted from such experiments is actually the pressure above which no significant increase in recovery will be achieved. The recovery mechanism is still effective at lower pressures.

ACKNOWLEDGEMENTS

The work on this project was supported by the United States Department of Energy, Gary Energy Co., and endowment funds at the University of Southern California.

REFERENCES

1. Beeson, D. M., and Ortloff, C.D., "Laboratory Investigation of the Water-Driven Carbon Dioxide Process for Oil Recovery", TRANS AIME (1959) 216, 388-91
2. Holm, L.W., "Carbon Dioxide for Solvent Flooding for Increased Oil Recovery", TRANS AIME (1959) 216, 225-231
3. Rathmell, J.J., Stalkup, F.I., and Hassinger, R.C., "A Laboratory Investigation of Miscible Displacement by CO_2", SPE 3483, 46th Annual Meeting of SPE of AIME (1971)
4. Holm, L.W., and Josendahl, V.A., "Mechanism of Oil Displacement by Carbon Dioxide", JPT (1974), 1417-1438.
5. Dunnyshkin, I.I., and Namoit, A., "Study of Conditions of Petroleum Miscibility with Carbon Dioxide", Neft. Khoz., (1978), v. 3, 59-61
6. National Petroleum Council, "Enhanced Oil Recovery - An Analysis of the Potential for Enhanced Oil Recovery from Known Fields in the United States - 1976 to 2000, Washington, D.C., (1976)
7. Yellig, W.F. and Metcalfe, R.S., "Determination and Prediction of CO_2 Minimum Miscibility Pressure", JPT, (1980), 160-168.
8. Gardner, J.W., Orr, P.M., and Patel, P.D., "The Effect of Phase Behavior on CO_2 Flood Displacement Efficiency", SPE 8367, 54th Annual Meeting of SPE of AIME, Las Vegas, (1979)
9. El Arabi, M., Ph.D. Dissertation, University of Southern California, June 1981.
10. Offeringa, J., and van der Poel, C., "Displacement of Oil From Porous Media by Miscible Liquids", TRANS AIME (1954) 201, 310-317
11. Warner, H. R., Jr., "An Evaluation of Miscible CO_2 Flooding in Waterflooded Sandstone Reservoirs", JPT, (1979), 1339-1347

12. Claridge, E.L., "Discussion of the Use of Capillary Tube Networks in Reservoir Peformance Studies", SPEJ (1972), 352-61

13. Glasstone, S., Text Book of Physical Chemistry, p. 713, D. Van Nostrand, New York, 1940.

14. Doscher, T. and Gharib, S., "Physically Scaled Models Simulating the Displacement of Resdiual Oil by Miscible CO_2 in Linear Geometry", SPE 8896, 50th Annual California Regional Meeting of SPE of AIME (1980)

15, Kane, A.V., "Performance Review of a Large Scale Carbon Dioxide-WAG Project, SACROC Unit-Kelly Snider Field, SPE 7091, SPE Improved Oil Field Recovery Symposium, Tulsa 1978

16. Gruy Federal, Inc., "Target Reservoirs for CO_2 Miscible Flooding", U.S.Department of Energy, Washington, D.C., (1980)

17. Kamath, K.I., Comberiati, J.R., and Zammerilli, A.M., "The Role of Reservoir Temperature in Carbon Dioxide Flooding", Paper N4, presented at the U.S.Department of Energy Symposium, Tulsa, Oklahoma 1979

18. Gerrard, W., Solubility of Gases and Liquids, A Graphic Analysis, Plenum Press, New York (1976). See also, Hildebrand, J.H., and Scott, R.L., The Solubility of Non-Electrolytes, Reinhold, New York (1950).

LABORATORY TESTING PROCEDURES FOR MISCIBLE FLOODS

S. G. SAYEGH and F. G. McCAFFERY*

Petroleum Recovery Institute, Calgary, Alberta, Canada T2L 2A6

ABSTRACT

The objective of this paper is to provide a state-of-the-art review and critique of laboratory testing procedures for miscible flooding for researchers in the field. An additional aim of the paper is to give reservoir and production engineers insight into those procedures, so that they may appreciate their potentials and limitations, and be better able to evaluate laboratory results in light of their field experience.

The topics treated include single- and multiple-contact phase behavior and physical properties measurements, and involve slim-tube and core displacement tests. General objectives for each type of test are listed, recommended practices are outlined, and many examples from the literature are referenced. In addition, general screening criteria are presented for the selection of suitable candidate reservoirs for miscible flooding.

INTRODUCTION

One of the principal enhanced recovery methods currently under consideration for light oil reservoirs is miscible flooding with carbon dioxide and/or hydrocarbon solvents. The process is complex and involves many parameters that have to be optimized so that a flood can lead to a technical and economic success. Some of the factors that have to be studied are the reservoir geology, oil and oil-solvent phase behavior, oil solvent displacement characteristics, waterflood performance, as well as reservoir engineering aspects such as solvent production and oil injection strategies, expected performance under both water and solvent flooding, and economics.

In this paper, laboratory testing procedures for miscible flooding will be discussed. These will include the measurement of the phase behavior and displacement data of reservoir crude oil-solvent systems, and how such data may be used in evaluating the suitability of a solvent flood for a particular application. The objective of this paper is to provide a state-of-the-art review and critique for researchers in the field. An additional aim of the paper is to give reservoir and production engineers insight into laboratory testing procedures so that they may appreciate their potentials and limitations and thus be better able to evaluate laboratory results in light of their field experience. For other

---

* Present address: Occidental Research Corporation, Irvine, Calif. 92713, U.S.A.

reviews of the miscible flooding process and its field applications, the reader is referred to the works by Holm[1], Stalkup[2], Dosher et al.[3], and Mungan[4,5]. Burnett and Dann[6] have reviewed screening tests for a variety of enhanced oil recovery processes.

PROCESS DESCRIPTION AND GENERAL SCREENING CRITERIA

In a miscible flood the solvent contacts the oil and a mixing zone is formed. In the mixing zone, there is a gradual change in composition from oil to solvent, without an interface. For economic reasons, the solvent is usually not injected continuously, but often in the form of a slug typically about 20-30% of the hydrocarbon pore volume (HCPV). The slug is then followed by a chase fluid, usually water or lean gas, to drive it through the reservoir towards the production wells. The slug may be injected in small portions alternating with water, commonly called the water-alternating-gas (WAG) process. Alternatively, water may be co-injected with the solvent. These latter injection modes help control the high mobility of the solvent.

It is rarely technically or economically feasible to inject a solvent that is directly miscible with the oil. Instead, miscibility is generally achieved through what are known as the multiple-contact miscibility (MCM) mechanisms[7-13]. Two such mechanisms can occur when gaseous or supercritical solvents are used: a condensation mechanism and a vaporization mechanism. When subcritical solvents are used at pressures above their bubble point, the process is one of liquid-liquid extraction[14,15].

The high oil recovery in miscible floods is attributed to the following factors:

- high microscopic displacement efficiency
- oil extraction by solvent
- low interfacial tension
- oil swelling
- oil viscosity reduction
- blowdown recovery

The solvents used (CO_2 and hydrocarbons) are generally less dense and viscous than the oils. This causes the solvent to override the oil and finger through it. These are adverse factors in horizontal floods and lead to early solvent breakthrough, poor sweep efficiency, and low oil recovery. In general, a good candidate reservoir for horizontal miscible flooding should have the following characteristics:

- thin pay zone, up to 5 m
- good horizontal continuity
- relatively homogeneous
- low vertical-to-horizontal permeability ratio
- not fractured
- contains undersaturated oil
- contains no free gas saturation
- contains no mobile water

The solvent should be chosen such that it:

- achieves miscibility with the oil at reservoir conditions
- is cheap
- is readily available

For vertical downward displacements, the requirements are somewhat less constraining:

- the reservoir should not contain permeability barriers to vertical flow
- the displacement should be carried out at a suitable rate such that the flood is gravity stable

PHASE BEHAVIOR MEASUREMENTS

Phase behavior measurements are carried out for several purposes:

- to characterize the oil-solvent system
- to determine the mechanism by which miscibility is achieved
- to fine-tune the phase behavior packages in compositional simulators

In general, the phase behavior studies involve the following measurements:

- solubilities
- multiple phase formation, including both liquid and solid phases
- densities
- oil swelling
- viscosities

Phase Behavior Measurement Equipment

High pressure phase equilibrium experimental techniques for a variety of applications have recently been reviewed by Eubank et al.[16] Apparatuses used in connection with CO_2 and hydrocarbon systems were described by other researchers.[17-26]

A common type of apparatus consists of a windowed cell whose volume may be manipulated by means of a piston or mercury from a positive displacement pump. The cell is placed in a thermostated oven for temperature control. The desired components of the mixture are loaded into the cell and then mixed. Mixing is usually done with a magnetically-coupled stirrer, by rocking the cell, or by circulating the fluids. Once equilibrium has been reached, visual observations of the coexisting phases may be carried out. Samples of these phases may also be withdrawn for density and viscosity measurements, and for compositional analyses. Constant composition expansions may also be carried out to determine bubble and dew points, and volumetric proportions of coexisting phases as functions of pressure.

The apparatus described by Lee[23] and Sayegh et al.[26] has two interconnected cells. This gives a greater flexibility of operation and permits the measurement of viscosity without using a separate viscometer. D. Robinson (personal communication) at the University of Alberta has the cell constructed entirely from sapphire. This permits unhindered visual observation of the entire contents of the cell. The apparatuses of Orr et al.[24], of Connor and Pope[25], and of D. Robinson have their sampling lines directly connected to gas chromatographs for analysis. The apparatus described by Orr et al.[24] differs from the others in that it resembles a continuously stirred tank reactor.

Phase Behavior Tests to Characterize the Crude Oil

Typical tests for the characterization of the crude oil involve the measurement of its composition, molecular weight, density, viscosity, compressibility, bubble point, formation volume factor, gas-oil ratio, distillation curve, differential liberation, constant volume depletion, and constant composition expansion characteristics.

These tests are generally carried out at reservoir temperature using, for example, ASTM standard procedures and are preferably carried out with bottom-hole samples. Testing of oil properties should be periodically repeated during the production lifetime of a reservoir, and be carried out on samples from the different producing zones or horizons of a pool to determine if there are any variations in oil properties. This is especially important where the reservoir pressure falls below the original bubble point of the oil.

Standing[17] and Henry et al.[28] presented descriptions of bottomhole sampling procedures. In general, the sampling well should be selected so that it is representative of the average reservoir conditions. The well should be produced at a slow rate during sampling to minimize pressure drawdown effects and the resultant phase changes. Also, sufficient sampling time should be allowed to ensure that the sample bomb is filled with fresh oil.

Large volumes of reservoir fluids are necessary to carry out a complete laboratory study of a miscible flood. Thus, it is unreasonable to use bottom-hole samples for all these tests. The normal procedure is to take large samples of separator oil and gas, then recombine them to match the properties of the bottomhole sample.[25]

Phase Behavior Tests to Characterize the Crude Oil-Solvent System

The general phase behavior of hydrocarbon fluids have been well reviewed.[17,28] Data for hydrocarbon floods of reservoir crudes were presented by several authors[7,8,25,29], while most of the recently published studies have dealt with the phase behavior of CO_2-oil systems.[11,12,14,18,23,24,26,30-38] This reflects the growing interest in using CO_2 as a miscible flooding agent.

The following discussion will concentrate on CO_2-reservoir crude oil systems since these are of most interest to the industry. The phase diagrams of CO_2-crude oil systems are often presented in the form of ternary phase diagrams.[9,12,14,24] Such a representation provides a convenient form for the visualization of the compositional path during a constant temperature and pressure displacement[11,37] and for determining the mechanism of achieving miscibility.[9] It should, however, be remembered that the ternary representation is not thermodynamically rigorous and hence should not be interpreted literally. More accurate predictions of the displacement path may be made using a quaternary diagram.[9,36]

A second type of test is the constant composition expansion.[7,23,32,33] This provides information on the phase behavior of the CO_2-oil system in the various locations of the reservoir where the pressure may vary. For example, at conditions where multiple liquid phases appear, the slug could break down, while asphaltene precipitation could lead to a reduction in reservoir permeability.

The density, swelling factor, and viscosity of the CO_2-saturated oil[18,26,31] are usually measured in parallel with the phase-envelope measurements described above. Connor and Pope[25] recently presented such data for hydrocarbon-oil systems. In general, as the pressure increases, more solvent gas dissolves into the oil causing it to swell and thus to reduce its density and viscosity. Carbon dioxide is generally more effective in this regard than hydrocarbon solvent gases.[36] At very high pressures, the density and viscosity curves could start increasing because the effect of pressure on the fluid properties predominates over the effect of solvent dissolution.

Phase Behavior Tests to Determine the Mechanism of Multiple-Contact Miscibility

The tests mentioned previously are all static, single-contact tests. The tests described in this section are designed to simulate the dynamic, multiple-contact process occurring in a reservoir between the injected solvent and the reservoir crude oil. These tests are carried out in a controlled manner in a PVT cell, thus the process parameters are well defined.

The first type of test is the generation of a Benham plot by a stagewise approximation of the continuous multiple-contact process.[12,19,24,25,39] In this procedure, a certain proportion of oil and solvent are mixed in a PVT cell and allowed to reach equilibrium. The proportions and properties of the resultant vapor and liquid are then measured. If a condensation process occurs, the vapor phase is then purged and a fresh batch of solvent is introduced into the cell. On the other hand, the liquid phase is purged if, based on changes in phase volume, a vaporization process is involved, and a fresh batch of oil is introduced into the cell. The entire process is repeated until only one phase appears in the cell, at which point MCM has been attained.

The drawback of this method is that it is a stagewise process, which only approximates the continuous contacts in a reservoir. As such, it is implicitly assumed that the oil and solvent in the reservoir have enough time to reach equilibrium. This is probably a reasonable assumption in many cases since reservoir flow rates are quite low, but if severe channelling, fingering, or gravity segregation occur in the reservoir, true equilibrium may not be attained and the prediction will be optimistic. Another problem associated with designing this type of batchwise experiment is the choice of volumetric ratios of gas-to-liquid contacted in each step. Reservoir parameters such as the mobilities of the phases and flow rates should be taken into account to determine a realistic ratio.[39]

The procedure described by Orr et al.[24] is a variation of the above method in that the multiple contacts are carried out continuously. In such an experiment, the rate of solvent injection into the cell would have to be carefully selected to obtain meaningful results.

LABORATORY DISPLACEMENT TESTS

Laboratory displacement tests provide important information on the behavior of reservoir fluid/solvent systems under dynamic displacement conditions. These tests are of two types: slim-tube and core displacements. It is important to carry out both types of tests in a laboratory study since each one provides different information necessary for the evaluation of a field application. Each type of test will now be discussed in further detail.

Slim-Tube Displacement Tests

Slim-tube displacement tests are laboratory tests that are carried out in an idealized porous medium. As such, they may be thought of as being an intermediate approximation to reservoir floods, lying between the more realistic core floods and the more idealistic multiple-contact PVT cell tests. A slim-tube test is carried out primarily to determine if a solvent achieves miscibility with an oil at a certain temperature and pressure. A laboratory investigation involving a series of runs could be done with either or both of the following objectives:

- minimum miscibility pressure (MMP) determination
- solvent screening

Orr et al.[24] have made a summary of slim-tube displacement apparatuses used by various investigators. The slim tube is normally constructed from horizontally coiled stainless steel tubing. The tube is 10-20 m long, about 5 mm internal diameter, and packed with fine glass beads or sands to a porosity of about 30% and to a permeability of 3-15 μm^2. The coil is first saturated with oil, then flooded with CO_2. The effluent from the slim-tube passes through a sight glass for visual observation, is sampled for analysis, and is then flashed to atmospheric pressure through a backpressure regulator. Produced liquid and gas phases are metered separately. The data obtained from the test include effluent color, number of phases, composition and gas-oil ratio, as well as oil recovery and pressure drop across the coil--each as a function of the volume of solvent injected.

The basic assumption in slim-tube tests is that the displacement is piston-like and that little or no fingering occurs. This is due in part to the uniformity of the packing and the dampening effect of the tube's walls. Accordingly, the criteria for miscibility being achieved in a carbon dioxide flood are:

- no appearance of a methane bank prior to breakthrough
- late solvent breakthrough (at around 0.8 pore volumes of solvent injected, or later)
- a smooth transition from oil to solvent in the mixing zone without the appearance of an interface
- high ultimate recovery (greater than 95% of the original oil-in-place, OOIP)

On the other hand, an immiscible displacement is characterized by:

- the appearance of a methane bank prior to solvent breakthrough
- early breakthrough
- the observation of an interface between the oil-rich and solvent-rich phases in the mixing zone
- low ultimate recovery

All of the above-noted symptoms of an immiscible displacement should appear if the pressure is well below the MMP. This also depends to some extent on the characteristics of the slim tube itself (tube diameter, uniformity of bead size and packing). It would be instructive to carry out two initial displacements to characterize the particular slim tube being used. The first flood could be conducted under definitely immiscible conditions using nitrogen, for example, as the flooding agent, while the second flood would involve first-contact miscible conditions using benzene, for example, as the displacing agent. For further discussions, the reader is referred to other published works.[24,31,40,41,42]

A variety of slim tube lengths have been used by various researchers.[24] It would appear that multiple-contact miscibility is achieved fairly early in the life of the displacement (within the first two meters), otherwise a high oil recovery would not be obtained. This is supported by the lower number of contacts (about 10) required in PVT cell, multiple-contact experiments[12,19,25] although, as mentioned previously, such experiments are open to interpretation. On the other hand, Yellig[15] concluded that longer lengths (2.5 - 5 m) were required to develop miscibility when carbon dioxide was in the liquid form. Thus, a slim tube length between 10-20 m is recommended. The rate at which slim-tube displacements are run affects the stability of the displacement front and the time allowed for contact between the oil and solvent. For this reason, displacement rates are best kept at less than 10 m/day. The use of relatively low rates also minimizes the pressure drop across the slim tube, which provides for good definition of the minimum miscibility pressure.

Benham et al.[8] have presented correlations for the minimum enrichment of dry gas (by LPG) required to achieve miscibility, while Jacobson[43] studied the contribution of acid gases to miscibility. Other researchers[10,31,40,41,42,44] have investigated the effect of the different process variables on the carbon dioxide MMP. In general, the MMP increases with decreasing oil gravity and its C_5 to C_{30} content, and with increasing temperature and molecular weight of the oil C_{5+} fraction. Hydrogen sulfide and LPG in the carbon dioxide decrease the MMP, while nitrogen and methane increase it.

In addition to studying dynamic miscibility conditions, the results of slim-tube experiments may be used to calibrate compositional simulators.[39,45,46] Wang and Loche[47] investigated the relative efficiency of different water-alternating-gas cycles and concluded that the total oil recovery was insignificantly affected by the injection sequence provided that the total amount of carbon dioxide injected remained the same.

In summary, slim-tube displacement tests are an extremely useful tool for studying the miscibility relationship between oil and solvent systems under controlled dynamic conditions. Caution must be exercised when transposing the results of such studies to reservoir systems since the effects of the reservoir rock properties (homogeneity, relative permeability, wettability, and pore geometry) have not been taken into account, hence displacement tests on reservoir rocks must follow. The following section deals with core displacement tests in an attempt to provide more detailed insight into the displacement behavior as it may occur in the reservoir in regions contacted by the solvent.

Core Displacement Tests

Following slim-tube displacement tests to confirm the establishment of miscibility with the oil for a given solvent at appropriate reservoir conditions of temperature and pressure, core flooding measurements are generally recommended. Such tests can be used to evaluate a variety of displacement phenomena that have bearing on the miscible flooding process. These include.

- recovery mechanisms[9,11,49]
- diffusion and dispersion coefficients, and dead-end pore volumes[49-56]
- miscible and compositional simulator tuning[48,64]
- chromatographic separation of components[11,48]
- water, oil, and gas relative permeabilities[57,70]
- rock interactions with gas and brine
- dynamic oil-solvent phase behavior[58]
- effect of the following factors on displacement efficiency or oil recovery:

. rock type[1,50,56]
. solvent type[14,60]
. water saturation (secondary or tertiary flooding mode)[59,60,63,65]
. phase behavior (multiple liquid and solid phases)[32]
. displacement pressure[11,15]
. solvent injection rate[15,61]
. flooding mode (continuous solvent injection, solvent slug size, WAG, water-solvent coinjection, CO_2-foam and CO_2-polymer injection)[62,63]
. blowdown
. low interfacial tension[66]

A core displacement apparatus consists of a core holder in which the core is placed under a confining pressure. The core is connected to reservoir oil and brine, injection water, and solvent containers. The core is flooded at reservoir temperature and pressure with these fluids in the proper sequence, and the fluid production and pressure drops are monitored. Visual observation of the core's effluents can be made through a sight glass.

It is recommended that core from the actual reservoir be used in the displacement tests. Although outcrop cores may also be used for certain mechanistic studies. The selection of reservoir cores for these tests is an important procedure which requires an understanding of the geology of the entire reservoir. The cores should be sampled from the pay zone of interest and chosen to properly represent the main rock types occurring in the reservoir. Cores with large heterogeneities such as fractures, vugs, and laminations would tend to give results that exaggerate the effects of the heterogeneities.[56] Studies of Rosman and Simon[66], and Batycky et al.[67] have, however, shown that the heterogeneity exhibited by individual core segments decreases when the segments are butted together to form a longer core assembly.

Full diameter, vertical cores may be used for evaluating vertical floods while, for horizontal floods, horizontal plugs have to be drilled out of the full diameter core. These plugs are typically 2-3 cm in diameter and 8-10 cm long. About 20 plugs should be butted together in a core holder to give a sufficiently long assembly for the displacement test, particularly if the development of multiple contact miscibility is involved. To achieve good capillary contact between the cores, the core faces can be machined square on a lathe, and there is the option of placing filter paper between the core faces prior to mounting them in a tiraxial core holder. It is recommended that the plugs be chosen such that they come from the same facies in the reservoir, and that they have similar and representative porosity-permeability characteristics. Combining plugs from different facies and with widely varying properties makes the interpretation of the displacement results difficult and of questionable value as input data for simulator predictions of field performance.

The cores available for testing may be in the preserved state or, more likely, are in an aged condition. If preserved, the cores can be used directly in the displacement experiments. Non-preserved core needs to be cleaned thoroughly by extraction or displacement with solvents such as toluene-methanol[68], mounted dry in a core holder, and then have its wettability and initial oil saturation re-established by contact with the reservoir fluids.

A typical test procedure utilized with cleaned, non-preserved core involves evacuating, saturating with reservoir brine, and then flooding with crude oil until the water saturation approaches the connate water saturation. If this procedure cannot provide a sufficiently low initial water saturation, then methods utilizing gas flow and/or evaporation can be used.[64,69] Following placement of crude oil in the core, it is left to age for several days for the purpose of re-establishing the original wettability[69]. After aging, the core is water-flooded with injection water down to residual oil saturation. The water-oil relative permeability may be calculated from the pressure drop and production history of the waterflood. Finally, the core is solvent flooded. If the solvent flood is to be a secondary one, the waterflood step is then naturally ommitted.

The distinct advantages of using non-preserved core are its ease of handling during the drilling of plugs, and the ability to examine the cores and measure their properties (such as air permeability and porosity) prior to the flood tests. The disadvantage of using aged core is that one is seldom sure of the adequacy of the measures taken to restore the core to its original state.

A prime reason for attempting to restore the reservoir wetting condition in the core relates to the reported trapping or shielding of oil by mobile water in water-wet sandstone[59,60]. It is generally believed that mixed or intermediately wet systems provide optimum tertiary recovery efficiencies with solvent floods.

RECAPITULATION

The first step in the implementation of a field-scale miscible flood is the selection of suitable candidate reservoirs and solvents. A set of technical screening criteria has been provided to aid in the selection. These should be augmented by other limitations and/or incentives (e.g. economic) specific to each locale.

Once the preliminary selection has been made, laboratory tests can be carried out to reduce the technical and economic uncertainties associated with field tests. The laboratory tests should be supplemented with geological (reservoir description) and computer simulation studies.

Laboratory tests have been categorized into static and dynamic measurements, and different types of tests that may be carried out under each category have been listed. Static phase behavior tests enable the measurement of the properties of the oil, solvent, and their mixtures under controlled conditions. Slim tube tests determine the dynamic miscibility characteristics of the oil-solvent system. Finally, core displacement tests help determine the effect of the process conditions and rock properties on the displacement efficiency in the swept zone of the reservoir.

ACKNOWLEDGMENTS

The authors wish to express their thanks to P.M. Sigmund for consultations, and to B. Moore for typing the manuscript.

REFERENCES

1. HOLM, L.W.; "Status of CO_2 and Hydrocarbon Miscible Oil Recovery Methods", J. Pet. Tech. (January 1976) 76.

2. STALKUP, F.I.; "Carbon Dioxide Miscible Flooding. Past, Present and Outlook for the Future", J. Pet. Tech. (August 1978) 1102.

3. DOSCHER, T.. et al; "Carbon Dioxide for the Recovery of Crude Oil. A Literature Search to June 30, 1979 - Final Report", U.S. Dept. of Energy Publication No. DOE/BETC/5785-1 (1980).

4. MUNGAN, N.; "Carbon Dioxide Flooding - Fundamentals", Pet. Soc. of CIM paper no. 80-31-04, presented at 31st Annual Technical Meeting of the Pet. Soc. of CIM, Calgary, Alberta (May 25-28, 1980).

5. MUNGAN, N.; "Carbon Dioxide Flooding - Applications", Pet. Soc. of CIM paper no. 81-31-22, presented at 31st Annual Technical Meeting of the Pet. Soc. of CIM, Calgary, Alberta (May 25-28, 1980).

6. BURNETT, D.B. and DANN, M.W.; "Screening Tests for Enhanced Oil Recovery Projects", paper SPE 9710, presented at the 1981 Permian Basin Oil and Gas Recovery Symposium of the Soc. of Pet. Eng. of AIME, Midland, Texas (March 12-13, 1981).

7. HUTCHINSON, JR., C.A. and BRAUN, P.H.; "Phase Relations of Miscible Displacement in Oil Recovery", A.I.Ch.E. J. (1961), $\underline{7}$ (1), 64.

8. BENHAM, A.L., DOWDEN, W.E., and KUNZMAN, W.J.; "Miscible Flood Displacement - Prediction of Miscibility", Trans. AIME (1960) $\underline{219}$, 229.

9. RATHMELL, J.J., STALKUP, F.I., and HASSINGER, R.C.; "A Laboratory Investigation of Miscible Displacement by Carbon Dioxide", paper SPE 3483, prepared for 46th Annual Fall Meeting of the Soc. of Pet. Eng. of AIME, New Orleans, Louisiana (October 3-6, 1971).

10. HOLM, L.W. and JOSENDAL, V.A.; "Effect of Oil Composition on Miscible-Type Displacement by Carbon Dioxide", paper SPE 8814, presented at the First Joint SPE/DOE Symp. on Enhanced Oil Recovery, Tulsa, Oklahoma (April 20-23, 1980).

11. METCALFE, R.S. and YARBOROUGH, L.; "The Effect of Phase Equilibria on the CO_2 Displacement Mechanism", Soc. Pet. Eng. J. (August 1979), 242.

12. GARDNER, J.W., ORR, F.M., and PATEL, P.D.; "The Effect of Phase Behavior on CO_2 Flood Displacement Efficiency", paper SPE 8367, presented at the 54th Annual Fall Technical Conference and Exhibition of the Soc. of Pet. Eng. of AIME, Las Vegas, Nevada (September 23-26, 1979).

13. WANG, G.C.; "Microscopic Investigation of CO_2 Flooding Process", SPE/DOE paper no. 9788, presented at the 1981 SPE/DOE Second Joint Symposium on Enhanced Oil Recovery of the Soc. of Pet. Eng., Tulsa, Oklahoma (April 5-8, 1981).

14. HUANG, E.T.S. and TRACHT, J.H.; "The Displacement of Residual Oil by Carbon Dioxide", paper SPE 4735, presented at Improved Oil Recovery Symposium of the Soc. of Pet. Eng. of AIME, Tulsa, Oklahoma (April 22-24, 1974).

15. YELLIG, W.F.; "Carbon Dioxide Displacement of a West Texas Reservoir Oil", paper SPE/DOE 9785, presented at the 1981 SPE/DOE Second Joint Symp. on Enhanced Oil Recovery of the Soc. of Pet. Eng., Tulsa, Oklahoma (Apr. 5-8, 1981)

16. EUBANK, P.T., HALL, K.R., and HOLSTE, J.C.; "A Review of Experimental Techniques for Vapor-Liquid Equilibria at High Pressure", Proceedings of 2nd International Conference on Phase Equilibria and Fluid Properties in the Chemical Industry, West Berlin (March 17-21, 1981), Dechema, Frankfurt/Main, 675.

17. STANDING, M.B.; "Volumetric and Phase Behavior of Oil Field Hydrocarbon Systems", Soc. of Pet. Eng. of AIME, Dallas, Texas, Eighth Printing (1977).

18. WELKER, J.R. and DUNLOP, D.D.; "Physical Properties of Carbonated Oils", J. Pet. Tech. (August 1963) 873.

19. MENZIE, D.E. and NIELSEN, R.F.; "A Study of the Vaporization of Crude Oil by Carbon Dioxide Repressuring", J. Pet. Tech. (November 1963), 1247.

20. SCHNEIDER, G., ALWANI, Z., HORVATH, E., and FRANCK, E.U.; "Phasengleich-wichte und kritische Erscheinungen in binaren Mischsystemen bis 1500 bar CO_2 mit n-Octan, n-Undecan, n-Tridecan und n-Hexan", Chemie-Ing.-Tech. (1967) 39 (11), 649.

21. JACOBY, R.H. and YARBOROUGH, L.; "PVT Measurements on Petroleum Reservoir Fluids and their Uses", Ind. Eng. Chem. (October 1967), 48.

22. YARBOROUGH, L. and VOGEL, J.L.; "A New System for Obtaining Vapor and Liquid Sample Analyses to Facilitate the Study of Multicomponent Mixtures of Elevated Pressures", Chem. Eng. Prog. Symp. Ser. (1967) 63 (81), 1.

23. LEE, J.I.; "Viscosities of Carbon Dioxide-Sulfur Dioxide Mixtures and the Effect of Methane on Sulfur Dioxide-Oil Miscibility", Research Note RN-9, Petroleum Recovery Institute, Calgary, Alberta (March 1979).

24. ORR, JR., F.M., SILVA, M.K., LIEN, C.L., and PELLETIER, M.T.; "Laboratory Experiments to Evaluate Field Prospects for CO_2 Flooding", paper SPE 9534 presented at 1980 Soc. of Pet. Eng. of AIME Eastern Regional Meeting, Morgantown, West Virginia (November 5-7, 1980).

25. CONNOR, H.J. and POPE, A.E.; "Development of Experimental Phase Equilibria for Miscible Hydrocarbon Enhanced Oil Recovery", Pet. Soc. of CIM, paper no. 81-32-41, presented at 32nd Annual Meeting of the Pet. Soc. of CIM, Calgary, Alberta (May 3-6, 1981).

26. SAYEGH, S.G., NAJMAN, J., and HLAVACEK, B.; "Phase Equilibrium and Fluid Properties of Pembina Cardium Stock-Tank Oil-Methane-Carbon Dioxide-Sulfur Dioxide Mixtures", Pet. Soc. of CIM, paper 81-32-14, presented at 32nd Annual Technical Meeting of the Pet. Soc. of CIM, Calgary, Alberta (May 3-6, 1981).

27. HENRY, R.L., FEATHER, G.L., SMITH, L.R., and FUSSELL, D.D.; "Utilization of Composition Observation Wells in a West Texas CO_2 Pilot Flood", paper SPE/DOE 9786, presented at the 1981 SPE/DOE Second Joint Symp. on Enhanced Oil Recovery of the Soc. of Pet. Eng., Tulsa, Oklahoma (April 5-8, 1981).

28. McCAIN, JR., W.D.; "The Properties of Petroleum Fluids", Petroleum Publishing Company, Tulsa, Oklahoma (1973).

29. RUTHERFORD, W.W.; "Miscibility Relationships in the Displacement of Oil by Light Hydrocarbons", Soc. Pet. Eng. J. (December 1962) 340.

30. HOLM, L.W.; "Carbon Dioxide Solvent Flooding for Increased Oil Recovery", Pet. Trans. AIME (1959) 216, 225.

31. HOLM, L.W. and JOSENDAL, V.A.; "Mechanisms of Oil Displacement by Carbon Dioxide", J. Pet. Tech. (December 1974) 1427.

32. SHELTON, J.L. and YARBOROUGH, L.; "Multiple Phase Behavior in Porous Media During CO_2 of Rich-Gas Flooding", J. Pet. Tech. (September 1977) 1171.

33. SIMON, R., ROSMAN, A., and ZANA, E.; "Phase Behavior Properties of CO_2-Reservoir Oil Systems", Soc. Pet. Eng. J. (February 1978) 20.

34. LEE, J.I. and SIGMUND, P.M.; "Phase Behavior and Displacement Studies of Carbon Dioxide-Hydrocarbon Systems", Research Report RR-37, Petroleum Recovery Institute, Calgary, Alberta (September 1978).

35. LEE, J.I.; "Miscibility of Carbon Dioxide-Sulfur Dioxide-Hydrocarbon Systems", Research Note RN-8, Petroleum Recovery Institute, Calgary, Alberta (August 1978).

36. LEE, J.I.; "Effectiveness of Carbon Dioxide Displacement Under Miscible and Immiscible Conditions", Research Report RR-40, Petroleum Recovery Institute, Calgary, Alberta (March 1979).

37. ORR, JR., F.M., YU, A.D., and LIEN, C.L.; "Phase Behavior of CO_2 and Crude Oil in Low Temperature Reservoirs", paper SPE 8813, presented at the First Joint SPE/DOE Symp. on Enhanced Oil Recovery, Tulsa, Oklahoma (April 20-23, 1980).

38. TUREK, E.A., METCALFE, R.S., YARBOROUGH, L., and ROBINSON, JR., R.L.; "Phase Equilibria in Carbon Dioxide-Multicomponent Hydrocarbon Systems: Experimental Data and an Improved Prediction Technique", paper SPE 9231, presented at the 55th Annual Fall Technical Conference and Exhibition of the SPE of AIME, Dallas, Texas (September 21-24, 1980).

39. METCALFE, R.S., FUSSELL, D.D., and SHELTON, J.L.; "A Multicell Equilibrium Separation Model for the Study of Multiple Contact Miscibility in Rich Gas Drives", Soc. Pet. Eng. J. (June 1973), 147.

40. YELLIG, W.F. and METCALFE, R.S.; "Determination and Prediction of CO_2 Minimum Miscibility Pressures", paper SPE 7477, presented at the 53rd Annual Fall Technical Conference and Exhibition of the Soc. of Pet. Eng. of AIME, Houston, Texas (October 1-3, 1978).

41. METCALFE, R.S.; "Effect of Impurities on Minimum Miscibility Pressures and Minimum Enrichment Levels for CO_2 and Rich Gas Displacement", paper SPE 9230, presented at the 55th Annual Fall Technical Conference and Exhibition of the Soc. of Pet. Eng. of AIME, Dallas, Texas (September 21-24, 1980).

42. JOHNSON, J.P. and POLLIN, J.S.; "Measurement and Correlation of CO_2 Miscibility Pressures", paper SPE 9790, presented at the 1981 SPE/DOE Second Joint Symposium on Enhanced Oil Recovery of the Soc. of Pet. Eng. of AIME, Tulsa, Oklahoma (April 5-8, 1981).

43. JACOBSON, H.A.; "Acid Gases and Their Contribution to Miscibility", J. Can. Pet. Tech. (April-June, 1972) 56.

44. CRONQUIST, C.; "Carbon Dioxide Dynamic Miscibility with Light Reservoir Oils", presented at 4th Annual DOE Enhanced Oil Recovery Symposium, Tulsa, Oklahoma (August 29-31, 1978).

45. SIGMUND, P.M., AZIZ, K., LEE, J.I., NGHIEM, L.X., and MEHRA, R.; "Laboratory CO_2 Floods and Their Computer Simulation", presented at the World Petroleum Congress, Bucharest, Rumania (September 1979).

46. WILLIAMS, C.A., ZANA, E.N., and HUMPHRYS, G.E.; "Use of the Peng-Robinson Equation of State to Predict Hydrocarbon Phase Behavior and Miscibility for Fluid Displacement", paper SPE 8817, presented at the First Joint SPE/DOE Symposium on Enhanced Oil Recovery, Tulsa, Oklahoma (April 20-23, 1980).

47. WANG, G.C. and LOCKE, C.D.; "A Laboratory Study of the Effects of CO_2 Injection Sequence on Tertiary Oil Recovery", Soc. Pet. Eng. J. (August 1980) 278.

48. LEACH, M.P. and YELLIG, W.F.; "Compositional Model Studies – CO_2 Oil – Displacement Mechanisms", Soc. Pet. Eng. J. (February 1981), 89.

49. BLACKWELL, R.J.; "Laboratory Study of Microscopic Dispersion Phenomena", Soc. Pet. Eng. J. (1962) 1, 69.

50. PERKINS, T.K. and JOHNSTON, O.C.; "A Review of Diffusion and Dispersion in Porous Media", Soc. Pet. Eng. J. (1962) 1, 77.

51. VAN DER POEL, C.; "Effect of Lateral Diffusivity on Miscible Displacement in Horizontal Reservoirs", Soc. Pet. Eng. J. (1962) 1, 93.

52. COATS, K.H. and SMITH, B.D.; "Dead-End Pore Volume and Dispersion in Porous Media", Soc. Pet. Eng. J., (March 1964) 73.

53. BRIGHAM, W.E.; "Mixing Equations in Short Laboratory Cores", Soc. Pet. Eng. J. (February 1974) 91.

54. BAKER, L.E.; "Effects of Dispersion and Dead-End Pore Volume in Miscible Flooding", Soc. Pet. Eng. J. (June 1977) 219.

55. YELLIG, W.F. and BAKER, L.E.; "Factors Affecting Miscible Flooding Dispersion Coefficients", Pet. Soc. of CIM paper no. 80-31-06, presented at 31st Annual Technical Meeting, Pet. Soc. of CIM, Calgary, Alberta (May 25-28, 1980).

56. SPENCE, JR., A.P. and WATKINS, R.W.; "The Effect of Microscopic Core Heterogeneity on Miscible Flood Residual Oil Saturation", paper SPE 9229, presented at the 55th Annual Fall Technical Conference and Exhibition of the Soc. of Pet. Eng. of AIME, Dallas, Texas (September 21-24, 1980).

57. SCHNEIDER, F.N. and OWENS, W.W.; "Relative Permeability Studies of Gas-Water Flow Following Solvent Injection in Carbonate Rocks", Soc. Pet. Eng. J. (February 1976) 23.

58. HENRY, R.L. and METCALFE, R.S.; "Multiple Phase Generation During CO_2 Flooding", paper SPE 8812, presented at the First Joint SPE/DOE Symposium on Enhanced Oil Recovery, Tulsa, Oklahoma (April 20-23, 1980).

59. STALKUP, F.I.; "Displacement of Oil by Solvent at High Water Saturation", Soc. Pet. Eng. J., (December 1970) 337.

60. SHELTON, J.L. and SCHNEIDER, F.N.; "The Effects of Water Injection on Miscible Flooding Methods Using Hydrocarbons and Carbon Dioxide", Soc. Pet. Eng. J. (June 1975) 217.

61. WATKINS, R.W.; "A Technique for the Laboratory Measurement of Carbon Dioxide Unit Displacement Efficiency in Reservoir Rock", paper SPE 7474, presented at the 53rd Annual fall Technical Conference and exhibition of the Soc. of Pet. Eng. of AIME, Houston, Texas (October 1-3, 1978).

62. HELLER, J.P., TABOR, J.J. and LOCKE, C.D.; "Mobility Control for CO_2 Floods – A Literature Survey", U.S. Dept. of Energy Publication no. DOE/MC/10689-3 (1980).

63. CHRISTIAN, L.D., SHIRER, J.A., KIMBLE, E.L., and BLACKWELL, R.J.; "Planning a Tertiary Oil Recovery Project for Jay-Little Escambia Creek Field Unit", paper SPE/DOE 9805, presented at the 1981 SPE/DOE Second Joint Symposium on Enhanced Oil Recovery of the Soc. of Pet. Eng., Tulsa, Oklahoma (April 5-8, 1981).

64. RANDALL, T.E., WANSLEEBEN, J. and SIGMUND, P.M.; "Physical Model, West Wilmar Rich Gas Pilot", Pet. Soc. of CIM paper no. 81-32-16, presented at the 32nd Annual Technical Meeting of the Pet. Soc. of CIM, Calgary, Alberta (May 3-6, 1981).

65. ROSMAN, A. and ZANA, E.; "Experimental Studies of Low IFT Displacement by CO_2 Injection"; paper SPE 6723, presented at the 52nd Annual Fall Technical Conference and Exhibition of the Soc. of Pet. Eng. of AIME, Denver, Colorado (October 9-12, 1977).

66. ROSMAN, A. and SIMON, R.; "Flow Heterogeneity in Reservoir Rocks", J. Pet. Tech. (December 1976) 1427.

67. BATYCKY, J.P., MIRKIN, M.I., JACKSON, C.H., and BESSERER, G.J.; "Miscible and Immiscible Displacement Studies on Carbonate Reservoir Cores", J. Can. Pet. Tech. (1981) 20 (1), 104.

68. GRIST, D.M., LANGLEY, G.O., and NEUSTADTER, E.L.; "The Dependence of Water Permeability on Core Cleaning Methods in the Case of Some Sandstone Samples", J. Can. Pet. Tech. (April-June 1975) 48.

69. CUIEC, L., LONGERON, D. and PACSINSZKY, J.; "On the Necessity of Respecting Reservoir Conditions in Laboratory Displacement Studies", paper SPE 7785, presented at the Middle East Oil Technical Conference of the Soc. of Pet. Eng., Manama, Bahrain (March 25-29, 1979).

70. HARVEY, JR., M.T., SHELTON, J.L., and KELM, C.H.; "Field Injectivity Experiences With Miscible Recovery Projects Using Alternate Rich-Gas and Water Injection", J. Pet. Tech., (September 1977) 1051.

COMPLEX STUDY OF CO₂ FLOODING IN HUNGARY

SÁNDOR DOLESCHALL, GÁBOR ÁCS, ÉVA FARKAS,
TIBOR PAÁL, JÁNOS TÖRÖK
Hungarian Hydrocarbon Institute

VALÉR BÁLINT
General Contracting and Designing Office for the Oil Industry, "Olajterv"

ZOLTÁN BIRÓ
Transdanubian Oil and Gas Production Company

ABSTRACT

A systematic program of carbonated natural gas flooding has been carried out in Hungary, based on laboratory PVT and displacement studies, followed by compositional mathematical simulation and field experiment on depleted reservoir.

PVT studies have proved that gas containing 81 mole % carbon dioxide can be used for EOR purposes. The studies covered the volumetric and phase behaviour of carbonated natural gas flooding under field conditions and the results proved that such flooding was efficient even if the gas is not pure carbon dioxide. Based upon theoretical considerations a technological scheme has been developed to increase the sweep efficiency.

A ten-component, three-phase mathematical model developed to simulate carbon dioxide flooding is suitable for treating single- and multi-phase systems. The difference equations handle the systems with different number of phases in a uniform way, thus the generation and disappaerance of phases can be followed by the model without difficulties.

The computer model was used to simulate partially miscible carbonated natural gas flooding in the western area of the Budafa oil field. The production history match and prediction agreed well with the field data.

INTRODUCTION

The oil resources of Hungarian reservoirs cover only a small part of the country's demand, and the import of crude oil imposes a considerable economic burden on a country developing its industry. Apart from the need to search for new oil fields, it became evident as long ago as the fifties that it was important to consider secondary and later the tertiary recovery methods. Among the other possibilities the effect of carbon dioxide was also studied, and

very soon most attention focused on the questions of CO_2 flooding because
in Hungary the occurrence of natural carbon dioxide in high carbon dioxide
content natural gases is more often found and to a greater extent than the
world average. Some results of CO_2 flooding in Hungary can be found in Ref. 1.

PVT AND PHASE BEHAVIOUR MEASUREMENTS

Feasibility studies of the application possibilities of carbon dioxide and
carbonated natural gases started in 1955 with a series of PVT measurements.
The very first PVT studies proved that carbonated natural gas alters the
viscosities and volumetric properties of crudes with very different densities
in a favourable way compared with the effect of lean or wet natural gases
under the same conditions, mainly if the carbon dioxide content of the dissolved
gas is above 60 mole %. Based upon the results of more detailed PVT measurement, sets of curves have been developed to predict the solubility,
swelling and viscosity of monophase reservoir oil--carbonated natural gas
systems. The actual PVT properties of the original gas saturated oil were
chosen as a reference state to eliminate the possible large errors coming
from the unknown parameters of such very complex systems, and only the change
of the given properties was correlated with the dissolved carbon dioxide content. In this way simple, easy to use equations with good accuracy have been
developed. For example, the prediction of viscosities of saturated and
undersaturated crudes under different conditions is possible with the use
of only one measured viscosity value.

It has been proved that in the case of Hungarian crude oils, bearing in mind
the actual reservoir conditions, that no complete miscibility occurs even if
the dissolved gas is pure carbon dioxide.

In the course of the thorough examination of the PVT data "unusual" behaviour
was observed. Repeated measurements in a windowed PVT cell revealed the presence of a carbon dioxide rich second liquid phase which exists within a
definite pressure-temperature range above a certain gas--oil ratio. This region
depends upon the total composition of the system and the phenomenon is connected with the restricted solubility of carbon dioxide in reservoir oils.
Partition of light and intermediate hydrocarbons between the reservoir oil
and the second liquid phase has been proven - in agreement with other experience. In the case of certain Hungarian crude oils reversible precipitation
of semi-solid particles has also been observed but mostly under such circumstances which cannot be realized in actual reservoirs. It is interesting that
these phenomena occur in the presence of carbonated natural gases, too, even
if they are relatively rich in light hydrocarbon fraction. The existence of
the mentioned multiphase systems had to be considered in planning vapour--liquid
equilibrium studies.

The aim of these studies is to determine exact K values according to the need
of compositional mathematical simulation. Equilibrium ratios had been determined for characteristic reservoir oil--carbonated natural gas as well as
reservoir oil--water--carbonated natural gas systems and a method for estimation was developed. As a result of additional measurements and comparison
of experimental with computed data using different equations of state it is
concluded that further improvements are necessary both for the development
of generalized K functions and equations of state together with the improvement
of interaction coefficients.

Judging by the results of other studies, the interfacial tension decreases with increasing carbon dioxide content in gas--oil--water systems.
Volumetric and phase behaviour as well as water content and hydrate forming conditions of carbonated natural gases in Hungary were also studied and the resulting data used to formulate generalized relationships.

Experimental data on solubility, swelling and viscosity of typical reservoir waters - saturated with carbonated natural gases having different composition, even in the presence of calcium carbonate and reservoir rocks containing clay minerals - together with vapour--liquid equilibrium ratios supplied further information enabling a better understanding of the mechanism of carbonated natural gas flooding. It has been pointed out that because of the interaction of carbonated water and reservoir rocks certain clay minerals contract and this may improve the efficiency of the process in practice.

A PVT model was used to follow the change of the volumetric and phase behaviour and the equilibrium composition of phases in the course of flooding. This model contained water-, oil- and gas-phases under reservoir conditions with a ratio corresponding to the actual saturation at a given, depleted field. The pressure was increased to the original reservoir pressure directly by the injection gas. In another set of experiments the final pressure was reached step by step, the vapour phase gradually being changed by the injection gas at each inter- mediate pressure until equilibrium composition was approached. These experiments were repeated for different field conditions using injection gases with different carbon dioxide content. Interpretation of the results revealed the importance of the dynamic pressure-increase process being applied for carbonated natural gas flooding, the role of the light hydrocarbon fraction present and supported the conclusion previously drawn on the basis of PVT studies of mono- and two- -phase systems.

Taking into consideration the composition of the gradually displaced gas, it has been concluded that it is not possible in practice to replace all the free and dissolved gas by a carbon dioxide slug of reasonable size. It has also been found that despite the dilution of the slug by hydrocarbon gas in the pores, relatively significant vaporization of the oil takes place if the carbon dioxide content of the free gas phase is above a certain critical concentration. This critical value, which depends upon the pressure, temperature and the characteristics of the oil, can also be exceeded by using carbonated natural gases for the injection. These observations confirmed indirectly the idea about the probable formation of a miscible front in the reservoir under dynamic conditions during carbonated natural gas flooding.

As to the volumetric properties and viscosities of the equilibrium liquid phases no substantial difference could be found on comparing the effect of a carbon dioxide slug and a larger volume of carbonated natural gas with higher carbon dioxide content.

LABORATORY DISPLACEMENT STUDIES

Following encouraging PVT results dynamical laboratory studies were carried out to examine the efficiency of CO_2 displacement processes. The linear model used for the measurements was 1 m long and 25 mm in diameter. Nonconsolidated reservoir sandstone cores and reservoir fluid were used for these displacement tests. This technique is suitable for studying production histories, as well as various forms of CO_2 flooding and the actual mechanism of the process.

As a first step the effect of carbonated water was examined. Carbonated water saturated at reservoir pressure and temperature was injected into the previously water flooded core. Carbon dioxide appeared in the effluent after injecting one pore volume of saturated water. To reach the injected equilibrium concentration of the carbonated water in the effluent 4-8 pore volumes of saturated water were necessary. Consequently, the additional oil was produced with a rather high water cut. The additional oil was 5-7 % of the original oil an place. Because of the injection of a large volume of water and the modest additional oil recovery, this method is uneconomic.

To increase the amount of injected carbon dioxide, oversaturated water was used in the next series of experiments. The additional oil reached 10 % of o.i.p. and favourable effects of free gas saturation were observed, too. However even in this case, 3-5 pore volumes of carbonated water were used to obtain this result.

Gaseous carbon dioxide was injected into the model when studying tertiary recovery methods for depleted reservoirs. Two different initial saturation conditions were used as average reservoir conditions for modelling production histories:
- the depleted reservoir has a high gas saturation, ~25-35 %;
- the depleted reservoir has a low gas saturation and high water saturation, ~50-60 %.

The pressure was increased to the original reservoir pressure by injecting carbon dioxide gas. After the pressure build-up, different sizes of CO_2 slugs were injected and followed by reservoir water flooding. The additional oil recovery as a function of slug size was studied. The probable optimal slug size was about 0.2 PV. Using this, the additional oil recovery was 12-16 % of the original oil in place for systems having a high initial gas saturation and 8-12 % for the case of high initial water saturation. The additional oil recovery was always related to the residual oil saturation of traditional water flooding.

All of the dynamic displacement tests, mentioned above were performed with practically pure carbon dioxide. Tests were conducted using carbonated natural gases, too. The results showed that the use of carbonated natural gases having a CO_2 content above 80 mole %, give not worse, but better results in most cases if the proper displacement technology is used.

Complex flow conditions and physico-chemical processes exist in reservoir oil--reservoir water--carbon dioxide--reservoir rock systems. The parameters influencing the effectiveness of CO_2 flooding must be individually determined for each project. If the wettability of reservoir rock changes from water-wet to oil-wet the favourable effects of free gas saturation and the later water flooding are reduced. This change of wettability depends upon many factors - among others, on the quantity of injected CO_2 (2).

If carbon dioxide is injected into the depleted oil reservoir it interacts with the reservoir fluid and component mass transfer starts among the phases. As a result of this process the oil phase will be richer in components having higher molecular weights. In extreme cases some of the components with interfacial active characteristics may adsorb on the rock surface, thereby changing the wettability properties of the system and leading to the rock becoming more oil-wet. Although the carbon dioxide content of the oil phase decreases the viscosity of crude rich in high molecular components and swells the oil-phase a possible increase in the oil-wet character counteracts these favourable effect.

Relative permeability curves for saturated carbonated water systems were also measured. The character of relative permeability curves justified the effect mentioned above. Under some circumstances the porous medium became more oil-wet. Decreasing oil and increasing water permeabilities could be observed in certain saturation ranges, depending upon the CO_2 content of the gas used. The increase in residual oil saturation was also observed with increasing CO_2 content. The bases of comparison were the relative curves of hydrocarbon gas saturated systems.

COMPUTER MODEL

A three-phase, ten-component mathematical model has been developed to study carbon dioxide displacement experiments and to predict performances (3, Part I.). The governing differential equations of the compositional model written in a usual form are as follows:

$$\frac{\partial}{\partial t} \sum_{j} \phi \varrho_j c_{j,i} S_j =$$

$$= \text{div} \left[\underline{\underline{K}} \sum_{j} \frac{k_j \varrho_j}{\mu_j} c_{j,i} (\text{grad } p_j + \varrho_j g \text{ grad } z) + \phi \sum_{j} S_j \varrho_j D_{j,i} \text{grad } c_{j,i} \right] +$$

$$+ q_i \qquad\qquad\qquad i = 1,2,\ldots,10$$
$$\qquad\qquad\qquad\qquad\qquad j = \text{gas, oil, water}$$

The basis of the calculations is the assumption that local thermodynamic equilibrium exists during displacement. In this way, the relationships correlated with laboratory PVT and equilibrium measurements can directly be employed. In accordance with the laboratory measurements, the formation fluid of Budafa oil field was considered as a ten-component system. The components are: seven hydrocarbon components /C_1, C_2, C_3, C_4, C_5, C_6, C_{7+}/, nitrogen, carbon dixide and water.

As the water phase exists everywhere in the formation, and during the water injection a great amount of carbon dioxide is to be transported by water, the solution of the carbon dioxide component in the water phase cannot be neglected. Besides three-phase regions, two-, moreover one-phase regions occur during the processes, thus a method has been developed that allows one to easily calculate a change in the number of phases. The three-phase equilibrium was interpreted as the simultaneous existence of two two-phase equilibriums.

To simplify the equilibrium calculations the following assumptions were made:
- the gas and oil phases do not contain a water component,
- the dissolved gas in the water phase consists of carbon dioxide only. /When checking the calculations the dissolved gas in the water phase contained methane, as well, but the little influence of this on the phase equilibrium made it reasonable to neglect it./

When calculating the phase equilibrium, flash calculations are used to determine the mole fractions of the phases; however, the calculation of three-phase equilibrium make it necessary to solve a coupled system of two nonlinear algebraic equations. Occasionally, mainly when the number of phases changes, convergence problems of iterative techniques occur. The system was transformed into one nonlinear algebraic equation, and a numerical procedure combining the Newton-method and the method of halving, ensure fast convergence in every case. The density of the gas phase is calculated using the Redlich-Kwong equation of state. When determining densities of the fluid phases the laboratory correlations are applied. In accordance with these correlations the formation volume factor is calculated as a function of the dissolved gas/fluid ratio for both fluid phases. Thus the quantity of the dissolved gas has to be known. Because the composition of the phases is known, the dissolved gas/oil ratio can be determined from the composition of the oil phase by normal flash calculation. As for the dissolved gas/water ratio, it was assumed that water in its normal state is free of gas.

In order to check the PVT and equilibrium calculations laboratory pressure-build-up measurements were simulated by a one-volume element model. Very good matches could be achieved by modifying the molecular weight of the C_{7+} component by 5 %.

FIELD EXPERIMENT

After some pilot tests the first large-scale process was started in the western area of the Budafa oil field in 1972. The area is a section of the Lower-Pannonian /Lower-Pliocene/ Budafa reservoir which consists of four separable sequences of strata of the same hydrodynamic system. The formations are heterogeneous vertically and horizontally. The effective formation thickness varies from 1-2 m to 30 m. The average porosity is 21 %, the average horizontal permeability 0.1 μm^2.

The sandstone formations occurring at an average depth of 850 m have a temperature of 64 °C. The initial pressure level judging by the hydrostatic condition at the beginning of production was 9800 kPa. The produced crude is of an intermediate-paraffin character, its average density at 20 °C being $0.817 \cdot 10^3$ kg/m³. The reservoir oil was initially saturated, the two upper sequences originally had an extensive gas cap. This accumulation was unfavourable from the point of view of tertiary recovery because the oil zones of the two lower layers were situated under the gas caps of the two upper layers.

Production was begun in July of 1937. Following the rapid increase in the number of wells, crude production amounted to 89,800 m³/year in 1941 which was the peak production of this area. The energy of the formation decreased because of the high production level and restricted edge water drive. In order to overcome the energy reduction, 139 million m³ hydrocarbon gas was injected into the reservoir from 1942 to 1958. During the primary and secondary displacements the solution gas drive, the energy of gas caps and, to a slight extent, edge water drive worked while the formation pressure decreased to an average level of 2900 kPa, which was considered as an abandon pressure. A total of 1 million m³ oil and 600 million m³ gas was produced. The average recovery efficiency was 22.6 %.

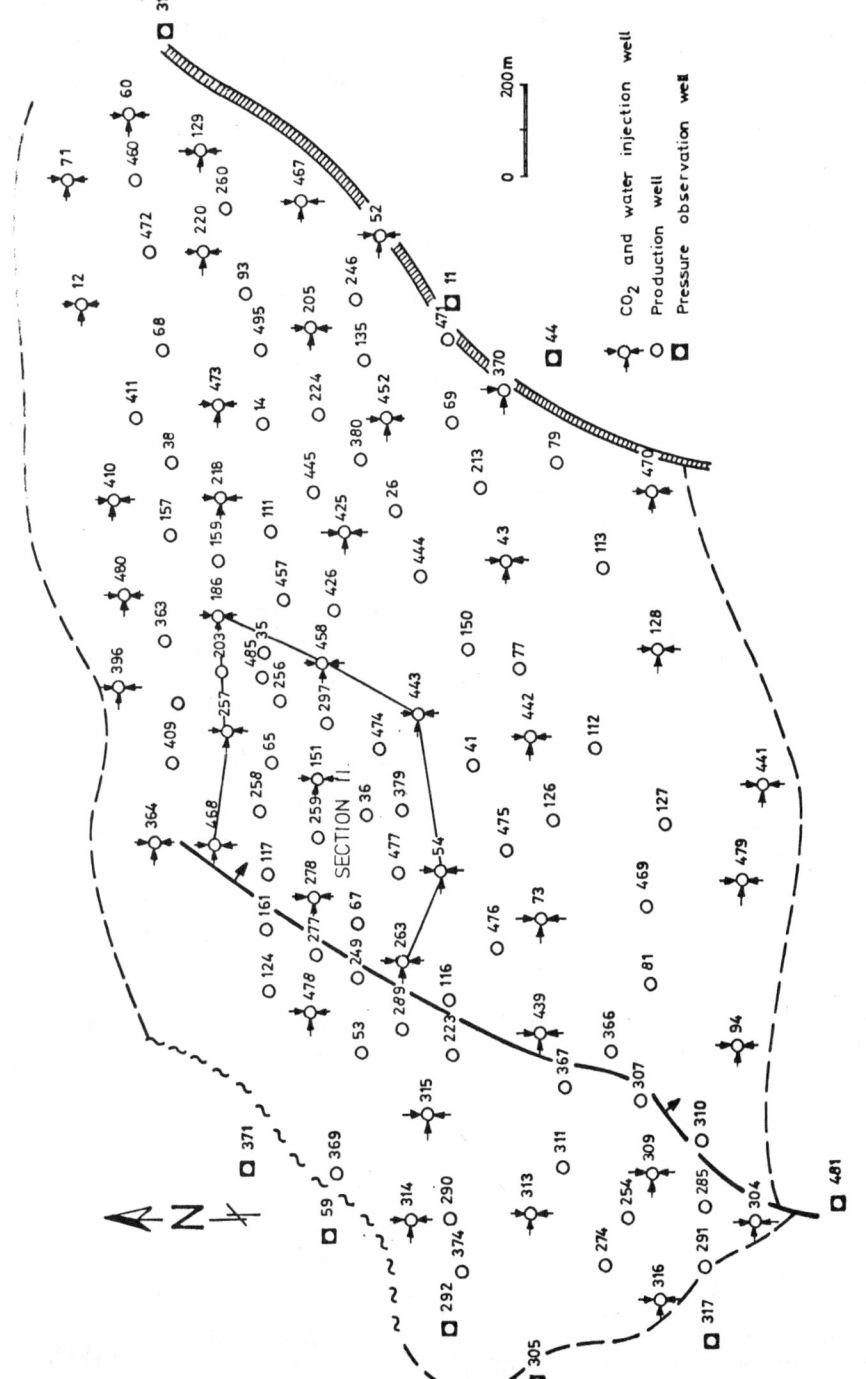

FIGURE 1. Budafa-West CO_2 injection pattern

At the beginning of tertiary recovery the oil zones of the area had a high gas saturation. Tertiary recovery by carbonated natural gas was realized by using 41 injection, 71 production and 9 observation wells. When designing the technology, the existing wells in the area were taken into account, and the system could be characterized by an irregular multi-spot pattern. The well pattern used is shown in Fig. 1.

In the first phase of the tertiary recovery, carbonated natural gas was injected into the formation during which controlled production was realized. The carbonated natural gas used was produced from a high pressure reservoir discovered in the actual area of Budafa. This gas - having a carbon dioxide content of 81 mole % and light hydrocarbons - was injected into the low pressure oil reservoir by means of natural energy. The carbon dioxide appeared in the production wells 1-2 months after the beginning of injection. Data relating to injection and production rates /Fig. 2/ show that the GOR amounted to a very high level /3000-5000 m^3/m^3/ during the injection. This disadvantageous effect was caused by the high gas saturation dating back to the primary and secondary recovery. No oil bank formation could be observed in any of the production wells. Gas and liquid flow always occurred simultaneously in the layers. In order to diminish the high GOR value of the produced fluid, water injection was started at the gas-oil contact of the two upper layers in the autumn of 1974, and the whole area was water flooded from the summer of 1975. At that time the average pressure of the reservoir was 10,900 kPa, the water injection rate 1500 m^3/day.

GOR response to water flooding was observed from the end of 1974 when the character of production changed remarkably. Along with increasing oil production rate, the gas/oil ratio decreased from the previous years' level of 5000 m^3/m^3 to about 600 m^3/m^3. The changed conditions can be seen in Fig. 2.

The carbon dioxide content of the produced gas remained above 65 mole % during the water injection, which made it evident that injection of additional carbonated natural gas was not necessary. Until 1st January 1981, 694 million m^3 carbonated natural gas and 3.013 million m^3 water had been injected into the formation. It should be mentioned that the greater part of the injected gas was used to fill up the gas caps. By January 1981, 173,000 m^3 oil and 1.072 million m^3 water had been produced and the average recovery efficiency had been 27.5 %, thus tertiary recovery resulted in additional oil of 3.9 %. This amount of additional oil is, however, an average value. For example, the additional oil from Section II. quite considerable in that the earlier value was 12.7 % o.i.p. The method has proved to be successful for one-layer, relatively homogeneous sections having low water saturation, and the effectiveness was poor, about 1-2 % for the multi-layers formation under the gas caps. The displacement is still continuing. The final amount of additional oil expected is 5.7 %. The production of additional oil proved to be economically worth while.

HISTORY MATCH AND PREDICTION

The field experiment was analysed by simulation of performance history (3, Part II.). The reservoir is thin, heterogeneous, laminated and nearly horizontal, thus an areal model was used and the effects of capillarity and gravitation were neglected. Because of the complex petrographic and heterogeneous saturation conditions, the Budafa-West multi-layer reservoir

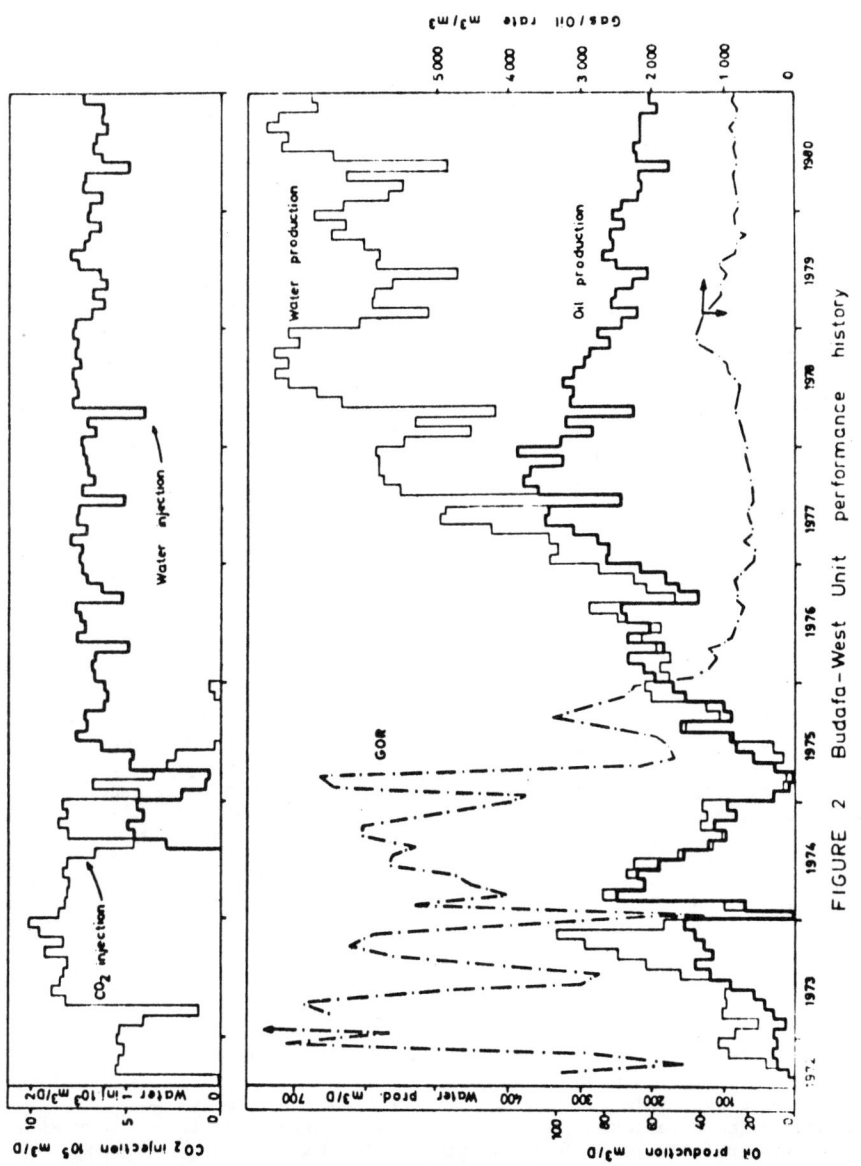

FIGURE 2   Budafa-West Unit performance history

constitutes a complicated system. For this reason an easily separable, one-
-layer section of the reservoir was examined belonging to that area where
the highest amount of oil originated from. /Primary and secondary displacement
resulted in 45.2 % for this section./ The section is shown is Fig. 1 as
Section II.

Because of computer restriction /an ICT 1905 computer with a memory of 32 Kwords
was used/, we could not describe all the injection and production wells of the
section; our intention was to obtain an overall picture of the process. /It is
to be noted that detailed data on formation parameters were also inaccessible./
The section was considered to be of constant thickness, horizontal, and the
average rock parameters and initial saturation referring to the beginning of
the tertary recovery were used. We wished to make use of all the measured
data, therefore on the bases of averaging the distances of the injection and
production wells of the section an eighth of a five-spot element was constructed. The injection and the production data of the model were calculated
from the cumulative data of the section using the pore volume ratio of the
section and those of the eighth of the five-spot element.

Relative permeability curves for three-phase carbonated systems were not
available. Based upon laboratory measurements and published data a simple form
of parametric relative permeability curves were constructed, and parameters
of the curves were determined by history matching. Pressure and production
data of 5.5 years /2.5 years of gas injection, 3 years of water injection/
were used. It seemed that no parameter group can be chosen to simulate early
breakthrough of carbon dioxide. Analysis of horizontal permeability distribution into vertical direction using continuous core samples of the reservoir
examined showed that 20 % of the permeability data were above 0.31 μm^2 which
differed remarkably from the average value. The flooding process is very
strongly influenced by the presence of high permeability zones. The heterogeneity was taken into account in a simple way, the thickness was divided into
a good and a poor permeability layer.

The results of the history match can be seen in Fig. 3. The computed average
pressures differed from the measured ones by only about 5 %.

After having good results on the history match for Section II, the model was
applied to the other 5 sections of the area. 15-25 simulations were used to
reach the final results for each case. We had to assume in the modelling,
that no flow boundaries existed between the sections though, as is to be
expected, this is not the case. This fact was proved by the calculations. To
some extent we had to modify the injected gas to get the good pressure history
match. However, these modifications were equalized from the viewpoint of the
whole area, and a calculated gas loss of only 6 % resulted.

The results of history matching are summarized in Fig. 4. The calculations
were performed in 1978. The figure shows predictions until 1983 together with
the actual production parameters of the last three years.

CONCLUSIONS

After thorough and extensive studies economic field-wide tertiary displacement
by carbon dioxide carried out in Hungary. Volumetric and phase behaviour of
the three-phase system can be modelled with good accurarcy using the
laboratory correlations. The CO_2 displacement has proved to be successful for
one-layer, relatively homogeneous sections having low water saturation.

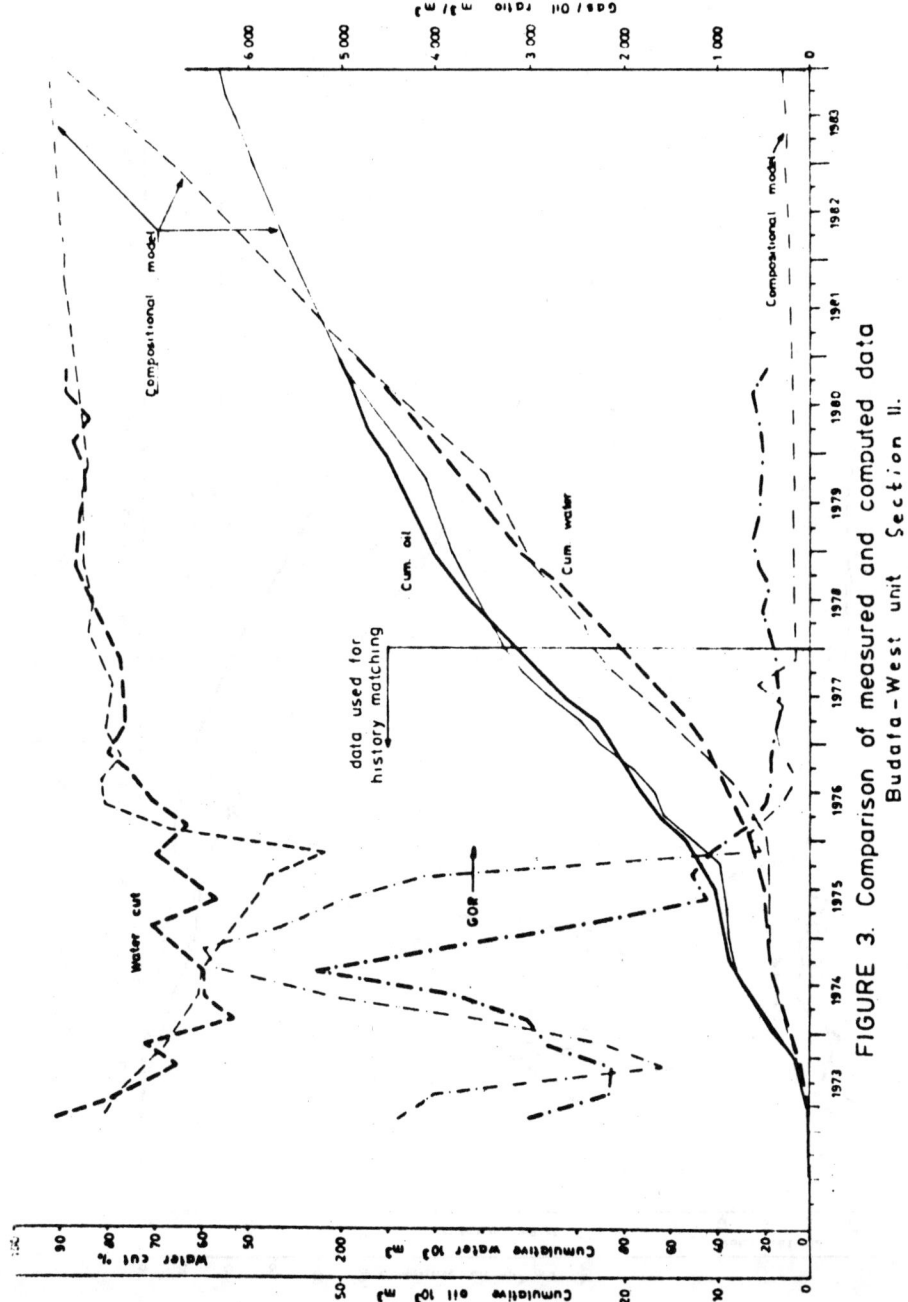

FIGURE 3. Comparison of measured and computed data
Budafa-West unit Section II.

310

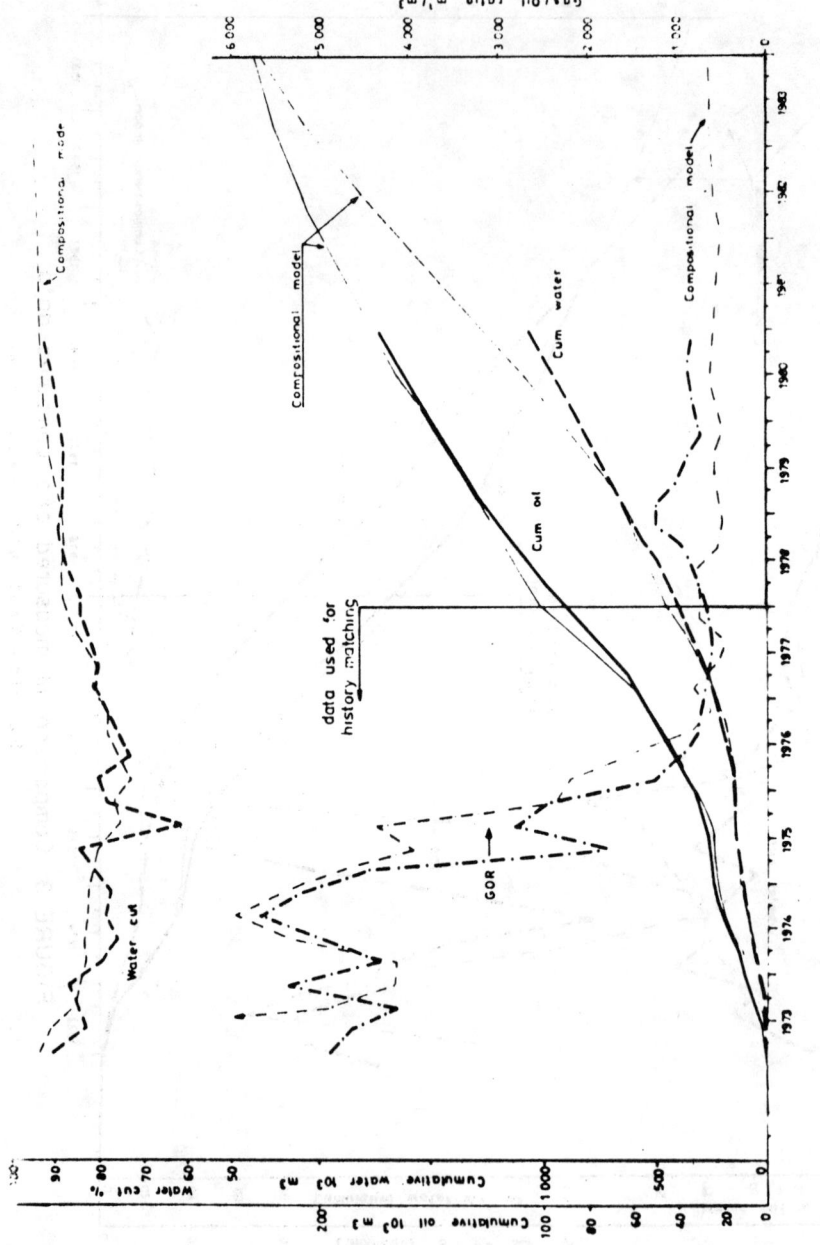

FIGURE 4. Comparison of measured and computed data
Budafa–West unit

311

Utilization of the local potential proved, in this case, to be a substantial
factor in achieving economic additional oil production thereby overcoming the
effects of unfavourable reservoir conditions.

NOMENCLATURE

C   mass concentration

D   diffusivity

g   gravitational acceleration

$\underline{\underline{K}}$ permeability tensor

k   relative permeability

p   pressure

q   mass sink per unit volume per unit time

S   saturation

t   time

z   depth

μ viscosity

ϱ density

Φ porosity

Subscripts

i   refers to ith component

j   refers to jth phase

REFERENCES

1. Bán, Á., Bálint, V., Doleschall, S., Zabrodin, P. I., Török, J.:
"Primenenije uglekislovo gaza v dobiche nefti" /"Application of carbon
dioxide in oil production"/, Nedra Publ. Co., Moscow, 1977

2. Bálint, V., Paál, T.: "A nedvesitési állapot és az áramlási jellemzők
változása CO_2-dal telitett fluidum-rendszerek porózus közegben való ára-
moltatásakor" /"Changes of wettability conditions and flow characteristics
for flowing carbon dioxide saturated fluid system in porous media"/, Kőolaj
és Földgáz, Nov. 1979

3. Ács, G., Doleschall, S., Biró, Z., Farkas É.: "Háromfázisu, kompoziciós
modell és alkalmazása a Budafa-nyugat telep szén-dioxidos müvelésének le-
irására" /"A three-phase, compositional model and its application for
describing CO_2 displacement of the Budafa-West reservoir"/, Part I,
Kőolaj és Földgáz, Jan. 1981; Part II, Kőolaj és Földgáz, Feb. 1981

AN ITERATIVE METHOD FOR PHASE EQUILIBRIA CALCULATIONS WITH PARTICULAR APPLICATION TO MULTICOMPONENT MISCIBLE SYSTEMS

NIKOS VAROTSIS, ADRIAN C. TODD, GEORGE STEWART

Petroleum Engineering Department,
Heriot-Watt University

ABSTRACT

An equation of state based method is used to establish phase behaviour and properties for mixtures of injection gases and reservoir fluids with specific application to multicomponent miscible systems including CO_2.

The modified Soave-Redlich-Kwong or the Peng-Robinson or a version of the Redlich-Kwong equation of state can be selected to be used in the model. The iteration method used requires a minimum number of variables for which simultaneous iteration is required and an algorithm based on the Broyden's modification of the full Newton step gives consistent phase properties and rapid convergence even near the very sensitive for a miscible displacement critical point area.

The model has been tested against published data including simple binaries, ternaries and multicomponent mixtures of reservoir oil and CO_2 injection gases. Good agreement between the predicted and the experimental values has been found together with a minimum number of iterations required to solve each problem.

The paper discusses briefly the specific use of the model in an experimental phase behaviour study for UK oil-CO_2 systems and as an integral part of a compositional reservoir simulator.

INTRODUCTION

One of today's more promising oil recovery techniques is miscible CO_2 flooding. The use of CO_2 to improve oil recovery is not a new idea since CO_2 has been investigated for miscible displacement, for immiscible displacement of reservoir oil, for producing well stimulation and for carbonated water flooding.

The current industry interest in CO_2 flooding is mainly concentrated on the mass transfer effect that takes place between the injected CO_2 phase and the reservoir oil inside the reservoir. The CO_2 extracts hydrocarbons from the oil phase and at the same time CO_2 is absorbed into the liquid phase up to the moment that miscibility is achieved. The study and prediction of oil recovery involving injection of CO_2 requires a knowledge of the vapour-liquid equilibria especially at the very sensitive critical point. A method is needed, first to calculate the saturation conditions for the mixtures of the injected gases and reservoir oils from which a prediction of the miscible pressure can be made and second to carry out the isothermal flash calculations for different pressures so that the phase behaviour of the system can be studied in detail. Such a model will be described which using any of the Peng-Robinson, modified Soave-Redlich-Kwong and

314

a version of the Redlich-Kwong equations of state can give predictions of the vapour-liquid equilibria of multicomponent mixtures and especially good and rapid convergence in the critical point region where most of the methods according to the literature fail to converge.

An extrapolation technique is used to improve the initial estimates for the consequative calculations of the saturation pressure of a reservoir oil-CO_2 mixture across the phase envelope and up to the critical point. Although the model has been specifically applied to CO_2-oil systems is obviously applicable to any injected gas or flowing system.

MISCIBILITY MECHANISMS - DIFFERENT MODELLING APPROACHES

Two of the most important and promising gas injection enhanced oil recovery practices are CO_2 flooding and lean gas injection.

The major mechanisms to improve the oil recovery in a carbon dioxide flooding are vaporization and condensation. Mass transfer takes place between the CO_2 rich phase and the oil rich phase and the initially immiscible phases gradually become miscible as they are enriched in intermediate and even heavy hydrocarbons and CO_2 respectively. The extraction of hydrocarbons by CO_2 and its condensation into the reservoir fluid results finally in an one phase miscible fluid. The development of miscibility can be visualised conceptually with a ternary diagram (Figure 1). This representation although not quantitative demonstrates how

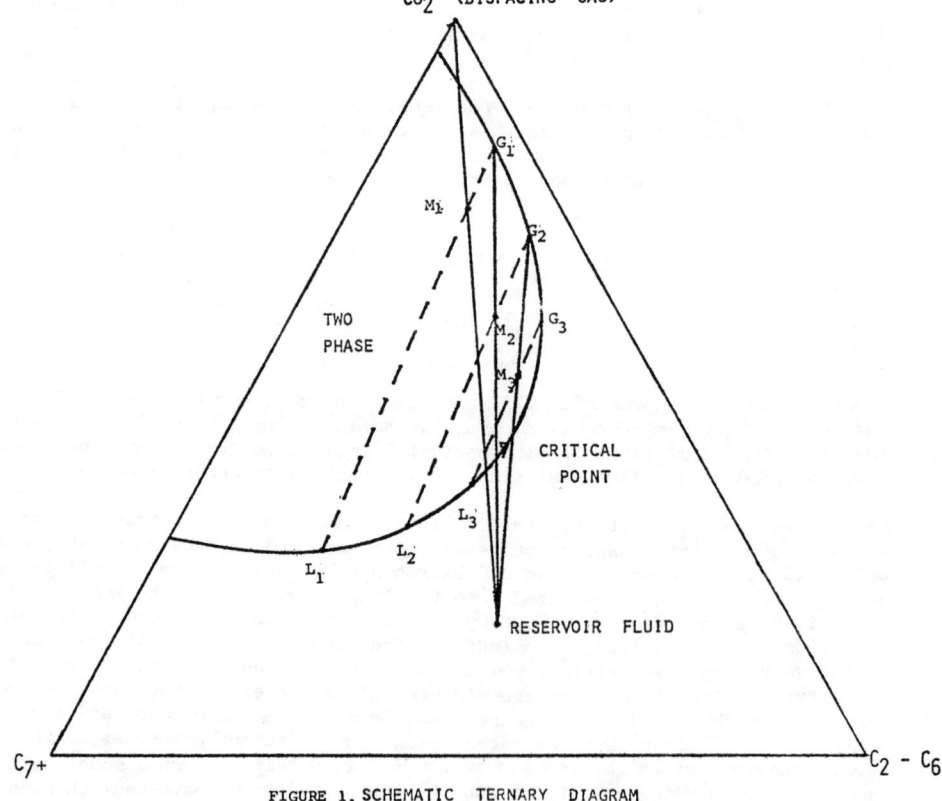

FIGURE 1. SCHEMATIC TERNARY DIAGRAM

important it is to be able to predict the critical point of a mixture for a multiple contact miscible process. The miscibility path passes through the critical point and it is its relative position in respect of the point that represents the reservoir fluid composition that defines whether under certain conditions the mixture of the injected gas and the reservoir fluid can obtain miscibility (Figure 2). The requirement for the generation of a miscible displacement is that the reservoir fluid composition must lie either to the right of the extension of the tangent to the phase boundary curve at the critical point or above the critical point in the single phase region.

The same remarks apply more or less for a lean gas injection flooding where the vaporization of the light hydrocarbons from the reservoir fluid to the gas phase controls the whole process.

There are also some minor mechanisms to improve the enhanced oil recovery by injection of CO_2. These are: oil swelling, reduction of oil viscosity, increase in oil density, high solubility of CO_2 in water which reduces the water density and therefore the overriding of the CO_2-water mixture and the acidic effect on the rock which increases the permeability of the reservoir.

The theoretical study of a miscible displacement experiment or of a miscible reservoir flooding requires accurate and reliable phase behaviour data. The phase envelope of the mixture at different conditions is required to determine the minimum miscibility pressure and the equilibrium lines (tie-lines) in order to study in detail the distribution of the different components in the two-phases.

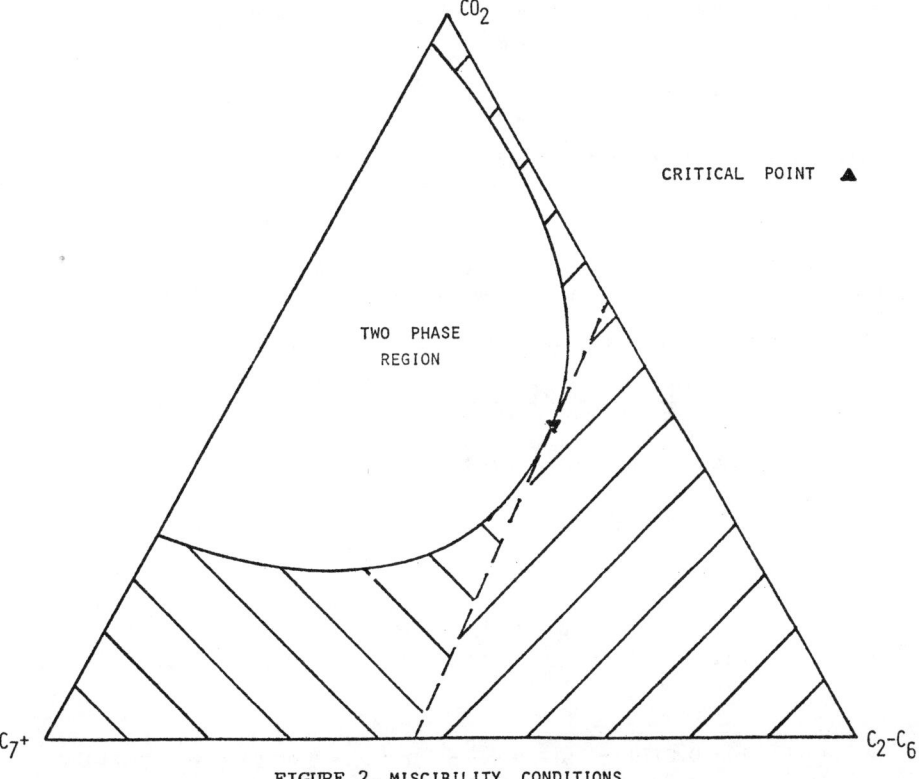

FIGURE 2. MISCIBILITY CONDITIONS

Either an equation of state based method is used to establish phase behaviour and properties or equations are used which have been obtained by curve fitting experimentally derived data. Due to inconsistent phase properties near the critical point and the requirement for comprehensive experimental data for each oil composition of the latter, the equation of state based method is now widely preferred. Most of the current published equation of state based methods appear to suffer from requiring a great number of iterations or do not converge at all in the critical point area, the key area for any miscible displacement.

PHASE EQUILIBRIA MODEL FOR A MISCIBLE OIL RECOVERY PROJECT

The technique being presented here for calculating vapour-liquid equilibria using an equation of state includes a system of non-linear equations and an iterative sequence to solve the equations. The system of equations consists of:

(i) An overall material balance equation

$$L + V = 1$$

(ii) Component material balance equations

$$Lx_i + Vy_i = z_i \qquad i = 1,n$$

(iii) Restrictive equations on the phase compositions

$$\sum_{i=1}^{n} x_i = 1, \qquad \sum_{i=1}^{n} y_i = 1$$

(iv) Thermodynamic phase equilibria equations

$$f_{iL} = f_{iV} \qquad i = 1,n$$

Three different equations of state can be used to provide values for the compressibility factor of the vapour and liquid phase. These are:

(1) The Peng-Robinson equation of state

$$P = \frac{RT}{v-b} - \frac{a(T)}{v(v+b) + b(v-b)} \qquad \text{(P-R)}$$

or in terms of the compressibility factor:

$$Z^3 - (1-B)Z^2 + (A-3B^2-2B)Z - (AB-B^2-B^3) = 0$$

where:

$$a(T) = 0.45724 \frac{R^2 T_c^2}{P_c} \sqrt{1+m(1-T_r^{\frac{1}{2}})}, m = 0.37464 + 1.54226W - 026992W^2$$

$$b = 0.0778 \frac{RT_c}{P_c} , \quad A = \frac{aP}{R^2 T^2} , \quad B = \frac{bP}{RT} \quad \text{(for pure components)}$$

(2) The modified Soave-Redlich-Kwong equation of state:

$$P = \frac{RT}{v-b} - \frac{a(T)}{v(v+b)} \qquad \text{(M-S-R-K)}$$

or in terms of the compressibility factor

$$z^3 - z^2 + (A-B-B^2)z - AB = 0$$

where:

$$a(T) = 0.42727 \frac{R^2 T_c^2}{P_c} \sqrt{1+m(1-T_r^{\frac{1}{2}})}, \quad m = 0.48508 + 1.55171W - 0.15613W^2$$

$$b = 0.0867 \frac{RT_c}{P_c}, \quad A = \frac{aP}{R^2 T^2}, \quad B = \frac{bP}{RT} \quad \text{(for pure components)}$$

(3) Modified Redlich-Kwong Equation of State

$$P = \frac{RT}{v-b} - \frac{aT^{-\frac{1}{2}}}{v(v+b)} \qquad \text{(M-R-K)}$$

or in terms of the compressibility factor

$$Z = \frac{1}{1-h} - \frac{A^2}{B}\left(\frac{h}{1+h}\right)$$

where:

$$a = \Omega_A \frac{R^2 T_c^{2.5}}{P_c}, \quad b = \Omega_B \frac{RT_c}{P_c}, \quad h = \frac{bP}{RTZ}, \quad A = \frac{aP}{R^2 T^2}, \quad B = \frac{bP}{RT}$$

(pure components).

Ω_A, Ω_B are supposed to be functions of temperature and of the nature of each component. The values of these parameters are calculated from generalised correlations applicable over a wide range of temperature. In Table I values of the parameters Ω_A, Ω_B calculated by our model are compared against those obtained by Coats and Fussell for a ternary mixture of $C_1 - nC_4 - nC_{10}$ at $160°F$ (344.3K).

TABLE I - Ω_A, Ω_B PARAMETERS

	Ω_A	Ω_B	Ω_A COATS	Ω_B COATS	Ω_A FUSS	Ω_B FUSS
C_1	0.4265	0.0862	0.42617	0.086173	0.4251	0.0859
nC_4	0.4198	0.0794	0.419367	0.0794	0.4154	0.0759
nC_{10}	0.4638	0.0734	0.451875	0.070452	0.46512	0.07259

For the same mixture and for composition (mole fraction) :

CH_4 : 0.253

n-Butane : 0.661

n-Decane : 0.086

the K-values and the saturation pressure estimated using the Modified Redlich-Kwong Equation of State compared to the values predicted by Coats and to the experimental ones are given in Table II.

TABLE II - K VALUES - SATURATION PRESSURE

	K-val. COATS	K-val OUR MODEL	K-val. EXPER.
C_1	3.173	3.174	3.174
nC_4	0.297	0.2969	0.297
nC_{10}	0.008	0.00806	0.013
Satur. Press.	972.7 psia	975.1 psia	1000 psia

For multicomponent mixtures the following mixing rules proposed by Soave are used:

$$a = \sum_{i=1}^{n} \sum_{j=1}^{n} x_i x_j a_{ij} \quad , \quad a_{ij} = a_i^{0.5} a_j^{0.5} (1-k_{ij})$$

$$b = \sum_{i=1}^{n} x_i b_i \quad , \quad K_{ij} = \text{interaction parameter}$$

The fugacity coefficients of component i in a mixture are calculated using the following equations

For the liquid phase:

$$f_{iL} = \frac{Px_i \exp\{b_{iL}(Z_L-1)\}}{(Z_L-B_L)\{1+\frac{B_L}{Z_L}\}^{\frac{U_i}{}}} \quad 1 \leq 1 \leq n$$

For the vapour phase:

$$f_{iv} = \frac{Px_i \exp\{b_{iv}(Z_v-1)\}}{(Z_v-B_v)\{1+\frac{B_v}{Z_v}\}^{\frac{W_i}{}}} \quad 1 \leq i \leq n$$

where: $U_1 = A_L(2a_{iL}-b_{iL})/B_L$, $W_i = A_v(2a_{iv}-b_{iv})/B_v$

The equilibrium ratios K_i are defined as:

$$K_i = Y_i/x_i = (f_i^L/x_iP)/(f_i^V/Y_iP) = \phi_i^L/\phi_i^V$$

DESCRIPTION OF THE PROGRAMME

The programme is written in Fortran IV language and is implemented as a conversational time-share package. The user is guided through the data input and calculation options by a question and answer sequence at the visual display unit.

There are four modes of calculations.

  (i)   Isothermal flash calculation

 (ii)   Bubble or dew-point calculation

(iii)   K-values prediction

 (iv)   Binary coefficient optimization

The programme stores physical properties for the pure components as molecular weights, critical temperatures and pressures and accentric factors.

The results printed out by the programme comprise the following items:

  (i)   Liquid and vapour phase mole fractions, L and V, compressibilities and densities

 (ii)   Composition of each phase by mole fractions x_i, y_i and k-values for each component

(iii)   Saturation pressure or temperature

 (iv)   Liquid and vapour phase enthalpies and liquid yield

  (v)   Mass ratio of vapour to feed and volume ratio of vapour to feed

 (vi)   Homogeneous mixture density

(vii)   Binary interaction coefficient (if requested)

SOLUTION TECHNIQUE

Iteration Method

A minimum variable iteration method is used to reduce the size of the correction step by eliminating as many unknowns as possible. The size reduction of the correction step is accomplished by dividing the unknown variables into two groups. The first group contains iteration (independent) variables which are the unknowns to be corrected. The second group contains dependent variables and there is an equal number of equations to define them.

The iteration sequence is a four step process:

- Select the iteration variables and assume values for these variables

- Use the defining equations to calculate the dependent variables

- Use the error equations to calculate the error

- Use a correction step to update the variables

Initial Estimates of the Iteration Variables

For the prediction of the saturation conditions a first estimation of the unknown phase composition has to be taken using the component K-values calculated by the empirical equation

$$K_i = \exp \{5.37(1+W_i)(1-\frac{1}{T_{ri}})\}/P_{ri} \ , \ \ 1 \le i \le n$$

For the isothermal flash calculations the corresponding saturation conditions are used as initial estimates.

Correction Step

The correction step used is a modified full Newton step and it requires the calculation of an approximation to the Jacobian obtained by numerical differentation of the function $f(\bar{x})$ for the first iteration and the Broyden's updating technique to improve the matrix for the rest number of iterations. This technique avoids analytical differentation of very complex functions and requires only one numerical differentation of the function $f(\bar{x})$ per calculation. The value of the iteration variable $\bar{x}$ at the K + 1 iteration is given by:

$$\bar{x}^{k+1} = \bar{x}^k + \Delta(\bar{x}^k)$$

$$\Delta(\bar{x}^k) = -B^{(k)-1} * f(\bar{x}^k)$$

where: $B^{(k)}$ is the Broyden's approximation to the Jacobian.

The array $\bar{x}$ contains n elements (the number of components present in the system). For a saturation calculation these n iteration variables are n-1 compositions plus the saturation pressure or temperature. For an isothermal flash calculation are n-1 compositions plus the vapour of liquid phase fraction (V or L).

Error Equations

The Euclidean norm of the residuals of the thermodynamic phase equilibria equation

$$M_i = f_{iL} - f_{iv}, \ \ 1 \le i \le n$$

must be less than the error tolerance.

321

EXTRAPOLATION TECHNIQUE TO IMPROVE INITIAL ESTIMATES

There is an option available in the programme to calculate the whole phase
envelope of a certain mixture of reservoir oil and injection gas starting from
the saturation conditions of the reservoir fluid and ending at the critical
point of the mixture where usually the injection gas composition is relatively
high and the two fluids are becoming first contact miscible. The step for each
successive calculation changes as the physical properties of the two phases are
approaching each other. As the overall composition approaches the critical
one, the step is being reduced to a minimum because in this region of the phase
diagram very small changes in composition cause very significant and radical
changes in the phase properties. Two different approaches have been tried to
carry out this series of calculations. Either the estimated phase compositions
and saturation conditions of the former step are used as initial estimates of
the iteration variables in the next step or these values are extrapolated
using a combination of quadratic and linear extrapolation to the new composition
step. The second approach improved the method drastically by reducing more
than 60% the number of iterations required to achieve convergence. Table III
indicates the total number of iterations required for a complete bubble-point
curve calculation. (At 18 different compositions of an eleven components
synthetic mixture ranging from 0% CO_2 up to 81% CO_2 which corresponds to the
critical composition). It also presents the number of iterations required
for bubble point calculation for two different compositions of the synthetic
oil-CO_2 mixture.

TABLE III - NUMBER OF ITERATIONS FOR A
SYNTHETIC MIXTURE

	No Extrapolation	Extrapolation
Total No. Iterations Bubble-point curve (18 points)	100	46
No. iterations for bubble-point 30% CO_2 - 70% Oil	5	2
No. iterations for critical point 82% CO_2 - 18% oil	5	1

Using the extrapolation technique and the step by step approach to the critical
region the final calculation for the critical point itself usually requires only
one or two iterations. Table IV gives the Euclidean norms for the same
synthetic mixture with and without extrapolation. Table V demonstrates how
close to the actual value, the extrapolated from the previous calculation
initial estimates of the vapour phase CO_2 compositions, are. The extrapolated
values are also compared with those that would be the initial estimates if the
extrapolation technique has not been applied.

There is also an option incorporated in the model to plot the pressure-composition data in an X - Y diagram.

An attempt was made to calculate the matrix used for the iteration sequence only
once in the beginning of each series of calculations and then to update it
continuously all across the saturation curve avoiding the recalculation of the
approximation to the Jacobian at each composition step. It has been found that
this method can be applied only when the composition step is very small, something that seems to be time consuming and not practical at all.

TABLE IV - EUCLIDEAN NORMS FOR A SYNTHETIC MIXTURE

	No. Iteration	No Extrapolation	Extrapolation
	1	0.4784	0.000502
Euclidean norms for bubble point 30% CO_2	2	0.04629	0.0000033
	3	0.00697	
	4	0.000397	
	5	0.0000257	

TABLE V - EXTRAPOLATED INITIAL ESTIMATES

Synthetic oil CO_2 mixture calculation steps	Extrap. initial estimates CO_2 mole fraction	Non extrap. initial estim- ates CO_2 mole fraction	Actual CO_2 mole fraction next step
$30\%CO_2$-$40\%CO_2$	0.5358	0.4176	0.5329
$40\%CO_2$-$50\%CO_2$	0.6393	0.5329	0.6356
$50\%CO_2$-$60\%CO_2$	0.7298	0.6356	0.7241

APPLICATIONS - DISCUSSION OF RESULTS

In order to test the accuracy of our computer model various calculations have
been performed for hydrocarbon/CO_2 mixtures for which the phase behaviour data
has been published.   These tests include binaries, ternaries, synthetic oils
and mixtures of injection gases and reservoir oils.

Isobutane - CO_2 system

The phase behaviour of this mixture has been measured by Besserer and Robinson and theoretical predictions reported by Peng and Robinson. The phase envelope was calculated at $100^{\circ}F$ and is illustrated in Figure 3. The fitting of the predicted phase boundaries to the experimental data is almost perfect. A binary coefficient of 0.105 was used in the modified Redlich-Kwong equation of state as the iteraction parameter for the mixture.

N-Butane-Decane-CO_2 system

The experimental data for this ternary mixture has been published by Metcalfe and Yarborough. The phase envelopes were calculated for two different pressures 1700 psia and 1500 psia at $160^{\circ}F$ (Figure 4).

Synthetic oil CO_2 system

This mixture has been studied by Metcalfe and Yarborough. The composition of the synthetic oil and the phase envelope were calculated at $150^{\circ}F$. Figure 5 and Figure 6 present the experimentally obtained pressure-composition data and the predicted data derived using the M-R-K and the M-S-R-K equations of state. The M-R-K equation of state seems to fit perfectly well the dew-point curve and the critical point. The maximum deviation between the predicted and the experimental points is 3.6%. The interaction coefficient used for CO_2-hydrocarbons is 0.1 and for methane $-C_{6+}$ is 0.04-0.05.

Rangely field oil - injection gases 1 & 2

The experimental data has been published by Graue et al. The calculated bubble point curve using the Peng-Robinson equation of state is plotted in Figure 7. The maximum observed deviation between the predicted and the experimental points is 3.5% and occurs at the critical composition.

Reservoir oil-CO_2 mixture

This oil has been studied by Simon et al. The C_{7+} cut has been divided into three pseudo components and the bubble point curve was calculated at $255^{\circ}F$ using the M-R-K equation of state (Figure 7). Once again in the near the critical point region the fitting of the predicted curve is almost perfect.

The theoretical model described above, is a part of a research programme on dynamic contact CO_2 miscible studies. The various calculations that have already been performed are for U.S. oils and reservoir conditions. The next stage will be the generation of experimental phase behaviour data for CO_2-oils for a whole range of North Sea crudes using a rig for multiple contact equilibrium experiments which has now already started to operate. The experimental results will be a test for our compositional simulator and especially for the assumptions we used. A valid and reliable theoretical model can decrease dramatically the cost of a miscible flooding because only a few experimental results, which are very expensive and time consuming, have to be obtained in order to establish a complete view of the phase behaviour of the reservoir fluids.

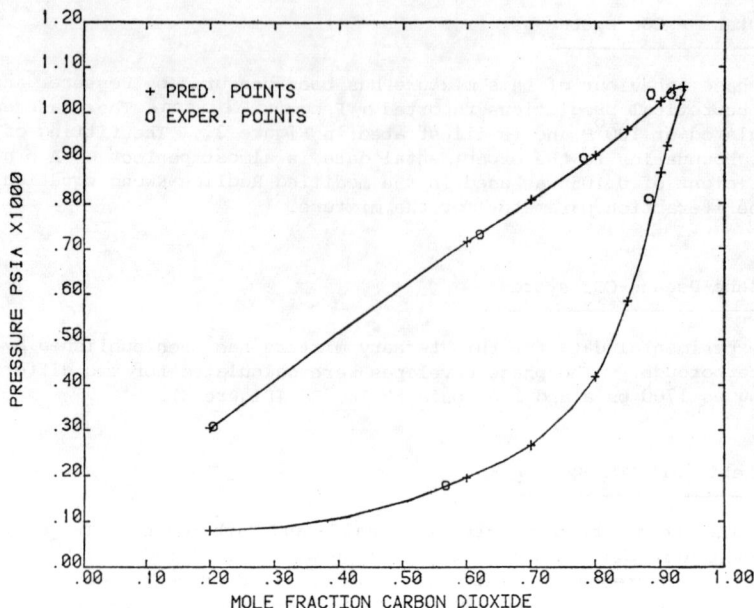

FIGURE 3. PHASE ENVELOPE FOR CO2- ISO-BUTANE MIXTURE TEMPERATURE=100F R-K EOS

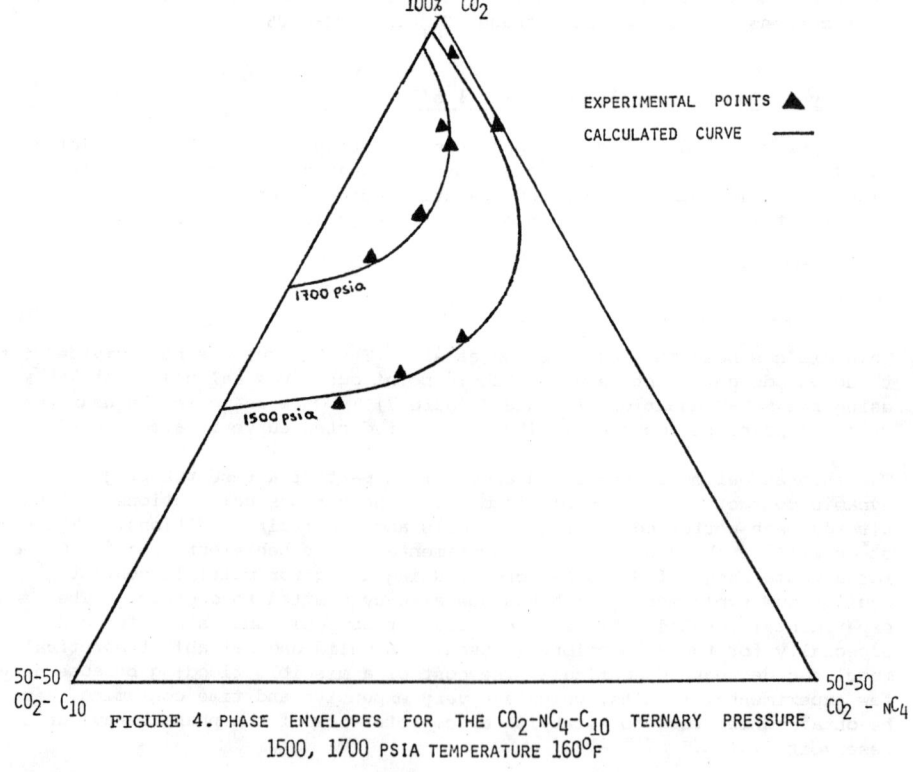

FIGURE 4. PHASE ENVELOPES FOR THE CO_2-nC_4-C_{10} TERNARY PRESSURE 1500, 1700 PSIA TEMPERATURE 160°F

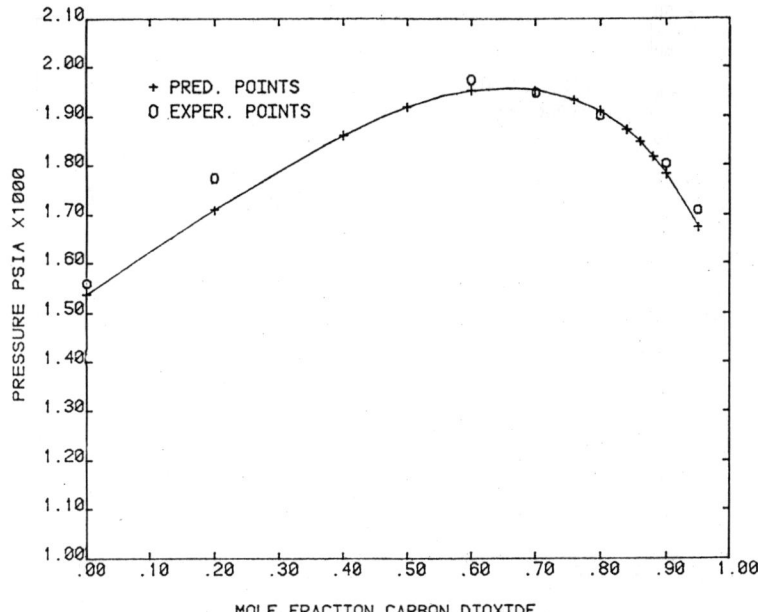

FIGURE 5. PHASE ENVELOPE FOR CO2−SYNTHETIC OIL MIXTURE
TEMPERATURE=150F R−K EOS

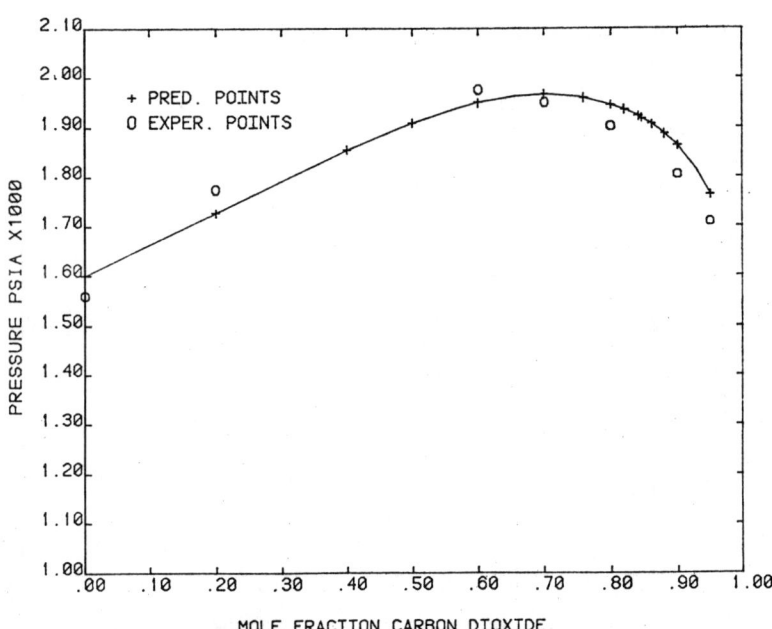

FIGURE 6. PHASE ENVELOPE FOR CO2−SYNTHETIC OIL MIXTURE
TEMPERATURE=150F MOD.S−R−K EOS

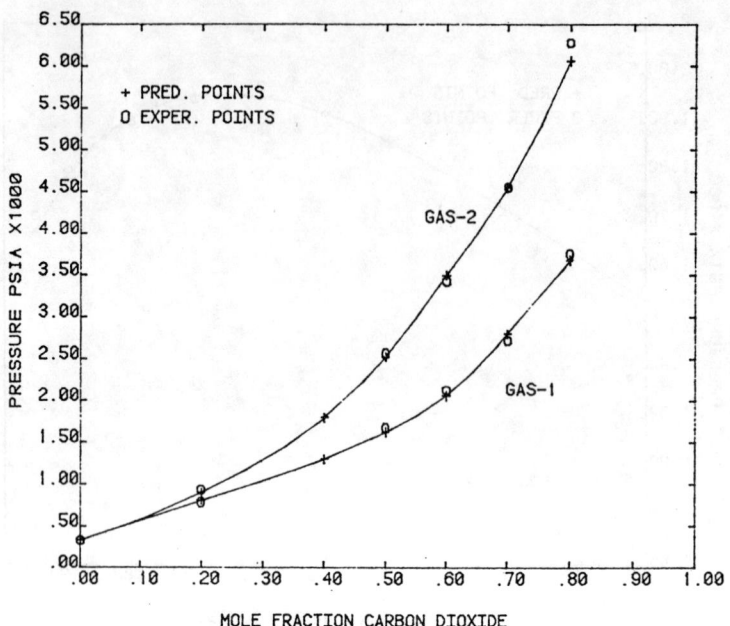

FIGURE 7. BUBBLE POINT CURVE FOR RANGELY FIELD OIL—CO2 MIXTURES
TEMPERATURE=160F P—R EOS

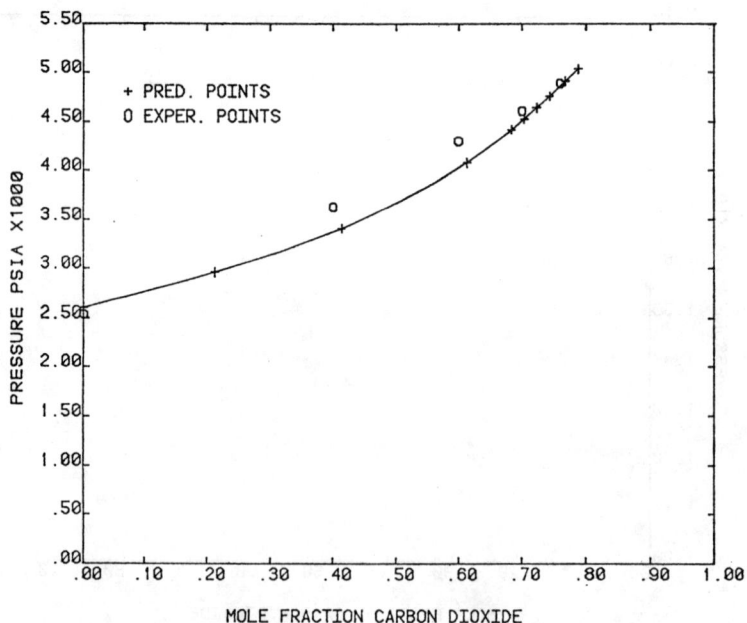

FIGURE 8. BUBBLE POINT CURVE FOR CO2—RESERVOIR OIL MIXTURE
TEMPERATURE=255F R—K EOS

CONCLUSIONS

A robust computer programme for isothermal flash, bubble and dew point calculations using one of the Peng-Robinson, Soave-Redlich-Kwong, Modified Redlich-Kwong equation of state has been developed applicable to the severe reservoir conditions encountered in miscible gas flooding enhanced oil recovery schemes.

The demonstrated examples show very good and rapid convergence at any point across the phase boundaries of the CO_2/hydrocarbon mixtures and particularly the sensitive critical point region. A case has not yet been found where the proposed scheme does not converge although we have, however, found situations where the MSRK or PR equations of state converge to unrealistic solutions.

Various tests indicated that the MSRK equation appears in some cases to give closer approximation to the experimental data than the PR equation. The MRK equation appears to fit better the dew point curve than the bubble point one. The treatment of the pseudo components in the oil mixtures is under more investigation.

NOMENCLATURE

a, b   =   Temperature, pressure and composition dependent parameters

f_{iL} =   Fugacity of component i in liquid phase, psia

f_{iV} =   Fugacity of component i in vapour phase, psia

K_i =   Equilibrium ratio y_i/x_i

L   =   Mole fraction of liquid phase

P   =   Pressure

P_c =   Critical pressure

P_r =   Reduced pressure P/P_c

R   =   Gas constant

T   =   Temperature

T_c =   Critical temperature

T_r =   Reduced temperature T/T_c

V   =   Vapour phase mole fraction

v   =   Molar volume

x_i =   Mole fraction of component i in liquid phase

y_i =   Mole fraction of component i in vapour phase

z_i =   Global mole fraction of component i

z_L =   Compressibility factor of vapour phase

W   =   Accentric factor

REFERENCES

1. McGLASHAN, R.S.; "A Compositional Phase Equilibrium Model Applied to Pressure Drop Prediction in North Sea Oil Wells", Ph.D Thesis, Heriot-Watt University, 1980

2. COATS, K.H.; "An Equation of State Compositional Model", SPEJ, October 1980, p.363

3. GARDNER, J.W., ORR, F.M., PATEL, P.D.; "The Effect of Phase Behaviour on CO_2 Flood Displacement Efficiency", SPE 8367, 1979

4. NGHIEM, L.X., AZIZ, K.; "A Robust Iterative Method for Flash Calculations using the Soave-Redlich-Kwong or the Peng-Robinson Equation of State", SPE 8285, 1979

5. FUSSELL, D.D., YANOSIK, J.L.; "An Iterative Sequence for Phase Equilibria Calculations Incorporating the Redlich=Kwong Equation of State", SPEJ, June 1978

6. PENG, D.Y., ROBINSON, D.B.; "A New Two-Constant Equation of State", Ind. Eng. Chem. Fundam. Vol. 15 No. 1, 1976

7. GRABOSKI, M., DAUBERT, T.; "A Modified Soave Equation of State for Phase Equilibrium Calculations", American Chemical Society Journal, 1978

8. BESSERER, G., ROBINSON, D.; "Equilibrium Phase Properties of i-Butane-CO_2 System", J. Chem. Eng. Data, Vol. 18, No. 3, 1973, p. 298

9. OLDS, REAMER, SAGE, LACEY; "Phase Equilibria in Hydrocarbon Systems N-Butane CO_2 System", Ind. & Eng. Chem., 1949, p. 475

10. SIMON, ROSMAN, ZANA; "Phase Behaviour Properties of CO_2 – Reservoir Oil Systems", SPEJ, Feb. 1978, p. 20

11. GRAUE, D.J., ZANA, E.; "Study of a Possible CO_2 Flood in the Rangely Field, Colorado", SPE 7060, 1978

12. METCALFE, R.S., YARBOROUGH, L.; "Effect of Phase Equilibria on the CO_2 Displacement Mechanism, SPE 7061. 1978

PHASE EQUILIBRIUM CALCULATIONS
IN THE NEAR-CRITICAL REGION

RASMUS RISNES

Norsk Agip A/S

VILGEIR DALEN, JAN IVAR JENSEN

Continental Shelf Institute

ABSTRACT

The present paper addresses the problem of phase equilibrium calculations in the critical point region. The approach is based on an equation of state, and both the Soave-Redlich-Kwong and the Peng-Robinson equations are considered. An accelerated and stabilized successive substitution method is presented. A procedure for disappearing phases is included, making the method convergent also in the single phase region. The accelerated successive substitution method has been compared with Newton type methods like Powell's method. Maps of an error norm which measures the fugacity deviations, are presented to illustrate how the different solution techniques perform. The general conclusion is that the accelerated successive substitution method is faster and much more stable than the Newton type methods considered.

INTRODUCTION

During recent years, hydrocarbon phase equilibrium calculations based on cubic equations of state has received considerable attention, partly because of the demand for accurate and consistent phase predictions encountered in connection with enhanced oil recovery techniques like gas miscible flooding. Both the Soave-Redlich-Kwong (SRK) equation /1/ and the Peng-Robinson (PR) equation /2/ have been extensively used. They both perform well on hydrocarbon mixtures, the PR equation being slightly better in predicting liquid densities. The basic solution procedure for flash calculations is the successive substitution method (SSM). It has however been reported to show poor convergence, or even no convergence, close to saturation pressures. To overcome the convergence problem, Fussel and Yanosik /3/ introduced Newton type iteration methods, and since then the trend has been towards such refined numerical solution techniques. Newton type methods are however dependent on good initial estimates. In a recent paper Nghiem and Aziz /4/ presented an algorithm using Powell's method which is a combination of a Newton method and the steepest decent method. They also presented a method to detect single-phase states. Their method was extended to three- and four-phase systems by Mehra /5/. Also Mott /6/ has presented a two-phase algorithm based on Powell's method. An important problem in equilibrium calculations is to avoid false trivial solutions where the vapor and the liquid phase are identical. This aspect has been discussed by Maddox and Erbar /7/.

The present work is part of a research project concerned with the development of numerical simulation models for enhanced oil recovery processes. It was directed towards the development of a thermodynamic simulator capable of predicting the phase behaviour of mixtures of hydrocarbon reservoir fluids and possible injection gases. The resulting computer program is called COPEC. The program is based on an equation of state approach, and both the SRK and the PR equations are included. Several solution options are available including Powell's method, but the basic solution method is an accelerated and stabilized successive substitution method (ASSM). The method is designed to converge also in the single-phase region, and it contains a bring-back procedure that brings the solution back to the two phase region if it reaches the single-phase region too soon. The acceleration routine employs an Aitken type formula for correcting K-values.

EQUILIBRIUM CONDITIONS

If we consider N moles of mixture or feed of composition z_i which separate into L moles of liquid of composition x_i and V moles of vapor of composition y_i, we have an overall material balance and a component balance equation for each component:

$$L + V = N \tag{1}$$

$$Lx_i + Vy_i = Nz_i \tag{2}$$

As the compositions are given in mole fractions we have the following constraints:

$$\Sigma z_i = \Sigma x_i = \Sigma y_i = 1 \tag{3}$$

Eqs. (1), (2) and one of the restricting equations (3) constitute a system of n+2 equations in the 2n+2 unknowns, L, V, x_i, y_i. n is the number of components. The remaining n equations needed are provided by the thermodynamic criterion stating that the fugacities in the liquid and the vapor phase must be equal:

$$f_{iL} = f_{iV} \tag{4}$$

These 2n+2 equations define the two-phase equilibrium problem.

In an equation of state approach, the fugacities can be calculated from the equation of state. The fugacity will depend on temperature, pressure, composition and the type of phase considered,

$$f_i = f_i (T, P, x_i, type)$$

With cubic equations of state, the same equation is used both for the liquid and the vapor phase. A cubic equation may give 3 solutions in volume. The distinction between liquid and vapor phase is then made by choosing the smallest volume for the liquid phase and the greatest volume for the vapor phase.

Formulas for the Soave-Redlich-Kwong (SRK) and the Peng-Robinson (PR) equations of state are given in Table 1.

BASIC SUCCESSIVE SUBSTITUTION METHOD

The successive substitution method is based on the concept of equilibrium constants K_i defined by:

$$K_i = y_i/x_i \tag{5}$$

Table 1   Summary description of the Soave-Redlich-Kwong and Peng-Robinson
equations of state.

Soave-Redlich-Kwong:	Peng-Robinson:
$P = \dfrac{RT}{v-b} - \dfrac{a}{v(v+b)}$	$P = \dfrac{RT}{v-b} - \dfrac{a}{v(v+b) + b(v-b)}$
$b = 0.08664 \dfrac{RT_c}{P_c}$	$b = 0.07780 \dfrac{RT_c}{P_c}$
$a(T) = 0.42747 \dfrac{R^2 T_c^2}{P_c} \alpha$	$a(T) = 0.45724 \dfrac{R^2 T_c^2}{P_c} \alpha$
$\alpha^{0.5} = 1 + m(1 - T_r^{0.5})$	$\alpha^{0.5} = 1 + m(1 - T_r^{0.5})$
$m = 0.480 + 1.574\omega - 0.176\omega^2$	$m = 0.37464 + 1.54226\omega - 0.26992\omega^2$
The cubic equation for the compressibility factor $Z = Pv/RT$ is:	The cubic equation for the compressibility factor $Z = Pv/RT$ is:
$Z^3 - Z^2 + (A-B-B^2)Z - AB = 0$	$Z^3 -(1-B)Z^2+(A-3B^2-2B)Z-(AB-B^2-B^3) = 0$
where $A = \dfrac{aP}{R^2 T^2}$ and $B = \dfrac{bP}{RT}$	where $A = \dfrac{aP}{R^2 T^2}$ and $B = \dfrac{bP}{RT}$
With mixing rules as given below, the fugacity coefficient of component k is given by:	With mixing rules as given below, the fugacity coefficient of component k is given by:
$\ln \Psi_k = \dfrac{b_k}{b}(Z-1) - \ln(Z-B) -$ $\dfrac{A}{B}\left(\dfrac{2\sum_i x_i a_{ik}}{a} - \dfrac{b_k}{b}\right) \ln(1+\dfrac{B}{Z})$	$\ln \Psi_k = \dfrac{b_k}{b}(Z-1) - \ln(Z-B) -$ $\dfrac{A}{2\sqrt{2}B}\left(\dfrac{2\sum_i x_i a_{ik}}{a} - \dfrac{b_k}{b}\right) \ln\left(\dfrac{Z+(\sqrt{2}+1)B}{Z-(\sqrt{2}-1)B}\right)$

The mixing rules employed for both equations are:

$$b = \sum_i x_i b_i$$

$$a = \sum_i \sum_j x_i x_j a_{ij}$$

$$a_{ij} = (1 - \delta_{ij}) a_i^{0.5} a_j^{0.5}$$

where δ_{ij} are binary interaction coefficients.

where y_i and x_i are mole fractions in equilibrium. If values for the equilibrium constants are assumed, and the fugacity equations (4) are replaced by the K_i-equations (5), the resulting set of equations can easily be solved for the unknowns L, V, x_i and y_i. With these compositions improved K_i-values can be obtained, and the cycle repeated.

Introducing the fugacity coefficients ψ_i, the fugacities can be written

$$f_{iL} = x_i \, \psi_{iL} P$$
$$f_{iV} = y_i \, \psi_{iV} P \tag{6}$$

In equilibrium the fugacities are equal and hence the equilibrium constants are given by

$$K_i = \psi_{iL}/\psi_{iV} \tag{7}$$

This is an important relation as it allows the definition of K-values also outside the two-phase region.

During the iteration process when the fugacities are not yet equal, the improved K-value estimates are obtained by

$$K_i^{j+1} = \frac{\psi_{iL}}{\psi_{iV}} = K_i^j \cdot R_i^j \tag{8}$$

where j is the iteration number and R_i is the fugacity ratio f_{iL}/f_{iV}. The criterion for acceptance of a solution is based on the fugacity ratios. To comply with other solution methods the following error norm is used

$$\rho = \Sigma(R_i - 1)^2 < \varepsilon \tag{9}$$

When the equilibrium constants are given, the system of flash equations (1), (2), (3) and (5) can most conveniently be solved by introducing the g(V) function following Nghiem and Aziz /4/. If we consider one mole of feed, N=1, eliminate L in Eq. (2), and then sum up all the equations we obtain

$$\Sigma x_i + V\Sigma(y_i - x_i) = 1 \tag{10}$$

The g(V) function is defined by

$$g(V) = \Sigma(y_i - x_i) = \Sigma \frac{(K_i-1)z_i}{1+(K_i-1)V} \tag{11}$$

From Eq. (10) we see that V is determined as the root in the equation

$$g(V) = 0 \tag{12}$$

This equation is readily solved by Newton's method. As the g(V) function always has a negative slope, there will only be one root of interest. When the value of V is determined, the compositions can be calculated in a straight-forward manner.

When the root of Eq. (12) gets outside the interval [0,1], this indicates a single-phase state. We then calculate the non-existing phase as if the system were at the saturation pressure:

If V<0, then V is set equal to 0, $x_i = z_i$, $y_i = \dfrac{K_i x_i}{\Sigma\, K_i\, x_i}$ (13)

If V>1, then V is set equal to 1, $y_i = z_i$, $x_i = \dfrac{y_i/K_i}{\Sigma\, y_i/K_i}$ (14)

The normalization is necessary as this is not automatically assured outside the two-phase region.

A common factor in the K-values has no effect on the composition of the non-existing phase, and when the K-values are corrected according to the fugacity ratios, the compositions are corrected only to the extent that the fugacity ratios deviate from the average value. The system will converge to a definite composition, and the fugacity ratios will converge to a common constant value. This limiting value will be different from unity except if the system is at the saturation pressure.

K-VALUE ESTIMATION

The set of initial K-values is the starting point for the iteration procedure. There are 3 conditions these estimates should meet:

1. The estimates should be as close as possible in order to obtain a rapid solution.

2. The estimates should assure that the calculations start in the two-phase region in order to avoid false single-phase solutions.

3. The estimates should have sufficient spread to avoid false solutions where all K-values become equal to one.

The often quoted empirical formula from Wilson /8/

$$K_i = \frac{1}{P_{ri}}\ \exp\left[5.3727\ (1+w_i)\ (1 - \frac{1}{T_{ri}}\ )\right] \tag{15}$$

normally meets these requirements well.

An alternative to the empirical formula can be based on the equation of state. The basic idea is the following. The mixture or feed is assumed to be liquid at the temperature and pressure given, and the fugacities are calculated. We then assume a gas phase to be formed by evaporation from this liquid. The evaporation rate for each component is assumed to be proportional to the fugacity of that component, the proportionality constant being the same for all components. The evaporation must be stopped before we run out of any component in the liquid phase, but if possible, the evaporation should proceed until only half of the liquid remains. From the resulting compositions, the fugacity coefficients are calculated, and from these we obtain the K-value estimates.

These fugacity based K-values work very well in the near critical region. The reason is probably that in addition to start with K-values consistent with the equation of state, we start in the middle of the two-phase region with both L and V equal to one half. The method also works well along the bubble point curve and in most of the two phase region. It may however break down along the lower dew point curve. There, a restriction of the liquid Z-factor to say $Z_L < 0.3$, may be needed in order to assure that the assumed liquid behaves liquid-like.

THE ACCELERATION PROCEDURE

When the system is close to the critical point, the convergency may be very slow. A method similar to Aitken's accelerating formula can then be used to speed up the convergence rate.

The equilibrium constants can be regarded as long products, starting with the initial estimate K_i^o and then multiplied by the fugacity ratios R_i which approaches unity as the number of iterations increases. This can be written

$$K_i = K_i^o \cdot R_i^1 \cdot R_i^2 \cdot R_i^3 \cdots R_i^j \cdots \tag{16}$$

Taking the logarithm we obtain

$$\log K_i = \log K_i^o + \log R_i^1 + \log R_i^2 + \log R_i^3 + \ldots + \log R_i^j + \ldots \tag{17}$$

In the first part of an equilibrium calculation the fugacity ratios may change from values smaller than one to values greater than one, causing alternate signs in the series above. But after say 20 to 50 iterations, the situation is characterized by a monotone and steady approach towards the solution.

The process can now be accelerated by replacing the remaining part of the series by a geometric series where k is the ratio between the terms

$$\log K_i = \log K_i^j + \log R_i^j (1 + k + k^2 + ..) = \log K_i^j + \frac{\log R_i^j}{1-k} \tag{18}$$

The quotient k is calculated as the ratio between the last two consecutive terms (omitting the subscript i)

$$k = \frac{\log R^{j+1}}{\log R^j} \approx \frac{\log R^j}{\log R^{j-1}} \approx \frac{R^j - 1}{R^{j-1} - 1} \tag{19}$$

and the resulting accelerating step is

$$K_i = K_i^j \cdot R_i^j \, (\frac{1}{1-k}) \tag{20}$$

where $1/(1-k)$ is exponent to the fugacity ratio.

When this acceleration step is used, it must always be tested that it leads to an improved solution in the way that it brings the fugacity ratios closer to unity. If not, it must be rejected and replaced by a single step. Between each accelerated step there must be a single step in order to determine the quotient k in the exponent in Eq. (20).

A simplified approach is to assume a constant value for the exponent to the fugacity ratio. If we wait until the system is well on the right track, an exponent of 2 may normally safely be employed. We routinely apply this exponent after 10 iterations, also in conjunction with proper acceleration.

THE BRING-BACK PROCEDURE

When the system is in a single-phase state, the composition of the non-existing phase is calculated by formula (13) or (14). As a common factor in the set of equilibrium constants has no influence on the composition of the non-existing phase, this gives the possibility to adjust the K-values to make the $g(V)$-function zero, or in other words, to keep the non-existing phase at the edge of the two-phase region while we test for its existence.

If we consider a single-phase liquid, V equals zero, and the multiplication factor γ is determined from

$$\Sigma(\gamma K_i - 1)\, z_i = 0 \tag{21}$$

which gives a new set of equilibrium constants

$$K_i^{new} = \gamma K_i = \frac{K_i}{\Sigma\, K_i\, z_i} \tag{22}$$

on which we apply the normal correction factors (f_{iL}/f_{iV}).

Physically this corresponds to creating a nucleus gas bubble and see if it will grow or disappear. A gas bubble will grow spontaneously if the fugacity is lower in the gas phase than in the liquid phase. This corresponds to having correction factors (f_{iL}/f_{iV}) mostly greater than one. The K-values will then be further increased and the system will return to the two phase state. If, however, the fugacities are higher in the gas phase than in the liquid phase, the bubble will disappear spontaneously. When the correction factors (f_{iL}/f_{iV}) are mostly less than one, the K-factors are reduced bringing the system back to the single-phase state.

If the single phase is gas, we may reason in a similar way. As V equals one, the multiplication factor is in this case determined from

$$\Sigma(1 - \frac{1}{\gamma\, K_i})\, z_i = 0 \tag{23}$$

NEWTON-TYPE METHODS

As a supplement to the acceleration procedure described above, we have also implemented the Newton-type method described by Powell /9/. This algorithm has been used by several other workers during recent times /4, 5, 6/ and is claimed to be a robust and efficient tool for the problem at hand. In summary, Powell's algorithm is based on the classical Newton method, but differs from that method in two important ways. First, the robustness is increased by combining the Newton method with the more stable steepest descent method. Secondly, inversion of matrices at each iteration step is avoided by using an approximate matrix-updating scheme directly on the inverse of the Jacobian.

Using Powell's method, the equilibrium problem is stated as a system of n nonlinear equations in n independent variables by

$$\frac{f_{iL}}{f_{iV}} - 1 = 0 \qquad \text{for all i} \tag{24}$$

and the solution is accepted when the residual norm ρ of Eq. (9) gets below a selected tolerance as before. Independent variables are selected from the following options:

LX-iteration:	L and x_i for all but the last component.
LY-iteration:	V and y_i for all but the last component.
NL-iteration:	n_{iL} for all components.
NV-iteration:	n_{iV} for all components.

n_{iL} and n_{iV} are liquid and vapor fractions per component, i.e. $n_{iL} = Lx_i$ and $n_{iV} = Vy_i$.

Following Nghiem and Aziz /4/, the strategy adopted here is to start with successive substitutions and then switch to Powell's method if the convergence is slow. A switch after j iterations requires all of the following conditions to be fulfilled:

$$\varepsilon_L < \rho^j < \varepsilon_U \quad \text{and} \quad \frac{\rho^j}{\rho^{j-1}} > \varepsilon_R$$

$$0 < v^j < 1 \quad \text{and} \quad \left| v^j - v^{j-1} \right| < \varepsilon_V \tag{25}$$

Default values of ε_L, ε_U, ε_R, ε_V and ε of Eq. (9) are set to 10^{-5}, 10^{-3}, 0.5, 0.01 and 10^{-8}, respectively.

The flash equilibrium problem may also be formulated as a pure minimization problem, and some tests of this approach has been made utilizing the general-purpose minimization routines E04JAF and E04FDF contained within the NAG library /10/.

APPLICATIONS

The above algorithms have been implemented into a computer program COPEC, and some applications on relatively simple fluid systems are discussed in the following. All cases are run in double precision on a VAX-11/780 computer. Relevant component properties are summarized in Appendix.

Mixtures of Isobutane and Carbon Dioxide

Tests with this binary system has been made with the PR version, and all results shown here are obtained for a temperature of 311 K (100°F). The phase behaviour of this system for a wide range of CO_2 contents is depicted in Fig. 1 and is in good agreement with previous calculations and experimental data /2, 6/.

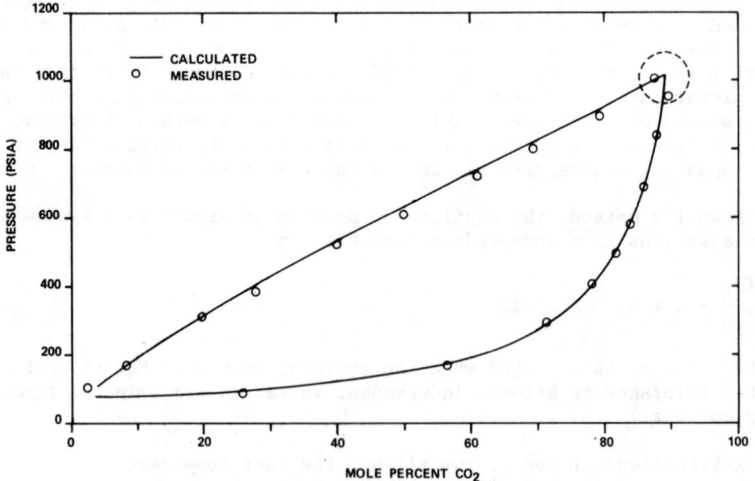

Figure 1   Phase envelope for binary mixtures of isobutane and carbon dioxide at 311 K (100°F).

In most parts of the two-phase region of Fig. 1, solutions are obtained easily even with pure successive substitutions, whereas the region indicated by a circle may pose more difficult problems. More detailed results in this region are given in Figs. 2 and 3. With the refined successive substitution method presented here, we encountered no serious problems obtaining saturation points as indicated in Fig. 2. These points are obtained by repeated flash calculations and not by direct saturation point calculations, and they demonstrate thus the ability to perform flash calculations very close to a critical point. Details of the volumetric behaviour in this near-critical region are shown in Fig. 3. A critical CO_2 content between 89.1 and 89.2% is predicted.

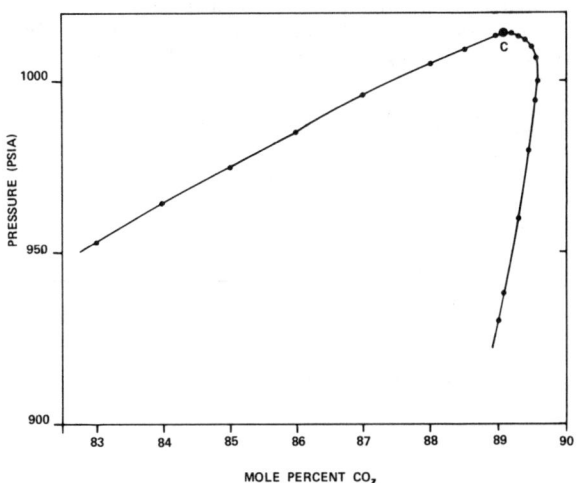

Figure 2   Same system in the nearcritical region (calculated points are indicated by dots).

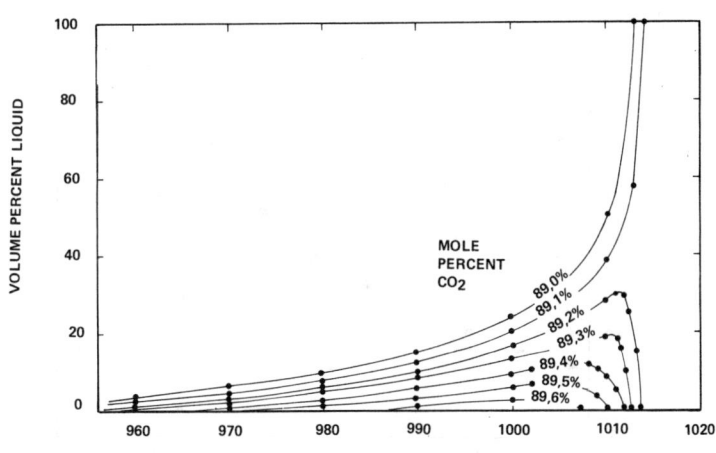

Figure 3   Volumetric behaviour of $I-C_4 - CO_2$ at 311 K (100°F) in the near-critical region (calculated points are indicated by dots).

338

In Fig. 4, the predicted flash behaviour for a CO_2 content of 89 mole percent is considered in detail, and several observations can be made. First, it can be seen that a relatively high tolerance level on the fugacity residuals (ε in Eq. (9)) tends to widen the two-phase region. $\varepsilon = 10^{-8}$ also gives a smooth curve, but is obviously a too high tolerance this close to a critical point. If such a condition is suspected, the sensitivity of the solution with respect to the tolerance level should always be investigated.

A more serious matter is the observation that for pressures above 6.977 MPa, the Powell method converges to a false or, more precisely, trivial solution. Such solutions are characterized by all K-values approaching unity and may appear when the feed composition yields one and only one root in the cubic equation for the compressibility factor. The problem of avoiding such trivial solutions is probably the biggest problem encountered with flash calculations near to a critical point.

The failure of Powell's method is investigated further in Figs. 5 and 6. These are plots of the fugacity residual norm ρ as function of the independent variables corresponding to a VY-iteration and depict in detail the performance of the different solution alternatives. Contour values refer to the logarithm of ρ ($\log_{10}\rho$). The vanishing norm for a $I-C_4$ vapor mole fraction of 0.11 corresponds to the trivial solutions of $K_i = 1.0$.

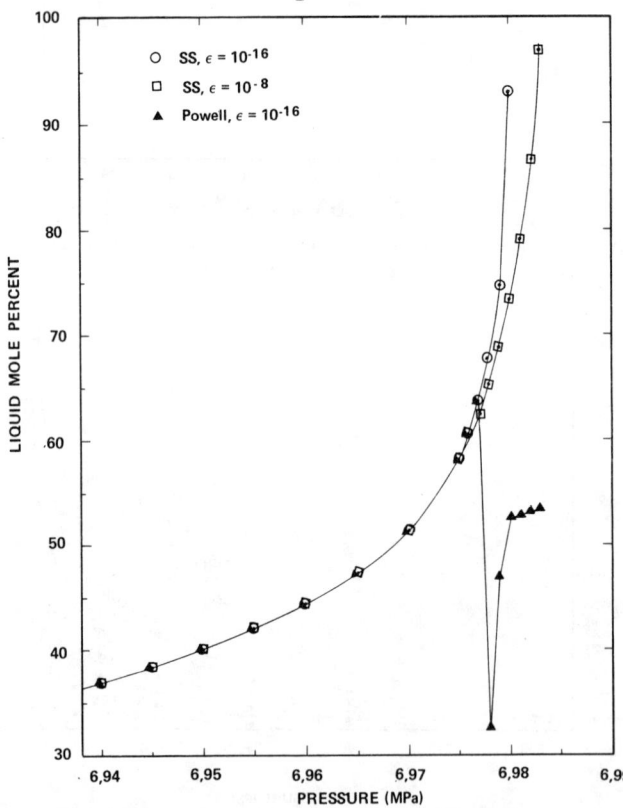

Figure 4 Flash behaviour of a $I-C_4$ - CO_2 mixture (89 mole percent CO_2) at 311 K ($100°F$) for different solution alternatives.

Fig. 5 shows the solution space at a relatively large scale and depicts the performance of the first 15 pure successive substitutions. K-value estimates both by formula (15) and the fugacity approach are included. The fugacity approach starts off somewhat better, but in this case this makes very little difference when some 15 iterations have been performed. After 15 iterations the residual norm is approximately 10^{-5}, and Fig. 6 depicts what happens if acceleration or Powell's method is started at this point. As may be seen, the accelerated successive substitution method proceeds in rather large steps towards a true solution whereas Powell's method rather quickly finds a trivial solution. The figure also shows what happens when Powell's method is employed at a norm of 10^{-7} and 10^{-9}. In the former case, a trivial solution is rapidly found, while after 300 iterations in the latter case, the solution is stuck at what appears to be something like a saddle point. It should be noted also, that it takes 172 pure successive substitutions to reach a norm of 10^{-9} whereas the accelerated version reaches a norm of 10^{-23} in just 31 iterations.

Fig. 5 clearly shows that 10^{-8} is not an appropriate tolerance level in the present case. A large region in the solution space satisfies this tolerance. Moreover, a maximum norm of only 10^{-12} separates the true solution from the trivial ones. It should be understood that the small-scale contour features in the lower left-hand parts of Fig. 6 are artifacts originating from that map (and all the other maps shown here) being contoured from a 101 by 101 network of points.

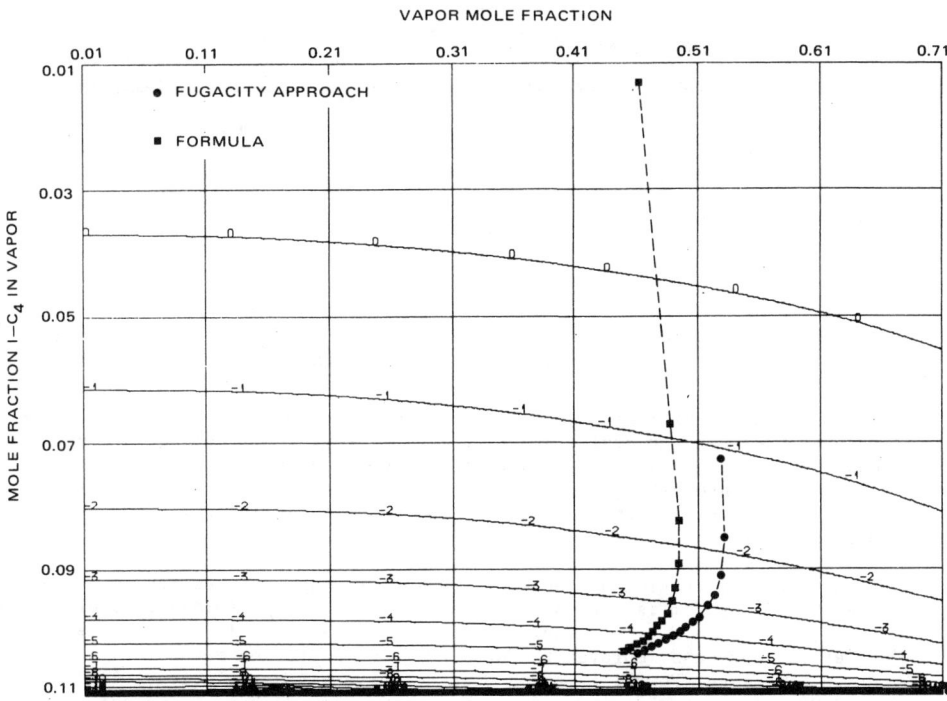

Figure 5   Residual norm plot for a 11% $I-C_4$ - 89% CO_2 mixture (P = 6.980 MPa, T = 311 K).

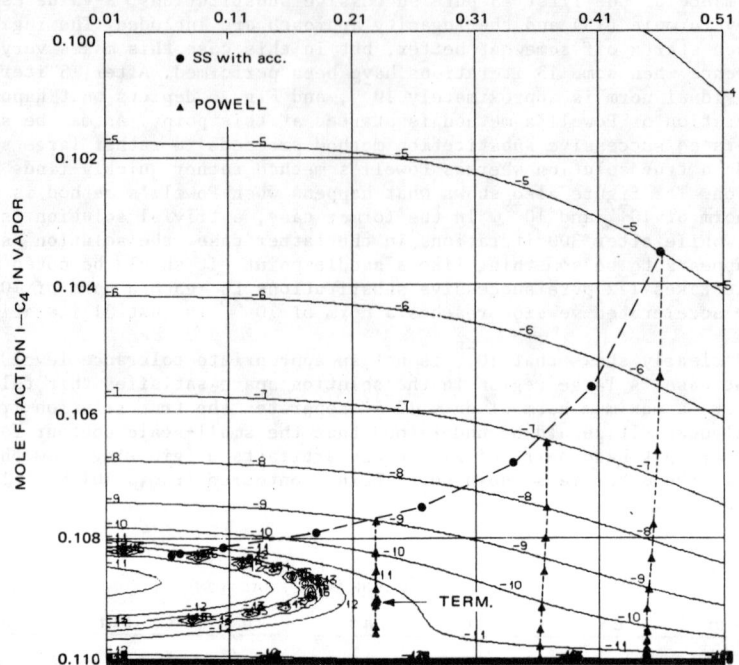

VAPOR MOLE FRACTION

MOLE FRACTION I-C₄ IN VAPOR

Figure 6   Residual norm plot for a 11% I-C$_4$ - 89% CO$_2$ mixture showing the
behaviour of different solution alternatives (P = 6.980 MPa,
T = 311 K).

Fig. 7 shows similar fugacity residual norm plots for a decreasing pressure.
P = 6.981 MPa corresponds to a single-phase liquid. The others show how the
delineation between true two-phase solutions and trivial solutions is gradually improved as the pressure drops off from the bubble-point.

The independent variables selected for the norm plots are probably not the
natural ones for successive substitution. If anything, successive substitution
must be considered as a process taking place in a space spanned by the K-values.
It would be interesting to see the performance in such a space, and it might
be an idea to employ the K-values as independent variables in Newton-type methods as well.

A further illustration of the acceleration process is given in Fig. 8. Again
the 89% CO$_2$ system is considered, and the iteration performance is plotted for
two pressures in Fig. 4, namely P = 6.94 and 6.98 MPa. It is seen that the acceleration process improves the successive substitution method dramatically.
As we have seen, Powell's method does not give a true solution at all at
6.98 MPa. At 6.94 MPa a proper solution is obtained, but at approximately
twice the no. of iterations required with acceleration. In addition, a Powell
iteration generally takes more computer time than a successive substitution
iteration.

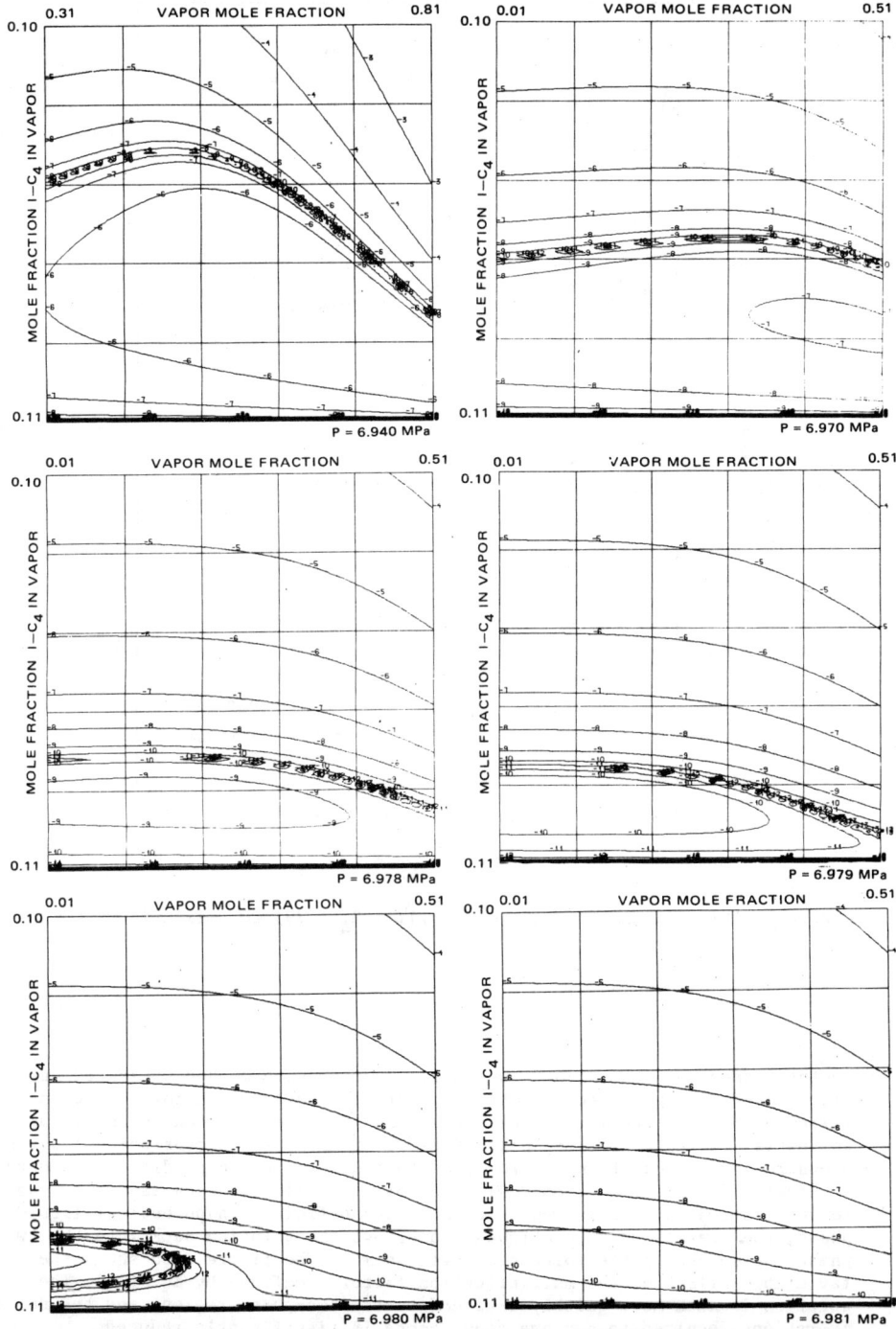

Figure 7 Residual norm plots for a 11% I-C$_4$ - 89% CO$_2$ mixture at 311 K.

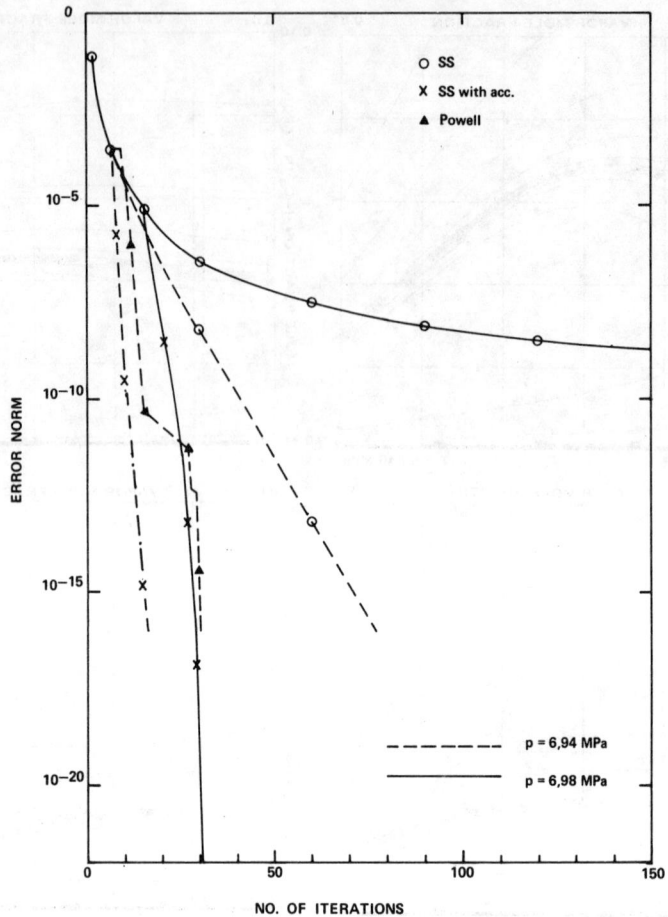

Figure 8   Iteration performance for a 11% I-C$_4$ - 89% CO$_2$ mixture at 311 K.

Another aspect of the refined successive substitution method presented here is illustrated in Fig. 9. It depicts the iteration performance for a step in a series of flash calculations where resulting K-values from one point is used as initial estimates for the next. A 89.3% CO$_2$ mixture is considered and the pressure step in question is from 6.82 to 6.62 MPa, corresponding to a decrease in liquid mole fraction from 9.01 to 0.46%. The solution escapes the two-phase region after 3 iterations, but with Eq. (14) defining a hypothetical liquid phase, the iteration is continued and brings the solution back into the two-phase region before the tolerance level is met. The figure also shows the favourable effect of the multiplication factor γ defined by Eq. (23). The solution is much more quickly returned to the two-phase region, and the no. of iterations required to achieve convergence is significantly reduced.

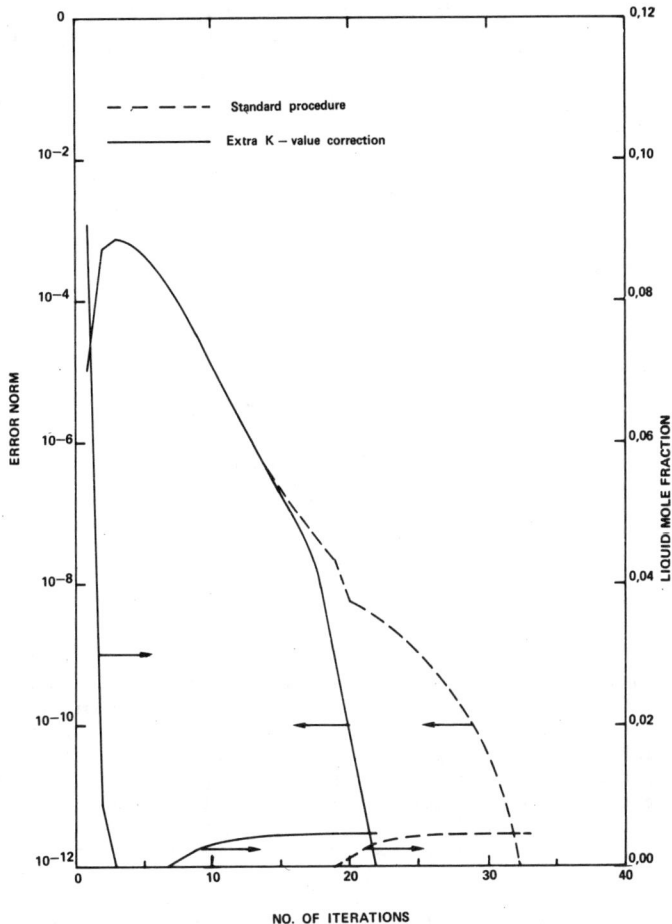

Figure 9   Effect of bring-back procedure.

A Ternary Mixture

Figs. 10 and 11 show some results obtained with a ternary mixture of 40% ethane, 40% propane and 20% n-butane. The SRK equation of state and the accelerated successive substitution method is used throughout this example. This mixture has also been studied by Gundersen /11/. Using a stepping procedure towards the critical region and a special treatment of Z-factors, he was able to perform flash calculations up to some 0.02 MPa from the critical point.

The predicted flash behaviour at some temperatures in the vicinity of the critical temperature is shown in Fig. 10. We had no difficulties obtaining convergence even more close to the critical point then indicated in the figure.

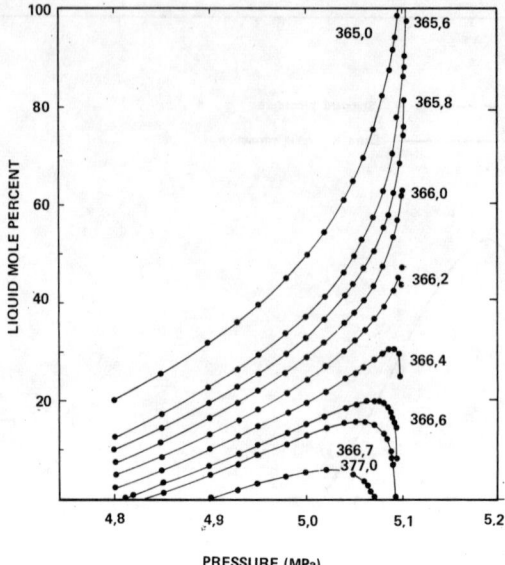

Figure 10 Flash behaviour of the ternary test mixture at different temperatures.

However, with a "normal" tolerance level of 10^{-8} some of the isotherms were found to become irregular as the saturation pressure was approached. The previous example clearly shows that such irregularities are to be expected, and that the tolerance has to be gradually reduced to get accurate results as a critical point is approached. In most practical applications, consistency near a critical point is probably more important than to pursue solutions very close to this critical point. In Fig. 10 an attempt is made to define a critical point vicinity where a phase separation is not insisted on and the solution is interpreted as a single-phase "critical" mixture. Specifically, the calculations are terminated when $\Sigma\, z_i (K_i - 1)$ gets less than 0.01, and this criterion is felt to be well adapted to the tolerance level of 10^{-8}. Nghiem and Aziz /4/ indicated a similar approach.

The effect on the isotherms is to create a discontinuity from two-phase to single-phase. In Fig. 10, all isotherms between 365.6 and 366.7 K experience this discontinuity (points beyond this discontinuity are not plotted), and the corresponding effect on the P-T phase diagram is to cut a top off the two-phase region as shown in Fig. 11.

Yarborough Mixture No. 8

The algorithms considered in this paper have been extensively tested also on systems consisting of a larger numbers of components, and some results obtained for a 6-component synthetic oil mixture commonly referred to as Yarborough mixture no. 8 /12/ will be presented here. These results will concentrate on the solution performance. However, to set a background, the flash behaviour obtained at several temperatures is plotted in Fig. 12. Fair agreement with experimental results is obtained at 200°F, and the critical temperature is estimated to approximately 55°F.

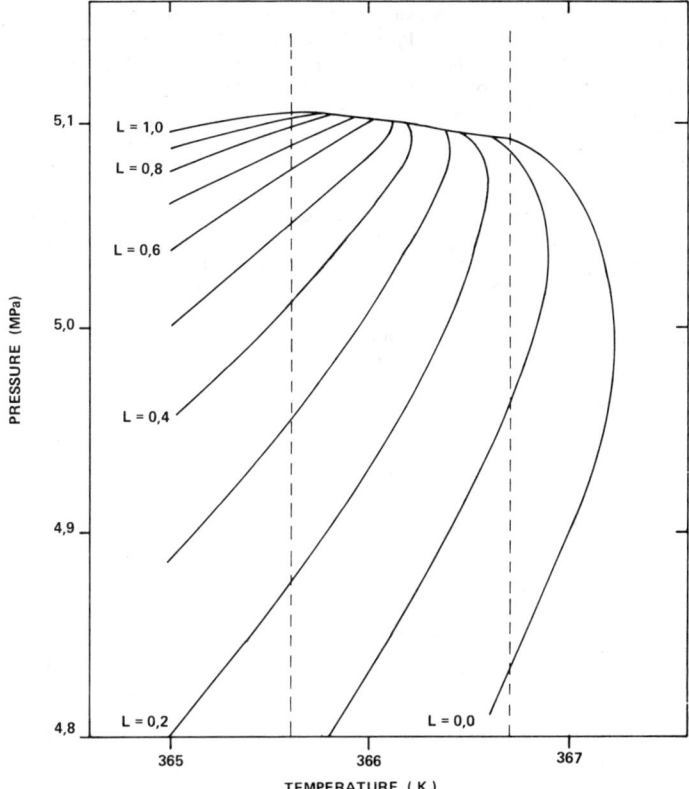

Figure 11   Phase diagram of the ternary test mixture.

The isotherm of 75°F is sufficiently close to the critical point to yield relatively hard flash equilibrium problems, and results of some testing of different solution alternatives for this temperature is given in Tables 2-4. Seven pressures are considered, and when converging, all the alternatives yield essentially the same results. A tolerance of 10^{-12} has been used in the present context. Resulting liquid mole fractions are included in Table 2.

Table 2   No. of iterations (and CPU-time) for different versions of SSM, Yarborough mixture no. 8.

Pressure (psia)	Liquid mole fraction	Pure SSM		SSM with overshoot		SSM with acceleration	
2000	0.2621	19	(0.23)	13	(0.16)	13	(0.16)
2500	0.3188	32	(0.35)	20	(0.23)	20	(0.27)
2750	0.3537	47	(0.48)	28	(0.31)	31	(0.37)
2875	0.3740	66	(0.67)	37	(0.42)	39	(0.41)
3000	0.3953	147	(1.41)	224	(2.45)	28	(0.34)
3050	0.3987	282	(2.71)	293	(2.98)	34	(0.45)
3075	0.3878	NC		NC		41	(0.48)

NC - Not converged within 300 iterations

346

Table 2 compares different version of the successive substitution method. The pure version yield a prohibitively high no. of iterations for the higher pressure values. With overhoot the fugacity ratios are raised to the power of 2 in Eq. (8) after a fixed no. of iterations (10 in the present case). This may be seen to function well for the lower pressure values, but not so well more close to the saturation point. With the acceleration procedure, a maximum of 41 iterations is used for all the pressure values considered.

The reason why the overshoot feature in some instances fails is probably that it is too uncritically employed, and the testing step included in the acceleration procedure should therefore be emphasized. The acceleration procedure proceeds in pairs of steps. First a simple iteration is done in order to determine k and the exponent of Eq. (20), and thereafter an accelerated step is made in accordance with Eq. (20). An important detail is, however, that the

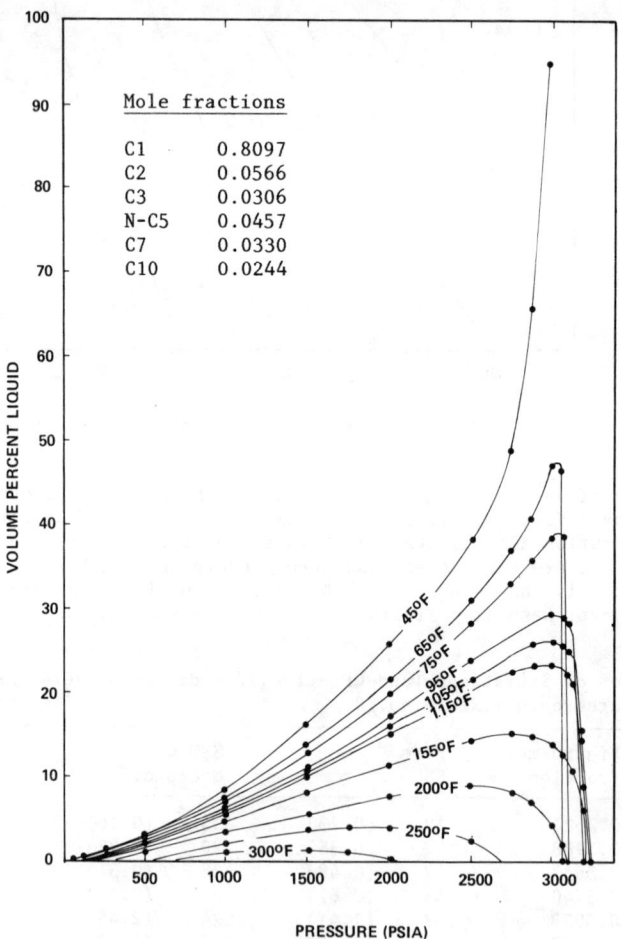

Figure 12 Volumetric behaviour of Yarborough mixture no. 8 at different temperatures.

accelerated step is rejected if the fugacity residual norm fails to be decreased by this step. In this case just the simple step is taken and is followed by a new pair of steps. The acceleration performance recorded at 3050 psia illustrates this point:

Iteration No.	Fugacity ratio exponent
22	17.764
24	-2.544 (rejected)
26	27.007
28	0.940 (rejected)
30	47.860 (rejected)
32	27.125

In essence we have applied the same criterion for start of acceleration as for switch to Powell's method, see Eq. (25). Table 3 compares different alternatives for the most important parameter in this criterion, namely ε_U, and illustrates that some caution should be used when setting this parameter. If it is too high, too many acceleration steps are rejected. If it is too low, too many iterations are made before acceleration is attempted. A value of 10^{-4} has been found to be suitable in most cases and yields typically some 15-30 iterations before acceleration is attempted.

Comparisons with Powell's method are made in Table 4. Here, NL-iterations are somewhat more efficient than VY-iterations, but both alternatives are somewhat slower than the accelerated successive substitution method.

Table 3   No. of iterations as function of start of acceleration, Yarborough mixture no. 8

Pressure (psia)	Residual norm at start			
	10^{-3}	10^{-4}	10^{-5}	10^{-7}
2000	13	13	13	13
2500	20	20	20	20
2750	31	31	31	31
2875	39	39	39	39
3000	29	28	36	51
3050	NC	34	44	NC
3075	NC	41	53	NC

NC - Not converged within 100 iterations.

Table 4   No. of iterations (and CPU-time) for different solution alternatives, Yarborough mixture no. 8

Pressure (psia)	SSM with acceleration		SSM + Powell VY-iteration		SSM + Powell NL-iteration	
2000	13	(0.16)	13	(0.16)	13	(0.16)
2500	20	(0.21)	20	(0.21)	20	(0.21)
2750	31	(0.37)	23 + 28	(0.70)	23 + 17	(0.52)
2875	39	(0.41)	32 + 26	(0.72)	32 + 18	(0.59)
3000	28	(0.34)	17 + 27	(0.58)	17 + 19	(0.45)
3050	34	(0.45)	21 + 30	(0.67)	21 + 26	(0.59)
3075	41	(0.48)	23 + 39	(0.83)	23 + 30	(0.69)

We also did some tests with general-purpose, Newton-type minimization routines /10/. Both the routines E04JAF and E04FDF were found to be less efficient than the other solution alternatives considered here, but one observation is worth mentioning. Working with the object function only, a more direct expression for Gibbs free energy is much better than a fugacity residual norm.

CONCLUSIONS

An accelerated and stabilized successive substitution method (ASSM) has been formulated for flash calculations of multi-component systems and has been especially designed for applications in the near-critical region. The method is made convergent also in the case of a disappearing phase and will therefore detect single-phase solutions automatically. The acceleration procedure is based on an Aitken type formula for correcting the K-values, but acceleration steps are never taken unless they lead to improved solutions. In the examples presented, the ASSM method has been shown to be a highly stable and efficient method. As special saturation pressure calculations are not needed to delineate the two-phase region, the method is well adapted for incorporation in compositional simulators.

Compared with Powell's method and other Newton type methods, the greatest advantage of the ASSM method is its stability close to saturation pressures. Generally, it is also faster than Newton type methods.

The method presented is based on the Soave-Redlich-Kwong and the Peng-Robinson equation of state. However, it can easily be adapted to other equations of state.

NOMENCLATURE

a, A	=	Equation of state coefficients
b, B	=	Equation of state coefficients
f_{iL}, f_{iV}	=	Liquid and vapor phase fugacities
G	=	Gibbs free energy
k	=	Acceleration parameter
K_i	=	Equilibrium constants, $K_i = y_i/x_i$
L	=	Liquid moles or liquid mole fraction
n	=	No. of components
N	=	Total no. of moles
P	=	Pressure
R, R_i	=	Gas constant and fugacity ratios
T	=	Temperature
v	=	Molar volume
V	=	Vapor moles or vapor mole fractions
x_i	=	Mole fraction of component i in liquid
y_i	=	Mole fraction of component i in vapor
z_i	=	Mole fraction of component i in system
Z	=	Compressibility factor
γ	=	K-value multiplication factor
ε	=	Tolerance
ρ	=	Fugacity residual norm
ψ_i	=	Fugacity coefficients
ω	=	Acentric factor

Subscripts

c	=	Critical
i, j	=	Component no.
j	=	Iteration no. (as superscript)
L	=	Liquid phase
r	=	Reduced
V	=	Vapor phase

ACKNOWLEDGEMENT

This research is part of a joint project between Norsk Agip A/S and the Continental Shelf Institute (IKU). The project is fully financed by Norsk Agip A/S. The authors wish to thank Norsk Agip A/S for permission to publish this paper.

REFERENCES

1. SOAVE, G.: "Equilibrium Constants from a Modified Redlich-Kwong Equation of State", Chem. Eng. Sci., Vol. 27 (1972), pp. 1197-1203.

2. PENG, D.-Y. and ROBINSON, D.B.: "A New Two-Constant Equation of State", Ind. Eng. Chem. Fundam., Vol. 15, No. 1 (1976), pp. 59-64.

3. FUSSEL, D.D. and YANOSIK, J.L.: "An Iterative Sequence for Phase-Equilibrium Calculations Incorporating the Redlich-Kwong Equation of State", Soc. Pet. Eng. J., Vol. 18, (June 1978), pp. 173-182.

4. NGHIEM, L.X. and AZIZ, K.: "A Robust Iterative Method for Flash Calculations Using the Soave-Redlich-Kwong or the Peng-Robinson Equation of State", SPE Paper 8285 presented at the 54th Annual Fall Meeting of SPE of AIME, Las Vegas (1979).

5. MEHRA, R.K. et al.: "Computation of Multiphase Equilibrium for Compositional Simulation", SPE Paper 9232 presented at the 55th Annual Fall Technical Conference and Exhibition of SPE of AIME, Dallas (1980).

6. MOTT, R.E.: "Development and Evaluation of a Method for Calculating the Phase Behaviour of Multi-Component Hydrocarbon Mixtures Using an Equation of State", AEE Winfrith Report 1331, Dorchester (1980).

7. MADDOX, R.N. and ERBAR, J.H.: "Equilibrium Calculations by Equations of State". Oil and Gas Journal, (Feb. 2, 1981), pp. 74-78.

8. WILSON, G.: "A Modified Redlich-Kwong Equation of State, Application to General Physical Data Calculations", paper no. 15C presented at the AIChE 65th National Meeting, Cleveland, Ohio, May 4-7, 1969.

9. POWELL, M.J.D.: "A FORTRAN Subroutine for Solving Systems of Non-Linear Algebraic Equations", in RABINOWITZ, P. (ed.): "Numerical Methods for Non-Linear Algebraic Equations", Gordon and Breach Science Publishers, London (1970).

10. NAG Library Manuals, Numerical Algorithms Group Ltd., Oxford (1978).

11. GUNDERSEN, T.: "Numerical Aspects of the Implementation of Cubic Equations of State in Flash Calculation Routines", to appear in Comp & Chem. Eng.

12. YARBOROUGH, L.: "Vapor-Liquid Equilibrium Data for Multicomponent Mixtures Containing Hydrocarbon and Non-Hydrocarbon Components", J. Chem. Eng. Data, Vol. 17 (1972), pp. 129-133.

13. McCAIN, W.D.Jr.: "The Properties of Petroleum Fluids", Gulf Publ. Comp., Tulsa (1973).

APPENDIX - COMPONENT DATA

The critical properties used in the computer program COPEC are taken from McCain /13/, and those used in the example calculations are given in Table 5.

For the binary test system considered in this paper, the PR equation with a binary interaction coefficient of 0.13 has been used.

For the tertiary test system, the SRK equation has been used with binary interaction coefficients as follows:

$$
\begin{aligned}
C_2 - C_3 &: \quad 0.001 \\
C_2 - N\text{-}C_4 &: \quad 0.009 \\
C_3 - N\text{-}C_4 &: \quad 0.012
\end{aligned}
$$

For the six-component Yarborough mixture the PR equation is used with all binary interaction parameters equal to zero.

Table 5   Component properties

Comp.	Mole weight	P_c (MPa)	T_c (K)	Acentric factor
CO2	44.010	7.387	304.21	0.2250
C1	16.043	4.606	190.58	0.0104
C2	30.070	4.882	305.42	0.0986
C3	44.097	4.251	369.82	0.1524
I-C4	58.124	3.650	408.14	0.1848
N-C4	58.124	3.799	425.18	0.2010
N-C5	72.151	3.370	469.65	0.2539
C7	100.205	2.737	540.26	0.3498
C10	142.286	2.096	617.65	0.4885

THE EFFECT OF SIMULATED CO_2 FLOODING ON THE PERMEABILITY OF RESERVOIR ROCKS

GRAHAM D. ROSS, ADRIAN C. TODD and J. ANDREW TWEEDIE

Department of Petroleum Engineering,
Heriot-Watt University

ABSTRACT

Both formation damage and stimulation effects have been experienced during "miscible" carbon dioxide field and laboratory tests in the USA. While the stimulation effects have been attributed to dissolution of the reservoir rock by carbon dioxide enriched flood water no work has been done to identify and quantify this phenomenon. Nor has any established theory for the formation damage been identified, although it seems likely that in some instances formation damage may be caused by formation fines, released by dissolution and subsequently migrating into pore throats.

This paper describes a laboratory investigation into the effects of rock - fluid interaction under simulated reservoir conditions, and in particular the carbonated water - carbonate mineral reaction in sandstones during a CO_2 enhanced recovery process. The design and operation of experimental equipment for flowing CO_2-water mixtures through linear rock cores are described, together with the analytical methods used to assess changes in core characteristics. The paper presents results from initial tests on four different carbonate containing core materials.

INTRODUCTION

(1) General

Successful laboratory investigations of miscible, carbon dioxide, flooding have been well documented in the literature. Field experience, however, has only recently begun to accumulate. All the projects reported have begun since 1972 (mostly in the United States), thus, only limited empirical data is currently available. Although encouraging, field results to date have been sufficient to identify several major problems and opportunities with the carbon dioxide technique.

One of the problems is that of reduced injectivity experienced in some reservoirs on injecting carbon dioxide. While many have reported this to be due to the deposition of high molecular weight materials upon mixing of crude and carbon dioxide[1,2,3] in situ plugging tests have not proved the occurrence of this type of precipitation[4]. Observed reductions in injectivity can probably therefore be attributed to other mechanisms, one of which may be the disintegration of carbonate cements in the reservoir rock, and movement of particulate matter into the throats of interstitial pores.

Conversely, increases in injectivity have also been experienced in the course of carbon dioxide field tests[5]. These were in turn attributed to dissolution of carbonate minerals in carbon dioxide enriched floodwater (carbonated water), causing increased permeability.

In view of the lack of data and uncertainty in the published results relating to carbonate dissolution on carbon dioxide flooding, a research programme has been initiated to study the phenomenon. The objectives of the programme are to evaluate the dissolution effects of carbonated water on formation carbonates, and to determine how formation permeability characteristics are likely to be altered during a carbon dioxide flood. This paper presents the first phase of the study, the development and operation of apparatus for flowing CO_2-water mixtures through linear rock cores, together with the results of experiments undertaken to establish the mechanism(s) of carbonate dissolution in porous media.

(2)   Carbonate Dissolution in Reservoir Rock

Many producing formations contain carbonates in some form. In the case of limestone and dolomite reservoirs, carbonates constitute the bulk of the formation rock. In sandstones, carbonates are commonly found as pore filling and replacement cements consolidating the sand grains, although varying, but usually minor amounts of detrital carbonate grains may also be present. Since the cementing material in sandstone is located between sand grains adjacent to flow channels, a relatively small change in the pore framework due to carbonate dissolution may significantly affect the total permeability.

Upon injection, carbon dioxide, mixing with either injection water or connate water, will form carbonic acid. One characteristic of carbonic acid is that at very low carbon dioxide partial pressure, the pH is reduced considerably. Thus, carbonated water will retain its acid nature with very little CO_2 in solution.

The carbonates most commonly found in reservoir rocks are those of calcium (calcite), combinations of calcium and magnesium (dolomite) and iron (siderite). These minerals have a low solubility in pure water at atmospheric conditions, but become increasingly soluble with increasing water carbonation (or CO_2 concentration) and pressure. The carbonate form is converted to that of the soluble bicarbonate, the following equation representing the chemical reaction for calcium carbonate:

$$H_2CO_3 + CaCO_3 \rightleftarrows Ca(HCO_3)_2$$

Similar chemical reactions take place with the other carbonates.

The solubility trends of calcium carbonate in carbonated water as a function of pressure and temperature are presented in Figure 1. Although no work has been carried out in the 0 to 100°C temperature range at pressures above 100 bars, indications from other studies[6-11] are that calcite solubility:

   (1)   increases with increasing temperature at constant total
             pressure and CO_2 concentration,

(2) increases with total pressure at constant temperature
     and CO_2 concentration, and

(3) increases up to a maximum at five weight per cent CO_2
     concentration before falling again at higher CO_2
     concentrations at constant temperature and total
     pressure.

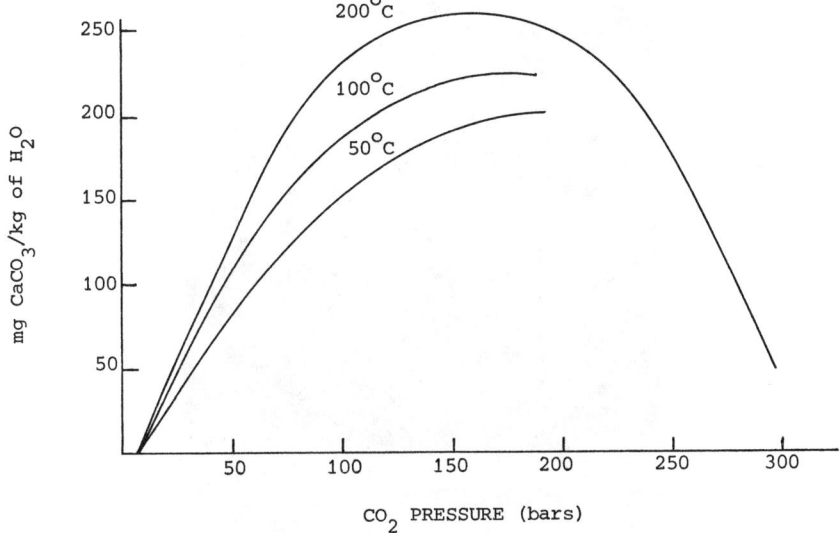

Figure 1   Solubility of calcite in carbonated water

Carbonated water, formed upon injection of carbon dioxide into a well, will
react with the carbonate minerals in the rock and transport the dissolved
products through the reservoir. This dissolution effect will be more
pronounced in the vicinity of the wellbore since the carbonated water
solution will approach total bicarbonate saturation as the water moves away
from the well. However, whether the reaction effects a reduction in
permeability in the reservoir by releasing particles which then migrate and
plug flow channels, or an increase in permeability, is not apparent from
tests undertaken to date.

EXPERIMENTAL

(1) Equipment

A high pressure, high temperature permeameter was designed and constructed to
permit an examination of carbonated water dissolution effects. The apparatus,
shown in Plate 1, is capable of operation in moderately corrosive liquid
environments under controlled conditions, of temperature, pressure and flow
rate. A process flow scheme of the core flooding apparatus is presented in
Figure 2.

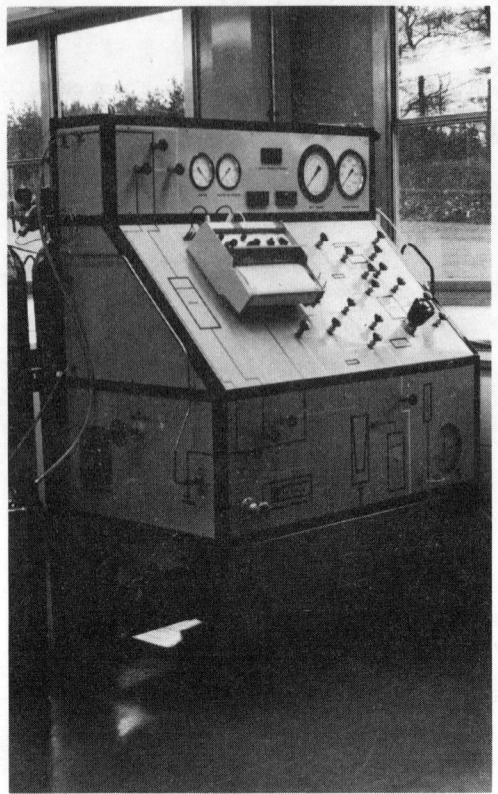

Plate 1   Front view of experimental apparatus

A detailed description of the major equipment components follows:

(a) Core Holder: The core holder cell was designed for high
      pressure core flooding in corrosive liquid environments.
      It consists of a thick-walled stainless steel outer
      cylinder with removable lid, fitted internally with a
      sleeve core holding assembly.

     The sleeved core is secured between the cell lid/inlet end
     plate and the outlet end plate by three tie rods. The
     end plates serve as distributor and receptor respectively
     for the fluid flowing through the core. Both end plates
     are scored with lines radiating from the centre and also
     with concentric circles about the centre. These lines
     allow even fluid and pressure distribution across the ends
     of a core. The outlet end plate can be precisely adjusted
     on the tie rods to enable short cores (down to 1.5 cm long)
     to be fitted in the cell.

     The cylindrical shell has four entry ports or taps, one in the
     side-wall for the core sleeve confining pressure and the
     others in the lid; one each for the core influent, the core
     effluent and a thermocouple probe. The cell lid is secured to

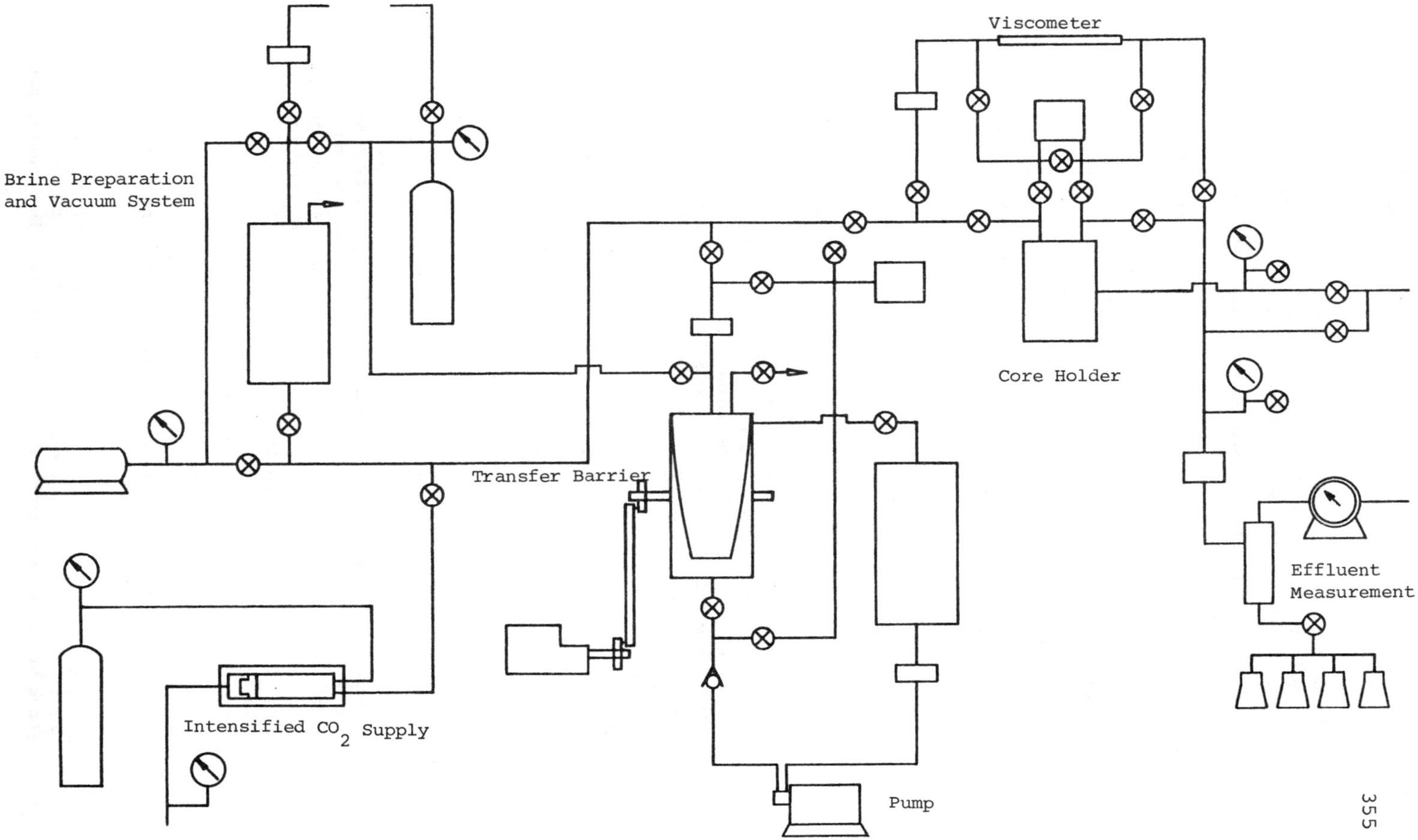

Figure 2 Experimental Flow Apparatus

the base by twenty high tensile bolts and sealed by
an O-ring. Water from a hydraulic pump is used to
supply the core sleeve confining pressure.

The core holder has been tested and certified for use up
to a maximum working pressure of 6,000 psi.

(b) Viscosity Measurement System: Required data on carbonated
brine viscosity are not reported in the literature.
Consequently an "in line" capillary tube viscometer was
incorporated in the flow apparatus to enable liquid
viscosity measurements to be made under test conditions.
The general arrangement of the viscometer is shown
diagrammatically in Figure 2. The main elements are (1)
a 20 cm length of 0.2 mm precision bore stainless steel
tube (secured by epoxy resin inside a length of support
tubing) and (2) a differential pressure transducer.

From the capillary tube dimensions and measurement of the
pressure drop across the tube at known constant flow rate,
the required viscosities can be calculated from the Hagen-
Poiseuille Equation.

(c) Transfer Barrier: The transfer barrier unit is a fluid pressure
transfer device, comprising an open-ended rubber bladder or
membrane enclosed in a 5 litre capacity cylindrical steel
pressure vessel. It serves as a mixing vessel during carbonated water preparation and as a fluid separator in which
pressure and volume changes between the drive fluid
(hydraulic oil) and the core flooding fluid (brine or
carbonated brine) are transmitted through the flexible
rubber membrane.

(d) Intensified CO_2 Supply: Carbon dioxide pressures greater
than cylinder pressure (830 psi) are obtained using a gas
booster. Intensification is obtained by a large area
reciprocating piston pushing a small CO_2 compression piston
with a ratio of 100 to 1 between the piston areas. A
compressed air driven hydraulic pump drives the large area
piston.

(e) Transfer Barrier Rocking Mechanism: To enable efficient and
rapid preparation of equilibrium solutions of carbon dioxide
in water, a rocking mechanism was attached to the transfer
barrier. The drive for the mechanism is supplied by a Kopp
variable speed motor, connected through a drive arm and
couplings to a steel cradle holder bolted to the transfer
barrier. The drive arm length is fixed to give a rocking
angle of 30 degrees, and the rocking rate from 15 to 90
cycles per minute, is controlled manually by a remotely
controlled adjuster from the variable speed motor.

The fluid lines to and from the transfer barrier are spiralled
around the axis of rocking. The spirals help to maintain
the integrity of various connections, by offering resistance
to the jerks caused by the rocking mechanism.

(f) Displacement System: The flow rate was determined in all cases
by employing an Eldex Precision Pump in conjunction with the
back pressure regulator. The Eldex positive displacement pump

delivers a steady flow that can be varied from 0 to
4.5 cc per minute. The flow rate within this range is
adjusted by a micrometer screw on the pump, which sets
the length of stroke. The pump is capable of delivery
pressures in excess of 5000 psi.

A non-corrosive fluid (hydraulic oil) was used as a drive
fluid to displace the core flood liquid from the membrane
in the transfer barrier. The drive oil was drawn from a
perspex reservoir by the Eldex pump and delivered at
constant volume to the base of the transfer barrier.

(g) Pressure Measurement System: As shown in Figure 2, the
     flow apparatus is equipped with four pressure gauges
     and two pressure transducers. The gauges are as
     follows:

   (1) 0 to -1.0 bar vacuum gauge, connected in
        the line to the vacuum pump - used during
        initial vessel and pipework evacuation.

   (2) 0 to 10 bar gauge, connected in the compressed
        air supply line to the gas booster - to
        monitor the air pressure to the gas booster
        and hence the level of gas intensification.

   (3) 0 to 600 bar gauge, connected to the core
        sleeve confining pressure line - to monitor
        core sleeve pressure.

   (4) 0 to 400 bar precision gauge with a stainless
        steel measuring element, connected immediately
        upstream of the back pressure regulator - to
        monitor system back pressure.

Both pressure transducers are S.E. Labs. 21/V models:

   (1) 0 to 5000 psi absolute pressure transducer,
        connected to the transfer barrier to measure
        the system "upstream" pressure. It is
        electrically connected to an Analogic digital
        unit for visual observation.

   (2) 0 - 50 psi differential pressure transducer,
        connected across the core holder and viscosity
        measurement capillary tube. It is linked to a
        strip chart recorder to provide a continuous
        record of the pressure differential data.

(h) Temperature Control System: The temperature control system
     consists of three independent sub-systems:

   (1) to heat the contents of the transfer barrier,

   (2) to maintain the core and fluids entering the
        core at the desired temperature level, and

   (3) to maintain the viscometer temperature.

Heat to the transfer barrier and core holder is supplied electrically by
close fitting mesh elements and controlled in each case within ($\pm$ 1°C) by
a thermostat/thermocouple controller. Insulation for the vessels is
provided by 4 cm thick layers of rock wool encased in aluminised glass

cloth jackets. To ensure that all fluid entering the core is at test temperature, the fluid line immediately upstream of the core holder is coiled tightly around the core cell lid. The capillary tube viscometer is enclosed in a water bath where it is maintained at the desired temperature by hot water circulation.

A series of chromel-alumel thermocouples are used to monitor temperature throughout the flow system. These are linked via a selector unit to a digital thermometer for visual display and recording.

(i) Effluent Collection and Measurement System: Core effluent, reduced to atmospheric pressure on discharge from the back pressure regulator, enters the gas/liquid separator. The separator is a sealed perspex cylinder with a capacity of 300 ccs. It has an inlet for the core effluent near the top and outlets in the lid and base for the separated gas and liquid respectively. The volume of carbon dioxide produced is measured by a wet-type volumetric meter connected directly to the gas outlet from the separator. The meter is a precision device provided with a 150 mm dial of 100 divisions and a six digit revolution counter form of totaliser. Liquid from the base of the separator flows via a five-way selector valve to sealed glass collection vessels for measurement and analysis.

(2) Experimental Procedure

Initial testing consists of flowing base water (i.e. brine or distilled water) through the core to establish the initial or reference (stabilised) permeability. Subsequently brine, carbonated to the desired level inside the rubber membrane of the mixing vessel, is injected into the core at constant rate by hydraulic oil displacement. The carbonated water and core temperatures are carefully controlled to represent oil reservoir conditions. A back pressure above the carbonation pressure is maintained throughout the test to ensure that only liquid phase exists at all points in the flow system. The permeability of the core is measured as a function of time, and all core effluent is collected for chemical analysis.

Following a core flood experiment a series of analyses are performed on the core and effluent liquid. The effluent liquid is analysed for content of calcium and magnesium by EDTA titration, and the core is divided into a series of segments. The permeability, porosity, pore size distribution and overall dissolution effect in each of the core segments is then assessed.

(3) Porous Media

To enable study of the carbonated water-carbonate mineral reaction in sandstone without interference from other effects such as clay or mica alteration, it was necessary to choose material with a relatively simple mineralogical composition. Thus, a relatively pure quartz-carbonate sandstone, a calcareous grit, was chosen for the initial stage of the study. The selection of this particular sandstone, from a quarry in the Yorkshire Jurassic, was also partly based on its high carbonate content.

However, to gain a more complete understanding of the reaction effect of carbonated water on carbonate mineral in reservoir rock, two other sandstones and a limestone were tested in this initial study. These materials are listed and described in Table 1. The analytical procedures used in the description of these materials, both before and after core flooding were:

TABLE I   Summarised descriptions of core materials

Formation	Rock Type	Description	Mineral Content		Physical Properties	
Yorkshire Jurassic	Calcarenaceous Sandstone	Composed of subrounded detrital quartz grains and detrital carbonate debris cemented by micritic calcite	Quartz Ferroan Calcite	80% 20%	Permeability Porosity	100mD 16%
Fife Carboniferous	Dolomitic Sandstone	Composed of angular to subrounded quartz grains partially cemented by secondary dolomite. The dolomite is evenly distributed, occurring as rhomb shaped crystals and crystalline masses in the voids between sand grains	Quartz Dolomite Felspar and Clay	90% 10% less than 1%	Permeability Porosity	200mD 10%
Rotliegende Sandstone	Calcitic Sandstone	Composed of subrounded to rounded quartz grains with patchy calcite pore fill and clay	Quartz Calcite Felspar and Clay	95% 2.5% 2%	Permeability Porosity	300mD 15%
Oxfordshire Jurassic	Oolitic Limestone	Composed of ooliths and shell fragments cemented by micritic calcite	Calcite Quartz	98% 2%	Permeability Porosity	60mD 15%

(1) thin section petrographic analyses,

(2) differential dye staining for carbonate identification

(3) scanning electron microscope analyses,

(4) porosity, pore size distribution and permeability
     measurements.

For each series of experiments a number of 2.5 cm diameter X 7.5 cm length
cylindrical cores were drilled and trimmed from the same block of rock, so
that variation in the properties would be kept to a minimum. As a precaution
against collapse on dissolution, the cores were coated on the cylindrical
surface with epoxy resin.

RESULTS

(1) The initial series of experiments were carried out on the Yorkshire
Jurassic calcitic sandstone. First tests with distilled water and brine
(no carbonation) were aimed at establishing a stabilised or reference
permeability, prior to any carbonated water flood. The results of two such
tests, R7 and R9, are presented in Figure 3. Significant increases in
permeability were obtained in the tests, with little apparent levelling off
in the rate of permeability increase and attainment of a reference value,
upon injection of up to 500 pore volumes. Chemical analysis of the core
effluents for calcium showed the permeability increases to be attributable
to the dissolution of calcite cement in the flood liquids.

As a comparison with the base liquid experiments, a series of tests with
carbonated brine were then undertaken. As shown in Figure 4, much greater
permeability increases were obtained, although again there was little
indication of any fall in the rate of permeability increase. The results
of the core effluent analysis as compared with those from a brine flood (R9)
are presented in Figure 5. As expected, the calcium concentrations in the
effluent samples from the carbonated brine tests were far higher than in
the brine test, although there was a significant difference between the
results for carbonated brine tests R20 and R21. Examination of the flooded
cores showed this was because in R20 a thin band was preferentially dissolved,
whereas in R21 a more uniform dissolution took place (Plate 2). Presumably
in R20, as the flood progressed, the main flow was through the thin
permeability "streak", resulting in lower total dissolution than in R21.

To gain an understanding of the variation in local permeability, the cores
from the various tests were retrieved after flooding and cut into three
2.5 cm long segments. The permeability of each segment was then measured
and plotted as a function of axial position in the core. The plots for R9
and R20 are presented in Figure 6. The profiles obtained show the
permeability at the inlet end of the cores was increased considerably more
than that at the outlet end. Also, the fact that the profiles have approximately the same shape, infers that the location of each profile is simply
determined by the level of carbonation of the brine. Since constant flow
rate was used in the experiments, this result implies that a zone of
increasing permeability, which can be considered as a front, was moving
through the cores. The velocity of the permeability front migration is in
turn a function of the liquid flow rate through the core and carbonation
level of the brine.

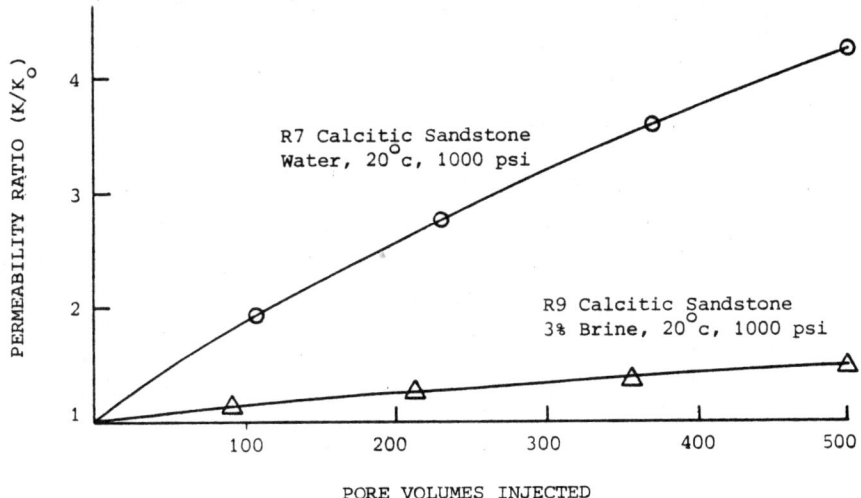

Figure 3   Permeability changes during runs 7 and 9

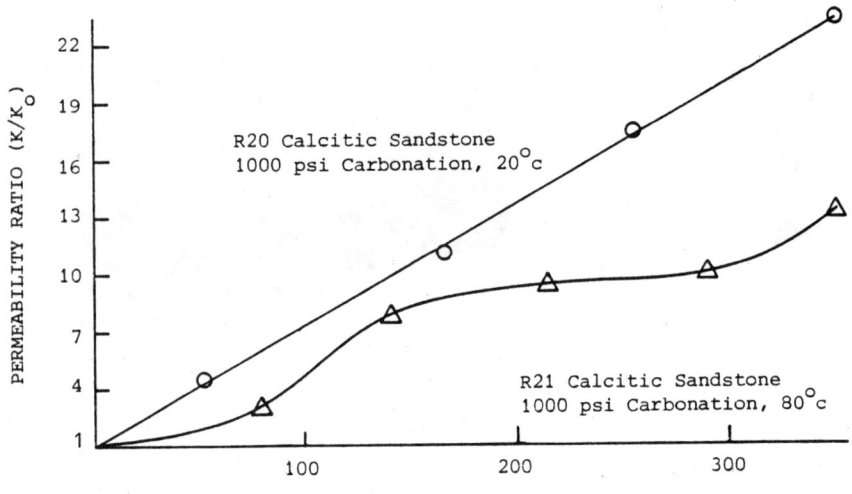

Figure 4   Permeability changes during runs 20 and 21

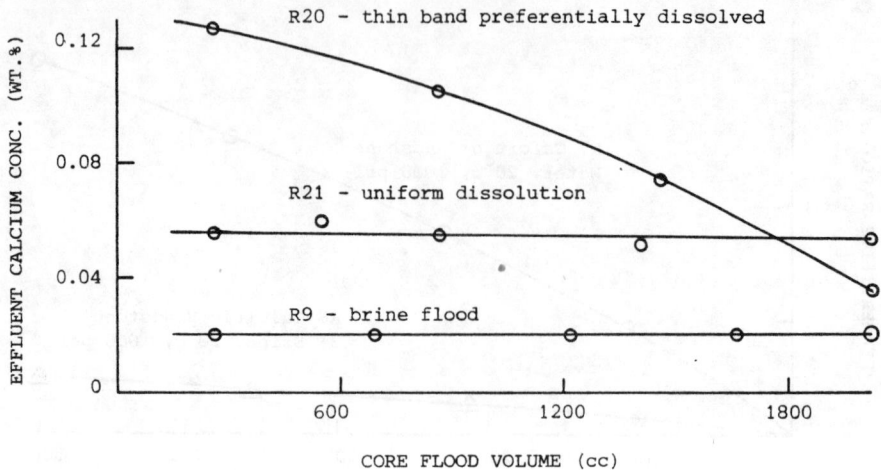

Figure 5   Comparison of effluent calcium concentration
profiles for runs 9, 20 and 21

Plate 2   Comparison of Yorkshire Jurassic sandstone cores
before (left) and after (right, run 21) a
carbonated water flood

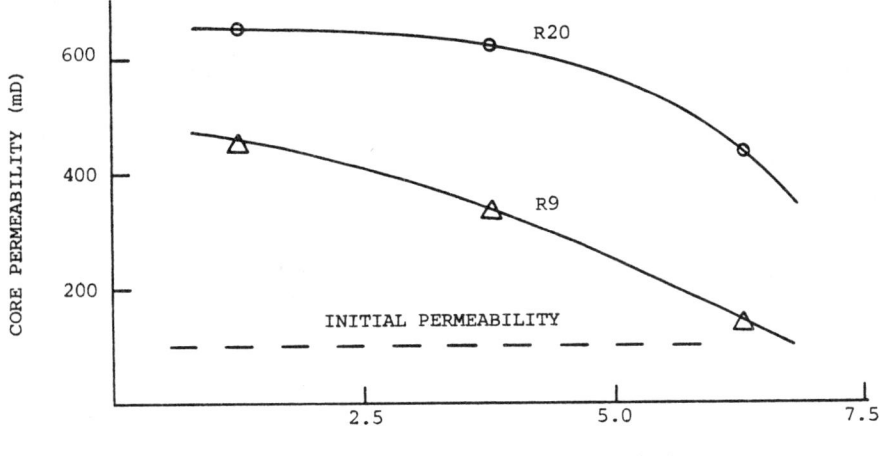

Figure 6   Permeability profile for runs 9 and 20

The porosity and pore size distribution of each of the 2.5 cm long segments from all the above tests were measured and compared to initial whole core values. Generally it was found that although large increases in the permeability had occurred, the porosity had changed little. This result illustrates that the main mechanism for the increase in permeability is probably not the uniform dissolution of carbonate cement, but rather the removal of constrictions in the larger pores. This is confirmed by the mercury porosimeter pore size distribution results, which show that it was primarily the diameters of the larger pores which were increased during the tests.

(2) To further, and more realistically, test the permeability front migration phenomenon, a series of tests were initiated on material with a much lower carbonate concentration than the Yorkshire Jurassic sandstone. Difficulty in acquiring calcite cemented sandstone led to a dolomitic material from the Fife Carboniferous being used at this stage. It was hoped that it would be possible to dissolve out all the dolomite cement from this sandstone and thus eventually achieve constant permeability. However, the low reaction rate of dolomite in carbonated water, compared to that of calcite, effectively ruled out this possibility.

The permeability profiles of two tests, R22 and R23, on the dolomitic sand-stone are presented in Figure 7. The very slow reaction rate of dolomite under ambient temperature conditions meant virtually no dissolution effects were observed in R22, while in R23, although a fairly significant permeability increase was obtained, chemical analysis of the core effluent showed that only a small proportion of the dolomite cement was leached out.

(3) Some tests were then carried out on a calcitic Rotliegende Sandstone from a Southern North Sea gas field, but a series of core collapses, caused by weakening on wetting, meant abandonning the use of this material and continuing the search for other sources of calcitic sandstone.

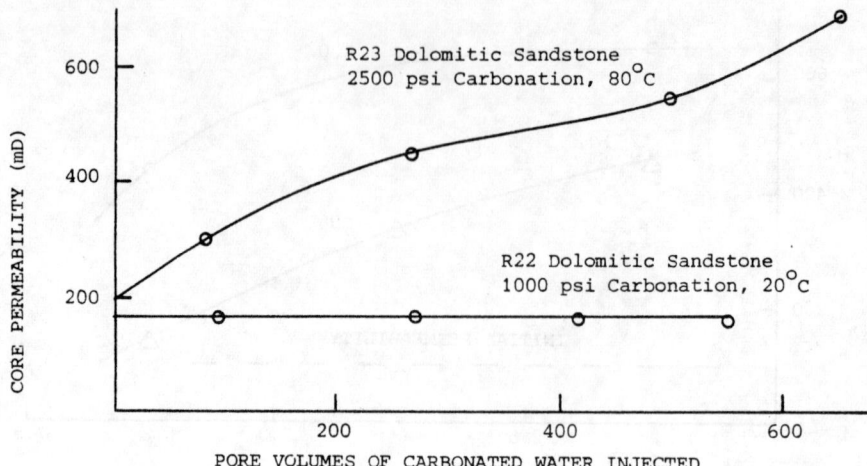

Figure 7   Permeability changes during runs 22 and 23

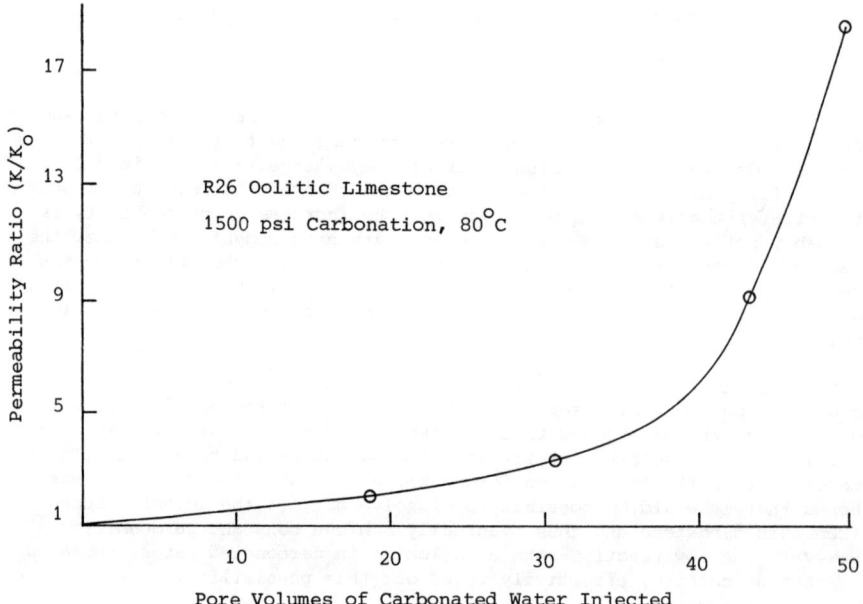

Figure 8   Permeability change during R26

(4) A test was run on an oolitic limestone from the Oxfordshire Jurassic, the permeability profile of which is presented in Figure 8. A very rapid increase in permeability was obtained, with the differential pressure across the core falling to almost zero at maximum flow rate, after injection of only 50 pore volumes. Examination of the flooded core showed this was because two 1.5 mm diameter "wormholes" of roughly circular cross section had formed over the length of the core.

Pore size distribution analysis of the limestone indicated an extremely wide pore diameter distribution and, as expected, it was the selective enlargement of the large pores at the upper extreme of the distribution that contributed significantly to the increase in permeability in this test.

CONCLUSIONS

(1) The high pressure, high temperature carbonated water permeameter constructed to investigate carbonate dissolution effects on carbon dioxide flooding is providing new insight into the variables that control the dissolution process.

(2) Only increases in core permeability from dissolution of carbonate minerals were experienced. No evidence for fines migration or particle plugging was obtained in the experiments.

(3) The dissolution of carbonate mineral from cores produces a change in local permeability which travels as a front through the cylindrical core.

(4) The dramatic increase in permeability of a core during a carbonated water flood is probably due to removal of constrictions and selective dissolution of the larger pores.

REFERENCES

1. NEWTON, L. E. and McCLAY, R. A.; "Corrosion and Operation Problems, CO_2 Project, SACROC Unit", Paper SPE 6391, presented at the SPE-AIME Permian Basin Oil and Gas Recovery Conference, Midland, TX, March 10-11, 1977

2. PONTIOUS, S. B. and THAM, M. J.; "North Cross Unit CO_2 Flood - Review of Flood Performance and Numerical Simulation Model", Paper SPE 6390, presented at the SPE-AIME Permian Basin Oil and Gas Recovery Conference, Midland, TX, March 10-11, 1977

3. HANSEN, P. W.; "A CO_2 Tertiary Recovery Pilot, Little Creek Field, Mississippi", Paper SPE 6747, presented at the SPE-AIME 52nd Annual Fall Technical Conference and Exhibition, Denver, Oct. 9-12, 1977

4. DOSCHER, T. M. and KUUSKRAA, V. A.; "Carbon Dioxide for Enhanced Recovery of Crude Oil", paper presented at the European Symposium on Enhanced Oil Recovery, Edinburgh, July 5-7, 1978

5. CAMERON, J. T.; "SACROC Carbon Dioxide Injection - A Progress Report", paper presented at the API Production Department Annual Meeting, Los Angeles, April, 1976

6. MILLER, J. P.; "A Portion of the System $CaCO_3-CO_2-H_2O$, with Geological Implications", Am. Jour. Sci., (March 1952) 250, 161-203

7. ELLIS, A. J.; "The Solubility of Calcite in Carbon Dioxide Solutions", Am. Jour. Sci., (May 1959) 257, 354-365

366

8.   SEGNIT, E. R., HOLLAND, H. D. and BISCARDI, C. J.; "The Solubility of Calcite in Aqueious Solutions", Geochim. et Cosmochim. Act., (1962) <u>26</u>, 1301-1331

9.   SHARP, W. E.; "The System $CaO-CO_2-H_2O$ in the Two Phase Region Calcite and Aqueous Solution", PhD Thesis, Univ. of California, 1964

10. SHARP, W. E. and KENNEDY, G. C.; "The System $CaO-CO_2-H_2O$ in the Two Phase Region Calcite and Aqueous Solution", Jour. of Geol. (1965) <u>73</u>, 391-403

COMPUTER MODELLING OF EOR PROCESSES

KHALID AZIZ

Computer Modelling Group,
3512-33 Street, N.W.,
Calgary, Alberta T2L 2A6, Canada

ABSTRACT

This paper presents a rather personal view of recent developments, current problems and future prospects for the computer simulation of enhanced oil recovery
schemes. While substantial progress has been made over the past twenty years or
so, some problems of significant practical importance remain unresolved.

INTRODUCTION

This paper is neither a comprehensive review of past work on reservoir simulation
- also referred to as reservoir modelling - nor a complete catalogue of current
activities in this field. Instead it presents the author's view of (a) the status
of simulation technology, and (b) current and future problems. The paper is
intended primarily for individuals interested in using models rather than those
who are engaged in the development of models.

The contents of the paper are heavily influenced by work conducted by the author's
students at the University of Calgary and his colleagues at the Computer Modelling
Group (CMG). Important work underway at other institutions may not be mentioned
here primarily because of the lack of up-to-date information available to the
author. CMG is, however, a vehicle for cooperative research in reservoir simulation among universities, research organizations, government agencies and
industry. Currently 34 such organizations are members of CMG and these organizations have a rather direct and significant influence on its work. Hopefully,
because of this type of interaction, problems being investigated by CMG reflect
current industry needs.

Modelling is an iterative process consisting of the following major steps[1]:

1. Describe Reservoir
2. Describe Recovery Mechanism
3. Write Mathematical Model
4. Develop Numerical Model
5. Develop Computer Model (Program)
6. Validate Model
7. Match History
8. Predict Future Performance

Often during steps 6, 7 and 8 it becomes necessary to go back to steps 1, 2, 3 or 4
and alter some of the assumptions made earlier. Assumptions are necessary at
various stages to (a) allow simulation of processes where recovery mechanisms are

not fully understood, (b) make the problem tractable, and (c) reduce cost of simulation. Obviously the need for the assumptions is constantly changing with improved understanding of the physical and chemical aspects of the recovery processes, development of new numerical techniques, and hardware innovations. Steps 6 and 7 dealing with the validation and use of models will not be considered in this paper.

CLASSIFICATION OF MODELS

A large variety of models are in current use and the number is constantly increasing. New models are developed to (a) simulate new processes, (b) simulate behaviour of reservoirs with special characteristics, (c) reduce cost, (d) improve accuracy, (e) have access to a suitable model under acceptable conditions, or (f) understand reservoir simulation. Table 1 provides a classification based on Recovery Mechanisms, Reservoir/Well Characteristics, Numerical Approximations, Fluid/Rock Properties, Solution Techniques and Computer Type.

Table 1
Classification of Models

1. RECOVERY MECHANISMS

    1.1 WATERFLOOD OR PRIMARY DEPLETION
        1.1.1 Three component, three phase
        1.1.2 Two component, two phase

    1.2 GAS OR SOLVENT INJECTION
        1.2.1 Multicomponent, single phase
        1.2.2 Multicomponent, multiphase

    1.3 CHEMICAL FLOOD
        1.3.1 Four component, two or three phase (polymer)
        1.3.2 Multicomponent, multiphase (surfactant, caustic)

    1.4 THERMAL MODELS
        1.4.1 Three component steam
        1.4.2 Compositional steam
        1.4.3 Steam with additives
        1.4.4 In situ combustion

2. RESERVOIR/WELL CHARACTERISTICS

    2.1 RESERVOIR WELL COUPLING

    2.2 FRACTURES
        2.2.1 Static fracture
        2.2.2 Dynamic fracture
        2.2.3 Uniformly distributed fractures

    2.3 CONSOLIDATION OF RESERVOIR ROCK
        2.3.1 Sand flow
        2.3.2 Ground subsidence

3. NUMERICAL APPROXIMATIONS

    3.1 PRIMARY VARIABLES
        3.1.1 Selection of variables
        3.1.2 Selection of equations
        3.1.3 Alignment of variables and equations

Table 1 (Cont'd)

3.2 LINEARIZATION
   3.2.1 Newton's method
   3.2.2 Other methods

3.3 DECOUPLING
   3.3.1 Fully implicit
   3.3.2 Sequential
   3.3.3 Implicit Pressure Explicit Saturation (IMPES)
   3.3.4 Dynamic Implicit
   3.3.5 Band reducing techniques

3.4 INTERBLOCK FLOW
   3.4.1 Single point upstream
   3.4.2 Two point upstream
   3.4.3 Harmonic average
   3.4.4 Centralized upstream
   3.4.5 Other interblock mobility calculation methods
   3.4.6 Nine-point schemes

3.5 TRUNCATION ERROR
   3.5.1 Standard finite-differences
   3.5.2 Higher order finite-differences
   3.5.3 Variational
   3.5.4 Semi-analytical
   3.5.5 Location of grid point in a block
   3.5.6 Curvilinear grid
   3.5.7 Local grid refinement
   3.5.8 Moving grid

4. FLUID/ROCK PROPERTIES

4.1 RELATIVE PERMEABILITY CALCULATION
   4.1.1 Three phase model
   4.1.2 Temperature effect model
   4.1.3 Composition effect model
   4.1.4 Hysteresis model

4.2 FLUID PROPERTIES
   4.2.1 Empirical correlations
   4.2.2 Equation of state

5. SOLUTION TECHNIQUES

5.1 ORDERING OF EQUATIONS

5.2 GAUSSIAN ELIMINATION

5.3 ITERATIVE METHODS

6. COMPUTER TYPE

6.1 STANDARD

6.2 VECTOR PROCESSORS

6.3 INTERACTIVE DATA INPUT AND ANALYSIS OF RESULTS

This classification provides a suitable framework for comment on the status of some aspects of the technology.

RECOVERY MECHANISMS

The simplest models that can be used for primarily depletion and water or hydrocarbon gas injection studies are referred to as black-oil or β models. The models of this type have been in use for over twenty years and are based on the assumption that the reservoir fluids can be assumed to consist of only three pseudo-components - oil, water and gas at standard conditions. This rather gross assumption works well for systems that remain far from the critical or the retrograde region during the recovery process and where the injected fluids consist of the same components as in the in situ fluids. Even in this relatively simple case different models can yield different results for the same problem[2]. Most of these differences may be attributed to the numerical aspects to be discussed later.

Compositional models allow for the representation of oil and gas by a mixture of several more pseudo-components each. They can handle complex phase behaviour associated with, for example, the injection of CO_2. Chemical flood models are even more complicated compositional models with capabilities to handle important rock/fluid and fluid/fluid reactions. Each component or pseudo-component yields one conservation or mass balance equation to be solved for each grid point. Hence as the number of components increases, the number of equations to be solved increases in direct proportion.

Thermal models can vary in complexity from the simple three component steam model to the complex in situ combustion model. In addition to the conservation of mass we must also add the conservation of energy to our system of equations to be solved.

The problems in defining the recovery mechanism from the point-of-view of the modeller usually relate to the lack of experimental information for the selection of pseudo-components, to predict their physical and chemical properties, and to validate the assumptions of the mathematical model. Contrary to the belief held by some, reservoir simulation does not reduce or eliminate the need for experiments - it allows us to get the most out of laboratory and field experiments we can afford to run.

Examples of the phenomena that can not be handled in a satisfactory fashion at this time are (a) formation and flow of emulsions, and (b) flow of more than two liquid phases.

RESERVOIR/WELL CHARACTERISTICS

The intimate interaction between the reservoir and the flow in the wellbore (tubing or annulus) of both injection and production wells must be recognized for realistic simulation. While it is relatively easy to do single phase well flow calculations to any desired accuracy, the same is not true when two or three phases exist in the wellbore. The transient nature of the flow causes further complications. Modellers often underestimate the importance of the wellbore/reservoir coupling and overestimate the reliability of correlations for performing wellbore calculations. Errors of the order of 20% are possible even when "best" available methods are utilized. Since there are no clear schemes for the determination of what wellbore flow calculation method may be the best in each situation, even higher than 20% errors are possible.

Simulation of the initiation, extension and closing of a fracture requires the coupling of rock and fluid mechanics. This important field has only recently started receiving attention. Much work is required before this technology can be used to improve the design of massive-hydraulic fractures now being conducted in tight formations[4,5]. These models are also required to predict fracture orientation and size in unconsolidated oil sands, where fracturing is used to provide initial communication between the injectors and the producers.

Except for single well studies in cylindrical coordinates, economic constraints demand that blocks containing wells be orders of magnitude larger than the size of the well. The problem then is to relate the calculated conditions at the grid point in a block to the well that may be located anywhere in that block. Analytical solutions based on single phase flow theory are used to relate the well pressure to the block pressure[6]. Detailed simulation of the zone near the well through the use of small blocks and cylindrical coordinates is necessary when saturation and/or thermal effects become important. The information generated from such a local study of the well vicinity is used in the form of pseudo-functions for the simulation of the reservoir. A better solution of this problem would be to have the capability to do local grid refinement without placing small blocks where they are not needed. The multi-grid approach may offer a solution to this problem[7].

Recently it has been possible to generalize the well treatment to handle vertical fractures that go through a number of grid points[8]. For single phase flow, where it is possible to compare numerical and analytical solutions, the agreement is excellent. For multiphase flow, in addition to the problems encountered for wells, we also have the unresolved problem of multiphase flow in the fracture.

Reservoir rocks that are naturally fractured behave in a significantly different fashion from conventional reservoirs. They may be simulated through the concept of double-porous-media with separate equations for each system and appropriate transfer terms for interaction between the systems[9]. Some of the problems with the practical use of this concept are (a) determination of the value of the transfer terms, and (b) experimental verification, particularly for multiphase flow.

Some reservoirs are either unconsolidated or only partially consolidated. The flow of sand in such systems alters rock properties. In shallow reservoirs removal of fluids (and/or solids) may also cause ground subsidence. Little is known about these two mechanisms and their simulation[10].

NUMERICAL APPROXIMATIONS

The mathematical model of flow in a conventional reservoir consists of one partial differential equation for each pseudo-component. Furthermore, for thermal processes an additional partial differential equation for temperature is obtained from the conservation of energy. In addition several constraints and algebraic relations must also be satisfied. The equations of the mathematical model may be manipulated to obtain a set that is more amenable to numerical treatment. At this stage a set of primary variables, equal in number to the partial differential equations to be solved, is selected. Sometimes the selection is postponed until after the application of some technique to translate the partial differential equations to difference equations. For example in black oil simulation, the oil phase pressure, and gas and water saturations form a suitable set of primary variables. The selection of primary variables and the alignment of these variables with appropriate equations can have a substantial effect on the eventual performance of the model.

Numerical difficulties can result due to the appearance and disappearance of a phase during simulation. This happens, for example, in thermal and in variable bubble point black oil problem simulation. This problem may be circumvented by variable substitution or by using a technique that does not allow the phase to disappear completely[6,11,12,13]. The same result is obtained by both techniques, however program complexity and computer time can differ substantially. With the use of variable substitution it is possible to solve for one less equation and

372

thus save computer time. However, special care is necessary for a smooth transition from one set of variables to another set[12].

Most reservoir simulators utilize three-point finite-difference approximation for second order space derivatives associated with transport terms and two-point backward difference approximation for first order derivatives associated with the accumulation terms. The end result of such an exercise is a set of non-linear, coupled algebraic equations of the form:

$$F(X) = 0 \tag{1}$$

where X is the vector of unknowns (= number of primary variables * number of blocks) for a time step. Such a set of non-linear equations can only be solved by some iterative technique. Application of Newton's method yields:

$$A(X^{\nu+1} - X^{\nu}) = -F(X^{\nu}) \tag{2}$$

where (ν) is the level of iteration and A is the Jacobian with elements $\partial f_i / \partial x_j$. These elements can be evaluated either numerically or analytically, depending on the problem. The Jacobian matrix is sparse with the form shown in Figure 1. Each non-zero entry in the matrix is a NEQxNEQ block element where NEQ is the number of primary variables.

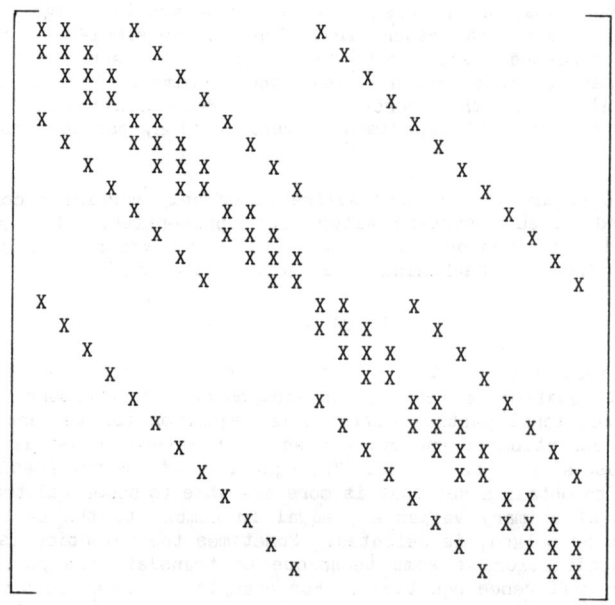

Figure 1
Structure of Matrix A for a 4x3x2 Grid
(Each X represents a NEQxNEQ block matrix)

Each time step usually requires 2 to 5 Newton iterations for the solution of (1). Hence for a typical problem, equation (2) must be solved many hundreds of times. As the number of blocks increases, the fraction of total computer time that is spent on (2) also increases. Recent research to reduce this effort will be discussed in the following section. The fully implicit method has unlimited stability, but Newton's method may not converge, or converge to an unreal solution if the initial guess (previous time step) is too far from the solution. Other variations of Newton's method like the semi-implicit or linearized implicit also work well for some problems. Non-linearities associated with the production/injection terms can have a significant influence on the stability and time truncation error of the model[14]. A problem of convergence to unreal solutions, which arises in the simulation of steam displacement with a non-condensable gas, has been eliminated through the addition of a "penalty source" term to the inert gas equation[12].

The size of the matrix equation to be solved can be reduced by suitable approximations that partially or fully decouple the equations (SEQUENTIAL METHOD) and reduce the number of implicit equations to one (IMPES). In the IMPES method the pressure is solved for implicitly while the saturations are treated explicitly. This results in a limitation on stability[1]. The decoupling can take place at the Jacobian level of the partial differential equations, the difference equations, or the matrix. Approximations of this type do, in some cases, increase the number of iterations necessary for convergence over the time step or fail to converge. The reliability of such methods is questionable for difficult problems.

The flow into and out of a block depends upon the permeability ($k_\ell = k \, k_{r\ell}$) values at the block boundaries. The value of absolute permeability at the boundary is computed as the harmonic average of the two adjacent blocks. The rules for the computation of relative permeability are not well defined. The most common approach is to use the relative permeability of the upstream block. Many different methods have been investigated with a view to reducing the grid orientation and truncation error[15]. Ko et al.[15] expressed transmissibilities for the two phase pressure and saturation equations as

$$\lambda_T \tag{3}$$

and

$$f_W \, \lambda_T \tag{4}$$

respectively. They found that the centralized upstream for f_W (CUF):

$$f_W \big|_{BB} = \tfrac{1}{4} f_{W_{uu}} - f_{W_u} + \tfrac{1}{4} f_{W_d} \tag{5}$$

and harmonic total mobility (HTM):

$$\lambda_T \big|_{BB} = \frac{2 \, \lambda_{Tu} \, \lambda_{Td}}{\lambda_{Tu} + \lambda_{Td}} \tag{6}$$

worked best. Even this method failed when the shock mobility ratio, M_s, exceeded 2. However their tests were for incompressible water flood problems. The behaviour of these schemes is different for compressible systems and when the saturation change is not monotonic.

Another approach to reduce the grid orientation effect is to allow flow in directions that are both parallel and diagonal to the grid. This can be accomplished through a nine-point (as opposed to five-point) scheme for two-dimensional problems[13,16] and a twenty-seven (as opposed to seven-point) scheme for three-dimensional problems. Abou-Kassem[13] has observed significant reduction in grid orientation with the nine-point scheme for a steam displacement problem where five-point shows substantial effect of the orientation of the grid.

Grid orientation is a major unresolved problem that raises some serious questions about the credibility of simulation for highly unfavourable mobility ratios. Although the nine-point formulation works, its use at this time is prohibitively expensive. For some situations curvilinear grid can be used to reduce both grid orientation and truncation error. However this approach is also not suitable for general applications.

Space and time truncation errors can be maintained at tolerable levels for conventional simulation. However, the effect of space truncation can mask the true phenomena in processes where block size is too large to define events in the reservoir. Examples of this situation are (a) miscible or chemical slugs, and (b) combustion front. In a multiple contact miscible drive process, the results can be especially sensitive to the size of the blocks in the zone where miscibility is being established. The accuracy of simulation could be improved by (a) adaptive grid refinement, (b) using higher order methods, or (c) using another (possibly analytical) model within the block to provide the necessary detail. None of these approaches have fully succeeded so far.

FLUID/ROCK PROPERTIES

Realistic prediction of fluid and rock properties for the changing conditions during simulation is of crucial importance in reservoir simulation. However, this aspect of the problem is not totally in the control of the simulator developer. Often lack of good experimental data and the need for answers within tight time constraints forces one to make assumptions that may or may not be justified. In situations of this type, it is the responsibility of the modeller to make the limitations of the results clear to the user of the information derived from the simulation.

Most of the properties required for the simulation of primary depletion or water flooding in crude oil reservoirs can easily be measured in the laboratory, and are usually available. One exception to this is data on three phase relative permeability, and on the effect of temperature and interfacial tension on relative permeability, and capillary pressure. Models are often used to predict three phase relative permeability from two phase data, and the effects of temperature, interfacial tension and hysteresis phenomenon. More data than what is currently available are required to validate and refine these models.

Relatively simple equations of state when properly tuned and used offer a powerful means of computing fluid properties in an accurate and consistent fashion[17]. These equations can be imbeded within a compositional model. Since computations with the equation of state are iterative, the modeller must ensure that the scheme will converge in difficult situations with relatively few iterations[17,18]. Several groups, including CMG appear to be making significant advances in this area. Investigations are also underway on methods of selecting an optimum number of components that can be used to represent the reservoir and injected fluids[19]. There is also concern that, at least for some processes like the injection of CO_2 in heavy oil, the assumption of thermodynamic equilibrium between phases in a block may not be valid.

As the processes become more complex the data requirements increase while the availability of data decrease. An example of this is the kinetics of low temperature oxidation for in situ combustion processes.

SOLUTION OF MATRIX EQUATIONS

The heart of a reservoir simulator is a program for the solution of a large set of linear equations that may be expressed as

$$A^\nu x^{\nu+1} = r^\nu \qquad (7)$$

where A is a sparse matrix with a well defined structure, $x^{\nu+1}$ is a vector representing change in the primary variables from ν to $\nu+1$ iteration, and r is the residual vector. Equations of this type may be solved directly by Gaussian elimination, or by some iterative method which involves the repeated solution of several sets of smaller matrix equations by Gaussian elimination. The work required for the direct solution of a system like this is given by

$$W_D = f \ I(NEQ \ J \ K)^3 \tag{8}$$

where I, J, and K are the number of grid points in the three directions. To minimize work I is chosen to be the direction with the largest number of grid points. The coefficient f=1 for standard ordering may be reduced to between .19 and .5 for D4 ordering[1].

Work required for iterative methods is difficult to predict since the number of iterations required depends on the problem. Another problem with iterative methods is their reliability in difficult situations. In general iterative methods that work are cheaper than direct elimination for larger problems. The cross-over point depends upon the methods and the problem.

Some rather powerful iterative methods have been developed recently. One such method, known as COMBINATIVE, has worked even for extremely difficult thermal problems[20]. This method became more economical than the direct elimination if

$$(J*K*NEQ + NEQ-1) \geq 50 \tag{9}$$

The combinative method involves the following steps: (a) decouple pressure equation by neglecting appropriate terms in (7), (b) solve the pressure equation by Gaussian elimination with D4 ordering and obtain initial estimate of pressure, (c) use this pressure estimate to form new residuals for (7), (d) do an LU factorization of the whole set and obtain an initial estimate of the remaining variables and an extra contribution to the initial pressure estimate obtained in step (a), and (e) apply ORTHOMIN[20] acceleration. This procedure is repeated until convergence is achieved.

Other iterative methods based on the multi-grid[21] approach now being developed show even greater promise. Iterative methods also require less storage than direct methods. In difficult problems it is necessary to treat the coupling between the well and the reservoir in a fully implicit fashion. If the well goes through more than one layer or block, additional terms are introduced. Unless properly handled, work required to solve the equations can increase substantially[22].

COMPUTER HARDWARE

In addition to the computers becoming faster with larger and larger memory, there are two other developments that are beginning to have a profound influence on reservoir simulation. These are (a) the development of pipeline and parallel processors, and (b) the development of display and interactive techniques.

The pipeline and parallel processors can perform a large number of operations (up to 5×10^8 floating point operations per second) very quickly provided the software is designed to take full advantage of the hardware. The current compilers can only go partways in achieving high efficiency with such processors. Program structure and solution algorithms are being developed for this class of computers. One disadvantage of this approach is that as efficiency on one machine increases, the program becomes less and less portable.

Interactive preparation of data and graphical display of results can make it much easier to run simulators and analyze results. This is particularly true of the new or infrequent users of a complex model. Within the next few years this is expected to become the normal procedure for conducting simulation studies.

CONCLUSIONS

The need for robust, economical, realistic and easy to use simulators is increasing as the oil recovery mechanisms being applied become more and more complex. Simulators are an essential tool for understanding and predicting reservoir performance. Their intelligent use can play a key role in optimizing oil recovery.

Along with the development of new numerical techniques, experimental studies must be continued to provide data and correlations for the prediction of fluid and rock properties. Model validation with carefully conducted experiments is also essential.

Significant new developments in numerical techniques, process understanding and hardware have taken place over the last few years, but much more needs to be done and will be done over the next few years.

ACKNOWLEDGEMENTS

The Department of Energy and Natural Resources of the Province of Alberta and the Department of Energy, Mines and Resources of the Government of Canada, provide partial funding for the work of CMG through the Alberta/Canada Energy Resources Research Fund. Additional support is provided by Associate Members of CMG through the membership fees. The work at the University of Calgary has been supported over the past sixteen years by the National Science and Engineering Research Council (previously National Research Council).

The author is indebted to these organizations for financial support and to his students and colleagues for the generation of ideas and for their implementation in practical simulations.

NOMENCLATURE

A   Jacobian matrix

f_W $\left(\dfrac{k\, k_{rw}}{\mu_w}\right)\bigg/\lambda_T$

I,J,K   Grid nodes along the three directions

k   absolute permeability

$k_{r\ell}$ relative permeability of phase ℓ

NEQ   Number of equations per grid block (= number of primary variables)

r   Residual vector

x   Vector of change in primary variables over an iteration

X   Vector of primary variables (unknowns)

λ_T $= \dfrac{k\, k_{rw}}{\mu_w} + \dfrac{k\, k_{ro}}{\mu_o}$

μ_ℓ viscosity of phase ℓ

Subscripts

BB   Block boundary

d   1 point downstream of block face in question

u   1 point upstream of block face in question

uu   2 points upstream of block face in question

Superscript

ν   Iteration level

REFERENCES

1.   AZIZ, K. and SETTARI, A.; "Petroleum Reservoir Simulation", Applied Science Publishers, London (1979).

2.   ODEH, A.S.; "Comparison of Solutions to a Three-Dimensional Black Oil Reservoir Simulation Problem", J. Pet. Tech. (January 1981) 33, 1, 13-25.

3.   FOGARASI, M., GREGORY, G.A. and AZIZ, K.; "Analysis of Vertical Two Phase Flow Calculations: Crude Oil - Gas Flow in Well Tubing", Cdn. J. Pet. Tech. (1980) 19, 1, 86-92.

4.   WADE, R. and AZIZ, K.; "Stimulating the Triassic Carbonates in the Foothills Gas Trend of Northeast British Columbia", CIM 81-32-35, presented at the 32nd Annual Technical Meeting of the Petroleum Society of CIM, Calgary, Alberta (May 1981).

5.   SETTARI, A.; "Simulation of Hydraulic Fracturing Processes", Soc. Petrol. Eng. J. (December 1980) 20, 6, 487-500.

6.   AU, A., BEHIE, A., RUBIN, B. and VINSOME, K.; "Techniques for Fully Implicit Reservoir Simulation", SPE 9302, presented at the 55th Annual Fall Technical Conference and Exhibition of the SPE of AIME, Dallas, Texas (September 1980).

7.   BRANDT, A.; "Multi-Level Adaptive Solution to Boundary Value Problems", Math. Comp. (April 1977) 31, 138, 333-390.

8.   NGHIEM, L.; "Modelling Infinite-Conductivity Vertical Fractures Using Source or Sink Terms", CMG.R8.02, (February 1981).

9.   GESHELIN, B.M.; "Static Fracture Model", CMG.R8.01, (January 1980).

10.   Ertekin, T. and Farouq Ali, S.M.; "Numerical Modelling of Reservoir Compaction and Associated Ground Subsidence under Non-Isothermal Two-Phase Flow Conditions", presented at the SIAM Fall Meeting, Houston, Texas (November 1980).

11.   CROOKSTON, R.B., CULHAM, W.E. and CHEN, W.H.; "A Numerical Simulation Model for Thermal Recovery Processes", Soc. Petrol. Eng. J. (February 1979) 19, 1, 37-58.

12.   FORSYTH, P.A. Jr., RUBIN, B. and VINSOME, K.; "Elimination of the Constraint Equation and Modelling of Problems with a Non-Condensable Gas in Steam Simulation", CIM 81-32-50, presented at the 32nd Annual Technical Meeting of the Petroleum Society of CIM, Calgary, Alberta (May 1981).

13.   ABOU-KASSEM, J.H.; "Investigation of Grid Orientation in a Two-Dimensional, Compositional, Three-Phase Steam Model", Ph.D. Thesis, University of Calgary (1981).

14.   FONG, D.K.S.; "Treatment of Nonlinearities and Production Allocation in a Fully Implicit, Three-Phase Coning Model", M.Sc. Thesis, University of Calgary (1980).

15.   KO, S.C.M., BUCHANAN, W.L. and VINSOME, K.; "A Critical Comparison of Finite-Difference Interblock Mobility Approximations in Numerical Reservoir Simulation", CIM 81-32-23, presented at the 32nd Annual Technical Meeting of the Petroleum Society of CIM, Calgary, Alberta (May 1981).

16. KO, S.C.M. and AU, A.; "A Weighted Nine-Point Finite-Difference Scheme for Eliminating the Grid Orientation Effect in Numerical Reservoir Simulation", SPE 8248, presented at the 54th Annual Fall Technical Conference and Exhibition of the SPE of AIME, Las Vegas, Nevada (September 1979).

17. NGHIEM, L. and AZIZ, K.; "A Robust Iterative Method for Flash Calculations Using the Soave-Redlich-Kwong or the Peng-Robinson Equation of State", SPE 8285, presented at the 54th Annual Fall Technical Conference and Exhibition of the SPE of AIME, Las Vegas, Nevada (September 1979).

18. MEHRA, R.K., HEIDEMANN, R.A. and AZIZ, K.; "Calculation of Multiphase Equilibrium for Compositional Simulation", SPE 9232, presented at the 55th Annual Fall Technical Conference and Exhibition of the SPE of AIME, Dallas, Texas (September 1980).

19. LEE, S.T., JACOBY, R.H., CHEN, W.H. and CULHAM, W.E.; "Experimental and Theoretical Studies on the Fluid Properties Required for Simulation of Thermal Processes", SPE 8293, presented at the 54th Annual Fall Technical Conference and Exhibition of the SPE of AIME, Las Vegas, Nevada (September 1979).

20. BEHIE, G.A. and VINSOME, K.; "Block Iterative Methods for Fully Implicit Reservoir Simulation", SPE 9303, presented at the 55th Annual Fall Technical Conference and Exhibition of the SPE of AIME, Dallas, Texas (September 1980).

21. BEHIE, G.A. and FORSYTH, P.A. Jr.; "Multi-Grid Solution of the Pressure Equation in Reservoir Simulation", CMG.R17.01 (July 1981).

22. GEORGE, A.; "On Block Elimination for Sparse Linear Systems", SIAM J. Numer. Anal. (June 1974) 11, 2, 585-603.

THREE-DIMENSIONAL NUMERICAL SIMULATION
OF STEAM INJECTION

P. LEMONNIER

Institut Français du Pétrole, Rueil Malmaison, France

ABSTRACT

A three-dimensional thermal model has been developed for simulating both cyclic steam injection and steam drive. The numerical model TWIST describes three-phase flow (oil, water and steam) heat flow in the reservoir and heat conduction in the surrounding formations. Wellbore heat losses between the surface and the reservoir are taken into account. The various reservoir heterogeneities and temperature dependent parameters (including relative permeabilities) are considered. Distillation effects are approximated through the decreased residual oil saturation when steam is present. Mass conservation and energy equations are solved simultaneously to improve stability. A semi-implicit method is used for time formulation. The oil phase equation is decoupled with a scheme of the type $p-T-S_g/S_w$. This formulation enables this thermal simulator to be very efficient in terms of computing time and stability.

Numerical results are presented showing a steam stimulation history of five cycles and the influence of steam quality, initial reservoir pressure and steam injection rate on steamflood performance in a five-spot pattern.

INTRODUCTION

TWIST (Tool When Injecting Steam) is a three-dimensional steamflood model, which describes three-phase flow (oil, water and steam) and heat flow in the reservoir. Vertical heat losses to overlying and underlying strata and wellbore heat losses between the surface and the reservoir are taken into account.

The literature on the simulation of steamflooding is extensive [1-7]. The efforts have been concentrated on methods of solution. The equations are solved sequentially or simultaneously and the formulation is explicit, weakly or highly implicit.

We use a formulation which requires significantly less computing time per grid block-time step than the implicit scheme, and is nevertheless highly stable, owing to the fact that the water and gas flow equation and the energy equation are solved simultaneously with implicit treatment of the gas transmissibility. This formulation has been mentioned in the literature [6] but not tested otherwise than in isothermal black-oil model.

We encountered no difficulties in simulating field cases with TWIST. Our experience includes steam stimulation and steam drive for pilots of various pattern shapes and various oil viscosities.

Numerical results are presented showing a steam stimulation history of five cycles and the influence of steam quality and reservoir pressure on oil recovery in steam drive.

MODEL DESCRIPTION

Simulator equations

The model consists of four equations expressing (1) conservation of mass for water and steam, (2) conservation of mass for oil phase, (3) conservation of energy and (4) equilibrium constraints. The four unknowns are oil pressure, temperature, steam and water saturations. We have three additional equations for obtaining oil saturation, gas and water pressures : (5) saturations constraint, (6) and (7) capillary pressures.

$$\frac{\partial}{\partial t} (\phi \rho_w S_w + \phi \rho_s S_s) + \nabla \cdot (\rho_w \bar{v}_w + \rho_s \bar{v}_s) + q_w + q_s = 0 \tag{1}$$

$$\frac{\partial}{\partial t} (\phi \rho_o S_o) + \nabla \cdot (\rho_o \bar{v}_o) + q_o = 0 \tag{2}$$

$$\frac{\partial}{\partial t} \left[\phi (\rho_w S_w U_w + \rho_o S_o U_o + \rho_s S_s U_s) + (1 - \phi) \rho_r U_r \right]$$

$$+ \nabla \cdot (\rho_w \bar{v}_w H_w + \rho_o \bar{v}_o H_o + \rho_s \bar{v}_s H_s) - \nabla \cdot (K_h \nabla T) + q_L + q_H = 0 \tag{3}$$

$$S_s = 0 \tag{4a}$$

$$\text{or } T = T_s (p) \tag{4b}$$

$$\text{or } S_w = 0 \tag{4c}$$

$$S_o + S_w + S_s = 1 \tag{5}$$

$$p_o - p_w = P_{cw} \tag{6}$$

$$p_s - p_o = P_{cg} \tag{7}$$

The phase velocity $\bar{v}_i$ is defined as

$$\bar{v}_i = - \frac{k \, k_{ri}}{\mu_i} \cdot (\nabla p_i - \gamma_i \nabla Z) \quad \text{for } i = w, o, s \tag{8}$$

The condensation term is eliminated by summing the water and steam mass conservation equations [3].

The **equilibrium constraints** are expressed by one of the three equations (4) for the following cases : no steam, saturated steam, superheated steam.

Additional assumptions

1 – The model can operate in one, two or three dimensions with cartesian or radial grid.

2 – Reservoir dip and gravity are taken into account.

3 – Reservoir rock and fluids are compressible.

4 – The model is not compositional. Distillation effects are approximated through decreased residual oil saturation in the presence of steam.

5 – Reservoir can be anisotropic, homogeneous or heterogeneous by layers or by cells.

6 – Temperature dependency of the physical and thermal parameters is accounted for.

7 – Three-phase relative permeabilities at each temperature value are calculated using Stone's method [8].

8 – Numerical simulations include steam drive and steam stimulation.

Heat loss to overburden and underburden

Heat loss by conduction to the overlying and underlying strata is calculated from the numerical solution of the heat conduction equation. The equation is approximated by the standard finite-difference approximation. We assume negligible effects of heat conduction in the horizontal directions [3]. The heat conduction equation for the surrounding rock is not solved simultaneously with the reservoir equations. At time step n+1 the equation is solved using the reservoir boundary temperature at the previous time step.

Well model

The wells have the following specifications :

1 – Bottom-hole pressure

2 – Water or steam injection rate

3 – Liquid production rate

4 – Oil production rate

5 – Shut in.

Each well can operate successively in injection and production modes (for huff and puff process for example).

Wellbore heat losses in the injection wells are calculated. The injection pressure and the steam quality are specified at surface or reservoir conditions. We use the method of Ramey[13], Satter[14] for wellbore heat losses computations [20]. The basic assumptions of the method are as follows:

– Steam is injected at a constant rate, wellhead pressure, temperature and quality.

– Any variation in steam pressure with depth is negligible.

Fluid and rock properties

The steam and water properties are expressed through correlations from the steam tables.

Oil and water enthalpies are expressed as polynomial functions of temperature. Steam enthalpy and all fluid densities are treated as functions of pressure and temperature.

Water and steam viscosities are expressed as functions of temperature. Oil viscosity is entered into the model as tabular function of temperature with exponential interpolation between adjacent entries in the table.

Residual oil and irreductible water saturations are represented as linear functions of temperature; the same assumption is made for the relative permeability end points. The influence of temperature on the relative permeability curves is described by shifting the curves without changing the curves shape[5].

The rock specific heat and thermal conductivity are represented as linear functions of temperature.

User facilities

User facilities have been developed in the simulator. Arrays are dimensioned automatically with appropriate values at the beginning of each run. All input cards are checked for validity and for inconsistencies. All errors encountered are listed at the end of the data processing. Various printing options are allowed for input data and output results. Array maps can be selected and printed with any orientation. Graphic plotting of input data (oil viscosity, relative permeabilities, ...) and well behaviour versus time (pressure, oil recovery, WOR, ...) are available, just as pressure, temperature or saturations contours or profiles.

SOLUTION METHOD

Discretization

The three equations (1) (2) (3) are discretized into finite - difference form with upstream densities, mobilities and enthalpies in the flow terms.

Semi-implicit approximations[9] are used for time discretization. Interblock transmissibilities or flow terms are treated as follows. Explicit (i.e. time level n) dating is considered for fluid viscosities, fluid densities and fluid enthalpies, on account of the weak sensitivity to implicit versus explicit dating encountered in numerical simulations. Explicit dating is used for water and oil relative permeabilities and semi-implicit formulation is used for steam relative permeability. Capillary terms are expressed explicitly in saturation.

The accumulation terms are written with implicit dating. The resulting formulation is then linearized as follows :

$$f\ (p^{n+1},\ S^{n+1},\ T^{n+1}) = f\ (p^n,\ S^n,\ T^n) + \frac{\partial f}{\partial p}\ \delta p + \frac{\partial f}{\partial S}\ \delta S + \frac{\partial f}{\partial T}\ \delta T \qquad (9)$$

where $\delta X = X^{n+1} - X^n$ with $X = p,\ S,\ T$.

With these approximations the four equations (1) (2) (3) (4) are expressed in terms of the four unknowns δp, δT, δS_s and δS_w. The first three equations can be represented by

$$A_{11} \, \delta p + A_{12} \, \delta T + A_{13} \, \delta S_s + A_{14} \, \delta S_w + F_{11} \, (\delta p) + F_{13} \, (\delta S_s) = R_1 \tag{10}$$

$$A_{21} \, \delta p + A_{22} \, \delta T + A_{23} \, \delta S_s + A_{24} \, \delta S_w + F_{21} \, (\delta p) = R_2 \tag{11}$$

$$A_{31} \, \delta p + A_{32} \, \delta T + A_{33} \, \delta S_s + A_{34} \, \delta S_w + F_{31} \, (\delta p) + F_{32} \, (\delta T)$$
$$+ F_{33} \, (\delta S_s) = R_3 \tag{12}$$

The variables A_{ij} and R_i denote coefficients while the variables F_{ij} denote differential operators.

The explicit treatment of oil and water relative permeabilities allows the water saturation unknown δS_w to be eliminated in equations (10) and (12) by means of equation (11). The elimination of δS_w results in a system of two coupled equations (13) (14):

$$A'_{11} \, \delta p + A'_{12} \delta T + A'_{13} \delta S_s + F'_{11} \, (\delta p) + F_{13} \, (\delta S_s) = R'_1 \tag{13}$$

$$A'_{31} \, \delta p + A'_{32} \delta T + A'_{33} \delta S_s + F'_{31} \, (\delta p) + F_{32} \, (\delta T) + F_{33} \, (\delta S_s) = R'_3 \tag{14}$$

$$A'_{ij} = A_{ij} - \frac{A_{i4}}{A_{24}} \times A_{2j} \qquad F'_{i1} = F_{i1} - \frac{A_{i4}}{A_{24}} \times F_{21}$$

$$R'_i = R_i - \frac{A_{i4}}{A_{24}} \times R_2 \qquad i = 1 \text{ et } 3, \qquad j = 1,3$$

The equations (4) (13) (14) involving the three unknowns δp, δT, δS_s are reduced to two equations in the two unknowns δp, δX by use of the equilibrium constraints (4). The unknown δX is equal to δT or δS_s. The definition of δX can vary from one grid cell to an other and from one time step to an other according to the equilibrium condition in the grid cell at time step n+1.

If no steam is present at time step n+1, $\delta X = \delta T$ and equation (4a) is used for eliminating δS_s.

If steam is present at time step n+1, $\delta X = \delta S_s$ and equation (4b) is used for eliminating δT.

If superheated steam is present at time step n+1, $\delta X = \delta T$ and equation (4c) is used for eliminating δS_s.

Explicit dating of saturation-dependent production terms can give saturation oscillations in grid cells near the wellbore. A semi-implicit formulation for production terms similar to that described by Spivak and Coats[10] is used for increasing computational stability.

Resolution procedure

The procedure of solution of the equations for the time step n+1 is as follows:

1 - Solve the two equations (13) (14) for δp and δX using Gaussian direct solution with D 4 ordering [11]. The substitution of the unknowns δS_s or δT is made with the equilibrium conditions at time step n.

2 - Check the validity of the solution for each grid cell. If not, choose an other equilibrium condition and solve again the two equations for δp and δX.

3 - Solve equation (11) for δS_w using the values δp, δT and δS_s.

EXAMPLES OF APPLICATIONS

We used the model for simulating a number of field cases. Numerical simulation studies of a steam drive pilot[12] were made with a two-dimensional grid. Simulations of the whole seven-spot pattern are now pursued with a three dimensional grid (12 × 9 × 5).

Two example cases are presented to illustrate the use of the model in the case of well stimulation and steam drive.

Well stimulation

The first application of the model consisted in simulating the production history of a well submitted to five successive steam stimulation cycles. The data are given in Table 1. The top of the reservoir is located at a depth of 228 m (748 ft). Oil viscosity at initial temperature of 26°C (79°F) is 4270 mPa.s (cP). Other values are given in Table 2. The run were made in a two-dimensional radial configuration (r, z) with a 12 × 6 grid.

TABLE 1 - DATA FOR WELL STIMULATION PROBLEM

Zone thickness	24 m (78.74 ft)
Exterior radius	200 m (656.16 ft)
Porosity	from 0.1 to 0.25
Horizontal permeability	from 0.8 to 2 μm^2 (800 to 2,000 md)
Anisotropy k_h/k_v	10
Irreductible water saturation	0.2
Residual oil saturation to water	0.43 at 26°C and 0.19 at 220°C
Residual oil saturation to steam	0.1
Rock compressibility	5.10^{-5} kPa^{-1} (3.4 10^{-4} psi^{-1})
Initial temperature	26°C (79°F)
Initial pressure	2800 kPa (406 psia)
Initial water saturation	0.2

TABLE 2 - OIL PHASE VISCOSITY

Temperature	Well stimulation problem	Steam drive problem
20°C (68°F)	8924 mPa.s (cP)	5159 mPa.S (cP)
50°C (122°F)	545 "	180 "
100°C (212°F)	36 "	19 "
150°C (302°F)	8.3 "	5.5 "
200°C (392°F)	4.1 "	2.6 "
260°C (500°F)	2.5 "	1.27 "

We specified a steam injection rate of 100 metric tonnes per day (629 B/D cold water equivalent) into the six layers. The steam quality was 100% at surface conditions and the injection temperature was 264°C (507°F). Wellbore heat losses computed by the model give at the end of injection a quality of 90% at the bottom of the well.

We simulated five cycles. One cycle involves 25 injection days, 5 soaking days and a producing period with a total fluid production rate of $20m^3/$ day (126 B/D). A new cycle is initiated when the oil rate has decreased to $5m^3/day$ (31 B/D). According to this criterion the durations of the successive production cycles have been 152, 99, 136, 130 and 133 days. The variations of oil rate and water cut with time for the successive cycles are shown in Fig. 1 and Fig. 2. The well bottom-hole pressure evolution is shown in Fig.3

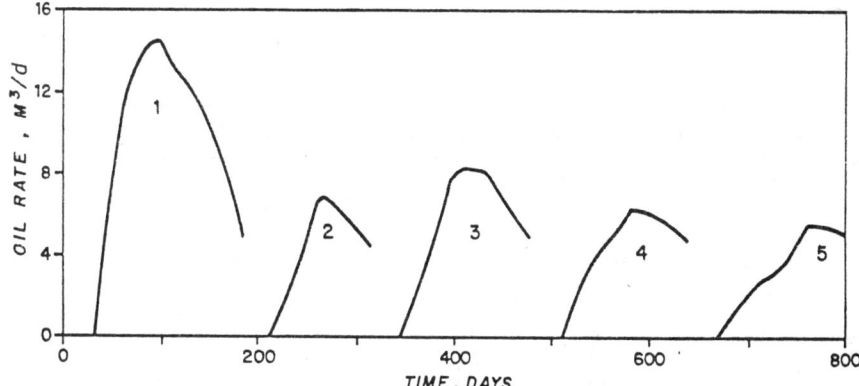

Fig. 1 - Oil production rate for five successive steam stimulation cycles

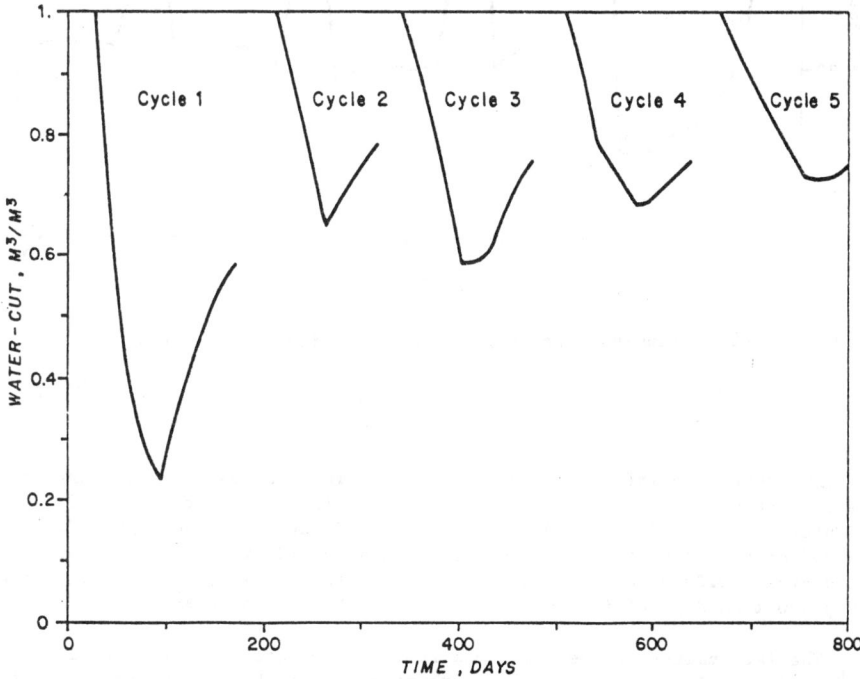

Fig. 2 - Water-cut versus time for five stimulation cycles

during the 800 days of history. The history includes 125 days of injection,
25 days of soaking and 650 days of production. The total oil recovery was
3,870m^3 (24,346 STB) for a total steam injection of 12,500 tonnes (78,616 STB
cold water equivalent) and a total water production of 8,127m^3 (51,113 STB).
The values of the oil/steam ratio for each of the five cycles are the following: 0.64, 0.18, 0.31, 0.23 and 0.19. The criterion chosen for the end of
the production phase leads to a greater length, a higher depletion and a better
performance for the first cycle and to a relatively poor performance for the
second cycle (Fig. 1).

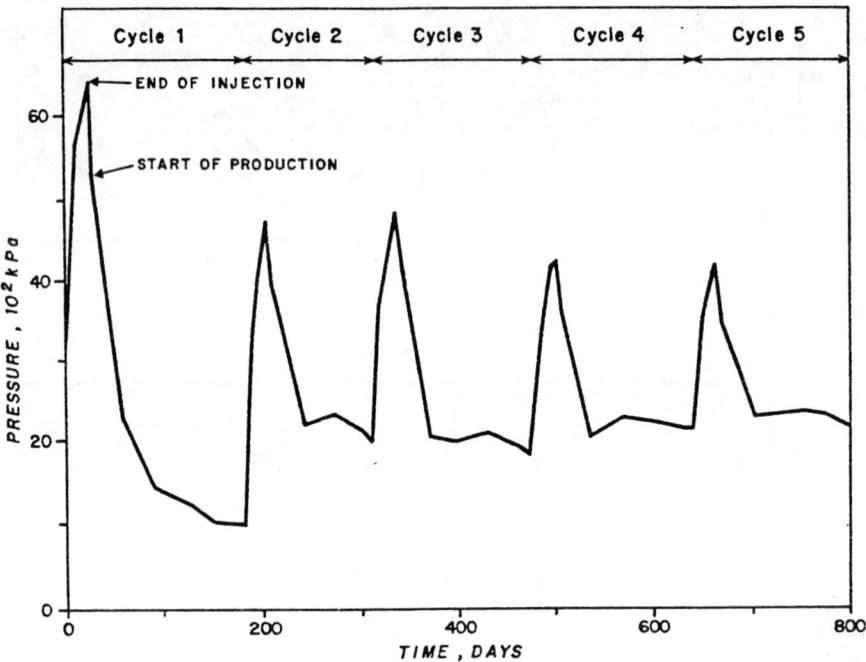

Fig. 3 - Well bottom-hole pressure versus time for five stimulation cycles

An other simulation performed with predefined values for the length of
the five successive production phases leads to the following values of the
oil/steam ratio: 0.4 after 90 producing days of the first cycle, 0.24 after
90 producing days of the second cycle, 0.32 after 120 producing days of the
third cycle, 0.29 after 150 producing days of the fourth cycle and 0.27 after
180 production days of the last cycle. The total oil recovery was the same.

The fair values of the oil/steam ratio may be essentially attributed to
the relatively low porosity of the reservoir and to the high viscosity of the
oil.

Steam-drive

Many studies have been devoted to the evaluation of the performance of steam flooding technique .[15-18]. However the influence of some operating parameters has not yet been clarified. The effects of steam quality and reservoir pressure on steamflood performance were investigated in a five-spot pattern with the simulator.

One eighth of a five-spot pattern was represented by a 6 × 3 × 5 grid (Fig.4) with $\Delta x = \Delta y$ = 14.14m (46.4 ft) and Δz = 4m (13.12 ft) . Table 3 summarizes the data for this problem and Table 2 shows the oil viscosity versus temperature. The relative permeabilities were temperature dependent. The water-oil relative permeability curves are shown in Fig. 5 for two temperature

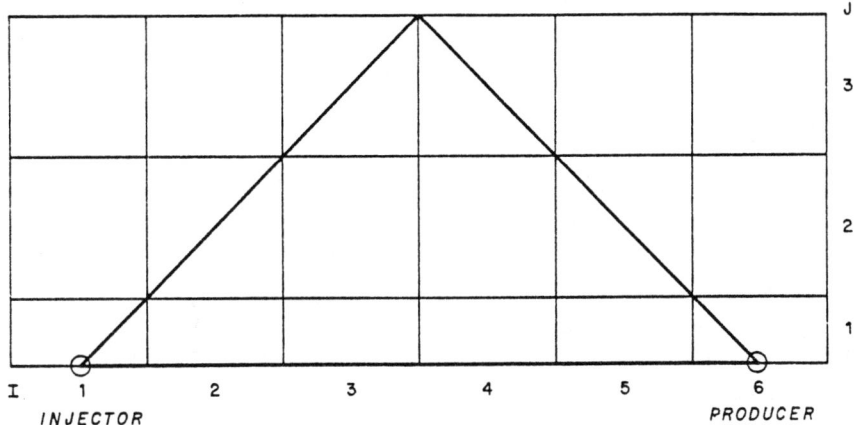

Fig. 4 - Simulation grid for one- eighth of five spot pattern

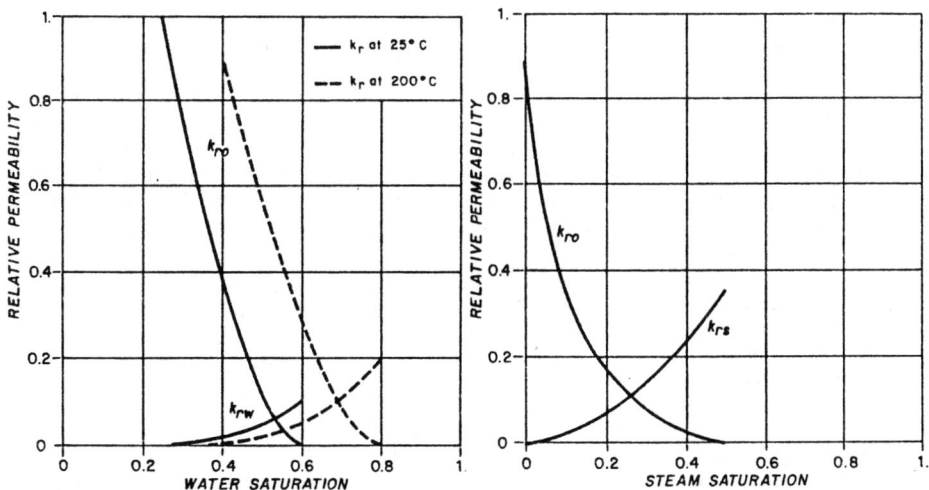

Fig. 5 - Water-oil relative permeability curves

Fig. 6 - Gas-oil relative permeability curves

TABLE 3 - DATA FOR STEAM-DRIVE PROBLEM

Area (5-spot)	10,000 m^2 (2.5 acres)
Reservoir thickness	20 m (65.6 ft)
Porosity	0.35
Horizontal permeability	2.5 μm^2 (2500 md)
Vertical permeability	1. μm^2 (1000 md)
Formation compressibility	1.10^{-5} kPa^{-1} (6.8 10^{-5} psi^{-1})
Specific heat of formation	2.35 J/cm^3 - °C (35 Btu/cu.ft -°F)
Specific heat of overburden and underburden	2.5 J/cm^3 - °C (37 Btu/cu.ft -°F)
Thermal conductivity of formation	2.3 W/m - °C (32 Btu/ft - day - °F)
Thermal conductivity of overburden and underburden	2.3 W/m - °C (32 Btu/ft - day - °F)
Oil compressibility	6.4 10^{-7} kPa^{-1} (4.4 10^{-6} psi^{-1})
Thermal expansion coefficient of oil	6.5 10^{-4} °C^{-1} (3.6 10^{-4} °F^{-1})
Specific heat of oil	2.1 J/g - °C (0.5 Btu/lb - °F)
Stock-tank oil density	0.95 g/cm^3 (60 lb/cu. ft)
Irreductible water saturation	0.25 at 25°C and 0.4 at 175°C
Residual oil saturation to water	0.4 at 25°C and 0.2 at 175°C
Residual oil saturation to steam	0.1
Initial temperature	25°C (77°F)
Initial water saturation	0.4
Initial pressure	500 kPa (72.5 psia) and 4000 kPa (580 psia)
Injection rate for full pattern (CWE)	50 m^3/day (314 B/D)
Production bottomhole pressure	300 kPa (43.5 psia) and 3800 kPa (551 psia)

values of 25°C (77°F) and 200°C (392°F). The gas-oil relative permeability curves are shown in Fig. 6.

We studied the effects of varying bottom-hole steam quality from 0 to 1 for two values of initial reservoir pressure, 500 kPa (72.5 psia) and 4000 kPa (580 psia) respectively. Specified injection rate for both set of cases was 50 tonnes/day (314 B/D cold water equivalent) for the full pattern; steam is injected only into the two bottom layers. Well injectivity and productivity indices as calculated according to Peaceman [19] were multiplied by two. Production wells produce from the five layers at deliverability against a bottom-hole pressure of 300 kPa (43.5 psia) and 3800 kPa (551 psia), for both cases respectively.

Fig. 7 shows the effect of steam quality on injection pressure. The high increase of injection pressure results from the formation of a high viscosity oil bank downstream from the condensation front. Injection pressure starts to decrease when the oil bank becomes mobile.

As shown in Fig. 8 the oil recovery for a given heat input is little sensitive to steam quality when quality is above 60%. The heat input is equal to the cumulative enthalpy of steam at sand face referred to initial reservoir temperature. Fig. 9 shows earlier steam breakthrough when steam quality increases and initial pressure decreases.

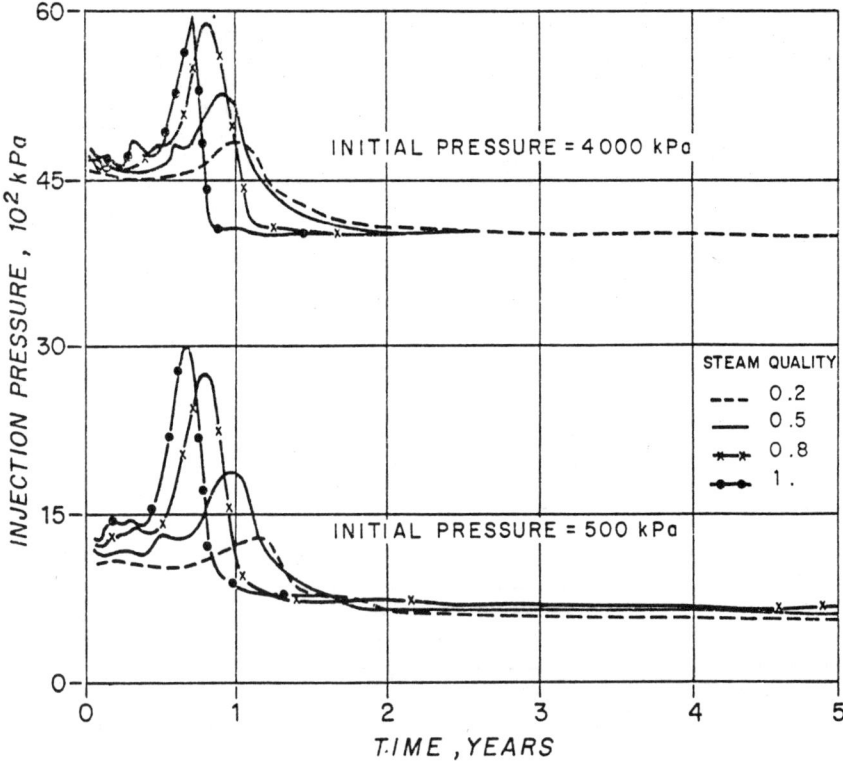

Fig. 7 - Effect of steam quality and initial reservoir pressure
on injection pressure

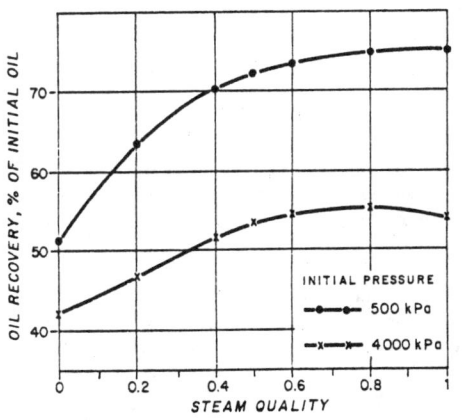

Fig. 8 - Effect of steam quality and
initial reservoir pressure
on oil recovery after 30 TJ
heat input.

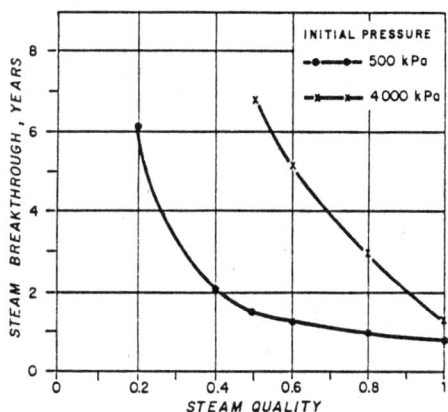

Fig. 9 - Effect of steam quality and
initial reservoir pressure
on steam breakthrough.

Vertical heat losses to overlying and underlying strata are relatively
independent of time after steam breakthrough. Heat losses after 6 years of
steam injection are presented in Fig. 10; less heat losses are achieved for
the lower initial pressure when quality is above 30%, due to lower steam
temperature and faster heating of the reservoir. For a steam quality of 60%
and an initial reservoir pressure of 500 kPa (72 psia) the vertical heat loss
is 31% of heat input. The heat injection rate for this case is 460 kJ/D-m^3 reservoir and the reservoir thickness 20 m. For these two values the vertical
heat loss curves of Gomaa[17] obtained with an initial pressure of 414 kPa give
the same value of 31%. The results in Fig. 10 show that the curves are also
reservoir pressure dependent. For a pressure of 4000 kPa (580 psia) the heat
injection rate is 505 kJ/D-m^3 res. and the heat loss is 38.5% of heat input,
instead of 29% in the low-pressure case considered by Gomaa.

Low heat loss to overburden strata and early breakthrough result in a high
amount of heat produced by the wells when operating at low initial reservoir
pressure (Fig. 11).

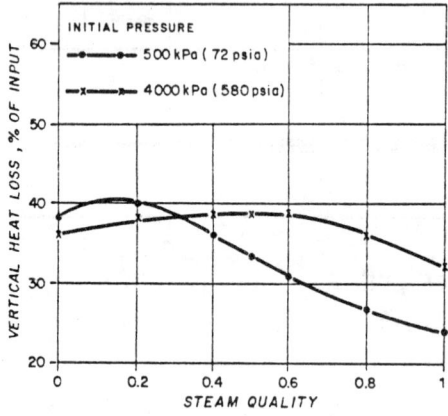

Fig. 10 – Effect of steam quality and
initial reservoir pressure
on vertical heat loss after
6 years

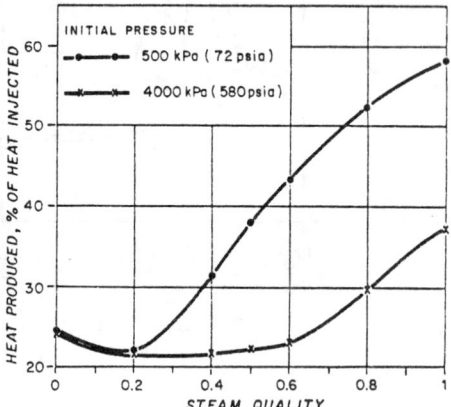

Fig. 11 – Effect of steam quality
and initial reservoir
pressure on heat produced

The cumulative oil/steam ratio is plotted in Fig. 12 versus steam quality
and initial reservoir pressure, after 6 years of injection. It appears that
the oil/steam ratio is improved when steam quality increases and initial reservoir pressure decreases.

The oil/steam ratio does not take into account the variation of heat input
due to the variation of steam quality. Hence we introduce an adimensional parameter, the energy yield EY, for comparing the performances of the tests. The
energy yield is defined as the ratio between the calorific value of the cumulative oil produced and the heat input previously defined. The value of EY is
equal to one when the energy content of the steam at sand face is equal to the
calorific value of the produced oil (calorific value of oil = 38 GJ/m^3). Fig. 13
shows the effect of steam quality and initial reservoir pressure on the energy
yield. An optimum steam quality value can be determined in Fig. 13, depending
on initial reservoir pressure and on the duration of steam injection.

The sensitivity to steam quality is much stronger at low pressure than at high pressure. This is related to the higher amount of heat transported by the produced fluids in the case of low pressure tests (Fig. 11). The same reason may explain the shift of the optimum steam quality towards lower values when time increases. As a matter of fact the heat produced after 6 years of injection is about twice the value obtained after 3 years.

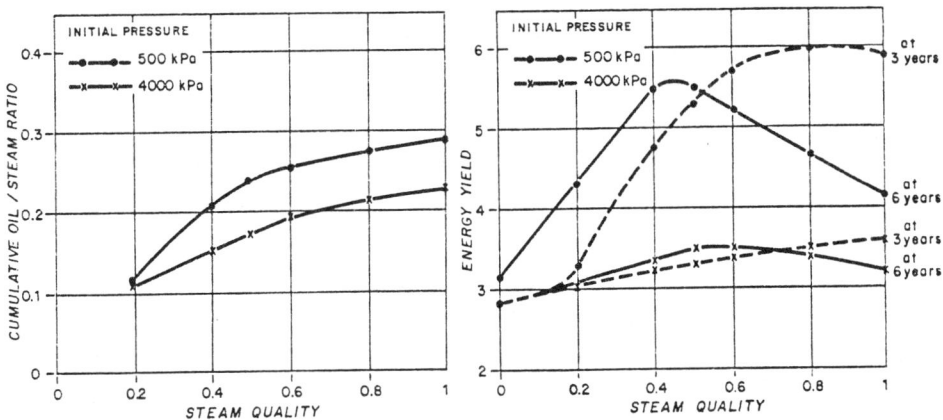

Fig. 12 - Effect of steam quality and initial reservoir pressure on cumulative oil/steam ratio after 6 years

Fig. 13 - Effect of steam quality and initial reservoir pressure on energy yield.

The optimum volumetric injection rate, after the project had reached a peak oil-production rate, was determined at Kern River from field results[16]. We made a similar study with the simulator in the case of 60% bottomhole steam quality and 500 kPa (72.5 psia) initial reservoir pressure. Fig. 14, as a result optimizes the rate of the instantaneous oil/steam ratio. An optimum steam injection rate of $2.5 \, 10^{-4}$ m^3/day/m^3 of reservoir volume (1.94 B/D - A-ft) was found (1.5 B/D -A-ft for Kern-River[16]). The curve in Fig. 14 has the same shape as the curve developed for Kern-River. The dispersion of the data is less than in the case of Kern-River since the simulations were carried out on a unique pattern whereas the correlation for Kern-River had been obtained from the field results of several pilot tests. This optimum steam rate corresponds to the value of 50 m^3/day (314 B/D) used in the sensitivity study for the full five-spot pattern.

Model running time

The formulation described above is very efficient in terms of computing time. A three-dimensional run with a 25 × 8 × 4 grid (800 cells) requires 0.002 seconds per grid block-time step on the CDC 7600 computer. The simulation of the steam drive pilot[12] with a three-dimensional grid 9 × 12 × 5 (405 active cells) requires 0.0018 seconds per grid block-time step. The case presented here of a three-dimensional steam-drive with a 6 × 3 × 5 grid (60 active cells) requires 0.0008 seconds per grid block-time step. The computation time for simulating a 9-year steam drive (582 time steps) with 4000 kPa initial pressure

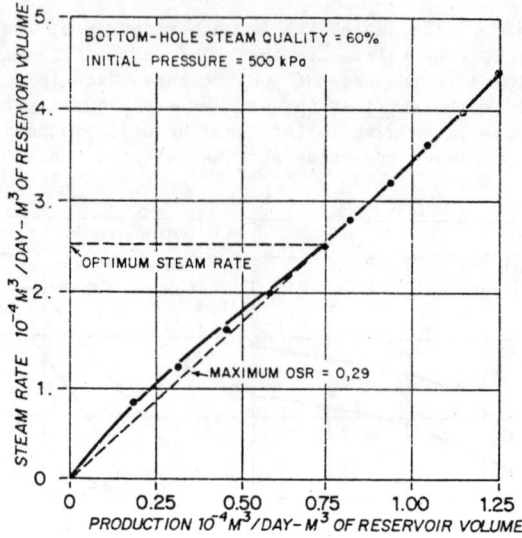

Fig. 14 - Optimum injection rate

and 60% downhole steam quality has been 28 seconds on the CDC 7600 computer. The ratio between computing times on a CDC 7600 and a vector computer CRAY 1 has been 5.5 for a 12 × 12 × 4 grid (576 cells).

CONCLUSIONS

1 - The semi-implicit formulation of the solution method used for the three-phase three-dimensional model TWIST enables the simulator to be very efficient in terms of computing time and stability. The model may be used for simulating a wide variety of thermal problems.

2 - Five successive steam injection cycles in a low porosity reservoir have been simulated to evaluate the decline of the oil/steam ratio from cycle to cycle.

3 - The influence of steam quality and initial reservoir pressure on steamflood performance has been investigated. It appears that these parameters affect the heat loss to the surrounding formations, the heat transported by the produced fluids and the performance of the process. Better performances and higher sensitivity to steam quality are observed at lower pressures.

4 - The analysis of the steamflood performances obtained for various steam injection rates at given quality and pressure indicates the existence of an optimum injection rate.

ACKNOWLEDGEMENTS

The author wishes to thank Institut Français du Pétrole for permission to publish this paper. He also expresses his appreciation to Mr. J.G. BURGER of Institut Français du Pétrole for his helpful and constructive discussions.

Partial financial support for the realization of the simulator used in this study was provided by Société Nationale Elf-Aquitaine (Production).

NOMENCLATURE

H = enthalpy (J/g)
k = absolute permeability (m^2)
k_r = relative permeability
K_h = thermal conductivity (W/m - °C)
p = pressure (kPa)
P_c = capillary pressure (Pa)
q = mass injection or production rate (Ton/day)
q_H = enthalpy production rate (J/day)
q_L = heat loss rate (J/day)
S = saturation
t = time
T = temperature (°C)
T_s = temperature of saturated steam (°C)
U = internal energy (J/g)
$\bar{v}$ = phase velocity
Z = depth, measured vertically downward (m)
γ = specific weight (kPa/m)
δ = time difference operator, e.g., $\delta X = X^{n+1} - X^n$
μ = viscosity (Pa.s)
ϕ = porosity
ρ = density (g/cm^3)

Subscripts and superscripts

g = steam
n = time level
o = oil
r = rock
s = steam
w = water

REFERENCES

1 - Shutler, N.C.: "Numerical, Three-Phase Model of the Two-Dimensional Steam-flood Process", Soc. Pet. Eng. J., (Dec, 1970) 405-417.
2 - Weinstein, H.G., Wheeler, J.A., Woods, E.G.: "Numerical Model for Thermal Processes", Soc. Pet. Eng. J, (Feb, 1977) 65-78.
3 - Coats, K.H, George, W.D., Chu, Chieh, Marcum, B.E.: "Three-Dimensional Simulation of Steamflooding", Soc. Pet. Eng. J. (Dec, 1974) 573-592.
4 - Ferrer, J. Farouq Ali, S.M.: "A Three-Phase, Two-Dimensional, Compositional Thermal Simulator for Steam Injection Precesses" - Paper 7613 presented at 27th Annual Technical Meeting of the Petroleum Society of CIM, Calgary, June 7-11 1976.
5 - Coats, K.H.: "Simulation of Steamflooding with Distillation and Solution Gas", Soc. Pet. Eng. J. (Oct. 1976) 235-247.
6 - Coats, K.H.: "A Highly Implicit Steamflood Model", Soc. Pet. Eng. J (Oct. 1978) 369-383.
7 - Grabowski, J.W., Vinsome, P.K., Lin, R.C., Behie, A. and Rubin, B.: "A fully Implicit General Purpose Finite-Difference Thermal Model for In-Situ Combustion and Steam", paper SPE 8396, presented at SPE 54th Annual Fall Meeting, Las Vegas, NV, Sept. 23-26, 1979.
8 - Stone, H.L.; "Estimation of Three-Phase Relative Permeability and Residual Oil Data", J. Can. Petr. Tech., V. 12, n° 4, (Oct. 1973).

394

9 - Nolen, J.S, Berry, D.W. : "Tests of the Stability and Time-Step Sensitivity of Semi-Implicit Reservoir Simulation Techniques", Soc. Pet. Eng. J (June 1972) 253-266.

10 - Spivak, A., Coats, K.H. : "Numerical Simulation of Coning Using Implicit Production Terms", Soc. Pet. Eng. J. (Sep. 1970) 257-267.

11 - Price, H.S., Coats, K.H. : "Direct Methods in Reservoir Simulation", Soc. Pet. Eng.J. (June 1974) 295-308.

12 - Sahuquet, B.C, Ferrier, J.J. : "Steam Drive Pilot in a Fractured Carbonated Reservoir Lacq Supérieur Field, "paper SPE 9453, presented at SPE 55th Annual Fall Meeting, Dallas, Texas, Sept. 21-24, 1980.

13 - Ramey, H.J, JR. : "Wellbore Heat Transmission", J. Pet. Tech. (April,1962) 427-435.

14 - Satter, A. :"Heat Losses During Flow of Steam Down a Wellbore", J. Pet. Tech. (July, 1965) 845-851.

15 - Chu, C., Trimble, A.E. :"Numerical Simulation of Steam Displacement Field Performance Applications", J. Pet. Tech. (June, 1975) 765-776.

16 - Bursell, C.G., Pittman, G.M. : "Performance of Steam Displacement in the Kern River Field", J. Pet. Tech. (August, 1975) 997-1004.

17 - Gomaa, E.E. :"Correlations for Predicting Oil Recovery by Steamflood", J. Pet. Tech. (Feb. 1980) 325-332.

18 - Nolan, J.B., Ehrlich, R., Crookston, R.B. :" Applicability of Steam-flooding for Carbonate Reservoirs", paper SPE 8821, presented at the First Joint SPE/DOE Symposium of Enhanced Oil Recovery, Tulsa, Oklahoma, April 20-23, 1980.

19 - Peaceman, D.W., :" Interprétation of Well-Block Pressures in Numerical Reservoir Simulation", Soc. Pet. Eng. J. (June, 1978) 183-194.

20 - Burger, J., Sourieau, P. :"Thermal Methods of Oil Recovery," Chapter 4. To be published by Editions Technip, Paris.

SPECIAL TECHNIQUES FOR FULLY-IMPLICIT SIMULATORS

J. R. APPLEYARD, I. M. CHESHIRE and R. K. POLLARD

Atomic Energy Research Establishment,
Harwell, Oxfordshire, England

ABSTRACT

This paper addresses some problems which arise when a fully-implicit black oil
simulator is allowed to take large time steps. It is shown that, by using a new
form of time averaged relative permeability, it is possible to reduce time
truncation errors to a very low level. The application of this technique also
reduces the non-linearities in the mass conservation equations which are solved
at each time step.

The solution of the linearised equations using iterative techniques becomes more
difficult as the time step is increased. A new 'nested factorisation' algorithm
for solution of these equations is described. The new method is shown to be more
efficient than existing techniques on a set of 2D test problems. Experience with
large 3D problems arising from North Sea applications has been most encouraging.

INTRODUCTION

The use of fully-implicit numerical methods in reservoir simulators is becoming
increasingly widespread[1,2,3]. This shift away from IMPES and semi-implicit
methods is motivated principally by the much greater stability and robustness of
fully-implicit methods when applied to problems involving strong gravity
segregation, high permeability contrasts, coning, bubble point crossing, etc.
As a direct result of this improved robustness, reservoir engineers are freed
from the need to consider the internal working of their simulator, and can
concentrate on more important issues. The wide applicability of fully-implicit
methods also reduces the need for special purpose simulators designed for
particular applications (e.g. coning).

It is often thought that fully-implicit simulators are less efficient in their
use of computer time than IMPES and semi-implicit alternatives. In our
experience[1], this need not be the case, as the strong stability of the method
allows the simulator to take much longer time steps than would otherwise be
possible. Indeed, for many problems, a fully-implicit simulator is the most
efficient, as well as the most robust alternative.

However, this gain in efficiency is realised only if the special problems
associated with long time steps can be overcome. Of these problems, the most
obvious is the increased numerical dispersion arising from time truncation
errors (as distinct from space truncation errors) resulting in additional smearing of flood fronts. The convergence of the non-linear equations which are
solved at each time step can also present difficulties for long time steps,
particularly if the relative permeability curves are highly non-linear. Finally,

solution of the linear equations using iterative techniques becomes more difficult as the time step is increased.

This paper addresses each of these problems in turn. Firstly, we show that it is possible to reduce time truncation errors significantly using a new technique for computing time averaged flows. The application of this technique also reduces the severity of non-linear convergence problems. Finally, we introduce a new and highly efficient technique for iterative solution of the linear equations, and present comparisons with other widely used methods.

TIME TRUNCATION ERRORS

The space discretised finite difference equations governing the flow of oil, water and gas can be summarised in the form

$$\frac{dM}{dt} = F \qquad (1)$$

where M and F are vectors. Elements of M represent the mass of a phase in a cell, and elements of F the sum of flows from neighbouring cells and wells. Integrating (1) over a time step Δt gives

$$\Delta M = \int_{t}^{t+\Delta t} F(t')dt' \qquad (2)$$

In the standard fully-implicit method the right-hand side of equation (2) is approximated by

$$\int_{t}^{t+\Delta t} F(t')dt' = F(t+\Delta t).\Delta t \qquad (3)$$

and the time discretised equivalent of the differential equation (1) is

$$\frac{\Delta M}{\Delta t} = F(t+\Delta t) \qquad (4)$$

Equation (4) is strongly stable which makes it possible to achieve high computing efficiency by taking large time steps. However, in practice, it is often necessary to limit the time step in order to prevent the growth of time truncation errors. An estimate of these errors can be obtained by comparing flow and well terms at the beginning and end of each time step.

$$E = F(t+\Delta t) - F(t) \qquad (5)$$

Using equation (5) it is possible to plot grid maps of the time truncation error and, as one might expect, the main time truncation errors occur in those cells where saturations are changing rapidly. This suggests that it may be possible to reduce time truncation errors significantly by performing a careful time integration in which special attention is given to the non-linear relative permeability terms.

Time Truncation Correction

The standard fully-implicit flow integral can be written as

$$\int_{t}^{t+\Delta t} F(t')dt' = \lambda(t+\Delta t)k_{r}(t+\Delta t)\Delta t \qquad (6)$$

where $k_r(t+\Delta t)$ is the value of the relative permeability at the end of the time step. To obtain a more accurate approximation we wish to replace the relative permeability by its time averaged value

$$\bar{k}_r = \frac{1}{\Delta t} \int_t^{t+\Delta t} k_r(t')dt' \qquad (7)$$

Relative permeabilities are functions of saturation, and it is therefore necessary to transform the time integration in (7) to an equivalent saturation integral. An approximate transformation can be obtained by assuming that
(a)   the flow is locally incompressible
(b)   capillary forces are negligible
(c)   the flow out of a cell depends only on the average saturation in the cell so that

$$\frac{ds}{dt} = \Theta(\delta - f(s)) \qquad (8)$$

where Θ is a constant, and $f(s)$ is the fractional flow curve for the phase under consideration. The constant, δ, is set to 1 for an invading phase and zero for a displaced phase to ensure that

$$\frac{ds}{dt} = 0 \qquad (9)$$

as the saturation approaches its limiting value. Equation (8) can now be used to transform the time integration in (7) to an equivalent saturation integral giving

$$\bar{k}_r \int_s^{s+\Delta s} \frac{ds}{\delta-f} = \int_s^{s+\Delta s} \frac{k_r ds}{\delta-f} \qquad (10)$$

Equation (10) forms the basis of a practical technique for computing average relative permeabilities during large time steps. If the relative permeability curves are approximated by piecewise linear functions, the integrals in equation (10) can be performed analytically, so that both $\bar{k}_r$ and $\frac{d\bar{k}_r}{ds}$ can be computed exactly at modest overall cost. Because the calculation takes detailed account of the shape of the relative permeability curve, rather than focusing attention at one or two point values, it is possible to take large time steps without losing accuracy.

It may be shown that the time averaged relative permeability is close to the time-centred value, $\frac{1}{2}(k_r(t+\Delta t) + k_r(t))$, if the saturation change is small, but approaches the implicit value, $k_r(t+\Delta t)$, as the saturation change increased.

The discretised equivalent of (1) may now be written as

$$\frac{\Delta M}{\Delta t} = \bar{F} \qquad (11)$$

where $\bar{F}$ is obtained by replacing relative permeabilities by their time averaged values, $\bar{k}_r$. All other terms are evaluated at the end of the time step.

The coupled non-linear equations (11) are solved for pressure and saturation in each grid block by Newtonian iteration:

$$-\left(\frac{1}{\Delta t}\frac{\partial M}{\partial x} - \frac{\partial \bar{F}}{\partial x}\right)\Delta x = \frac{\Delta M}{\Delta t} - \bar{F} \qquad (12)$$

where the solution variables (pressure and saturation changes) are represented by x. The Jacobian matrix, $\left[\dfrac{1}{\Delta t}\dfrac{\partial M}{\partial x} - \dfrac{\partial \bar{F}}{\partial x}\right]$, and residual vector, $\dfrac{\Delta M}{\Delta t} - \bar{F}$, are evaluated at the current best estimate of the solution, x.

In some cases, this iteration converges very slowly. For example, when studying the effects of water injection, engineers frequently use relative permeability curves for the injected phase which are set to zero below the Buckley-Leverett saturation. This technique helps to reduce numerical dispersion, but it also introduces a discontinuity into the fully-implicit equation (4), which makes the solution much more difficult to find. These problems are much reduced if time averaged relative permeabilities are used. This is because $\bar{k}_r$ is evaluated as an integral over the time step, and as such varies continuously in time, even if $k_r(t+\Delta t)$ does not. As a result there is no discontinuity in the time averaged equation (11).

Numerical Examples of Time Truncation Correction

The effect of the Time Truncation Correction (TTC) is illustrated using 1D, 2D and 3D examples.

The first is a standard Buckley-Leverett problem with equal oil and water viscosities. The fractional flow of water used is

$$f_w = \frac{k_{rw}}{k_{rw} + k_{ro}} = \frac{s^2}{s^2 + (1-s)^2} \quad \ldots\ldots\ldots\ldots\ldots\ldots \quad (13)$$

Figure 1 shows water saturation distributions after twelve cell pore volumes have been injected. All the results show considerable numerical dispersion due to space truncation errors, but our concern here is solely with time truncation errors. Results are displayed for one point upstream weighting using 32 time steps both with and without the time truncation error correction. A series of runs showed that the time truncation error is halved as the number of time steps is doubled and that the standard fully-implicit method eventually converges to the TTC result as the time step is refined. The convergence rate is shown in Table 1 for cell number 12.

TABLE 1

THE WATER SATURATION IN GRID BLOCK 12. RESULTS
OBTAINED USING SINGLE POINT UPSTREAM WEIGHTING

Number of time steps	16	32	64	128	256	512
Standard result	.659	.677	.687	.692	.695	.697
TTC result	.700	.698	.698	.698	.698	.698

Figure 1 also shows results using two point upstream weighting[4]. The effect of the time truncation error correction is similar to that observed with single point upstream weighting. The analytic Buckley-Leverett result is also shown for comparison. Finally, we note that if the relative permeability of water is set to zero below the Buckley-Leverett saturation $(\frac{1}{\sqrt{2}})$ then the fully-implicit two point upstream weighting result is virtually identical to the analytic result.

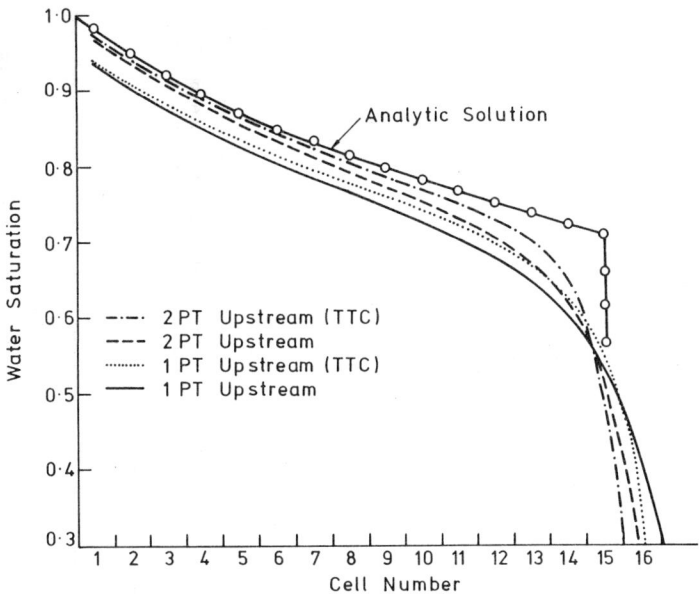

FIG.1.SATURATION PROFILES FOR ONE DIMENSIONAL TEST
OF TIME TRUNCATION CORRECTIONS

Problems 2 and 3 were run on PORES, a fully-implicit black oil simulator
described in reference 1. Time stepping in PORES is not controlled by maximum
permitted saturation and pressure changes as in most other simulators. Instead,
the estimate of time truncation error (equation 5) is converted to a local
material balance error by scaling with cell pore volumes and formation volume
factors. The time stepping algorithm attempts to keep the root mean square of
this local material balance error within specified bounds. Thus time stepping
is tied directly to the best available estimate of time truncation error.

Case 2 is a 38 x 8 cross section with a water injector in column 1 and an oil
producer in column 38. Both wells are completed in layers 1, 3-5 and 7-8. The
second layer is inactive and the reservoir is therefore in two sections which
communicate only through the wells. Further details of this problem are given
in reference 1.

Figure 2 shows the water cut as a function of time for Case 2 using
(a)   PORES default TTE controls ($\cong$ ΔSmax = 0.3)
(b)   PORES default TTE controls/100 ($\cong$ ΔSmax = 0.05)
(c)   PORES default TTE controls with time averaged relative permeabilities.

The results indicate that the time averaging technique virtually eliminates time
truncation errors.

Case 3 is a 3 dimensional gas/oil problem described by Odeh[5]. The results
shown in figure 3 again illustrate that time truncation errors are dramatically
reduced using time averaged relative permeabilities.

400

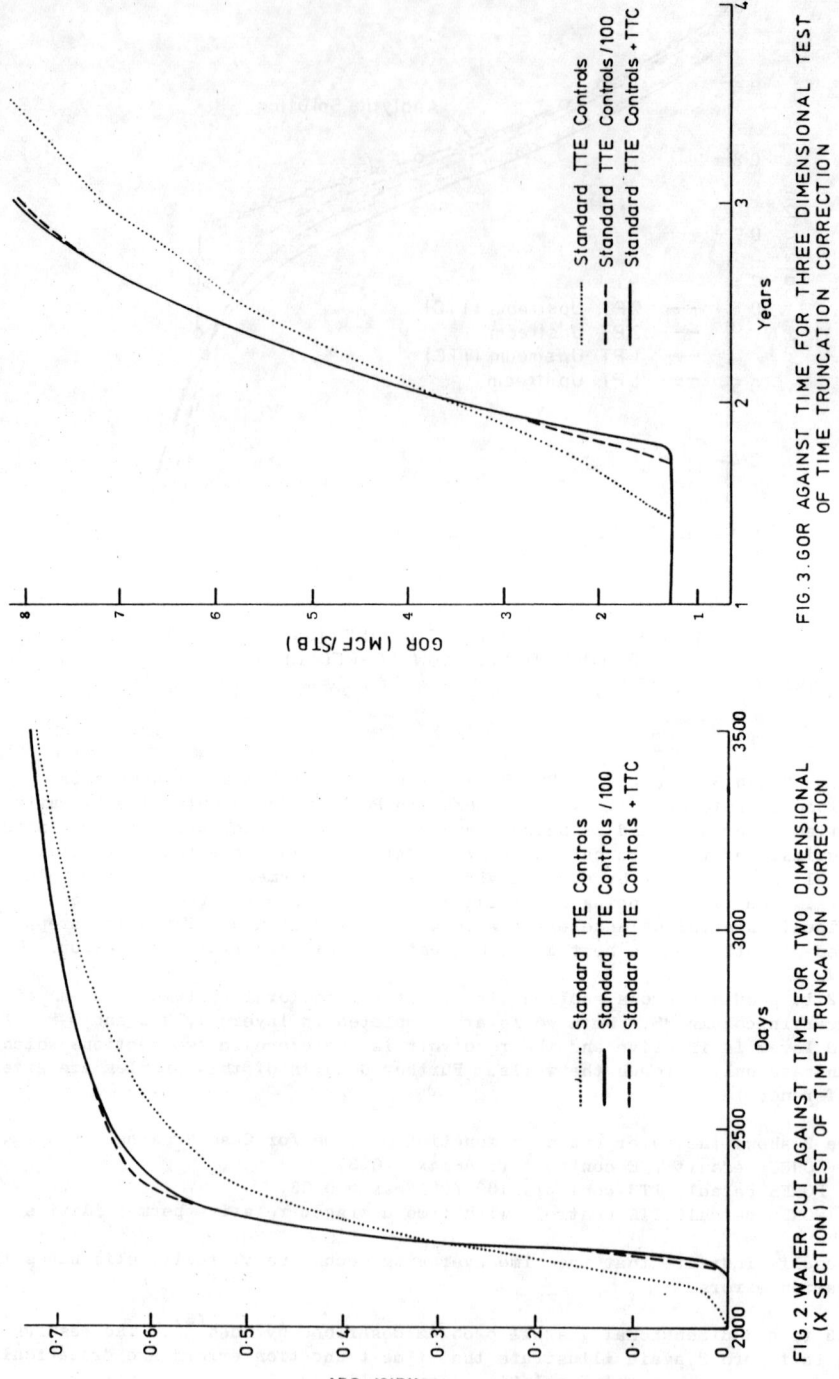

FIG.3. GOR AGAINST TIME FOR THREE DIMENSIONAL TEST
OF TIME TRUNCATION CORRECTION

.... Standard TTE Controls
– – – Standard TTE Controls /100
——— Standard TTE Controls + TTC

FIG.2. WATER CUT AGAINST TIME FOR TWO DIMENSIONAL
(X SECTION) TEST OF TIME TRUNCATION CORRECTION

.... Standard TTE Controls
——— Standard TTE Controls /100
– – – Standard TTE Controls + TTC

LARGE TIME STEPS AND THE SOLUTION OF THE LINEAR EQUATIONS

At each Newton iteration, it is necessary to solve the linearised equations (12), to obtain an updated estimate of the solution to the non-linear equations (11). For small problems, these equations may be solved by Gaussian elimination. However this is not practicable for large 3 dimensional studies, as storage and computing time increase very rapidly with the problem size. In such cases, some form of iterative solution method must be used. The efficiency and robustness of the resulting procedure depend critically on the effectiveness of the solution method adopted.

This observation is particularly true of fully-implicit simulations with long time steps, as the mass accumulation term in the Jacobian matrix (equation (12)) which makes it diagonally dominant (and therefore non-singular) is inversely proportional to the time step length, Δt. As a result, the linear equations are more difficult to solve if long time steps are taken. This effect is illustrated in Table 2, which shows how the number of linear iterations required for each Newton iteration increases with the time step size, for a fairly eventful period in a typical simulation.

TABLE 2

EFFECT OF TIME STEP SIZE ON CONVERGENCE

Time Steps for Total Simulation Period	Total Number of Newton Iterations	Total Number of Linear Iterations	Linear Iterations Per Newton Iteration
12	67	366	5.46
18	55	237	4.31
23	64	226	3.53
48	116	273	2.35

Because the linear equations are more difficult to solve with large time steps it is important to devise powerful iterative methods to make fully-implicit codes efficient. In the following section we describe the method used in PORES.

Nested Factorisation

All iterative methods for the solution of the linear equations

$$Ax = b \qquad \ldots\ldots\ldots\ldots\ldots\ldots\ldots\ldots\ldots\ldots\ldots (14)$$

depend on the existence of an approximation, B, to the coefficient matrix A such that $B^{-1}\phi$ is easily calculated for any vector ϕ. The rate of convergence of the iteration depends primarily on how well B approximates to A.

The best choice of B will, in general, depend on the structure of A. Five point finite difference methods give rise to the nested block tridiagonal structure shown in Fig. 4.

$$A = d + \ell_1 + u_1 + \ell_2 + u_2 + \ell_3 + u_3 \qquad \ldots\ldots\ldots\ldots\ldots (15)$$

In this paper, we present a way of approximating such a matrix by nested factorisation. Because the algorithm exploits the structure of the matrix to the full, it is not easily adapted to deal with general sparse matrices.

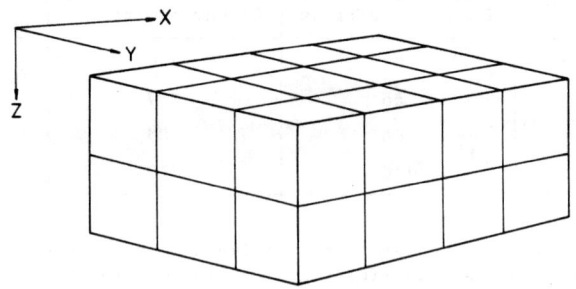

l_1, u_1 Bands connect Cells in X Direction
l_2, u_2 Bands connect Cells in Y Direction
l_3, u_3 Bands connect Cells in Z Direction

FIG.4.THE STRUCTURE OF THE COEFFICIENT MATRIX FOR A
FIVE POINT FINITE DIFFERENCE SIMULATION

The nested factorisation approximation for a 3D system may be summarised as

$$B = (\alpha + \ell_3)\alpha^{-1}(\alpha + u_3) \quad \text{..................................} \quad (16)$$

$$\alpha = (\beta + \ell_2)\beta^{-1}(\beta + u_2) \quad \text{................................} \quad (17)$$

$$\beta = (\gamma + \ell_1)\gamma^{-1}(\gamma + u_1) \quad \text{................................} \quad (18)$$

$$\gamma = d - \ell_1\gamma^{-1}u_1 - \text{colsum}(\ell_2\beta^{-1}u_2 + \ell_3\alpha^{-1}u_3) \quad \text{..........} \quad (19)$$

Here, colsum(X) is the diagonal matrix formed by summing the elements of X in columns.

By combining equations 16-19, we obtain the following expression for B

$$B = A + (\ell_2 \beta^{-1} u_2 + \ell_3 \alpha^{-1} u_3) - colsum(\ell_2 \beta^{-1} u_2 + \ell_3 \alpha^{-1} u_3) \quad \ldots\ldots \text{ (20)}$$

For two dimensional systems, $\ell_3 = u_3 = 0$, and the outermost layer of the factorisation (equation (16)) may be omitted, leaving

$$B = (\beta + \ell_2)\beta^{-1}(\beta + u_2)$$

$$= A + \ell_2 \beta^{-1} u_2 - colsum(\ell_2 \beta^{-1} u_2) \quad \ldots\ldots\ldots \text{ (21)}$$

The method reduces to an exact Cholesky decomposition for one dimensional problems

$$B = (\gamma + \ell_1)\gamma^{-1}(\gamma + u_1)$$

$$= A \quad \ldots\ldots\ldots\ldots\ldots\ldots\ldots\ldots\ldots\ldots\ldots\ldots \text{ (22)}$$

In PORES, this approximation is used as a preconditioning matrix for a truncated conjugate gradient algorithm of the type described by Vinsome[6]. For applications which give rise to a symmetric coefficient matrix, it would be more appropriate to use it as preconditioning for a symmetric conjugate gradient algorithm as described by Meijerink and Van der Vorst[7].

The procedure for evaluating $B^{-1}\Phi$ is hierarchical. At the outermost level, we solve block triangular matrix equations of the form

$$(\alpha + \ell_3)\mu = \Phi \quad \ldots\ldots\ldots\ldots\ldots\ldots\ldots\ldots \text{ (23)}$$

Because the matrix α is block diagonal, these equations can be solved a plane at a time, using

$$\mu = \alpha^{-1}(\Phi - \ell_3\mu) \quad \ldots\ldots\ldots\ldots\ldots\ldots\ldots \text{ (24)}$$

This is not recursive, as $\ell_3\mu$ involves the solution from the previous plane.

Within each plane, the equations are solved a line at a time using

$$\mu = \beta^{-1}(\Phi - \ell_2\mu) \quad \ldots\ldots\ldots\ldots\ldots\ldots\ldots \text{ (25)}$$

where once again, only known solution variables appear on the right-hand side.

Similar considerations apply to the evaluation of γ, which must be done before the iteration begins. The calculation proceeds a plane at a time and, within each plane, a line at a time. At each stage, the calculation involves elements of γ from the previous cell in the current line, the previous line in the current plane, and from the previous plane. The evaluation of $colsum(\ell_2\beta^{-1}u_2 + \ell_3\alpha^{-1}u_3)$ which is required in the calculation of γ, is achieved by multiplying a column vector whose elements are all unity by the transpose of $(\ell_2\beta^{-1}u_2 + \ell_3\alpha^{-1}u_3)$. The result vector contains the diagonal elements of $colsum(\ell_2\beta^{-1}u_2 + \ell_3\alpha^{-1}u_3)$.

It will be noted from equation (20) that γ has been set in such a way that the column sum of the error matrix, B-A, is zero. The object of this choice of γ is to force the sum of the residuals, which in a reservoir simulator corresponds to a total material balance error, to zero. If the iteration is started from the initial solution

$$x_o = B^{-1}b \quad \dots\dots\dots\dots\dots\dots\dots\dots\dots \quad (26)$$

then the initial residual r_o is given by

$$r_o = b - Ax_o$$

$$= (B - A)B^{-1}b \quad \dots\dots\dots\dots\dots\dots\dots \quad (27)$$

and the sum of the components of r_o is zero if colsum$(B - A) = 0$. It is easily shown that this condition, once established, remains valid throughout the iteration[1]. Moreover, it may be seen that because the error matrix is block diagonal, the residuals sum to zero independently within each plane (or, for 2D systems, within each line) and that this condition also holds throughout the iteration.

It is well known that in most iterative methods for solving linear equations, it is low frequency eigenvectors which are most persistent, and which ultimately determine the rate of convergence. Indeed, it was this observation which prompted the additive correction methods of Watts[8] and of Settari and Aziz[9], and which led to the recent upsurge of interest in other multi-grid methods[10]. The method described above avoids the worst of these problems by eliminating the lowest frequency eigenvectors (those with a non-zero residual sum) from the outset. Algorithms displaying similar properties have been described by Gustafsson[11] and Cheshire et al[1] although only the latter noted the importance of starting the iteration from an initial solution with zero residual sum (equation 26). The fact that residuals sum to zero within each plane may also be used as a test on the correctness of the program.

An important consideration in the implementation of this algorithm is the orientation of lines (cells connected by the ℓ_1 and u_1 bands) and planes (lines connected by the ℓ_2 and u_2 bands). The best strategy, deduced from a series of numerical experiments, appears to be to align the axes so that the largest off-diagonal elements are on the ℓ_1 and u_1 bands, the next largest on the ℓ_2 and u_2 bands, and the smallest on the ℓ_3 and u_3 bands. In the PORES implementation, this choice is made automatically.

Whilst it is convenient when describing the algorithm to view it as a series of nested LDU factorisations (equations 16-18) a significantly more efficient implementation is achieved by combining the D and U factors.

$$B = (\alpha + \ell_3)(I + \alpha^{-1}u_3) \quad \dots\dots\dots\dots\dots\dots \quad (28)$$

where

$$\alpha = (\beta + \ell_2)(I + \beta^{-1}u_2) \quad \dots\dots\dots\dots\dots\dots \quad (29)$$

and

$$\beta = (\gamma + \ell_1)(I + \gamma^{-1}u_1) \quad \dots\dots\dots\dots\dots\dots \quad (30)$$

Given this form of the algorithm, the evaluation of $B^{-1}\Phi$ for an arbitrary n-vector Φ requires about 22n floating multiplications for 3D problems (10n for 2D problems). The total for each preconditioned conjugate gradient iteration is 33n(19n) if A is symmetric, or somewhat more for a Vinsome type truncated conjugate gradient algorithm. The figure for 2D problems can be reduced to 16n by combining the calculation of $B^{-1}\Phi$ with that of $AB^{-1}\Phi$.

An important feature of the nested factorisation algorithm is its very modest storage requirements. Evaluation of $B^{-1}\Phi$ requires storage for only one diagonal matrix (in general γ^{-1}) in addition to the elements of A. Some storage is also required by the conjugate gradient algorithm.

Numerical Tests of the Nested Factorisation Algorithm

The nested factorisation algorithm was tested on a series of single phase 2D test problems described by Settari and Aziz[9]. Because these problems give rise to symmetric coefficient matrices, we have used a symmetric conjugate gradient algorithm to accelerate convergence. The results are shown in Table 3, together with results obtained on the same problem using SIP[12], ICCGO[7], a variant of ICCGO in which the column sum of the error matrix is forced to zero[1], and ICCG3[7]. In each case, we have shown the computational work required to reduce the largest normalised residual to 10^{-6}. The unit of work is taken as a SIP iteration (about 22n floating point multiplications). The corresponding figures for the other methods are 16n for ICCGO and ICCGO with colsum, 22n for ICCG3 and 19n for nested factorisation, (it would be possible to reduce this figure to 16n on these problems).

TABLE 3

COMPARISON OF NESTED FACTORISATION
WITH OTHER ITERATIVE METHODS

Problem Number	COMPUTATIONAL WORK (SIP ITERATIONS)				
	SIP*	ICCGO	ICCGO (WITH COLSUM)	ICCG3	NESTED FACTORISATION
1	28	32	19	20	13
2	20	26	20	29	6
3	38	36	22	22	13
4	28	31	20	20	16
5	>50	34	21	21	12
6	50	27	17	15	11

* These results are abstracted from reference 9.

The results show the nested factorisation method to be fastest of the methods tested on all the problems. Perhaps the most significant result is that on problem number 2, which is the only one which exhibits the strong directionality characteristic of cross section and 3D reservoir simulations. On this problem the nested factorisation algorithm is fastest by a factor of three.

The importance of forcing the sum of the residuals to zero by choosing B in such a way that colsum(B−A) is zero is shown by the results from ICCGO and ICCGO (with colsum). The latter converges significantly faster in every case, and competes effectively with ICCG3 which is more complex and requires more storage.

The 2D results discussed above correspond only to a single nesting (B = α) and do not demonstrate the full potential of the technique arising from the second nesting for 3D problems. The algorithm is implemented in PORES for 1, 2 and 3 phase simultaneous problems and our experience to date on large complex 3D simulations arising from practical North Sea studies has been very encouraging.

CONCLUSIONS

1. A new technique has been developed to control time truncation errors in reservoir simulation.

2. The method can be incorporated into existing simulators with relative ease and has been demonstrated on one, two and three dimensional test problems.

3. It is shown that, as larger time steps are taken, the linearised equations become more difficult to solve by iterative methods.

4. A new nested factorisation method is described for the solution of the linearised finite difference equations.

5. Because the nested factorisation method is highly recursive it makes minimal additional demands on computer storage.

6. By comparison with existing published results it is shown that the nested factorisation method is highly efficient for simple two dimensional problems.

7. The nested factorisation method has been incorporated in PORES and our experience to date indicates that the full power of the method is most apparent on large difficult three dimensional studies arising in North Sea applications.

NOMENCLATURE

A = Jacobian matrix arising in finite difference calculations

B = an approximation to the Jacobian matrix A

Colsum = diagonal matrix formed by summing the elements of a matrix in columns

b = right-hand side of the linear equation

d = diagonal elements of A, each element of d is a 3x3 matrix in simultaneous 3 phase simulations

E = estimate of time truncation error vector

F = flow vector, each element represents the sum of flows into a cell from neighbouring cells and wells

$\bar{F}$ = flow vector obtained by replacing all relative permeabilities at the end of a time step by time averaged relative permeabilities

f = fractional flow of oil, water or gas

f_w = fractional flow of water

k_r = relative permeability (oil, water or gas)

$\bar{k}_r$ = average relative permeability over a time step

k_{rw} = relative permeability of water

k_{ro} = relative permeability of oil

ℓ = lower band of the Jacobian matrix

M = mass accumulation vector, each element of M represents the mass of a phase in a cell

ΔM = change in M over a timestep Δt

n = number of elements in d

r = residual vector

s   = saturation

Δs   = change in saturation during a time step

t   = time

Δt   = duration of a simulation time step

TTC   = time truncation correction

u = upper band of the Jacobian matrix

x = solution of a linear equation, state vector

Δx = change of state vector during a Newton iteration

α = matrix for two dimensional systems

β = matrix for one dimensional systems

γ = diagonal matrix

δ = 0 or 1 corresponding to an invading or displacing phase

Θ = constant

λ = pressure dependent component of the flow vector

Φ = right-hand side of a linear equation

μ = solution of a linear equation

ACKNOWLEDGEMENTS

The authors wish to thank the United Kingdom Department of Energy, the British National Oil Corporation and the British Gas Corporation for permission to publish this work.

REFERENCES

1. CHESHIRE, I.M., APPLEYARD, J.R., BANKS, D., CROZIER, R.J., and HOLMES, J.A.; "An Efficient Fully Implicit Simulator", paper EUR 179 presented at the European Offshore Petroleum Conference and Exhibition, London, England, (Oct. 1980), 325-336.

2. BANSAL, P.P., HARPER, J.L., McDONALD, A.E., MORELAND, E.E. and ODEH, A.S.; "A Strongly Coupled, Fully Implicit, Three Dimensional, Three Phase Reservoir Simulator", paper SPE 8329 presented at the SPE-AIME 54th Annual Fall Meeting of the SPE, Las Vegas, Nev. (Sept. 1979).

3. AU, A.D.K., BEHIE, A., RUBIN, B., and VINSOME, P.K.W.; "Techniques for Fully Implicit Reservoir Simulation", paper SPE 9302 presented at the SPE-AIME 55th Annual Fall Meeting of the SPE, Dallas, Texas (Sept. 1980).

4. TODD, M.R., O'DELL, P.M. and HIRASAKI, G.J.; "Methods for Increasing Accuracy in Numerical Reservoir Simulators", Soc.Pet.Eng.J. (Dec. 1972), 515-530.

5. ODEH, A.S.; "Comparison of Solutions to a Three-Dimensional Black-Oil Reservoir Simulation Problem", J.Pet.Tech. (Jan. 1981) 33, 13-25.

6. VINSOME, P.K.W.; "Orthomin, an Iterative Method for Solving Sparse Banded Sets of Simultaneous Linear Equations", paper SPE 5729 presented at the SPE-AIME Fourth Symposium on Numerical Simulation of Reservoir Performance, Los Angeles, Ca. (Feb. 1976).

7. MEIJERINK, J.A. and VAN DER VORST, H.A.; "An Iterative Solution Method for Linear Systems of which the Coefficient Matrix is a Symmetric M-Matrix",Mathematics of Computation (Jan. 1977) 31, 148-162.

8. WATTS, J.W.; "An Iterative Matrix Solution Method Suitable for Anisotropic Problems", Soc.Pet.Eng.J. (March 1971) 11, 47-51.

9. SETTARI, A. and AZIZ, K.; "A Generalisation of the Additive Correction Methods for the Iterative Solution of Matrix Equations", SIAM J.Numer.Anal. (June 1973) 10, 506-521.

10. BRANDT, A.; "Multi-Level Adaptive Solutions to Boundary-Value Problems", Mathematics of Computation (April 1977) 31, 333-390.

11. GUSTAFSSON, I.; "A Class of First Order Factorisation Methods", BIT (April 1978) 18, 142-156.

12. STONE, H.L.; "Iterative Solution of Implicit Approximations of Multidimensional Partial Differential Equations", SIAM J.Numer.Anal. (Sept. 1968) 5, 530-558.

SOME CONSIDERATIONS CONCERNING THE EFFICIENCY OF CHEMICAL FLOOD SIMULATIONS

R. W. S. FOULSER

AEE Winfrith, Dorchester, Dorset, DT2 8DH

ABSTRACT

This paper discusses some aspects of improving computational efficiency and accuracy in surfactant flood calculations. The use of curvilinear grids in chemical flooding simulation can reduce grid orientation effects and speed up the calculations. For the low tension, low concentration surfactant processes being considered, the work of Martin and Wagner supports the application of conformal grids. The development at Winfrith of the PASL code, which applies the results of potential theory, has enabled suitable grid patterns to be generated for use in the CFTE code. This mesh generation code has also proved useful in visualising flow patterns and consequent mesh requirements for studying possible pilot applications of surfactant flood under real reservoir conditions.

A further improvement in the efficiency with which the CFTE code can be applied has been achieved by incorporating a line successive overrelaxation iterative solver (LSOR) into the code, as an alternative to the direct inversion method. Initially written for a scalar machine, its application on the CRAY vector machine has led to the development of alternative LSOR codes using different grid block orderings. These LSOR options run about twice as quickly as the direct inversion on the IBM scalar machine and, when vectorised on the CRAY, they run up to 65 times faster than the original code on the IBM. Thus the vectorised LSOR approach has proved to be very powerful.

INTRODUCTION

The Chemical Flood Ternary Equilibrium Simulator (CFTE Ref 1) is one of several EOR codes being used at AEE Winfrith. During the period for which CFTE has been available, a number of assessments studies have been performed, one of which has been reported elsewhere by Fayers et al (Ref 2). As more difficult assessment calculations have been undertaken, it has become necessary to develop supplementary programs to aid the generation of input for the studies. Computational time is also becoming a significant factor in the use of the CFTE code in large assessment calculations. This paper briefly presents some of the experience associated with the use of a special mesh generation code, PASL, that has been written as one of the support programs, and also the steps that have been taken to improve the efficiency of CFTE when used on the CRAY vector processor.

The PASL mesh generation code has been used as a tool in preliminary studies to determine broad characteristics of flow patterns within possible pilot areas of an oil field. This has been especially useful in examining alternative well patterns in a pilot flood and also as a means of studying pattern confinement strategies. The preliminary flood pattern studies provide a basis for identifying computing mesh requirements in subsequent CFTE calculations. The PASL code may also be used to generate curvilinear co-ordinate systems for use in CFTE. The use of curvilinear co-ordinates in surfactant flood calculations allows a considerable improvement to be obtained in calculation efficiency as well as a reduction in the errors caused by grid orientation effects. These improvements are quantified by comparing the results using parallel and diagonal Cartesian grids with curvilinear grids.

The CFTE code uses the IMPES formulation, and the resulting implicit pressure matrix problem was originally solved by direct methods. With about 400 grid blocks in three-dimensional calculations, the matrix inversion time begins to dominate the total execution time. Because of this, it was decided to incorporate an iterative solver into the code. The recent availability of a CRAY-1 vector processor in association with the IBM 3033 at Harwell, where the computing for this project is undertaken, emphasised the need to choose a method which could be readily vectorised. The LSOR method was chosen, partly because of its robustness and its relatively simple form for programming, but also because this iterative method is more readily vectorised than some of the more complicated strategies, such as SIP. The second part of this paper is concerned with an investigation of the performance of the LSOR method in chemical flooding calculations and also with the aspects of vectorisation of this particular method.

THE DETERMINATION AND USE OF CURVILINEAR GRIDS IN SURFACTANT FLOODING STUDIES

The generation of curvilinear grids and the study of flow fields for arbitrary arrangements of injection and production wells has been implemented in the PASL code. No-flow boundaries associated with linear fault lines have also been included in the formulation which uses the basic results of potential theory. The faults are dealt with by applying conformal transformations to obtain a continuous line fault and the method of images is then used to provide an equivalent system of wells in an infinite domain. The solution method then uses the classical equations for representing a distribution of wells in infinite space (Ref 3). For a system of wells the velocity potential and stream function are given by:

$$\Phi(x,y) = -\sum_i [\frac{m_i}{2} \log \{(x - x_i)^2 + (y - y_i)^2\}] \tag{1}$$

$$\Psi(x,y) = -\sum_i m_i \tan^{-1} \frac{y - y_i}{x - x_i} \tag{2}$$

where m_i is the strength of well i. Once the streamlines and equipotentials have been computed, inverse conformal transformations are used to map the results back to the original co-ordinate system. The code provides output of curvilinear mesh dimensions in a format suitable for direct input to CFTE.

One of the first examples studied using a grid generated by PASL, was concerned with the problem of mesh orientation effects during the calculation of a simple surfactant flood in a 5-spot geometry with a uniform residual oil saturation. The CFTE code was used to model a low concentration surfactant flood in a 1/8 th symmetry sector of the 5-spot pattern. Some investigations of this type of surfactant flood for North Sea applications have been discussed in Ref 2. The basic data adopted are similar to those described in that reference. A variation of relative permeability with capillary number similar to that used by Todd and Chase (Ref 1) has been assumed where, in this case, the relative permeability of the oil and water phases move progressively from their initial oil-water forms at low capillary numbers ($\sim 6 \times 10^{-6}$), to their limiting straight line forms as the capillary number increases to 2×10^{-3}.

The results of the calculations are shown in Figure 1 where the 5-spot region has been covered by 90 grid blocks for the parallel and curvilinear grid systems and 91 blocks for the diagonal grid system. These diagrams show the oil distribution after 0.5 PV of surfactant solution injection which is not long after breakthrough of the oil bank. All three calculations show low concentrations around the injection well with a fairly steep rise in oil saturation into the oil bank where the oil saturation reaches about 40%. All three calculations also show the oil bank pinching off a small region where the saturation has not yet risen above that left after waterflooding. As would be expected, the parallel grid shows a preferential flow along the direct path between the wells, while the diagonal grid allows the less accessible corner region to be swept more easily. The prediction of the curvilinear grid simulation lies between these two rectangular grid results. The cumulative oil production curves, not shown, also indicates that the curvilinear grid gives results intermediate to the Cartesian mesh arrangements.

From a computational point of view, the curvilinear calculation was by far the most efficient, running some 5 times faster than the others. This occurred because the connections with the wells in the curvilinear calculations was via 5 fairly large grid blocks, rather than via the single smaller grid block in the Cartesian systems. It is these connecting grid blocks which tend to control the timestep size that can be taken in the explicit solution of the concentration and saturation equations.

In the particular example studied the water in front of the oil bank has a mobility of 1.25 mD/cp, the total mobility ($\frac{Kk_{ro}}{\mu_o} + \frac{Kk_{rw}}{\mu_w}$) in the oil bank is 1.05, and behind the bank the mobility of the surfactant solution is somewhat greater than 5. Thus the equally mobile two phase mixture in the oil bank and the watered out zone ahead are being displaced by a much more mobile fluid. This is, therefore, an example of the unfavourable mobility ratio effects in surfactant flooding. Martin and Wagner (Ref 4) suggest that this type of problem is amenable to fixed stream tube methods, thus supporting use of curvilinear two-dimensional grids in CFTE calculations with this type of system.

The suitability of curvilinear grid methods for more general application is also being studied. Figure 2 shows flow patterns generated from the PASL code, when applied to an inverted 5-spot operated in isolation, but in the vicinity of two intersecting impermeable faults. Note that the faults are streamlines of the flow pattern and that the equipotentials cross the steamlines perpendicularly. The intersection of streamlines and equipotentials have been used in CFTE calculations as the grid block corners of a conformal curvilinear co-ordinate system. For maximum efficiency in the CFTE code the equipotentials are usually chosen to produce equal volume grid blocks along the shortest stream tube. The shaded region in the Western sector of this figure has been used in this way to generate co-ordinate systems for the problem on which CFTE performance estimates are reported in the second half of this paper. This is an example of a field problem for which selection of a satisfactory Cartesian mesh would have been very difficult. At this stage, it has only been possible to study problems for specially chosen sectors, because of limitations in the connectivity of mesh which can be handled by the CFTE code. Generalisation of the mesh connectivity arrangements will be studied in the future in connection with the LSOR method discussed in the next section.

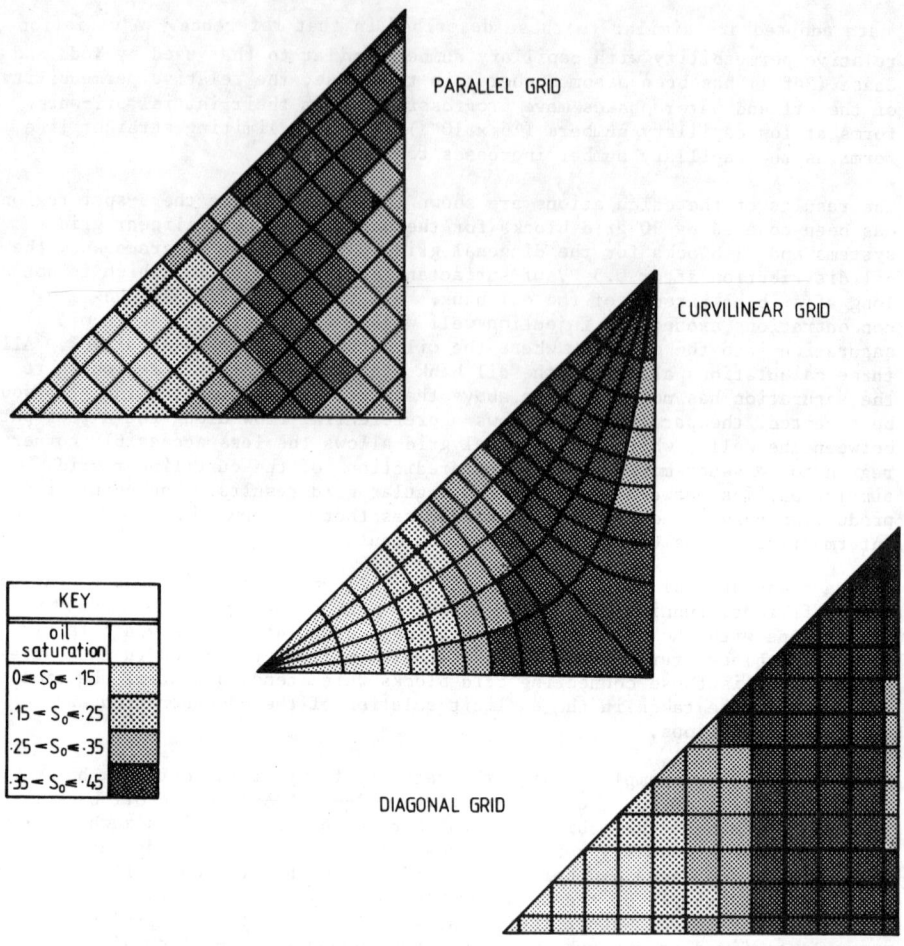

PARALLEL GRID

CURVILINEAR GRID

KEY	
oil saturation	
$0 \leqslant S_o \leqslant \cdot 15$	
$\cdot 15 < S_o \leqslant \cdot 25$	
$\cdot 25 < S_o \leqslant \cdot 35$	
$\cdot 35 < S_o \leqslant \cdot 45$	

DIAGONAL GRID

FIG.1 COMPARISON OF LOW TENSION SURFACTANT FLOOD COMPUTED OIL
DISTRIBUTIONS AFTER 0·5 PV OF FLUID INJECTION.

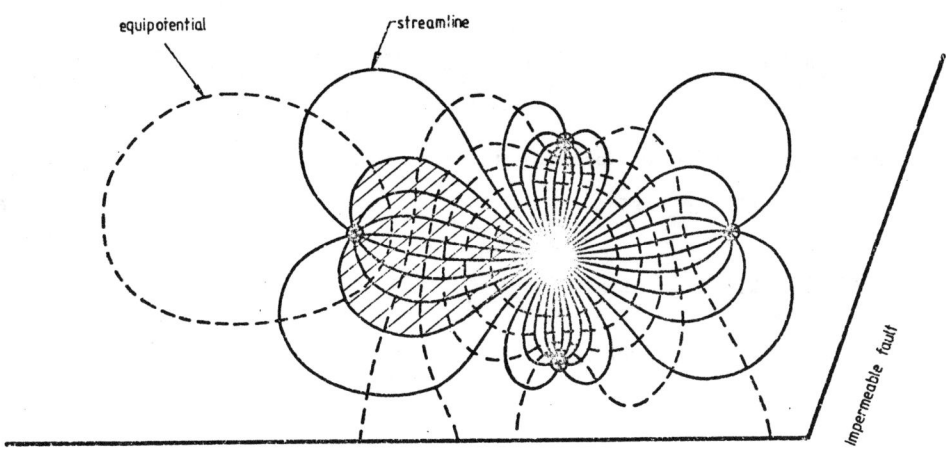

equipotential

streamline

Impermeable fault

impermeable fault

FIG. 2   EXAMPLE OF APPLICATION OF POTENTIAL FLOW THEORY TO FLOW PATTERN
VISUALISATION IN AN INVERTED 5-SPOT CONFINED BY INTERSECTING FAULTS.

THE PERFORMANCE OF SCALAR AND VECTOR VERSIONS OF THE LSOR METHOD IN CHEMICAL
FLOOD CALCULATIONS

In view of the need to speed up the inversion process for large problems with
curvilinear meshes, and the practical difficulties imposed by generalised
connectivities between meshes introduced by such schemes, it was decided to
investigate the application of the LSOR iterative procedures as an alternative to
the existing direct inversion method in the CFTE code. The robustness of the
convergence behaviour of LSOR, its ease of programming, and its simplicity
for vectorisation on a CRAY were further reasons for choosing this method.

Outline of LSOR

In block successive over relaxation the system of equations to be solved can
be written in the form:

$$\sum_i A_{ki} \, \underline{x}_i = r_k \qquad\qquad k = 1, \ldots, K \qquad\qquad (3)$$

where $\underline{x}_i$ is an n-tuple of variables (the block size) and A_{ki} and r_k are the
corresponding sub-matrix of co-efficients and the n-tuple right-hand side.
In line successive over relaxation, the blocks are chosen to correspond to
lines of grid blocks. An iterative sequence of vectors $\underline{x}_k^n$ may then be defined

by:

$$\underline{x}_k^{n+1} = (1-\beta)\, \underline{x}_k^n - \beta \left[\sum_{i=1}^{k-1} A_{kk}^{-1}\, A_{ki}\, \underline{x}_i^{n+1} + \sum_{i=k+1}^{n} A_{kk}^{-1}\, A_{ki}\, \underline{x}_i^n - A_{kk}^{-1}\, \underline{r}_k \right] \quad (4)$$

where β is a relaxation factor. This may be written in matrix form:

$$\underline{x}^{n+1} = E^{-1}F\, \underline{x}^n + E^{-1}A^{-1}\, \underline{r} \quad (5)$$

where $E = (E_{ki}) = \begin{cases} I_{kk} & \text{if } k=i \\ \beta A_{kk}^{-1}A_{ki} & \text{if } i<k \\ 0 & \text{if } i>k \end{cases}$

$F = (F_{ki}) = \begin{cases} (1-\beta)I_{kk} & \text{if } k=i \\ 0 & \text{if } i<k \\ -\beta A_{kk}^{-1}A_{ki} & \text{if } i>k \end{cases}$

I_{kk} is the block identity matrix.

The convergence of the iterative scheme described by equation (5) depends on the spectral radius ω of the iteration matrix $E^{-1}F$ (Ref 5). For a seven point approximation to the three-dimensional Laplacian operator it can be shown that, provided the blocks A_{ki} are ordered consistently, then ω is related to the spectral radius μ of the matrix $(A_{kk}^{-1})A-I$ by:

$$\mu = \frac{\beta - 1 + \omega}{\beta}\cdot\omega^{-\frac{1}{2}} \quad (6)$$

Since the matrix $(A_{kk}^{-1})A-I$ is dependent only on the original matrix and on the block selection (ie the direction of the lines in LSOR), μ depends only on the block arrangement of the matrix and on the relaxation factor β. As in point successive relaxation the optimal relaxation factor $\bar{\beta}$ can be shown to be given by:

$$\bar{\beta} = \frac{2}{1+(1-\mu^2)^{\frac{1}{2}}} \quad (7)$$

By putting $\beta=1$ in equation (6), it may be seen that $\mu^2 = \omega_1$ so that equation (7) may be rewritten as:

$$\bar{\beta} = \frac{2}{1+(1-\omega_1)^{\frac{1}{2}}} \quad (8)$$

This shows that the optimum relaxation factor can be determined by powering the iteration matrix $E^{-1}F$ assuming $\beta = 1$. ω_1 is the asymptotic ratio of successive iterates of the eigenvector elements (p287 Ref 6). The spectral radius of the iteration matrix with the optimal relaxation factor can be shown to be $\bar{\omega} = \bar{\beta}-1$ and this gives a value for the asymptotic rate of convergence of:

$$R_\infty = -\log(\bar{\omega}) \quad (9)$$

The essential points of LSOR are thus:

(i) To choose the direction of the lines (ie the A_{kk}) so as to maximise the rate of convergence given by equation (9). In reservoir situations this almost invariably means choosing the vertical direction for the lines.

(ii) To choose an order for revising the x_k that is consistent. A necessary and sufficient condition for an ordering to be consistent is given on page 245 of Ref 7.

(iii) To power the matrix $E^{-1}F$ with $\beta=1$ so as to determine ω and then use equation (8) to determine the optimum relaxation factor.

(iv) To use equation (5) to iterate on the unknown vector $\underline{x}^{n+1}$ until convergence is achieved.

The convergence test for the pressure distribution that has been used in this application is:

$$\max_{i}\ (|x_i^{n+1} - x_i^n|) < (1-\bar{\omega})\varepsilon\ (\max_{i}\ x_i^{n+1} - \min_{i}\ x_i^{n+1}) \tag{10}$$

and ε has been taken to be 0.001. This test ensures that successive iterates must be more tightly converged when the convergence is slow, ie ω is close to unity. Checks of the material balance in chemical flooding calculations have shown this to be an adequate test.

Scalar Code

Either the x, y or z direction may be chosen for the line direction in all three variants of the LSOR code discussed in this paper. However, for convenience it is referred to as the vertical direction. In the scalar version of the code, the lines of grid blocks are ordered in a natural order in the horizontal plane, starting in one corner of the area modelled and working up to the opposite corner. An ordering of a 7 x 7 arrangement of grid block lines is shown in Figure 3. This ordering is consistent. Although the grid block arrangement is displayed as a square, this in fact represents the curvilinear co-ordinate system for the curvilinear structure shown in Figure 2, where the wells are connected to the grid blocks along the left and right edges.

In order to determine the iteration matrix eigenvalue the matrix is powered. This involves the identification of the eigenvector corresponding to the maximum eigenvalue of the iteration matrix. This step in the procedure can be quite costly unless all the eigenvector is stored for use as an initial guess at subsequent time steps. Usually the eigenvector changes relatively slowly during a displacement calculation and only a few additional iterations are needed at each timestep. The eigenvector array represents about half the scratch storage requirement. The other half of the scratch storage is used to store the previous eigenvector iterate during the spectral norm calculation, and after this the previous iterate of the pressure distribution is stored during the pressure solution.

The pressure distribution is found using equations (5) and (8) and the convergence criterion in equation (10).

The total scratch storage space needed by the scalar LSOR code is compared in Figure 4 with the storage required by the D4 direct elimination code originally in CFTE. The advantage of iterative schemes in this respect is well known.

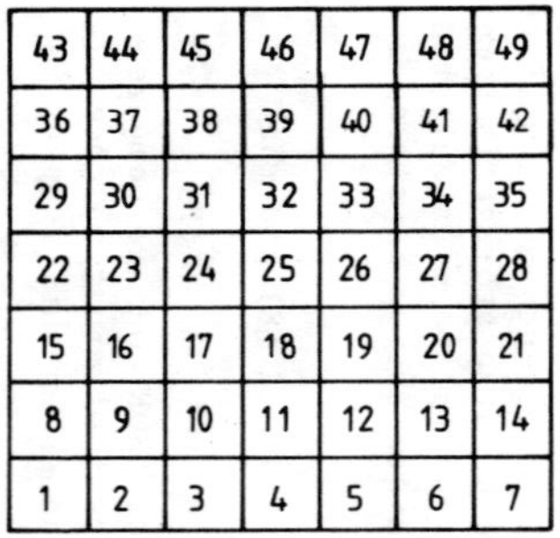

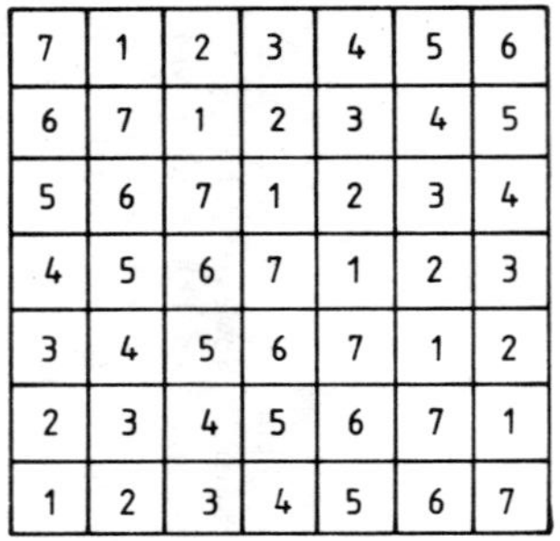

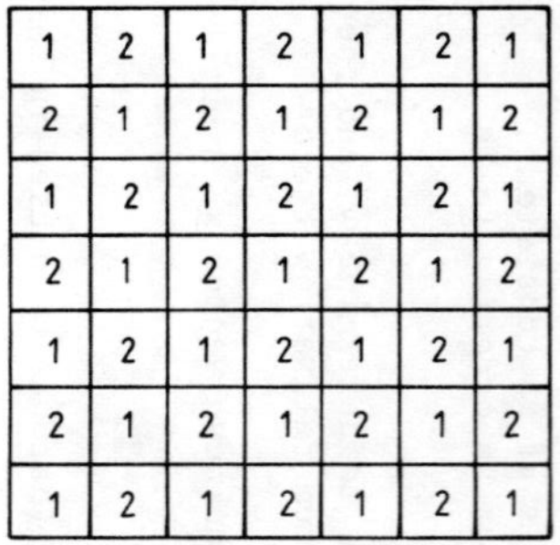

NOLEN'S RED-BLACK ORDERING
USED FOR VECTOR PROGRAM

STANDARD ORDERING
USED FOR SCALAR PROGRAM

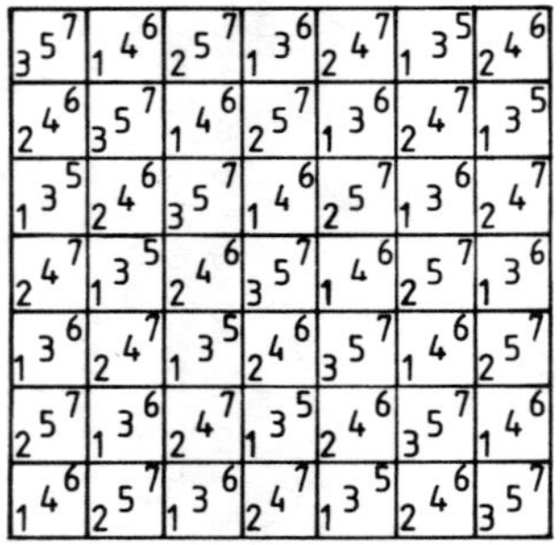

DIAGONAL GROUPING USED FOR VECTOR PROGRAM
PRINCIPLE OF ORDERING SHOWN ABOVE HAS VECTOR LENGTH OF 7
DEVELOPMENT BELOW HAS VECTOR LENGTH OF 21

FIG. 3 GRID BLOCK ORDERING SCHEMES USED TO INVESTIGATE LSOR PERFORMANCE.

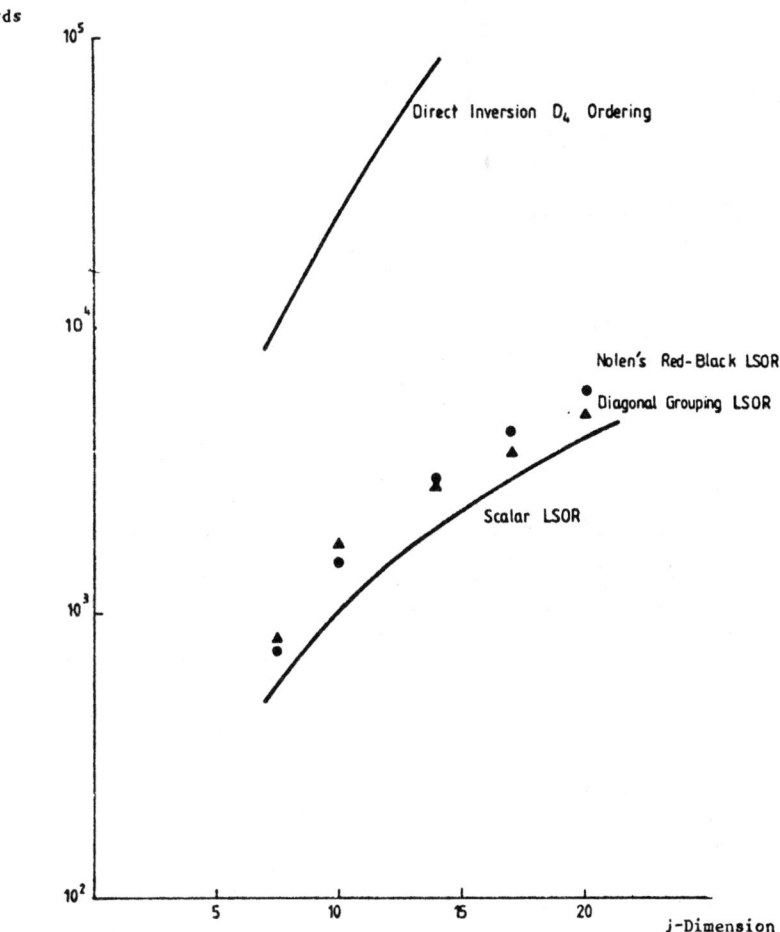

FIG.4 COMPARISON OF SCRATCH STORAGE REQUIREMENT FOR LSOR MATRIX
INVERSION STRATEGIES J x J x 5 PROBLEM.

A number of tests have been made using this version of the code. The number
of grid blocks varying from 7x7x5 (245 grid blocks), 10x10x5 (500 grid blocks)
and 14x14x5 (980 grid blocks). The same surfactant system as described in the
section on curvilinear grids has been used, and to introduce some axial
variations, the permeabilities of the 5 vertical layers have been varied with
values of 10,3,10,3,10 mD respectively. Pressure constrained well models were
assumed in which 15% of the total pressure drop at the initial conditions was
lost in the well completion factor. This influences the spectral norm of the
iteration matrix.

The computing time taken to invert the pressure matrix using LSOR varies with
time according to the difficulty of the problem (ie initial guess and the
matrix spectral radius). Figure 5 shows the time taken for the first 150
timesteps of the sample problem. Also shown for comparison purposes is the
time taken by the direct D4 matrix inversion routine available in the CFTE

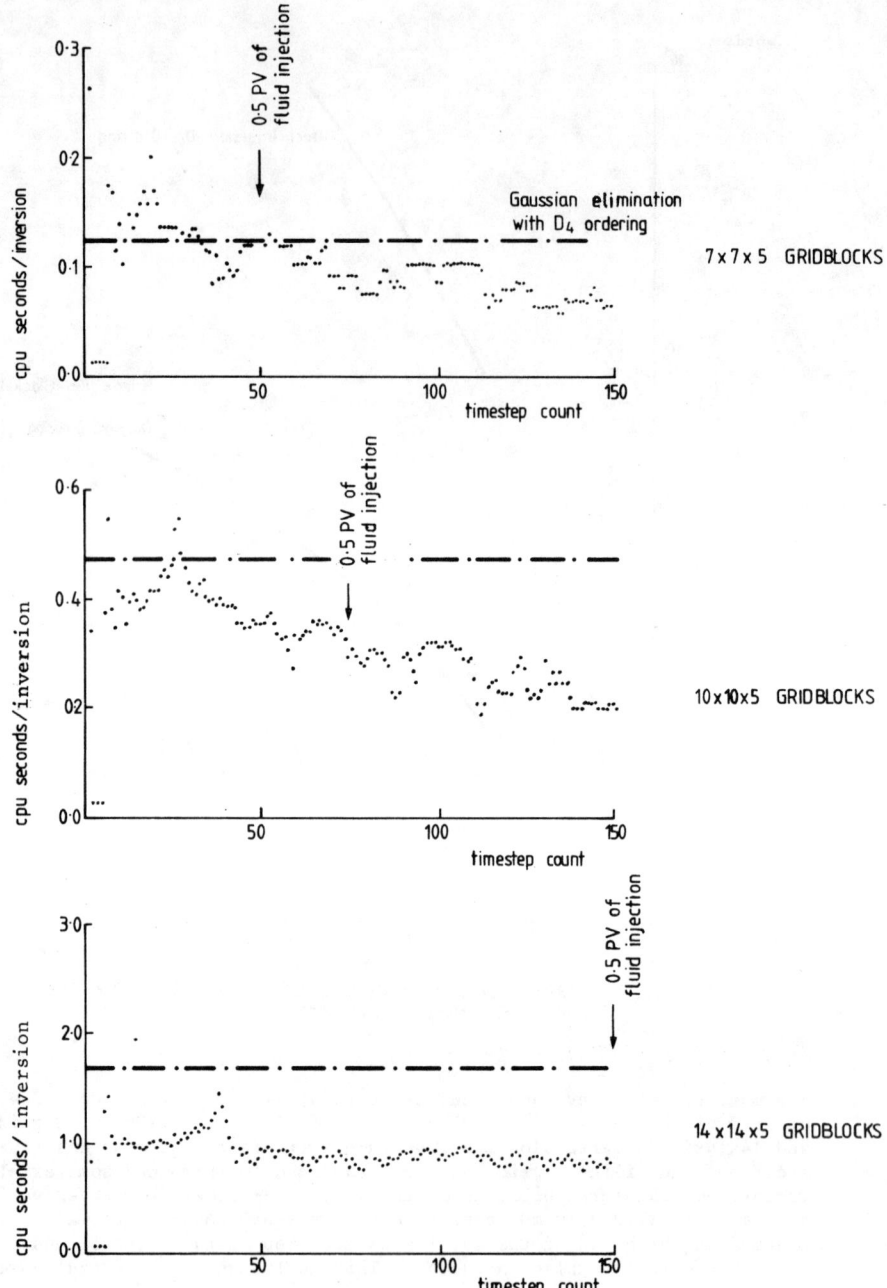

FIG. 5   EXAMPLES OF TIME SPENT IN MATRIX INVERSION USING SCALAR VERSION OF LSOR CODE AND CURVILINEAR GRID BLOCK REPRESENTATION OF WESTERN QUADRANT OF INVERTED 5-SPOT SHOWN IN FIG. 2

code. The overall pattern of behaviour in these calculations, most clearly seen in the largest problem, is that of a decrease in inversion time from the initial conditions, followed by an increase at about 0.2 PV of fluid injection and then a gradual decrease. Over the timescales represented, LSOR is more efficient than the direct method, even for the smallest problem (245 grid blocks).

At long timescales the calculational time in LSOR becomes insignificant as the pressure distribution changes very little during the timesteps. The increase in inversion time at 0.2 PV co-incides with the breakthrough of the oil bank at the producer well. This occurs because of the relatively rapid pressure changes that occurred making the eigenvector and pressure distributions at the previous timestep less good as initial guesses. In micellar/polymer floods the most pronounced peaking has been noted when a highly viscous micro-emulsion breaks through into the producer well.

It is of interest to compare the achieved convergence rates with the asymptotic rates given by equation (9). This is summarised in Table 1 below:

TABLE 1

COMPARISON OF THEORETICAL AND CALCULATIONAL CONVERGENCE RATES

Mesh size	7x7x5	10x10x5	14x14x5
Iteration matrix spectral radius	0.674	0.796	0.854
Asymptotic convergence rate (9)	0.39	0.23	0.16
Typical average convergence rate	0.30	0.18	0.12

As may be seen, the average convergence rate achieved is somewhat slower than the asymptotic rate, as would be expected. The increase in iteration matrix spectral radius, and thus the decrease in convergence rates, with increasing problem size is well known; however, also of interest is the fact that the well factors can considerably influence the convergence properties of the iteration scheme. Making the well factors very large, and thus increasing the coupling between the calculational mesh and the fixed well bore pressures, speeded up convergence by 20% in the examples discussed above.

Vectorisation of LSOR for CRAY - Algorithm Performance

The scalar version of LSOR described in the previous section was transferred to the CRAY without modification and some matrix inversion timings were made to compare the CRAY and IBM performance.

The CRAY calculations without vectorisation were about twice as fast as the equivalent IBM calculations. This is less than the factor of four expected on the basis of the relative clock times of the machines. It is believed that this results from the optimising capability of the IBM compiler used, which more than halved the running time relative to code generated by a non-optimising compiler.

It should also be noted that very little of the original LSOR coding could be vectorised automatically. This was due to three causes:

    (i)   the use of conditional testing to eliminate inactive grid blocks.

    (ii) the recursive nature of the Thomas' algorithm which is the kernel sub-program in the LSOR method.

    (iii) the extensive use of indirect addressing to reference neighbouring grid blocks.

Most of the conditional testing and indirect addressing could be removed by suppressing the elimination of inactive grid blocks. For Cartesian grid approximations to flow patterns in repeated patterns, the removal of this facility incurs a significant work penalty since unnecessary matrix solutions are undertaken for grid blocks which are isolated from the active region by imposing zero transmissibilities. However, for the preferred curvilinear co-ordinate method the inactive grid block facility would not normally be required.

Each line of grid blocks gives rise to a tridiagonal system of operations which cannot be vectorised by the CRAY due to the recursions that occur in performing the forward eliminations and back substitutions. Following Buzbee et al (Ref 8) the method employed to overcome this has been to solve for a number of lines simultaneously.

The choice of lines which are solved simultaneously is important for two reasons. Firstly, the ordering of the grid blocks must not degrade the basic solution strategy of the algorithm. Killough has described an attempt to vectorise the SIP algorithm (Ref 9); unfortunately the vectorisation method degraded the convergence of the algorithm so as to outweigh the gain in speed due to vectorisation. Secondly, the choice must lead to a grid block referencing system which allows the grid blocks in a line, and the grid blocks in neighbouring lines, to be easily referenced. The attraction of the LSOR method is that the lines for simultaneous solution can be chosen so that the asymptotic convergence behaviour is preserved. The required condition is that any chosen ordering must be consistent.

Nolen (Ref 8) has identified one possible ordering scheme in which the lines are ordered with a chequer board arrangement, the red lines are solved first and then the black lines. The concept has been generalised slightly from that reported by Nolen to allow an odd number of grid block lines to be used, as well as even. An example of the line ordering is shown in Figure 3. The vector length for the simultaneous tridiagonal solver is NX.NY/2 for problems with an even number of grid blocks, and (NX.NY+1)/2 and (NX.NY-1)/2 for the red and black groups respectively, when there is an odd number of lines. The generalisation to an odd number of grid block lines in the plane makes the arithmetic expressions for referencing neighbouring grid blocks more complicated than in the less general situation.

An alternative to the red-black ordering scheme has also been investigated. This was prompted by the observation that the red-black scheme resembles a two-step line over relaxed Jacobi scheme for the initial iterations, and therefore the average convergence properties might be less than for other possible ordering schemes. The alternative scheme devised corresponds to grouping the lines in diagonals so as to obtain a number of line problems that can be solved simultaneously. Each diagonal is considered in turn until all the grid block eigenvector elements or pressures have been revised. The basic arrangement is illustrated in the top middle diagram of Figure 3. This scheme allows neighbouring grid blocks to be referenced by simple arithmetic expressions.

To understand the possible differences in initial convergence behaviour further, it is necessary only to consider curvilinear calculations with a symmetric arrangement of NxN grid blocks in the horizontal plane. With the diagonal grouping, the influence of the well connected to the left hand edge of the grid pattern is transmitted in the first iteration to the lower triangular zone of $0.5 \ (N^2 + N)$ mesh points, but only the line of 1.0N grid points along the right boundary are directly coupled to the well on the right in this iteration. In the red-black arrangement, 0.5N grid points are coupled to wells on each side in the red sweep (for N even), and 1.0 N grid points in the black sweep. Thus the diagonal arrangement can on average transmit boundary effects across the area faster, but is asymmetric in its behaviour. Symmetry could be introduced in the diagonal scheme by reversing the sweep order on successive iterations, but this has not been investigated.

The asymptotic convergence rates of the two schemes must be identical with the standard ordering, since all three are consistently ordered. Thus for slowly convergent problems the schemes should behave identically.

On the CRAY the vector performance becomes more advantageous the longer the vector length, subject to the limitation of filling the 64 element registers. The diagonal grouping leads to vector lengths of NY for the tridiagonal solver, which for small problems may not be sufficient to take full advantage of the vector machine. Thus, for small problems, the diagonals have been combined to increase the vector length. The lower diagram in Figure 3 indicates one of these combinations, and the order in which the grid block line problems are solved in this strategy. The ordering remains consistent despite these combinations. In the particular case shown each pass of the program revised each line three times, so one pass is equivalent to three iterations.

The calculations already reported for the scalar code have been repeated using code modifications employing the red-black and diagonal grouping ordering. These calculations correspond to the initial steps in a chemical flood simulation and as such represent the calculation of an almost symmetric problem. The average rates of convergence achieved by the alternative strategies are shown in Table 2:

TABLE 2

COMPARISON OF CONVERGENCE RATES WITH VARIOUS ORDERINGS

Mesh size	7x7x5	10x10x5	14x14x5
Standard ordering	0.30	0.18	0.12
Red-black ordering	0.38	0.22	0.16
Diagonal line ordering	0.37	0.22	0.14

For some small problems, such as the 10x10x5 calculation, the red-black and diagonal group orderings co-incide due to the combining of groups to increase the vector length, and thus the convergence rates are identical. The diagonal grouping in the 7x7x5 problem leads to a slight asymmetry in the initial revision pattern, so the convergence in this case is slightly slower. In the 14x14x5 case, four diagonals are grouped together, and the asymmetric propagation is more marked, thus the reduced convergence compared with the red-black scheme. In all cases the standard ordering leads to an initial asymmetric behaviour even more pronounced than the diagonal grouping, and this explains why this scheme has the poorest average convergence rate.

Vectorisation of LSOR for CRAY - Coding Performance

While the overall strategy in the vectorised scheme is the same as in the original code, the implementation of the two alternative ordering schemes entailed completely rewriting the routines concerned with powering of the iteration matrix, and iterating the pressure distribution. Initial attempts to vectorise the above algorithms involved the extensive use of additional scratch storage associated with use of the CRAY SCILIB routines GATHER and SCATTER (Ref 10). However, program refinements eliminated the need for these routines and the additional scratch storage, with the resulting storage requirements shown in Figure 4.

Conditional testing, an inefficient computing task, has also been completely eliminated from the routines, except when testing for convergence.

Most of the execution time is expended in the iterating routines. Table 3 indicates the vector lengths achieved in the major sections of these routines when applied to the 14x14x5 sample problem.

TABLE 3

EXAMPLE VECTOR-LENGTHS ACHIEVED WITH DIFFERENT ORDERING SCHEMES

Operation	Red-Black Ordering	Diagonal Grouping
Updating previous iterate	980	980
Setting up right-hand sides for Thomas' algorithm	5	70
Setting up left-hand sides	490	70
Thomas' Algorithm	98	56
Updating solution (relaxation step)	490	70
Convergence test	980	980

Table 4 gives the CRAY cpu time required by the various vectorised options to perform one iteration in either the eigenvalue calculation or the calculation of the pressure distribution.

TABLE 4

CRAY CPU TIME IN MILLI-SECONDS TO EXECUTE ONE ITERATION

Mesh size	Eigenvalue calculation			Pressure distribution		
	7x7x5	10x10x5	14x14x5	7x7x5	10x10x5	14x14x5
Standard ordering	3.67	–	11.04	3.69	–	14.78
Red-Black ordering	0.73	1.47	2.85	0.77	1.49	2.88
Diagonal grouping	0.34	0.51	0.89	0.35	0.51	0.90

As may be seen the red-black coding is 4 or 5 times faster than the original code and the diagonal grouping is 10 to 16 times faster. The difference between the last two options is almost certainly due to the slightly more complex arithmetic needed in the red-black code to identify the neighbouring grid blocks, and also the shorter vector length associated with setting up the right-hand side column vector. By reverting to the restricted case where the number of grid blocks in the plane is known to be even, it may be possible to make the two vectorised options comparable.

A small part of the gains achieved above are associated with simply reducing the generality of the inversion programming and the generation of better fortran coding. This has been demonstrated by using the same inversion routines on the IBM machine, where enhancements by factors of 1.2 and 1.3 were observed.

CONCLUSIONS

This paper has discussed advantages which can be derived in utilising curvilinear mesh co-ordinate systems in surfactant flood calculations. This allows a computational geometry to be adopted broadly consistent with the anticipated flow patterns of a problem. Implementation of flow stream geometry utilising the code PASL has been illustrated for a generalised field problem with sealing fault lines. The curvilinear geometry gives advantages in choice of mesh blocks adjacent to wells, which in turn gives a superior time step capability in an IMPES formulation, such as that adopted in the CFTE code for simulation of surfactant floods. The reduction in mesh orientation errors resulting from the use of curvilinear co-ordinates has been demonstrated for a 5-spot pattern with surfactant flooding.

Direct inversion employed in the solution of the implicit formulation of the pressure equation in the CFTE program leads to computer speed limitations for large curvilinear mesh problems. To overcome this, the LSOR method has been programmed and tested using the standard ordering, as well as a red-black and a diagonal line consistent ordering. The last two arrangements have been shown to be amenable to selection of long vector lengths on a CRAY computer so that diagonal line formulation in the revised code runs up to 19 times faster than the standard ordering in LSOR. Relative to D4-direct inversion on the IBM 3033 machine, LSOR diagonal line inversion runs some 65 times faster on the CRAY. The total code performance is dependent on problem size and for the largest example discussed here of 980 mesh blocks, direct inversion required about 60% of the overall running time. Much larger problems are needed for real field studies where the inversion aspect becomes completely dominant. Thus these improvements have placed field computation closer to practical realisation in terms of computer costs.

Consideration of further generalisation of the LSOR method to curvilinear meshes with a wider range of connectivities, and with consequent difficult patterns of off-diagonal non-zero elements, needs to be considered in the future.

ACKNOWLEDGEMENT

The work reported in this paper has been funded by the UK Department of Energy.

REFERENCES

1   TODD M.R. and CHASE C.A., "A Numerical Simulator for Predicting Chemical Flood Performance". Paper SPE 7689, presented at the Fifth Symposium on Reservoir Simulation, Denver 1979.

2   FAYERS F.J., HAWES R.I. and MATTHEWS J.D., "Some Aspects of the Potential Application of EOR Processes in North Sea Reservoirs". Paper EUR 194, presented at the European Offshore Petroleum Conference and Annual Exhibition, 1980.

3   MUSKAT M., "Flow of Homogeneous Fluids Through Porous Media", McGraw Hill, 1937.

4   MARTIN J.C. and WAGNER R.E., "Numerical Solution of Multiphase Two-Dimensional Incompressible Flow Using Stream Tube Relationships". Soc. Pet. Eng. J. (October 1979) pp313-323.

5   PEACEMAN D.W., "Fundamentals of Numerical Reservoir Simulation", Elsevier, 1977.

6   VARGA R.S., "Matrix Iterative Analysis", Prentice Hall, 1962.

7   FORSYTHE G.E. and WASOW W.R., "Finite Difference Methods for Partial Differential Equations". John Wiley, 1960.

8   BUZBEE B.L., BOLEY D. and PARTER S.V., "Applications of Block Relaxation". Paper SPE 7672, presented at the Fifth Symposium on Reservoir Simulation, Denver 1979.

9   KILLOUGH J.E., "The Use of Vector Processors in Reservoir Simulation". Paper SPE 7673, presented at the Fifth Symposium on Reservoir Simulation, Denver 1979.

10   CRAY-1 Library Reference Manual. SR-0014.

CONTROL OF NUMERICAL DISPERSION
IN COMPOSITIONAL SIMULATION

D. C. WILSON, T. C. TAN, P. C. CASINADER

Department of Mineral Resources Engineering,
Imperial College, London SW7 2BP

ABSTRACT

This paper presents a technique, suitable for multidimensional application, for reducing numerical dispersion on fully implicit compositional simulators. Simple geometrical analysis of curve profiles and re-examination of various weighting schemes lead to the development of a dynamic weighting technique that exploits the optimum features of 1-point, 2-point and mid-point weighting schemes. This weighting scheme is general in its application in finite difference models for reducing dispersion in convective parameters such as saturation and concentration. We show how this scheme can be implemented on an implicit, equation of state, compositional model. Numerical examples including multiple contact (MCM) and near miscible (NM) problems are used to compare its performance with two published compositional simulators which utilise full upstream weighting. Our results show a significant reduction in the number of grid blocks required to achieve the same numerical accuracy.

1.0 INTRODUCTION

The first compositional simulators appeared in the late 1960's. Since then, tremendous progress has been achieved in the treatment of fluid properties, solution techniques and model generality.

In common with B-simulators, numerical dispersion remains a major problem. The most dominant aspect of numerical errors occur in the compositional field. As in B-models, albeit in a less obvious manner, saturation dispersion and errors in the pressure field remain.

Various techniques have been reviewed and are discussed below.

McFarlane et al (1) used smaller cells in the region of maximum compositional change, with larger cells elsewhere. This technique is not even robust enough for 1-D problems. Price et al (2) proposed the use of time and space discretisation in 1-D, such that the numerical diffusivity is of the same order as the physical diffusivity. This is too expensive for practical application. The works of Peaceman (3) and Lantz (4) on stability and truncation error analysis laid down the foundation for many subsequent works, including the present one. It is thus possible to use an artificial diffusion term to cancel out the numerical diffusivity in explicit backward (Chaudhari (5)), and implicit backward (Van Quy (6)) difference equations. They require severe time step, grid size limitations and are primarily

suitable for miscible displacement. Laumbach (7) developed a truncation
cancellation procedure which removed the time step and grid size limitations
by cancelling a portion of the error in the convection term with that in the
accumulation term. It is applicable for miscible, incompressible systems,
where compositions are the only variables solved. Field application,
however, requires both pressure and concentration solutions. It appears that
the early work of Gardner et al (8) using the method of characteristics is of
greater utility in miscible flooding. However, the compositional field is
decoupled from the conservation equation used for solving the pressure, and
so does not strictly observe the conservation principle. The explicit
2-point upstream weighting scheme is the most widely quoted dispersion
control technique (Todd et al (9)). It has recently been improved (Banks et
al (10)), and extended for implicit treatment (Wheatley (11)). Nghiem et al
(12), reported the use of 2-point upstream weighting scheme for an IMPES
compositional simulator.

This paper gives some simple analysis of various representative profiles of
convective parameters in the light of existing stability and truncation error
analysis. This leads to the identification of several weaknesses in the
2-point upstream weighting scheme, and the best method of exploiting
mid-point weighting. A variable time level, variable distance weighting
scheme has been developed which optimises the best features of the 1-point,
2-point and the hitherto unused mid-point schemes. The development is
empirical, in nature, but is within the constraints of numerical stability,
and is guided by the available knowledge on truncation errors. We report
successful applications on difficult numerical problems.

2.0   THEORETICAL DEVELOPMENT

The following sections discuss the theoretical basis for our model.

2.1   Interpolation Methods

Consider a convective paramter C, which is assumed to be continuous and
linear in space and time (Fig. 1).

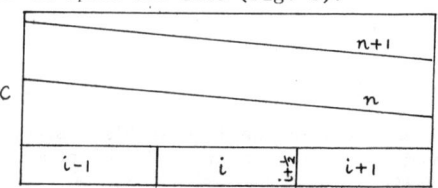

Figure 1 - C profiles at 2 time
levels, n and n+1

For interblock flow at $i+\frac{1}{2}$, the parameter C for the finite difference
equation must be evaluated at $i+\frac{1}{2}$. An observer at $i+\frac{1}{2}$ would notice a
continuous change in the value of C from C^n $i+\frac{1}{2}$ at the start of a time step,
to $C^{n+1}_{i+\frac{1}{2}}$ at the end of the time step. The integral average of C $i+\frac{1}{2}$ over the
time step is the arithmetic mean of C^{n+1} $i+\frac{1}{2}$ and C^n $i+\frac{1}{2}$. This analysis
agrees with the linearised truncation error analysis for the convection
equation in which mid-point weightings in space and time are found to be the
most accurate (Appendix A1).

In practice, however, C varies non-linearly, and may become discontinuous.
Consequently, the inter-block value $\bar{C}i+\frac{1}{2}$ is not a simple arithmetic mean
between t^n and t^{n+1}; instead, it must be found by a time-integration over this
range.

$$\bar{c}_{i+\frac{1}{2}} = \left[\int_{t^n}^{t^{n+1}} c(x_{i+\frac{1}{2}}, t) \, dt \right] \Big/ (t^{n+1} - t^n) . \qquad (1)$$

Since the parameter C propagates with time, it follows that the above integral must be equal to a distance integral between $x_{i+\frac{1}{2}}$ and some upstream point X Thus:-

$$\overline{C}_{i+\frac{1}{2}} = \left[\int_{x_{i+\frac{1}{2}}}^{X} C(x,t^n)\, dx\right] / (X - x_{i+\frac{1}{2}}), \qquad (2)$$

and, by the Mean Value Theorem, $\overline{C}i+\frac{1}{2}$ must correspond to some point in the interval $(xi+\frac{1}{2}, X)$. This shows that neither single point upstream nor midstream weighting is totally correct for the non-linear problem, and that some intermediate weighting factor must be determined. A description of our proposed theory now follows.

Consider the general linear interpolation formula:-

$$C_x^* = \theta\left[W C_i^{n+1} + (1-W) C_{i+1}^{n+1}\right] + (1-\theta)\left[W C_i^{n} + (1-W) C_{i+1}^{n}\right] \qquad \begin{array}{l} t^n \leq t^* \leq t^{n+1} \\ x_i \leq x \leq x_{i+1}, \end{array} \quad (3)$$

where θ and W are the time and distance weighting factors, with values between 0 and 1. Two conditions must be satisfied for linear interpolation to be valid.

(1) A continuous linear (or near linear) curve between the pivotal points.
(2) The pivotal points must be "mobile". By this we mean that C should be in the mobile range bounded by the maximum and minimum possible values. For example, the mobile range of water saturation is between Swc and (1-Sor), but does not include these actual values.

Our aim is to use equation (3) to predict the value of an interblock convective parameter such that it lies close to the true time-integral average value of C on the history curve at $i+\frac{1}{2}$. Before doing this, it is necessary to examine the representative curves which are to be interpolated.

2.2 Curve Analysis

Consider 4 arbitrary profiles at a fixed time level (Fig. 2). In the following description, equal grid spacing is assumed.

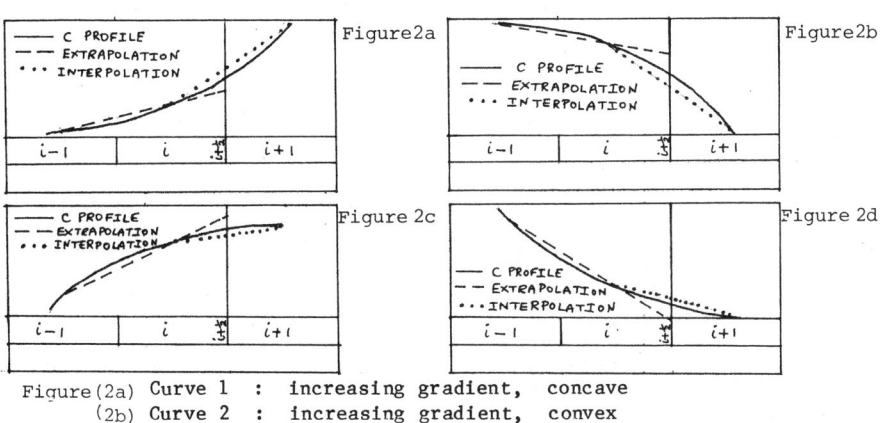

Figure(2a) Curve 1 : increasing gradient, concave
(2b) Curve 2 : increasing gradient, convex
(2c) Curve 3 : decreasing gradient, convex
(2d) Curve 4 : decreasing gradient, concave

We are interested in approximating the value of C at $i+\frac{1}{2}$. Both upstream extrapolation from i-1 and i, and interpolation between i and i+1, are possible. On Curves 1 and 2, the extrapolated values can be found on the upstream portion of the curves, while the interpolated values lie on the downstream part of the curves with reference to $i+\frac{1}{2}$. This implies that the use of upstream extrapolation is effectively upstream weighted $(\frac{1}{2} < w < 1)$, and hence stable. The use of interpolation is effectively downstream weighted ($0 < W < \frac{1}{2}$), and hence unstable. This situation occurs just behind an immiscible displacement, where a shock is present, or is building up. Thus Curve 1 can represent the oil saturation, and Curve 2 can represent the water saturation. It is well known that the use of mid-point weighting creates overshoot under such circumstances, while full upstream weighting results in an undershoot of the displacing phase (saturation dispersion).

On Curves 3 and 4, the reverse conditions occur. The use of upstream extrapolation is effectively downstream weighted. Thus on Curve 3, upstream extrapolation creates undershoot, and on Curve 4, it creates overshoot. Further analysis of upstream extrapolation and mid-point interpolation is given in the Appendix A2. Suffice it to say here that both extrapolation and interpolation have their advantages and limitations. They are complementary in their functions. When extrapolation is unstable, interpolation is stable, and vice versa. Under certain conditions they are both unstable. This occurs when the pivot(s) become "immobile". In such situations upstream weighting is the best stable approximation.

The same analysis can be extended to the history curves at a fixed point. We are interested in the history curve at $i+\frac{1}{2}$ over the time step. Due to the convective nature of C, a history curve at a fixed point over a time step is related to the portion of the distance profile immediately upstream of the fixed point at the beginning of the time step.

2.3 A Dynamic Weighting Scheme

The purpose of this development is to find a method of evaluating explicit local weighting factors, such that the linear interpolation formula, equation (3), can be incorporated into an implicit simulator to substitute for interblock convective parameters or their dependent functions.

The 4 basic curve types (Fig 2.) can be subdivided into 2 groups. Group 1 (curves 1 and 2) has increasing gradients in the flow direction, and Group 2 (curves 3 and 4) has decreasing gradients. Either Group 1 or Group 2 are present locally at a fixed time. Identification is possible through gradient testing.

$$\left[|G_i| - |G_{i-1}| \right] \begin{cases} > 0 \Rightarrow \text{Group 1} \\ < 0 \Rightarrow \text{Group 2} \\ = 0 \Rightarrow \text{linear} \end{cases}, \text{ where:} \begin{cases} G_i = \dfrac{C_{i+1} - C_i}{x_{i+1} - x_i} \\ G_{i-1} = \dfrac{C_i - C_{i-1}}{x_i - x_{i-1}} \end{cases} \quad (4)$$

The basic assumptions are:-

(1) Group 1 curves do not evolve into Group 2 curves over a time step. The same distance weighting factors can be used at two fixed time levels, n and n+1, in the linear interpolation equation.

(2) The history curve at the block interface $i+\frac{1}{2}$ over the time step is in the same curve group as the immediate upstream profile at time level n (see next section). Therefore, the interpolation factor on the history profile, θ, is assumed to be equal to the distance interpolation factor, W.

Based on the previous curve analysis, the following strategy is adopted.

When a Group 1 curve is detected, an upstream extrapolation is required to provide low numerical dispersion, while maintaining numerical stability. This can be invoked on the linear interpolation formula by choosing weighting factors between 0.5 and 1.0 (equal grid spacing). Simple geometrical constructions show how this can be done. (Figs. 3a, b).

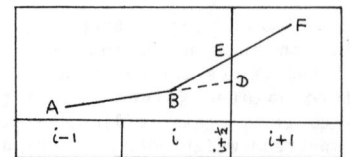

Figure 3a

2-point upstream extrapolation.
D can be found on the chord BF.

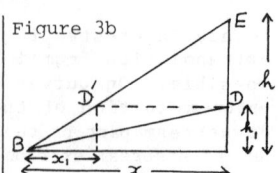

Figure 3b

A B is an upstream chord. BF is a downstream chord. It is required to find the extrapolated point D on BF. This point is D'. The necessary weighting factor for the interpolation formula, W, is derived below.

$$a_i = (x_{i+1} - x_{i+\frac{1}{2}})/(x_{i+1} - x_i) \; ; \tag{5a}$$

$$W_i = a_i + (1 - a_i)(x - x_1)/x \; ; \tag{5b}$$

$$R_i = \frac{h_1}{x_1}\Big/\frac{h_1}{x} = \frac{G_i}{G_{i-1}} \; ; \tag{5c}$$

substituting equation (5c) into equation (5b) gives

$$W_i = 1 + (a_i - 1)/R_i \tag{5d}$$

When a Group 2 curve is detected, interpolation is superior. The weighting factors are calculated by setting Ri=1. If the grid spacing is uniform, this gives Wi=ai=0.5.

Screening must be applied to exclude the use of the interpolation formula under 2 invalid conditions:-
(1) Gradient reversal (R i is negative)
(2) Either, or both, of the pivots (C_B, C_F) are "immobile"
Once these conditions are detected, full upstream weighting affords the best stable alternative available.
The interpolation scheme proposed is therefore dynamic in nature. Effective full upstream, 2-point upstream extrapolation, or mid-stream interpolation with varying degree of implicitness can be invoked locally via the same linear interpolation formula.

2.4 Relationship Between Time Weighting and Distance Weighting

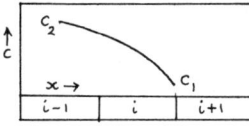

Figure 4a

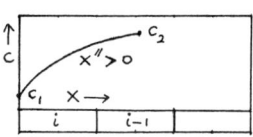

Figure 4b

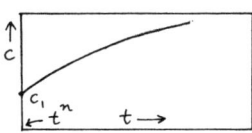

Figure 4c

Figure 4a shows the profile of a typical parameter C at time t^n , C_1 being its value at the interface between blocks i and i+1. Figure 4b shows the same profile with respect to X, the distance measured upstream from the interface. Assuming that the profile X(C) and the velocity of propagation V(C) are known, it is required to determine the shape of the history curve, t(c) (Figure 4c), at the interface, which is given by:-

$$t(c) = \frac{X(c)}{V(c)} . \tag{6}$$

Case 1 V(C) = constant = v
This assumption is approximately valid if the time step is small, so that the band of values (c_1, c_2), which crosses the interface over the time-step, is narrow. For such a case,

$$t(c) = \frac{X(c)}{V} , \tag{7}$$

and hence t "(C) = X "(C)/V.
Since V is positive, t" has the same sign as X", and hence t (C) belongs to the same group of curves as X (C).

<u>Case 2</u> V(C) is variable.
Differentiating (6) w.r.t. C, we obtain:

$$t' = (vx' - xv')/v^2,\tag{8}$$

and, differentiating yet again, we have

$$t'' = [v(vx'' - xv'') - 2v'(vx' - xv')]/v^3.\tag{9}$$

Now, since X' and V' have opposite signs, it follows that the second term of
the above expression $[-2v'\ (VX' - XV')]$ is always positive. Furthermore, if
X "> 0, then V"< 0, and the entire expression will be positive, and so X" and
t" will have the same sign, and the curves will belong to the same group
(i.e. Group 1). If, however, X"< 0 (i.e. Group 2), then, due to the second
term being positive, we cannot be certain whether or not t" will change
sign. Nevertheless, for practical purposes, we shall assume that, for all
cases, X and t belong to the same group.

3.0 OVERALL APPLICATION TO AN IMPLICIT COMPOSITIONAL SIMULATOR

There is a unique dependence of the overall component fractional flows on the
overall composition. For propagational stability, concentration velocities
at a fixed point in space and time are equal (Helfferich (13)). It is
therefore appropriate to find the dynamic weighting factors based on the
local overal concentration profiles. We further assume the local existence
of either the Group 1 curves, or the Group 2 curves. The concentration
profile of the most "sensitive" component is utilised to evaluate the
weighting factors. They are used for all components in both hydrocarbon
phases, if 2 phases exist. Selection of the most "sensitive" component is
important to avoid the need to choose the most stable weighting factors
evaluated from all the concentration profiles. The "immobile" conditions for
the selected components must also be defined. These ideas are illustrated in
the numerical examples. To account for phase discontinuity, full upstream
weighting is used if the upstream and downstream blocks do not have the same
number of hydrocarbon phases.

3.1 Construction of the Model
The compositional model used in this study is based on an equation of state,
and follows the implicit formulation presented by Coats (14). We will,
therefore, only discuss the model where it has been modified to take into
account the preceding discussion.

3.1.1 Temporal and Spatial Weighting

Figure 5

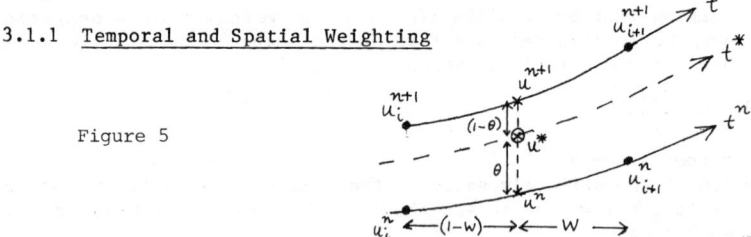

At any point in the system, the value of a variable u, at time t^* (where $t^n \le t^* \le t^{n+1}$), can be related to its values at t^n and t^{n+1} by:-

$$u^* = \theta u^{n+1} + (1-\theta)u^n = u^n + \theta \delta u^{n+1}\tag{10a}$$

or,

$$u^* - u^n = \delta u^* = \theta \delta u^{n+1}.\tag{10b}$$

so, for each iteration, this implies:-

$$\delta^{\ell} u^* = \theta \, \delta^{\ell} u^{n+1}$$

(10c)

The symbols $\delta^{\ell} u \; (= u^{\ell+1} - u^{\ell})$ and $\delta u \; (= u^{\ell+1} - u^{n})$ denote the change in u over the iteration l, and the cumulative change, respectively.

If, however, in addition to this intermediate time-level, u is also evaluated at a point other than the block-centres, then it has to be related to the block-centre values by means of the distance weighting formula. Thus:-

$$u^* = W u_i^* + (1 - W) u_{i+1}^* \; ; \qquad\qquad \left(\tfrac{1}{2} \leq W \leq 1 \right) \qquad (11a)$$

substitution then leads to:

$$\delta^{\ell} u^* = \theta W \delta^{\ell} u_i^{n+1} + \theta (1 - W) \delta^{\ell} u_{i+1}^{n+1}. \qquad (11b)$$

3.1.2 Expansion of the "Flow" term

The flow between two neighbouring blocks i and i+1 can be expressed in the form:-

$$T^* \left(P_{i+1}^* - P_i^* \right).$$

Increments in the transmissibility term T are calculated by means of partial derivatives w.r.t. the complete set of variables $(U_1, \ldots\ldots U_n)$.

$$\delta^{\ell}(T^*) = \sum_{k=1}^{N} \left(\frac{\partial T}{\partial u_k^*} \right)^{\ell} \delta^{\ell} u_k^* , \qquad\qquad (12a)$$

and, finally, using equations (10c), (11b) and (12a), we obtain the expansion of the flow term, as follows:-

$$T^* \left(P_{i+1}^* - P_i^* \right) = \left[T^* \left(P_{i+1}^* - P_i^* \right) \right]^{\ell} + (T^*)^{\ell} \left[\theta \, \delta^{\ell} P_{i+1} - \theta \, \delta^{\ell} P_i^{n+1} \right]$$

$$+ \left(P_{i+1}^* - P_i^* \right)^{\ell} \sum_{k=1}^{N} \left(\frac{\partial T}{\partial u_k^*} \right)^{\ell} \left[\theta W \left(\delta^{\ell} u_i^{n+1} \right)_k + \theta (1 - W) \left(\delta^{\ell} u_{i+1}^{n+1} \right)_k \right]. \quad (12b)$$

4.0   DISCUSSION OF RESULTS

Three different displacement problems were chosen, in order to demonstrate the application of the above theory in a variety of situations. The data for these runs were taken from Coats (14), Leach and Yellig (15), and Smith and Yarborough (16), respectively.

4.1   Displacement 1 (Coats (14)).

This is an MCM problem, involving components: C1,n-C4 and n-C 10. The system exists initially as an undersaturated liquid, and is displaced by a rich gas.

In the simulation, three zones can be identified: a downstream zone containing undersaturated oil, a middle zone comprising two phases whose compositions converge in the upstream direction, and finally an upstream miscible zone containing a single dense fluid whose composition changes from the critical composition to that of the injection gas. The boundary between the first two zones will be referred to as the gas front, while that between the latter two will be called the miscible front.

In MCM problems, the use of single point upstream weighting leads to severe compositional dispersion which causes substantial delay in the attainment of miscibility. This retardation of the miscible front is conspicuous in Coats'results, where the use of 20, 40 and 80 blocks show a progressive

432

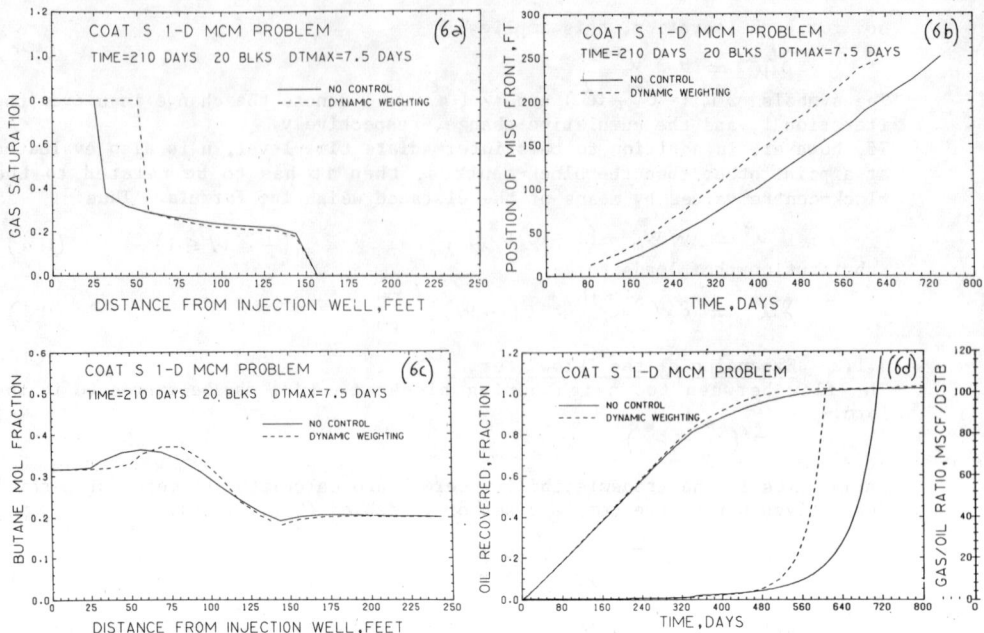

Figure 6 - Displacement 1 (MCM) : 1-D comparison of dynamic weighting with full upstream weighting.

(a) Sg profile. (b) Advance of miscible front. (c) C4 concentration profile (d) oil recovery and GOR vs time.

increase in its speed of propagation. In the absence of an analytical solution, it is justifiable to assume that the 80-block solution is the nearest to reality.

The introduction of the dynamic weighting scheme described in this paper produces a marked improvement, and has enabled us to obtain, with 20 blocks, answers which are of comparable accuracy to Coats' 40-block solution. Figures 6a, b, c, d show a comparison between the use of this technique and single-point upstream weighting. The use of the proposed technique clearly results in a faster advance of the miscible front, which is confirmed by the early and steep rise in GOR, following its breakthrough to the producer.

The scheme has also been tested in 2D, using a cartesian grid of 9X9 blocks, with the injection and production wells located in two diagonally-opposite corner blocks. (Figures 7a, b, c, d). Once again, an improvement in the size of the miscible zone can be observed, using our technique.

4.2   Displacement 2   (Smith and Yarborough (16)).

The system used in this displacement was a binary mixture of C1 and nC5, being displaced by dry gas (C1). In this case, evaluation of weighting factors can be carried out on either component. Thus C1 was arbitrarily chosen for this purpose. Two runs were performed on this system, the first of these being designed to simulate FCM displacement. This was achieved by assuming an initial composition of 50% C1 and 50% n-C5, and simulating the displacement in the super-critical region (at 3000 psi). Since this is a perfect piston-type displacement, the analytical solution consists of a step change in composition from the injection composition to the initial composition.

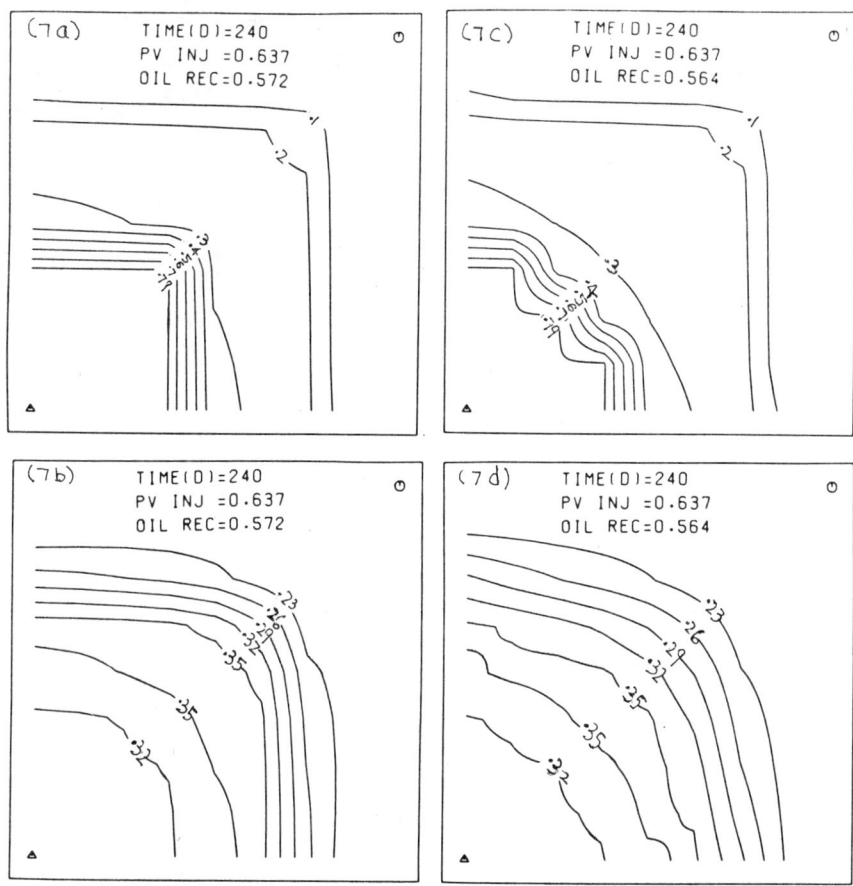

Figure 7 – Displacement 1 (MCM) : 2-D comparisons.

(a) Gas saturation map, dynamic weighting. (b) C4 concentration map, dynamic weighting. (c) Gas saturation map, full upstream weighting. (d) C4 concentration map, full upstream weighting.

Figures 8a and 8b show the Cl profile at 210 days, and the n-C5 concentration in the effluent as a function of time. The weighting technique shows better results than the "full upstream" case, although both show an appreciable compositional dispersion relative to the analytical solution.

In the second run, an initial composition of 87.5% Cl and 12.5% n-C5 was chosen, so as to yield an initial condensate liquid of 7% saturation at 1525 psi, which was also the pressure at which the simulation was conducted. Again, Cl was injected, and the problem was run in the 2-phase mode, with the liquid assumed to be immobile. The purpose of this run was to demonstrate that, for some problems (such as of this type) the amount of compositional dispersion is negligible.

This postulated absence of compositional dispersion is verified by the numerical results shown in Figures 8c and 8d, in both of which the results of using single point upstream weighting are virtually identical to those obtained with the present technique.

434

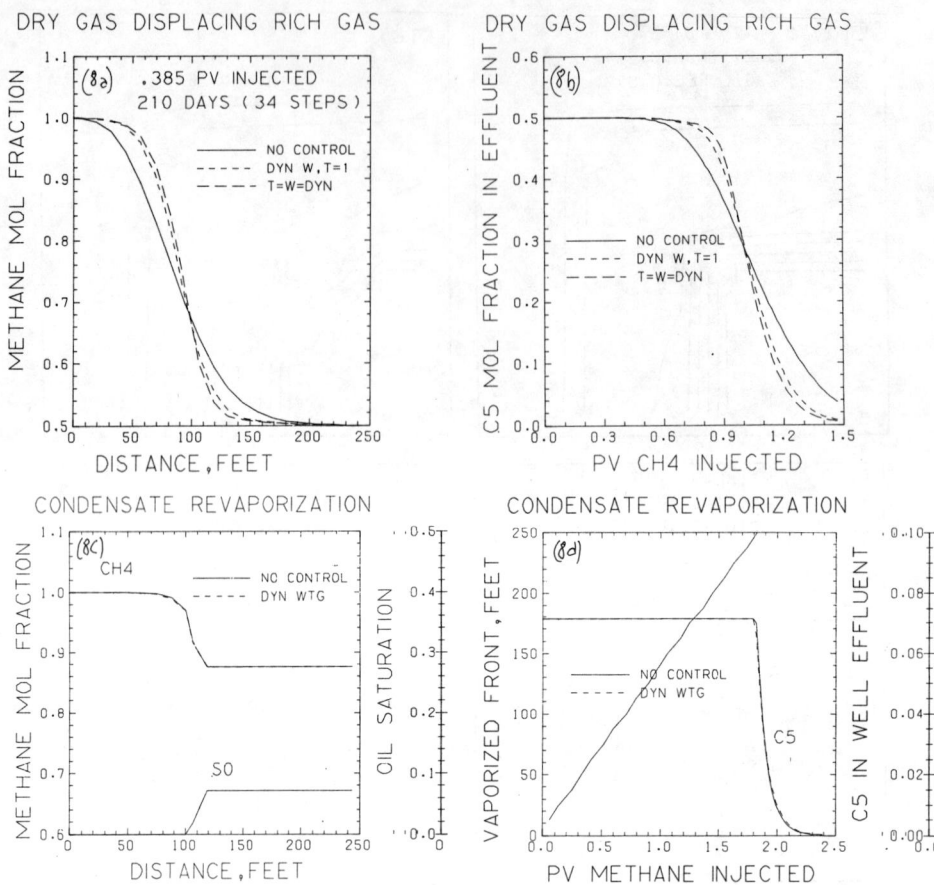

Figure 8 - Displacement 2 : 1-D comparison of dynamic weighting with full upstream weighting.

(a) C1 concentration profile (FCM). (b) C5 concentration in effluent vs time (FCM). (c) C1 and oil saturation profiles (re-vaporization). (d) C5 concentration in effluent, and advance of "dry front" vs time (re-vaporization).

4.3 Displacement 3 (Leach and Yellig (15)).

This was a study of the mechanisms involved in the displacement, by CO_2, of a synthetic crude oil. Leach et al (15) presented laboratory results covering the various displacement types (FCM, MCM and NM), and also simulated these on their compositional model, using 100 blocks.

To test our technique, two runs were chosen: an MCM drive (Run 6), and an NM drive (Run 7). The component which we selected for "gradient testing" was the one which had the least initial concentration - namely C6. The weighting technique has enabled us to match the laboratory results to a good accuracy, with merely 20 blocks. Considering the MCM results first, Figure 9a demonstrates the faster advance of the miscible front, and the steeper CO_2 profile resulting from the use of this technique. Figures 9b and 9c further support our dispersion control method, by showing the delayed breakthrough of CO_2, and the steep change in the GOR and the effluent composition.

The above features have also been verified in the NM run, perhaps to a greater extent, as can be seen, for example, in the significant sharpening

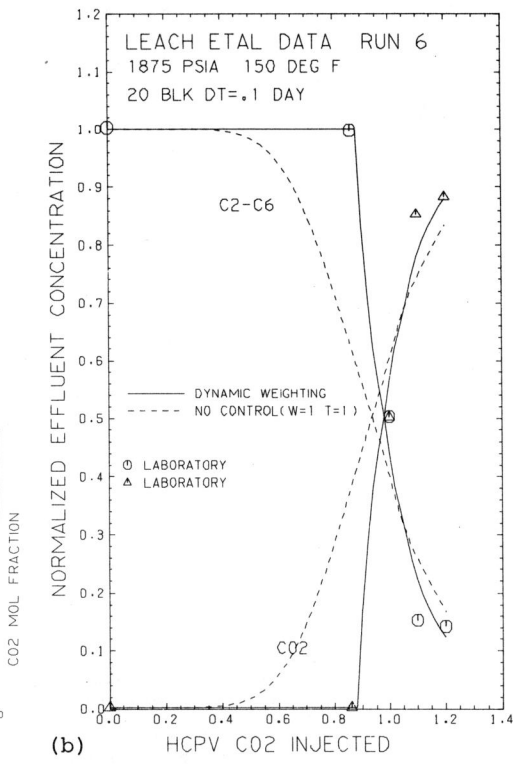

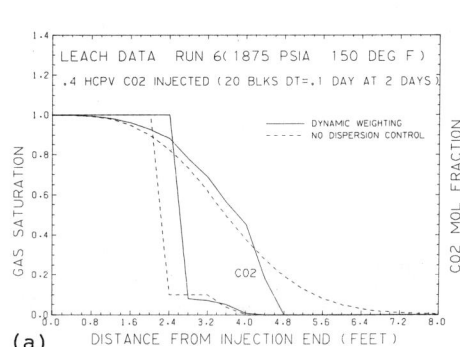

(a)

(b)

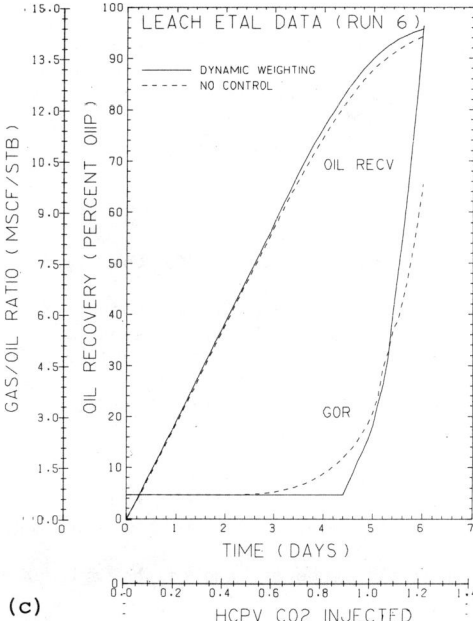

(c)

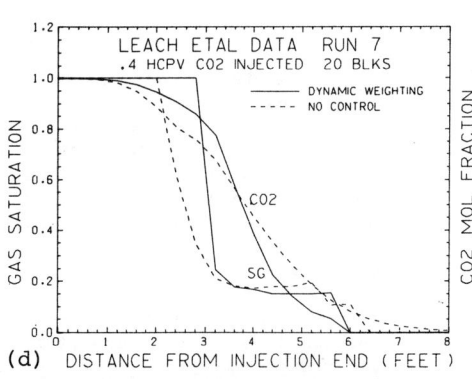

(d)

Fig.9 (a)-(d). Caption overleaf.

436

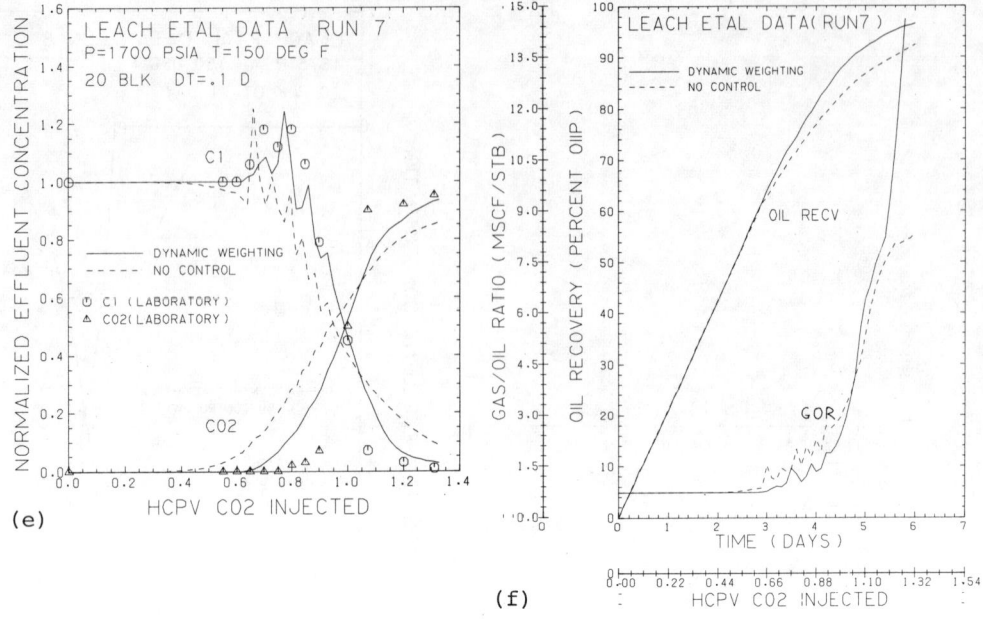

(e)

(f)

Figure 9 – Displacement 3 : 1-D comparison of dynamic weighting
with full upstream weighting. (a) Sg and CO2 concentration
profiles (MCM). (b) Normalized concentration of CO2, and C2-C6 in
effluent vs HCPV injected (MCM). (c) GOR and oil recovery vs HCPV
injected (MCM). (d) Sg and CO2 concentration profiles (NM). (e)
Normalised concentration of CO2 and C1 in effluent vs HCPV injected
(NM). (f) GOR and oil recovery vs HCPV injected (NM).

which occurs in the Sg and CO2 profiles, due to the weighting scheme (Figure
9d). The production history (Figure 9e) and the effluent C1 – and CO2 –
concentrations (Figure 9f) confirm the delayed arrival of the 2-phase zone,
and the consequent higher recovery resulting from the use of this technique.
It needs to be mentioned, however, that a critical gas saturation of 15% had
to be introduced to the relative permeability table, before the results of
Leach et al (15) could be successfully reproduced.

5.0   CONCLUSIONS

The work described in this paper leads us to the following main
conclusions:-
(1)   The use of single-point upstream weighting causes severe compositional
dispersion, particularly when simulating FCM, MCM and NM displacements.
(2)   A dynamic weighting scheme has been developed, which utilises the
profile of the variable concerned, to determine the optimum weighting factors
in time and space.   It exploits the classical features of mid-stream,
two-point upstream, and single-point upstream schemes, based on the
properties of the profile.
(3)   The technique has been successfully tested on MCM, FCM and NM
displacements, and yields results which, if their accuracy is to be
reproduced on a "fully upstream" model, would require several times as many
grid blocks.
(4)   The method is supported by geometrical arguments, and can be implemented
easily in multi-dimensional simulators.

ACKNOWLEDGEMENTS

The authors would like to thank the UK Department of Energy and Imperial College of Science and Technology for supporting this research, Professor C.G. Wall, Dr. R.A. Dawe for their continuing interest, and Miss M. Green of ERC for her patience in typing the various drafts of this paper.

REFERENCES

1. MCFARLANE, R.C., MUELLER, T.D., MILLER, F.G.;
"Unsteady-State Distributions of Fluid Compositions in Two-Phase Oil Reservoirs Undergoing Gas Injection", Society of Petroleum Engineers J. (March, 1967), $\underline{7}$, 61-74.
2. PRICE,H.S. and DONOHUE, D.A.T.,
"Isothermal Displacement Processes with Interphase Mass Transfer", Society of Petroleum Engineers J. (June 1967) $\underline{7}$, 115-130.
3. PEACEMAN, D.W.
"Fundamentals of Numerical Reservoir Simulation", Elsevier, Amsterdam, (1977) 65-82
4. LANTZ, R.B.
"Quantitative Evaluation of Numerical Diffusion (Truncation Error)", Society of Petroleum Engineers J. (1971), $\underline{11}$, 315-320; Trans. AIME, 251
5. CHAUDHARI, N.M.
"An Improved Numerical Technique for Solving Multi-Dimensional Miscible Displacement Equations", Society of Petroleum Engineers J. (1977), $\underline{11}$, 277-284; Trans., AIME, $\underline{251}$
6. VAN QUY, N., SIMANDOUX, P. and CORTEVILLE, J.;
"A Numerical Study of Diphasic Multicomponent Flow", Society of Petroleum Engineers J. (April 1972), $\underline{12}$, 171-184; Trans., AIME $\underline{253}$
7. LAUMBACH, D.D.;
"A High Accuracy, Finite Difference Technique for Treating the Convection-Diffusion Equation", Society of Petroleum Engineers J., (1975) $\underline{15}$, 517-531
8. GARDNER, A.O. and PEACEMAN, D.W. and POZZI, A.I.;
"Numerical Calculation of Multidimensional Miscible Displacement by the Method of Characteristics", Society of Petroleum Engineers J. (1964), $\underline{4}$, 26-36
9. TODD, M.R., ODELL, P.M., and HIRASAKI, G.J.;
"Methods for Increased Accuracy in Numerical Reservoir Simulators", Society of Petroleum Engineers J. (1972), $\underline{12}$, 515-530
10. BANKS, D., CHESHIRE, I.M., and POLLARD, R.K.;
"A Technique for Controlling Numerical Dispersion in Finite-Difference Oil Reservoir Simulation", Proceedings of BAIL Conference, Dublin (June 1980), 99-203
11. WHEATLEY, M.J.;
"A Version of Two Point Upstream Weighting For Use in Implicit Numerical Reservoir Simulators", paper presented at Society of Petroleum Engineers 5th Symp. On Reservoir Simulation, Denver, 1979; SPE Paper No. 7677
12. NGHIEM, L.X., FONG, D.K., and AZIZ, K.;
"Compositional Modelling with An Equation of State", SPE Paper 9306, SPE Annual Fall Meeting, Dallas, Texas (September 1980)
13. HELFFERICH, F.G.;
"General Theory of Multicomponent, Multiphase Displacement In Porous Media", Society of Petroleum Engineers J. (February 1981), $\underline{21}$, Trans., AIME, $\underline{261}$
14. COATS, K.H.;
"An Equation of State Compositional Model", Society of Petroleum Engineers J. (October 1980), $\underline{20}$, 363-377

438

15. LEACH, M.P. and YELLIG, W.F.;
"Compositional Model Studies: CO_2 - Oil Displacement Mechanisms", SPE Paper 8368, SPE Annual Fall Meeting, Las Vagas, Nevada (September 1979)
16. SMITH, L.R. and YARBOROUGH, L.;
"Equilibrium Revaporization of Retrograde Condensate by Dry Gas Injection," Trans. AIME, (1968), 243 87-94
17. PEACEMAN, D.W.;
"A Nonlinear Stability Analysis for Difference Equations Using Semi-Implicit Mobility", Society of Petroleum Engineers J. (February 1977), 17, 79-91; Trans., AIME 259

APPENDICES

A1. Stability, Truncation Errors and Numerical Dispersion

The nonlinear convection equation is:

$$-v\frac{\partial f(c)}{\partial x} = \frac{\partial c}{\partial t}. \tag{A1.1}$$

Truncation error analysis on the finite difference approximation of the linearised equation

$$-vf'\frac{\partial c}{\partial x} = \frac{\partial c}{\partial t} \tag{A1.2}$$

shows a leading truncation error term of the form:

$$D_{num}\frac{\partial^2 c}{\partial x^2}, \tag{A1.3}$$

$$\text{where } D_{num} = vf'\Delta x \left[\left(W-\tfrac{1}{2}\right) + vf'\tfrac{\Delta t}{\Delta x}\left(\theta-\tfrac{1}{2}\right)\right].$$

W and θ are the distance and time weighting factors for C in the difference equation. By solving the difference equation of equation A1.2, we are, in effect, solving a diffusion - convection equation of the form:

$$D_{num}\frac{\partial^2 c}{\partial x^2} - vf'\frac{\partial c}{\partial x} = \frac{\partial c}{\partial t}. \tag{A1.4}$$

This creates artificial diffusion of C, and is termed numerical dispersion. Linearised stability analysis shows that the numerical solutions are stable if the weighting factors lie in the range 0.5 to 1 (Equal grid spacing). Peaceman (17) showed that a nonlinear stability analysis gave the same practical criteria for a full upstream difference scheme (W=1). The results of the linearized stability analysis are summarised in the diagram shown below. The approximate stability subdomain in which the dynamic weighting scheme is operating is more restrictive than that permitted by the linearized stability analysis.

A Schematic Illustration of The Numerical Stability Domains

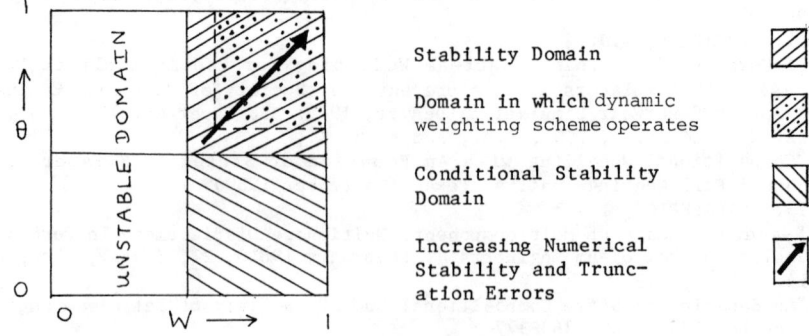

A.2   Additional Notes on 2-Point Upstream Weighting and Mid-Point Weighting
Schemes

The previous curve analysis shows that 2-point upstream weighting cannot be
applied on Group 2 curves.   Todd et al (9) showed 2 cases which, according to
our present analysis, belong to the Group 2 curves category.

Case 1: Todd et al showed an example of unit mobility, miscible displacement
of oil by solvent.   It is possible to calculate oil relative permeability
which is greater than 1 by 2-point upstream extrapolation, as shown. Todd et al
recommended setting the spurious extrapolated value to the maximum of the 2
bounding values.   Our curve analysis indicates that this is effectively
downstream weighting and could create an undershoot of the oil phase if it is
approaching zero saturation.   A midstream interpolation is the best
alternative.   It is effectively upstream, but not fully upstream weighted on
the actual curve profile.

Case 2: Todd showed that a spurious extrapolation error would occur near a
sharp WOC or GOC.   This is a Group 2 curve situation created by the use of
relative permeability (Kr) extrapolation, which is less consistent than
saturation extrapolation on the following grounds:-
(1)   Saturations at the block interface   dictate the interblock flow.
However, the extrapolated relative permeabilities will not correspond to a
total saturation of 1.
(2)   It creates, or accentuates the creation of, a Group 2 profile (which is
not amenable to linear extrapolation).   In this example, a Group 1 saturation
profile exists.   Had saturation extrapolation been used, Krw at i=2½ would be
0.7 instead of the spurious negative value.
(3)   Typically, for a water flooding problem, Group 2 saturation profiles
exist above the shock front saturation value.   The corresponding Krw profiles

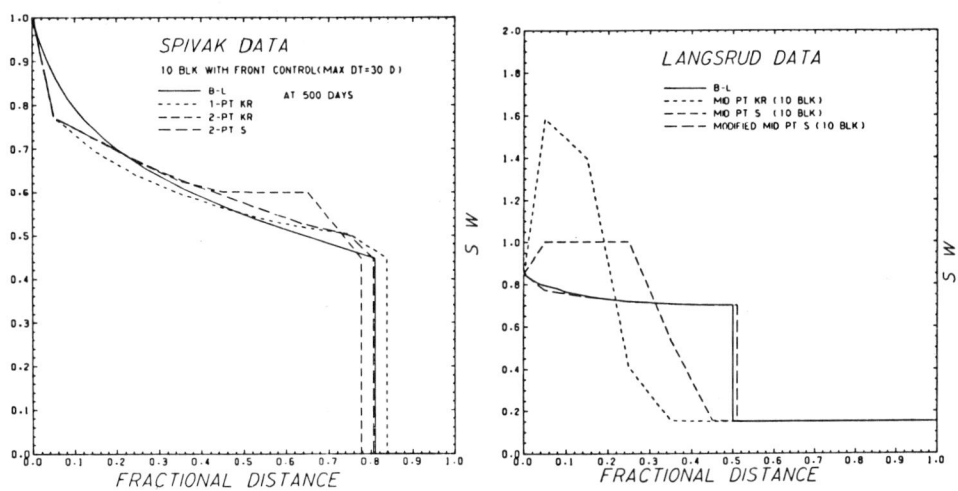

Figure 10 - Comparison of water saturation profiles for various
mobility evaluation schemes (black oil model).
(a) Spivak data (SPEJ, February 1977).   (b) Langsrud data (Nolen
and Berry, SPEJ June 1972).

have stronger Group 2 characteristics. The fact that explicit 2-point Kr extrapolation does not cause instability is probably due to the non-sharpening nature of this saturation range. On the other hand, Group 1 curves (which are not amenable to interpolation) are present around the flood front. The saturation range below the shock front saturation is self-sharpening, thereby aggravating the weakness of interpolation. This supports the evidence that mid-point weighting is unstable.

Figure 10a illustrates some weaknesses of using Kr extrapolation, normally not observable without imposing frontal control. An explicit version of the dynamic weighting scheme, using 2-point saturation extrapolation at the controlled front and midstream weighting everywhere else, is illustrated in Figure 10b.

INTERPHASE MASS TRANSFER EFFECTS IN IMPLICIT BLACK OIL SIMULATORS

D. BANKS and D. K. PONTING

Atomic Energy Research Establishment,
Harwell, Oxfordshire, England

ABSTRACT

Mass transfer may be described in black oil simulators by allowing oil and gas to exist in both liquid and vapour phases. An efficient fully implicit method of simultaneously modelling bubble and dew point is described. A subtracted total gas formulation is found to combine the advantages of the free and total gas approaches. A partial re-solution algorithm option is described which interpolates between total- and no- re-solution logic. The dispersion of dissolved gas and vapourised oil is discussed.

INTRODUCTION

Black oil simulators, characterised by the treatment of just two hydrocarbon components, have traditionally been more concerned with displacement mechanisms that the PVT dominated processes of EOR studies. Compositional effects are modelled simply by mass transfer between liquid and vapour phases. In this paper we discuss a number of aspects of mass transfer in black oil simulators, mainly from the standpoint of a fully implicit formulation.

Black oil models generally describe the concentration of dissolved gas in the reservoir liquid by the bubble point pressure, P_b, or the solution gas–oil ratio, R_s. The quantity of oil in the vapour is described by the dew point pressure, P_d, or the vapour oil–gas ratio r_s or R_v,[1,2]. The vapour oil–gas ratio is generally preferable, as it enables the vapour to be described in regions of low pressure and r_s, for which no dew point exists. The 'oil' and 'gas' may be any two groups of hydrocarbon components, or true stock tank oil and gas.

In a general black oil model there are therefore five independent variables per cell: P_o, P_b or R_s, P_d or r_s, S_w, S_g. The equations determining P_b and r_s would involve diffusion, convection and mass transfer rates. At present, the extra computational effort required to solve the fourth and fifth equations is prohibitive. The three equation picture may be restored by employing bubble and dew point models, making two of the variables dependent on the primary ones. A variable substitution method for simultaneously modelling of P_b and r_s variations in an implicit black simulator is described in Section 2. While this cannot yield the detailed description obtained from a true multi-component compositional simulator, the greater computational efficiency enables detailed full field studies to be performed. Facilities such as fault connections and directional relative permeabilities are then available for studies involving mildly volatile oils or dew point transitions, and numerical dispersion may be limited by the use of small grid blocks. In our experience variable substitution is the only method of modelling mass transfer which does not limit the ability of the simulator to take large time steps, although other methods are possible if the time step length is restricted.

The quantity of gas existing in solution may be typically 50–1000 times greater than that existing in the vapour phase. The question arises as to whether the mass conservation equation for gas should involve all the gas, or just the free component. Originally a free gas formulation was used in PORES [10]. In solving the material conservation equations, however, a column sum condition is imposed which attempts to zero the sum of errors on diagonal planes of cells within the reservoir model. This is particularly important in the sequential method of solving the linear matrix equations. For a free gas formulation, the column sum condition represents an attempt to conserve free gas, a conservation condition violated when interphase mass transfer occurs. A total gas formulation avoids this, but, due to the large dissolved gas contribution, leads to poorly conditioned equations which the sequential method frequently fails to solve. A subtracted total gas method which overcomes this is described in Section 3. Such methods may be important for compositional simulators as the increasing number of equations renders fully simultaneous solution methods impractically expensive.

Gas must come out of solution when the oil pressure crosses the bubble point, but the re-solution of gas depends on the presence of gas in contact with liquid oil, and the rate at which solution occurs. Experiment [7] indicates that, where an intimate gas-oil contact exists, equilibrium is established on a timescale short compared to those typically involved in reservoir engineering. The determining factor in gas solution is the rate at which gas diffuses through liquid oil. In PORES, and other black oil simulators, two alternatives are available for the treatment of gas solution. These are the no- re-solution and total- re-solution options. No- re-solution assumes that dissolved gas does not diffuse through oil, so that a layer of saturated oil will immediately build up at a gas-oil interface and prevent further solution. While this option is logically consistent, it is unrealistic for residual oil droplets, and will overestimate gas cap sizes. Using typical diffusion coefficients, it can easily be shown that a 1mm droplet will reach 99% of its ultimate dissolved gas concentration in less than 24 hours.

Total re-solution logic assumes that interphase equilibrium always exists in each cell, so that free gas may only exist with saturated oil. This assumption of instantaneous equilibrium is usually also made in compositional simulators, and essentially implies instantaneous flow of dissolved gas through oil. In practice, however, vapour invading oil is likely to finger or channel, resulting in the gas by-passing some of the oil. Free gas may then pass through a cell without completely saturating the oil. In Section 4 we describe a partial re-solution option which enables the engineer to set a re-solution or equilibrium fraction for each cell, the fraction of the liquid hydrocarbon in a cell in contact with the vapour. This can still be fully expanded in a three variable formulation, and is similar to the trapping fraction approach.

Simulators which permit gas solution have difficulties with the dispersion of dissolved gas and cell size dependence. These problems are, if anything, more severe for vapourised oil. This is due to the non-specification of the determining diffusion rates, so that changes in R_s and r_s are immediately propogated across cells. In a sense, the artificial cell boundaries introduced by the simulator prevent dispersion from being total, rather than causing it. No re-solution logic has an equivalent problem in that gas is evolved from undersaturated oil when flow occurs across an R_s gradient. It is possible to control this dispersion for the simple case of free gas invading oil by only allowing R_s to rise due to contact with free vapour. However, such methods run into trouble when a dry vapour or dissolved gas slug is propogated, as they modify the slug shape by sharpening the leading edge.

MODELLING BUBBLE AND DEW POINT VARIATIONS

Two main approaches exist to modelling mass transfer in black oil simulators: the variable substitution method [4,5], and one cell methods in which cell properties are modified to be consistent with the solution in terms of a fixed set of variables [3]. These correspond to saturation pressure and flash techniques in compositional simulation [5,6]. Both methods have been used in PORES, and variable substitution has proved superior, although it involves organisational difficulties in keeping track of whether the third solution variable is S_g, P_b or r_s. The cell by cell technique retains P_o, S_w and S_g as solution variables, and adjusts $R_s(P_b)$ to satisfy phase equilibrium. If gas injection occurs into undersaturated oil, for example, R_s is increased, as in the pseudo solution gas method. Unless this is done exactly, a negative gas saturation is obtained on the subsequent iteration and generally causes material balance errors. This can be overcome by using a one cell Newtonian iteration to exact material balance to fix the R_s change precisely. This yields a working scheme, but runs into convergence problems on long time steps, as the pressure changes which occur when free gas goes into solution disturb the interblock flows in a manner not incorporated into the Jacobian of the Newtonian iteration. In the undersaturated oil case the gas equation is being converged to a known solution, $S_g = 0$; more precisely, the bubble point is implicit, but not fully expanded.

Variable substitution does not attempt to retain gas saturation as the third variable at all times. Depending on the conditions in a cell, gas saturation, bubble point or vapour oil-gas ratio, r_s, may be the primary solution variable. In each case it is crucial that the functional dependence of the secondary variables, (such as P_b and r_s in a cell in which S_g is the primary variable), and of functions of these secondary variables, is known and included in the Jacobian. The omission of apparently minor terms from the Jacobian can limit convergence of the non linear equations unacceptably. However, the exact calculation of interblock and well flows at the advanced time level, which is obtained with increasing accuracy as the Newtonian iteration converges, prevents instabilities which can occur using first order approximations to the implicit flows [11].

Assuming interphase equilibrium, there are only three possibilities for the state of a cell:-

(i) Vapour only. P_o, S_w and r_s are solution variables, with $S_g=1-S_w$ and $P_b=P_o$

(ii) Liquid and vapour hydrocarbon present. P_o, S_w and S_g are solution variables, with $P_b=P_o$, $r_s=r_s^{sat}(P_o+P_{cog}(S_g))$

(iii) Liquid only. P_o, S_w and P_b are solution variables, with

$$S_g=0 \text{ and } r_s=r_s^{sat}(P_o+P_{cog}(0)) \dots\dots\dots\dots\dots\dots\dots\dots (1)$$

$r_s^{sat}(P_g)$ is the curve describing the oil-gas ratio for vapour in equilibrium with liquid oil.

The mass conservation equations take the form

$$R_j = \frac{1}{\Delta T}\left[m_j^{T\Delta T} - m_j^{T}\right] - q_j^{T\Delta T} - \sum_n f_{nj}^{T+\Delta T} = 0 \quad \ldots\ldots\ldots\ldots (2)$$

$j=1,..,N$, N the number of cells.

Elements of the residual, R, mass terms, well terms and flows have a three vector form:-

$$R_j = \begin{vmatrix} R_{oj} \\ R_{wj} \\ R_{gj} \end{vmatrix} \quad m_j = \begin{vmatrix} S_o b_{or} + r_s S_g b_{gr} \\ S_w b_{wr} \\ S_g b_{gr} + R_s S_o b_{or} \end{vmatrix} \quad f_{nj} = \begin{vmatrix} f_{nj}^{o} + r_{s_{nj}} f_{nj}^{g} \\ f_{nj}^{w} \\ f_{nj}^{g} + R_s f_{nj}^{o} \end{vmatrix} \quad q_{jw} = \begin{vmatrix} q_{jw}^{o} + r_{s_{jw}} q_{jw}^{q} \\ q_{jw}^{w} \\ q_{jw} + R_s q_{jw}^{o} \end{vmatrix}$$

$$\ldots\ldots\ldots\ldots (3)$$

where f_{nj}^{o}, f_{nj}^{w} and f_{nj}^{g} are the free oil, water and free gas flows given by Darcy's Law in the usual way, and g_{jw} are the corresponding well terms.

The equations given by $R(X^{T+\Delta T})=0$ are solved by Newtonian iteration, derivatives being taken with respect to the primary solution variables for each cell. Transitions may occur between the three states of (1), on the basis of the current approximation to the advanced time step solution, as follows:-

From state   (i), if $r_s > r_s^{sat}(P_o + P_{cog}(1-S_w))$. Set $r_s = r_s^{sat}$, $S_g = \epsilon$,
change to (ii)

From state   (ii), if $S_o < 0$. Set $r_s = r_s^{sat}(P_o + P_{cog}(S_g)) - \epsilon$, $S_g = 1 - S_w$,
change to (i)

From state   (ii), if $S_g < 0$. Set $S_g = 0$, $P_b = P_o - \epsilon$, change to (iii)

From state (iii), if $P_b > P_o$. Set $P_b = P_o$, $S_g = \epsilon$,   change to (ii)

$$\ldots\ldots\ldots\ldots\ldots (4)$$

This is essentially a combination of the methods proposed by Cook et al [2] and Spivak and Dixon [1].

The extra cost of modelling r_s variation is small, as cells in which r_s is   the solution variable would otherwise be repeatedly solved for a constant gas saturation of zero. When r_s is not the primary variable, the effect is merely to add extra terms to the Jacobian. This enables effects such as the vapourisation of residual oil into re-injected gas, an EOR type process expected to some extent in most reservoirs with gas injection, to be followed, as well as gas solution and a primary recovery waterflood.

THE SUBTRACTED TOTAL GAS FORMULATION

The residual in (3) includes terms for total oil, water and total gas. The corresponding free gas residual is $R_j{}^{fg} = R_j{}^o - R_{sj} R_j{}^g$. A free gas formulation was originally used in PORES. Both the iterative linear solver and sequential method use column sum methods to preserve zero residual sum—effectively material balance on diagonal planes of cells. This constraint, which generally speeds convergence, has little value if a free gas formulation is used, as free gas residual sum does not correspond to material balance if saturated oil is present.

It is possible not to differentiate the R_{sj} term in the residual, leading to a set of equations, which, if solved exactly, are equivalent to those obtained from a total gas residual. The Jacobian is then not the derivative of the residual, and some of the convergence properties of the full Newton method are lost.

The alternative is to use a straightforward total gas formulation. This is possible for simultaneous solution methods, but the sequential method fails completely. The dissolved gas contribution swamps the free component in the equation determining the gas saturation.

The sequential method of solving the linear equations, in the simple case of a non-condensate gas oil system, involves the matrix decomposition:-

$$\begin{vmatrix} J_{o,p} & J_{o,g} \\ J_{g,p} & J_{g,g} \end{vmatrix} \simeq \begin{vmatrix} 1 & D \\ 0 & 1 \end{vmatrix} \begin{vmatrix} J_{o,p} - DJ_{g,p} & 0 \\ J_{g,p} & J_{g,g} \end{vmatrix} \qquad \cdots\cdots\cdots (5)$$

The elements of R_s are given by $(R_s)_{ij} = S_{ij} R_{sj}$, and D is a diagonal matrix, such that $DJ_{g,g} \simeq J_{o,g}$. The change in ΔS_g over a Newtonian iteration is $X_g = \delta(\Delta S_g)$, defined by

$$J_{g,g} X_g = R_g - R_x^{tg} \qquad \cdots\cdots\cdots\cdots\cdots\cdots\cdots\cdots\cdots\cdots\cdots (6)$$

where

R_x^{tg} is approximate, given by $R_x^{tg} = J_{g,p}(J_{o,p} - DJ_{g,p})^{-1}(R_o - D(R_{fg} + R_s R_o))$

If a free gas formulation is used, X_g is given by

$$(J_{g,g} - R_s J_{o,g})X_g = (R_g - R_s R_o) - R_x^{fg} \equiv R_{fg} - R_x^{fg} \qquad \cdots\cdots\cdots (7)$$

Errors arise in the evaluation of the right-hand side of (6). In the free gas case R_x^{fg} consists of an approximate matrix acting on $R_o - D R_{fg}$, while in a total gas formulation R_x^{tg} is given by a approximate matrix acting on $R_o - D(R_{fg} + R_s R_o)$. If the errors in the liquid and vapour flows are comparable, $R_x^{tg} \simeq R_s R_x$, with $R_s \sim 0(10^2 - 10^3)$. The right-hand side of the equation defining gas saturations consists, in the total gas case, of two large cancelling terms, one of which is approximate, and the resulting values of ΔS_g are less accurate than in the free gas case

Errors in S_g are fed back into the oil residual via $S_o = 1 - S_w - S_g$, and the iteration often diverges.

The advantages of both formulations may be combined in a subtracted total gas formulation, in which the gas equation residual is

$$R_{stg} = R_g - R_s^{sub} R_o \quad\dotfill\quad (8)$$

The choice of R_s^{sub} is rather critical, and several alternatives have been tried. The best seems to be $\overline{R}_s$, updated by the predictor at the start of each time step. The use of R_s^{min} rather than $\overline{R}_s$ can increase run times by 50%. The continuous updating of R_s^{sub} also seems essential.

An advantage of this method is that isolated dissolved gas changes, such as those due to gas injection, stand out over the R_s subtraction, causing residuals which the simulator converges out accurately. All that is removed is the bulk of the initial dissolved gas which otherwise causes the gas equation to be a near multiple of the oil equation. The sequential method can still fail when a large initial R_s gradient exists across the reservoir, in which case a fully simultaneous solution method must be used.

THE SOLUTION OF GAS IN OIL

Black oil simulators have tended not to provide the engineer with very comprehensive facilities for investigating gas solution effects. Partly, this is due to a lack of knowledge concerning the processes involved. It seems clear that residual oil droplets will equilibrate quickly, but a metre diameter area of oil will take over a year to saturate if free gas channels past it. Providing no- and total- re-solution options enables the sensitivity of the problem to gas solution effects to be established. If this is a major effect, however, as in the Odeh test problem [9], there are no facilities for history matching. In particular, the degree of equilibrium between phases will be different for residual oil in a gas cap from that attained in the case of gas injection into undersaturated oil, and these processes may occur in the same study.

The PORES partial re-solution option allows the user to define a re-solution or equilibrium fraction, f, of the oil in a cell which is in close contact with vapour. For a time step from T to T+ΔT, the partial re-solution option may be summarised as:-

Undersaturated oil, P_b is solution variable

If $P_{b_{min}}^T < P_b^{T+\Delta T} < P_o^{T+\Delta T}$ then $S_g^{T+\Delta T} = 0$, $P_{b_{min}}^{T+\Delta T} = P_{b_{min}}^T$

If $P_b^{T+\Delta T} < P_{b_{min}}^T$ then $S_g^{T+\Delta T} = 0$, $P_{b_{min}}^{T+\Delta T} = P_b^{T+\Delta T}$

Saturated oil, S_g is solution variable

If $S_g > 0$ then $P_b^{T+\Delta T} = P_o^{T+\Delta T}$, $P_{b_{min}}^{T+\Delta T} = Min\left[P_{b_{min}}^T , P_b^{T+\Delta T} \right]$

Bubble point transition, gas appearing

$$\text{If } P_b^{T+\Delta T} > P_o^{T+\Delta T} \text{ then } S_g \text{ set to } \epsilon, P_b^{T+\Delta T} = P_o^{T+\Delta T}, P_{b_{min}}^{T+\Delta T} = \text{Min}\left[P_{b_{min}}^T, P_b^{T+\Delta T}\right]$$

$$\dots\dots\dots\dots\dots \quad (9)$$

P_b^{min} is the minimum bubble point attained by the cell during the simulation. The quantity of dissolved gas in the cell and interblock flows is obtained as a weighted average using

$$R_s(P_b) \text{ and } R_s(P_b^{min}), \quad P_b^{min}$$

acting as bubble point for the 'trapped' or non-equilibrium oil fraction. This model yields total and no-resolution as the limits of f=1 and 0 respectively. The r_s value can be followed for the vapour as described in Section 2. All functional derivatives can be expanded in the Jacobian in terms of the primary variables.

There are several respects in which this partial re-solution scheme is less than ideal:-

(i)   If pressure drops through the bubble point of the by-passed oil, and gas comes out of solution at less than mobile saturation, then it will not re-dissolve in the non-equilibrium fraction or re-pressurisation. This could be allowed, but it seems unrealistic to ascribe different behaviours to gas at just under and over critical saturation. In addition, discontinuous changes in the functional form of the residual can slow convergence.

(ii)   When fingering or channelling occurs, it would be expected that transverse saturation would occur behind the front. This does not occur in this model.

Other possibilities exist for partial re-solution options, such as re-solution in residual oil. However, when gas displacement is stable, due to gravity seg-regation, this is equivalent to total re-solution. For gas injection, residual oil saturation is rarely attained. There is also the difficulty of identifying the residual oil if the cell is subsequently flushed with mobile oil.

It would be possible to have a similar option for oil vapourisation. In most condensate studies, however, oil saturations remain less than critical, so that equilibrium is a reasonable assumption. The equivalent of gas injection rarely arises.

DISPERSION PROBLEMS IN MASS TRANSFER

Variations in R_s and r_s, due to convection and mass transfer usually show dispersion effects. When, for example, the bubble point rises in a cell due to gas solution this rise is assumed to occur evenly throughout the entire cell, and is communicated to its neighbours by oil flows. The resulting rise in R_s for the neighbouring cell is then passed, in the same time step, to the next cell. In cases of high throughput ratio, a considerable fraction of the reservoir may need to be saturated before free gas appears in the injection cell. This problem might be expected to be rather less severe in IMPES type simulators, which effectively impose an upper limit of $\Delta X/\Delta T$ on the speed at which dissolved gas or vapourised oil may propogate, as diffusion is limited to one block per time step. However, such a simulator is likely to take more steps than a fully implicit one to solve a given problem, off-setting the advantage of lower dispersion per time step.

For simple gas injection into undersaturated oil, it is fairly easy to prevent this dispersion by assuming that gas passing into a cell saturate oil only as it overrides it, so that an increased R_s value will not appear at the downstream interface until the cell is saturated. The results of such a technique on the case 2 Odeh problem are shown in fig. 1. However, this method of dispersion control runs into problems when the convection of slugs is considered. Particularly in the case of condensate reservoirs, where gas injection may be controlled by production rates and sales contracts, the resulting r_s distributions may not have simple shapes. Front sharpening methods will tend to distort such slugs by sharpening the leading edge. Interblock flow schemes other than upstreaming will yield unphysical results if flow occurs from a cell of zero r_s to one of finite r_s, but upstreaming causes unacceptable artificial diffusion.

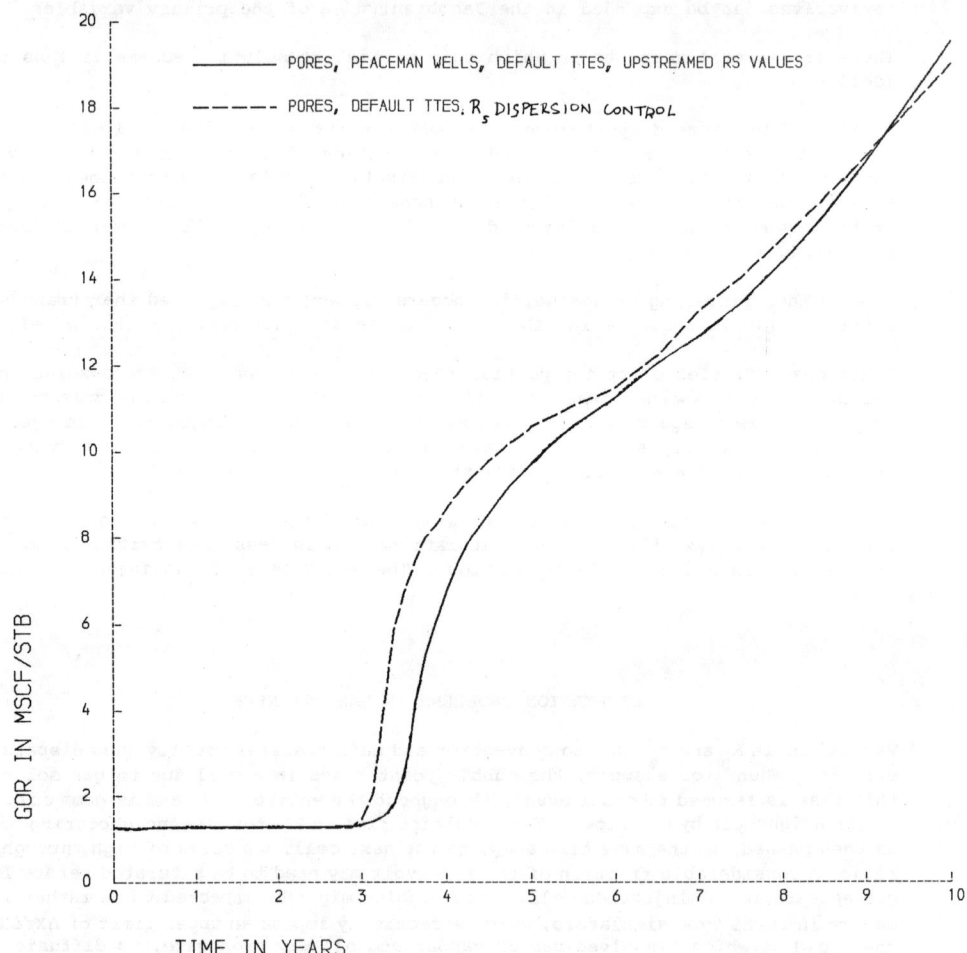

CASE 2, GOR VS TIME

——————— PORES, PEACEMAN WELLS, DEFAULT TTES, UPSTREAMED RS VALUES

– – – – – PORES, DEFAULT TTES. R_s DISPERSION CONTROL

GOR IN MSCF/STB

TIME IN YEARS

Figure 1.

We can present no simple solution to the dispersion problem, and it may be that it is inherent in the concept of using a dew or bubble point model rather than a separate equation. The basically convective nature of r_s and R_s transport suggests a point following algorithm, although those suggested to date are not fully implicit. It may be cheaper, however, to add an extra transport-type equation to the traditional black oil picture than to go to the number of cells required to reduce dispersion to an acceptable level.

CONCLUSIONS

(i) It is possible to model gas condensate and bubble point effects simultaneously in an efficient general purpose black oil simulator. Minor oil vapourisation and condensate effects will occur in many studies, and these can be included at little extra cost.

(ii) There is a need to provide engineers with a more flexible method of matching gas solution effects. A re-solution fraction approach enables oil by- pass and channelling effects to be included, and has a simple functional form which is particularly suitable for implicit simulators. No- and total- re-solution options are obtained as limiting cases.

(iii) For sequential methods, a continuously modified subtracted total gas formulation is preferable to either total or free gas formulations.

(iv) All bubble and dew point models are a poor excuse for solving a dissolved gas or vapourised oil equation. In particular, numerical dispersion of vaporised oil and dissolved gas can be significant. This can be controlled in simple cases, such as dry gas invading undersaturated oil, but dispersion control methods may distort the shape of convected slugs. This may be particularly important for gas injection into condensate reservoirs.

NOMENCLATURE

b_p : Inverse formation volume factor for phase p, defined as $\rho_p^{reservoir}/\rho_p^{surface}$, where ρ is fluid density

b_{pr} : $b_p \times b_{rock}$

f_{nj}^p : The flow rate, measured in terms of surface volume, of phase p, from cell n to cell j

N : The number of active cells in the reservoir

P_{bj} : The bubble point pressure of cell j

P_{dj} : The dew point pressure of cell j

P_{pj} : The pressure of phase p in cell j

S_{pj} : The saturation of phase p in cell j

q_{jw}^p : The rate of flow, measured in terms of surface volume, from well w to cell j

R_p^j : The residual of the mass convervation equation for cell j, phase p

$J_{p,v}^{ij}$: The element of the Jacobian, $\partial R_p^i / \partial X_{vj}$

X_{pj} : The pth primary solution variable for cell j

R_s^{min} :The minimum R_s value in the reservoir model

R_s : The mean R_s value in the reservoir model

REFERENCES

1. SPIVAK. A. and DIXON, T.N.; "Simulation of gas condensate reseervoirs", SPE4271, Proc. 3rd Symp. on Numerical Simulation of Reservoir Performance, Houston, 1973.

2. COOK, R.E., JACOBI, R.H. and RAMESH, A.B.; "A beta-type reservoir simulator for approximating compositional effects during gas injection", Soc.Pet.Eng.J., (Oct. 1974), 471-481

3. AU, A.D.K., BEHIE, A, RUBIN, B. and VINSOME, K.; "Techniques for fully implicit reservoir simulation", Proc. 55th Ann. Fall Conf. and Exhibition of SPE, Dallas, 1980.

4. BANSAL, P.P. et al; "A strongly coupled, fully implicit, three dimensional, three phase reservoir simulator, SPE8329, Proc. 54th Ann. Fall Conf. and Exhibition of SPE, Las Vegas, 1974.

5. COATS, K.H.; "An equation of state compositional model", Soc.Pet.Eng.J., (Oct. 1980), 363-376

6. NGHEIM, L.X., FONG, D.K. and AZIZ, K.; "Compositional modelling with an equation of state", SPE9306, Proc. 55th Ann. Fall Conf. and Exhibition of SPE, Dallas, 1980.
   NOLAN, J.S.; "Numerical simulation of compositional phenomena in petroleum reservoirs", SPE Reprint Series, No. 11, (1973), 269.

7. RAIMONDI, P. and TORCASO, M.A.; "Mass transfer between phases in a porous medium: A study of equilibrium", Soc.Pet.Eng.J. (March 1965), 51-59, Trans. AIME 234

8. SANDREA, R. and NIELSEN, R.; "Dynamics of petroleum reservoirs under gas injection", Gulf Pub.Co., 1974

9. ODEH, A; "Comparison of solutions to a three dimensional black-oil reservoir simulation problem", J.Pet.Tech., (Jan. 1981), 13-25

10. CHESHIRE, I.M. et al; "An efficient fully implicit simulator", EUR179, Proc. European Offshore Conference and Exhibition, London, 1980, 325

11. COATS, K.H.; "Reservoir simulation: A general model formulation and associated physical/numerical sources of instability", Proc. BAIL1 Conf., Dublin, June 1980, 62-76, ed. Miller, J.J.H., Bode Press.

A NOVEL DEVICE FOR CO$_2$ CORE FLOODING

VOLKER MEYN

Institut für Tiefbohrkunde und Erdölgewinnung der TU Clausthal

ABSTRACT

A newly developed core flooding apparatus is described. The apparatus permits the conducting of flood experiments with living oil
within a pressure range between 1 and 600 bar at flooding rates
from 1 to 50 cm^3.h^{-1}. During the experiment, the mass flow of CO$_2$
at the input is held constant. The following data are thereby measured:

oil production, water cut produced , number of moles of gas produced, gas chromatographical gas analysis up to C$_7$, analysis of
stock tank oil up to C$_{26}$, pressure difference.

Because the perfomance of core flooding experiments is feasible only within a certain length limit, whereas the development of the
transition zone near the minimal miscibility pressure requires a
flood distance of at least 1 m, a new experimental set-up has been
tested.

The flooding experiments are to be conducted with the transition zones previously established. This design is based on a publication
by WATKINS. The establishment of the transition zone during the flood
process is simulated in a three-stage mixing device, which consists of inclined pipes with a total length of about 6 m. The mixer
was tested with oil from German oil reservoir. The concentrations of
the components CO$_2$, as well as C$_1$ to C$_{26}$, were recorded gas-chromatographically at the output of the mixing device. With the help of
these experiments, it can be demonstrated that such a mixer is capable of preparing a phase whose composition simulates that in the
real transition zone, even in the vicinity of the minimal miscibility pressure.

INTRODUCTION

Laboratory investigations are being performed in the course of
a project /1/ concerning the possibilities of CO_2 flooding in
West Germany. For the flood experiments, a device which should
be suited for both slim tube tests and core flood experiments
has been developed. A main objective is to provide experimental
data for a simulation study.

A requirement for the use of black oil simulators is that the transition zone is restricted to a single cell. For this reason the
transition zone must be short. Such a requirement cannot be satisfied in the case of core flood experiments with pure CO_2, since
the construction of the transition zone requires a length of at least
1 m /2/.

WATKINS /3/ has demonstrated that the residual oil saturation
can be substantially reduced, even in "short" reservoir models,
by the use of a premixing vessel. Consequently, only a short
length is necessary for constructing the transition zone in this
case. Following this concept, the use of a premixer should permit displacement by a medium whose composition is similar to that
of the transition zone. In order to demonstrate such a possibility, mixer tests and comparable slim tube experiments have
been conducted.

In order to allow a measurement of the unit displacement efficiency, slim cores are being employed during an initial phase.
However, the embedding, especially of slim cores, imposes difficulties because of the high-pressure CO_2 and the temperatures up
to $120^{\circ}C$. In general, organic sealing materials tend to swell
and blister under these conditions.

A further difficulty arises from the invasion of the core by the
adhesive. For this reason a cell of the Hassler type consisting
only of Teflon (PTFE) and stainless steel has been developed.

EXPERIMENTAL SET-UP

The set-up is designed for a pressure up to 600 bar and a temperature up to $150^{\circ}C$. It comprises a pumping unit for injecting
the CO_2, the flood tube, and the analytical equipment.

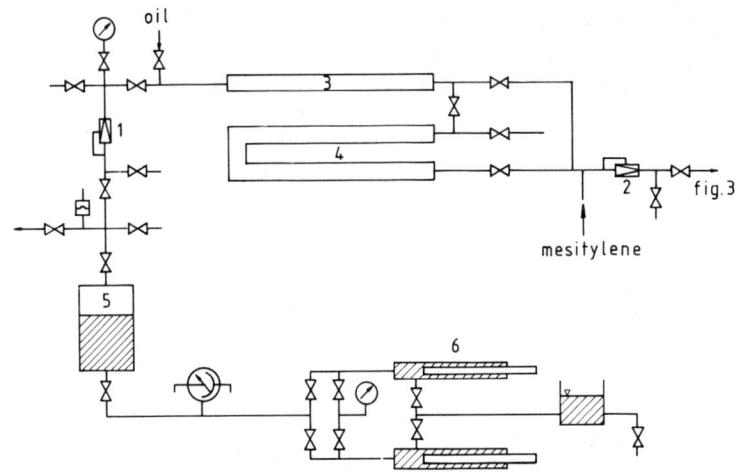

Fig. 1: Set up of the flood device:
1,2 back pressure regulator; 3 mixer tank; 4 flood tube;
5 CO_2-storage vessel; 6 displacement pumps

The set-up (fig. 1) is designed such that the CO_2 mass flow at
the inlet is maintained constant. For this purpose, CO_2 is displaced by mercury at a constant flow rate from a storage vessel
(5), in which the pressure and temperature are maintained constant.
The storage vessel is thermostated at a temperature below the
critical value, in order to keep the compressibility low.
The pressure in the storage vessel is held constant by means of
the back-pressure regulator (1). The pumping rate can be varied
within a range from 0.4 to 50 cm^3/h.
In the case of the slim tube tests, the flood tube (4) consists
of straight tube sections 2 m in length connected by elbows. For
analytical reasons a comparatively large diameter of 0.875 cm
was chosen for the tubing. The flood tube is immersed in a ther-
mostatic oil bath which can accomodate a total length of 30 m.
In order to facilitate the packing of the flood tube, tee fittings were installed after every 4 m of length. The tube bundles
can be easily emptied and therefore reused often.
For the core flood experiments of the first phase, a diameter
of about 11 mm was selected. The aim is to achieve a total
length of 2 m. According to previous experience, cores up to

200 mm in length and with a diameter of 11 mm can be drilled
without difficulty. The drilling of longer cores presents difficulties with the sandstone used here. The core sections are
inserted into a Teflon sleeve. This sleeve is installed in the
stainless steel body by means of packers (fig. 2). Since a single packer seals at two positions, it is possible to drill a hole
through the packer down to the core after the assembly, in order
to attach a pressure transducer.

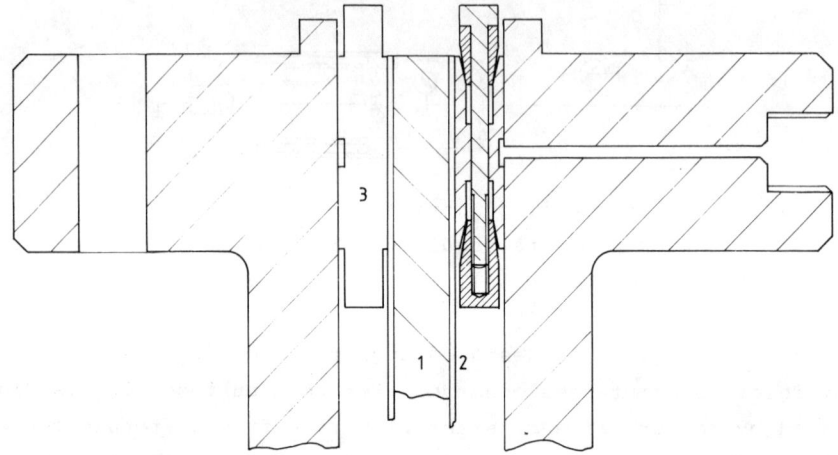

Fig. 2: Packer assembly:
1 core; 2 PTFE sleeve; 3 packer

In order to prevent leakage due to the flow of the Teflon at
elevated temperatures, the packer is designed as an automatic
gasket. To simplify the assembly, the tube consists of four
parts. Each flange includes a port for connecting a pressure
transducer.

Analytical equipment

The analytical equipment (fig. 3) is intended to collect such
data as oil production, brine production and gas production,
as well as composition of the gas up to C_7 and of stock tank
oil up to C_{26}.
Oil/brine is separated from the gas in a small packed column
(1); the gas is subsequently withdrawn into evacuated vessels.
The number of moles of gas is determined by means of a pressure

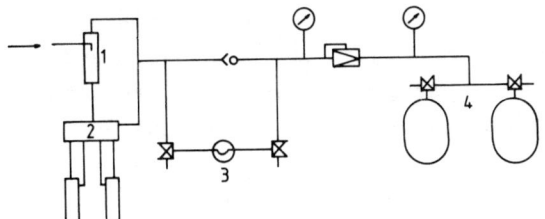

Fig. 3: Analytical equipment:
1 gas/oil separator; 2 brine/oil separator; 3 sample
valve; 4 evacuated vessels

measurement. In order to maintain the pressure in the separator
constant, a back-pressure regulator has been installed. The
advantage of such a set-up is the fact that the measurement is
largely independent of the production rate and that the method
of measurement is cumulative.

A bypass with a sample valve (3) to a gas chromatograph (Perkin
Elmer Sigma 1) is included.

In preliminary tests it became evident that emulsions can be produced with the oil employed. An electrostatic oil/brine separator was therefore installed (fig. 4). This separator has been
milled from Plexiglas. The channels are predominantly 5 x 5 mm
in size. The voltage of 60 V over the plates is sufficient for
separation. The water content in the stock tank oil produced was
less than 0.1 per cent after the separation.

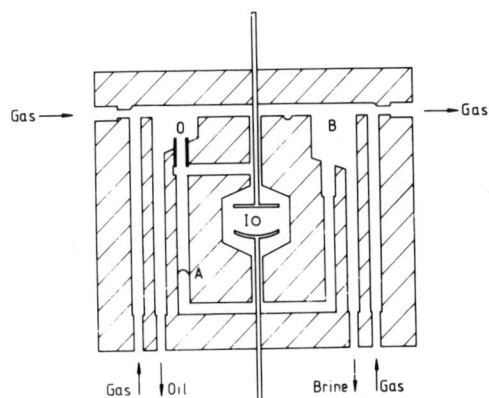

Fig. 4: Sketch of the oil/brine separator
0 oil/weir; B brine/weir, A location of the brine/
water contact

The quantities of oil and brine in the separator are governed
by the difference in heights of the weirs O and B, respectively.
In order to achieve sufficient accuracy, the quantities of brine
and oil in the separator must be kept constant within $\pm$ 0.1 cm^3.
That is, the height of the oil/brine contact at A must be held
constant within $\pm$ 0.2 cm. This is not feasible for a constant
setting of the weirs, since the bond stress between the media
and the wall material can change, for example. Therefore the
separator is automatically tilted when the oil/brine contact at
A deviates from the setpoint. The position is determined by
means of the change in conductivity when the oil/brine contact
passes an electrode. In order to circumvent the remaining problems with the bond stress (creeping of oil, forming of a hemis-
phere at the water weir) a Viton insert was used as oil weir,
and the edge of the brine weir was provided with a cellulose
rider.

The separators are thermostated at a temperature of 20°C. The
oil and brine which emerge from the separator are weighed in
collecting bottles.

In order to determine the overall concentration of the individual oil components, especially in the transition zone, gas chromatographical analyses are performed on both the gas and the oil.
The samples of stock-tank oil are withdrawn from the bottom of
the gas-oil separator by means of a syringe. Sampling behind the
oil-brine separator is not feasible because of the large dead
volume and of the resulting remixing. Gas and oil analyses are
carried out simultaneously in the gas chromatograph. For the gas
phase, a Porapack Q/S-packed 1/8" column 6 m in length is employed; the oil is analysed in a silicone rubber capillary column
40 m in length. This procedure implies that a compromise must be
reached between the analytical requirements imposed by the capillary column and the packed column. The simultaneous execution of
both analyses is necessitated by the long duration of 2 h. The
sample storage required for avoiding such a procedure appears
impractical.

ANALYTICAL PROCEDURE

A calculation of the local concentration under reservoir conditions from the measured data is possible only if the oil analysis is complete. However, the analysis extends only up to C$_{26}$.

Furthermore, the accuracy of injection does not suffice for calculating the molar flux of the components. Hence, a tagging compound, in this case mesitylene, is added to the outflow at constant rate before the pressure reduction. Thus it is possible
to calculate the molar flux of all detected components from the
gas chromatograms.

From the known flow rate in the reservoir model, the concentration
is obtained directly from the molar flux. According to the following equation, the flow rate can be calculated from the known
CO_2-mass flux and the measured pressure values:

$$\frac{dV}{dt} = \frac{k}{\rho(p)} - (V_1 \, x_1 \, \frac{dp}{dt} + V_o x_o \, \frac{d(p - \frac{\Delta p}{2})}{dt} \,)$$

The main purpose of this procedure is to correct for interference due to insufficiencies in the performance of the back-pressure regulator at the outlet.

SLIM TUBE EXPERIMENTS

The slim tube experiments were performed primarily for obtaining
an estimate of the minimal miscibility pressure. In order to determine the residual oil, the flood tube was flushed with a solvent, for example toluene. The flushing process was checked for
thoroughness by means of preliminary tests. For this purpose sand
from the inlet and outlet was analysed for total carbon. For the
oils used here, a mass content of carbon less than 0,1 per cent
was obtained after flushing, and drying by low pressure CO_2.

In order to determine the amount of residual oil in the effluent,
the major part of the solvent was distilled off, and the distillate was analysed gas-chromatographically. Of course, only a small
quantity of oil is present in the distillate. Since the major
share of the residual oil lies beyond C_{26} and therefore cannot be
analysed gas-chromatographically, n-hexane is added to the residue at a ratio of 1 : 1. From the gas chromatogram, the mass ratio of n-hexane to solvent is determined, and thus the residual
oil mass can be calculated with good accuracy.

MIXER TANK

WATKINS /3/ has used premixed media for displacement experiments.
He allowed CO_2 to bubble through the oil from below in an auto-

clave. With the use of such a procedure, a good mixing performance cannot be expected, because of the unfavourable diameter-to-
length ratio. Hence a different method was used. CO_2 is injected
into an oil-filled mixer tank consisting of slightly inclined
tubes (fig. 5).

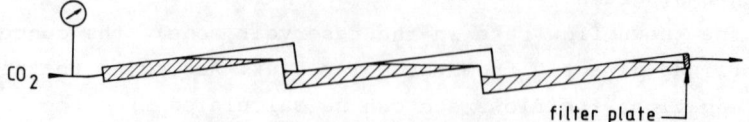

Fig. 5: Sketch of the mixer tank

The tubes have an inner diameter of 0.9 cm and a length of 2 m.
The vertical tube sections, in which the denser liquid phase is
displaced by a gas phase, subdivide the mixer into three stages.
Within the individual stages, the density difference between CO_2-
rich and CO_2-poor oil provides for adequate circulation, whereby
the diffusion paths are short because of the small diameter-to-
length ratio.

The flooding of the tank (fig. 10) shows that only pure oil flows
out at first. If the amount of oil displaced purely by swelling
is calculated from the PVT data for CO_2-oil mixtures, a value
of 170 cm^3 is obtained. The quantity which was produced prior
to the CO_2 break-through was 137 cm^3. After 161 cm^3 had been produced, the stock-tank oil was only slightly coloured. From these
two findings it can be concluded that the equilibration is rather
good in the individual stages.

RESULTS AND DISCUSSION

Exclusively recombined oils were used for the investigation.
For both slim-tube experiments described here, oil with a viscosity of 6 mPa.s (under reservoir conditions) was employed.

For the first slim tube test shown in fig. 6 a sand pack with a
permeability of 3.08 D was used.
For preparing the sand pack for the second test (fig. 8 and 9)
a 1 : 1 mixture of sand and silca powder was used (permeability: 3.03 D).

The first experiment (fig. 6 and 7) was performed at a mean pressure of 219 bar and a CO_2-mass flux of 4.77 g/h, with flood
length of 6.3 m, resulting in a velocity of 5.3 m/d.

Curve I shows the measured pressure difference over the entire
length, curve II shows the amount of stock-tank oil produced,
and curve III shows the quantity of gas produced (fig. 6).

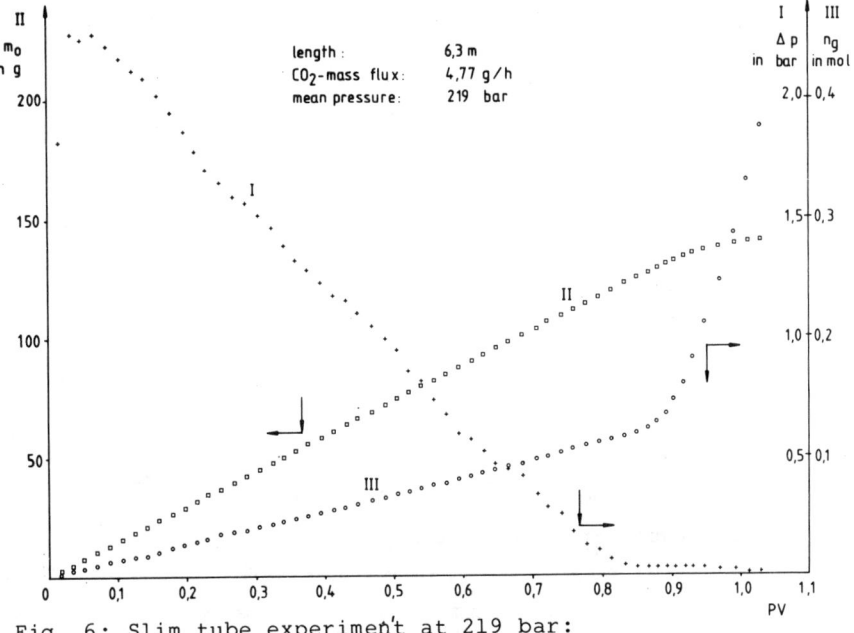

Fig. 6: Slim tube experiment at 219 bar:
Dead oil mass m_o produced, pressure difference Δp,
moles of gas produced n_g plotted versus pore volumes
injected PV

From the produced mass and the flow rate, the "density" of the oil $\frac{dm_o}{dV}$
under reservoir conditions can be calculated. The calculated va-
lues are shown by curve I (fig. 7). The mean "density" before CO_2
break-through amounts to 0.775 g/cm^3. The expected value is 0.805.
This deviation corresponds to a volume effect due to the dissolution of CO_2 in the oil. If it is assumed that the volume effect
due to dissolution during the flood process is comparable with the
effect observed in single-contact PVT measurements, and if the total volume shrinkage is estimated from the length of the transition zone, the following value is obtained for the "density":

$$\frac{dm_o}{dV} = 0.78 \text{ g/cm}^3$$

The decline of dm_o/dV (fig. 7 and 8) after the CO_2 break-through
remains linear over a certain range. Hence it appears plausible

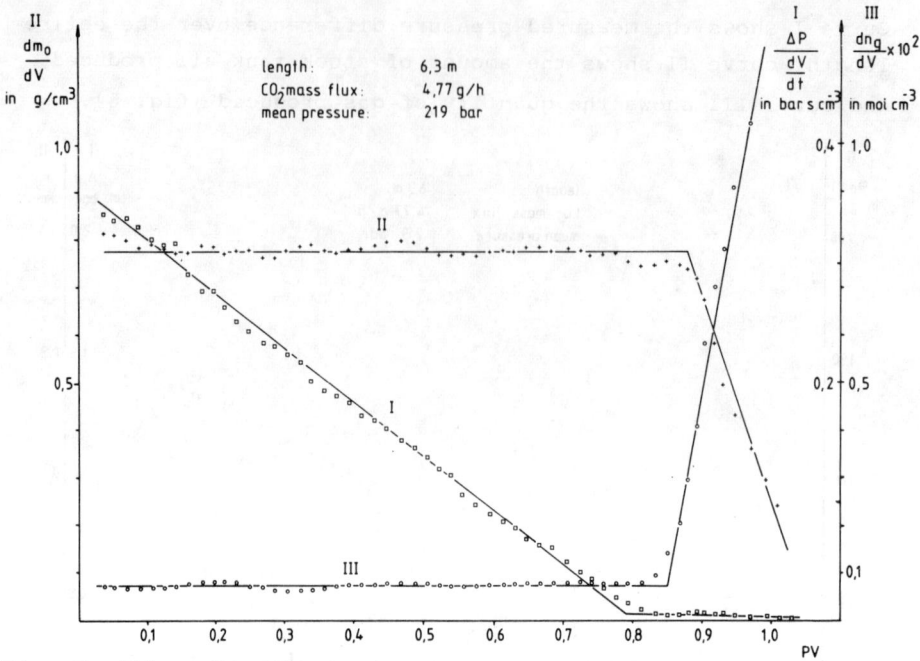

Fig. 7: Slim tube experiment at 219 bar:
"Density" $\frac{dm_o}{dV}$, gas concentration $\frac{dn_g}{dV}$,
flow resistance $\frac{\Delta p}{\frac{dV}{dt}}$ plotted versus pore volumes injected PV

to define the length of the transition zone by means of the inter-
cepts of the straight line with the abscissa and with the horizontal portion of the curve.

Figures 8 and 9 show a selected interval of an experiment at
189 bar and a CO_2 mass flux of 1.98 g/h. In figure 9 the overall
concentration dn_i/dV of C_1, C_2, $i-C_4$ and $n-C_4$ under reservoir
conditions are plotted. These values have been calculated from
the gas production dn_g/dt, the flow rate dV/dt, and the gas-chromatographically measured concentration.

It can be seen that very pronounced concentration maxima occur
for the lower alkanes. Moreover, the maxima are all situated at
the same position. The maximum of the normalized C_1-gas-concentra-
tion $\frac{c}{c_o}$ in the effluent amounts to 1.06 only.

The pressure is approximately equal to the MMP. According to the
literature /2,4,5,6/ the behaviour of methane should differ from
that of the other light alkanes. No such difference is recogni-
zable here. Methane does not show a lead, and the increase in me-

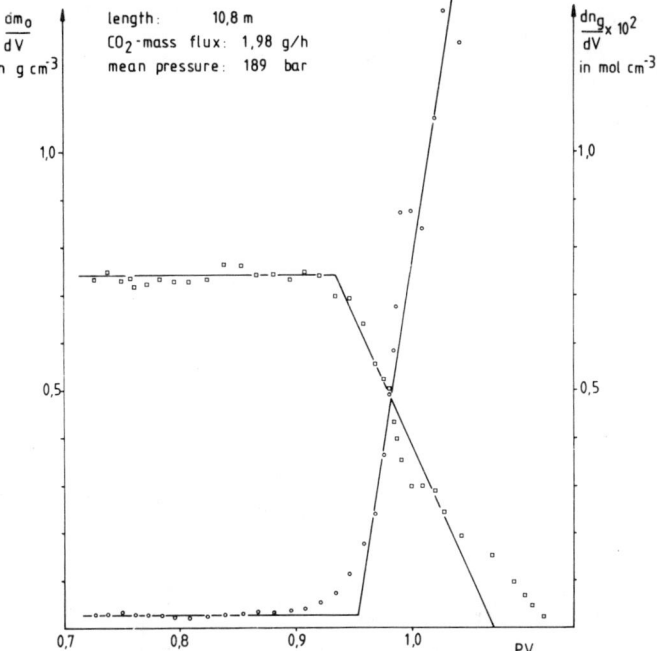

Fig. 8: Part of a slim tube experiment at 189 bar: Density $\frac{dm_o}{dV}$, gas concentration $\frac{dn_g}{dV}$ plotted versus pore volumes injected PV

thane concentration is of the same order of magnitude as that for the other components shown.

Fig. 10 shows the calculated CO_2 and C_1 concentration under reservoir-conditions. It can be seen that the start of concentration increase is at the same position for methane and carbon dioxide.

To demonstrate the similarity between the flooding results obtained with the mixer and with the slim tube, the concentration calculated in the same manner from the mixer outflow is plotted against injected PV in figure 11. This experiment was conducted at a CO_2 mass flux of 9.91 g/h and a pressure of 202 bar. The scale is so chosen that 1 PV corresponds to the mixer volume of 424 cm^3.

The observation that the concentration maxima for C_1, C_2, $i-C_4$ and $n-C_4$ occur at the same time for the mixer test too is especially striking.

462

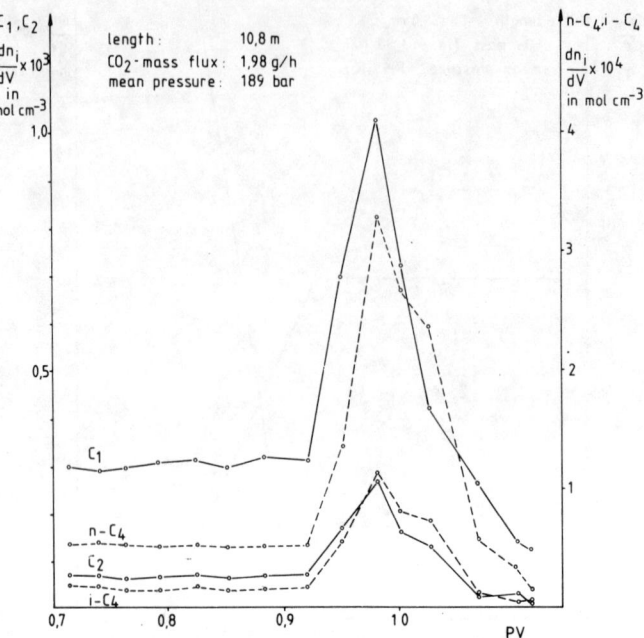

Fig. 9: Part of a slim tube experiment at 189 bar:
Concentration $\frac{dn_i}{dV}$ of C_1, C_2, $i-C_4$, $n-C_4$ under reservoir-conditions plotted versus pore volumes PV injected

The maximal concentrations for the slim tube and mixer tests are presented in the following table.

Table: Maximal concentrations in mol/cm^3

	C_1	C_2	$i-C_4$	$n-C_4$
Slim tube	$1.03 \cdot 10^{-3}$	$2.60 \cdot 10^{-4}$	$1.14 \cdot 10^{-4}$	$3.28 \cdot 10^{-4}$
Mixer	$1.38 \cdot 10^{-3}$	$3.42 \cdot 10^{-4}$	$1.61 \cdot 10^{-4}$	$5.13 \cdot 10^{-4}$

In view of the fact that the maximal value of the concentration is given only by one gas chromatogram, the agreement between the values is remarkably good. Furthermore, it must be taken into consideration that the relative change in flow rate due to dissolution of CO_2 in the oil is certainly different for the slim tube and mixer tests. For both tests it should be emphasized that increase in concentration by a factor of about 5 occurs for the lower alkanes considered here.

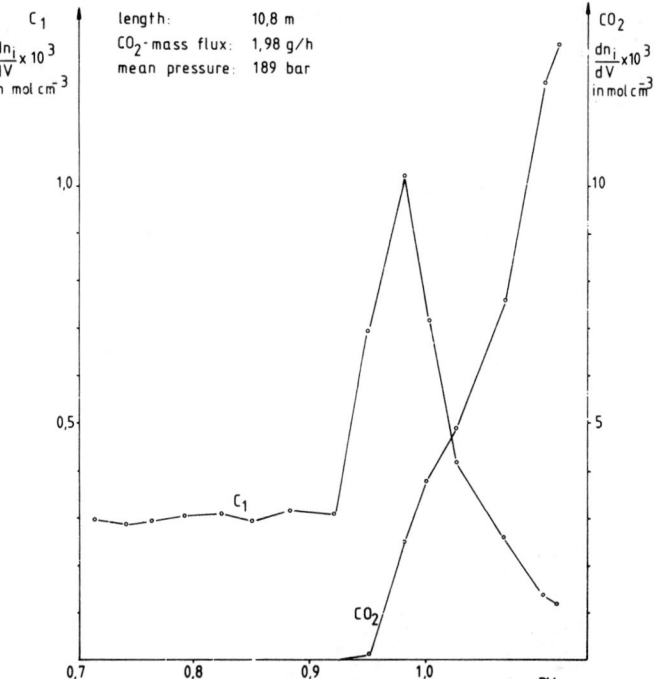

Fig. 10: Part of a slim tube experiment at 189 bar:
Concentration $\frac{dn_i}{dV}$ of C_1 and CO_2 under reservoir conditions plotted versus pore volumes PV injected

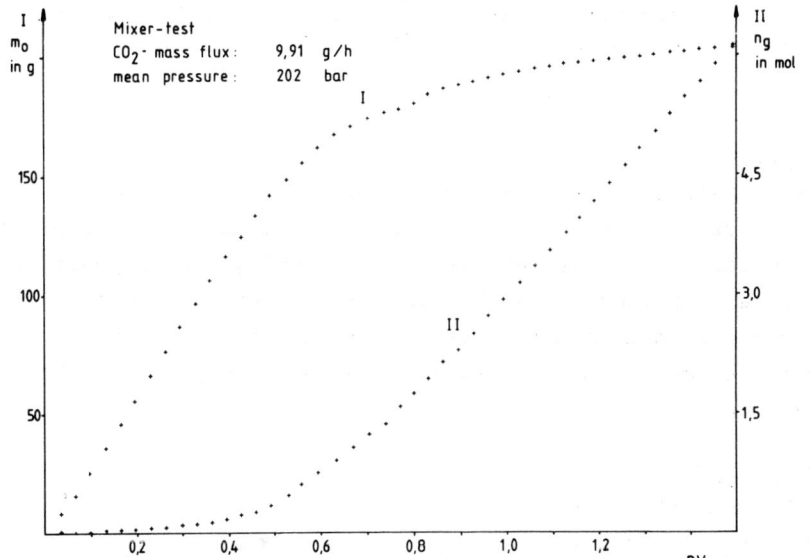

Fig. 11: Mixer test
Dead oil mass produced m_o, moles of gas produced n_g, plotted versus pore volumes PV injected

464

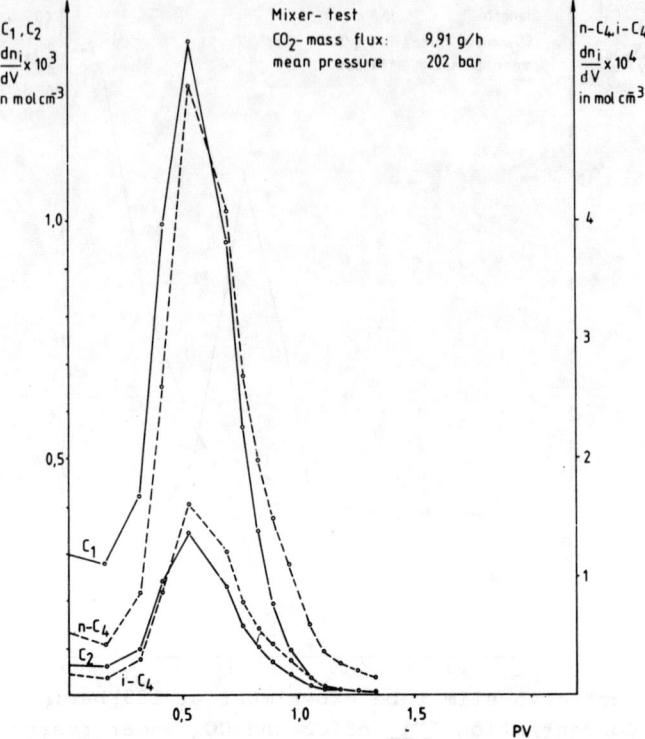

Fig. 12: Mixer test
Concentration $\frac{dn_i}{dV}$ of C_1, C_2, $i-C_4$, $n-C_4$ under reservoir conditions plotted versus pore volumes injected

Conclusions

On the basis of the results obtained so far, the apparatus developed appears to be suited for CO_2 core flooding experiments. In particular, it allows conclusions concerning the composition under reservoir conditions. The use of the three-stage mixer tank for core flooding experiments appears promising. By means of this device, cores can be flooded with media corresponding to a transition zone.

Nomenclature

p	= inlet pressure
$\rho(p)$	= CO_2-density
k	= CO_2-mass flux
V	= volume under reservoir conditions

V_2 = volume filled by CO_2
V_o = volume filled by oil
X_o = compressibility of oil
X_1 = compressibility of CO_2
m_o = mass of stock tank oil
n_g = moles of gas
t = time
C = gas-concentration of methane in the effluent

Acknowledgements

The autor is grateful to the Federal Ministry of Research and Technology (BMFT) for financial support of the project as well as to Prof. G. Pusch, the project leader, for his advice and support.

References

1. V. Meyn   Laboruntersuchungen zum Kohlendioxid-
   G. Pusch   Fluten (PVT-Verhalten und Flutversuche)
                    ET 3048 A

2. J.J. Rathmell   A Laboratory Investigation of Mis-
   F. J. Stalkup   cible Displacement by Carbon Dioxide
   R. C. Hassing   SPE 3483 (1971)

3. R. W. Watkins   A Technique for the Laboratory Measure-
                    ment of Carbondioxide Unit Displace-
                    ment Efficiency in Reservoir Rock
                    SPE 7474 (1978)

4. R.S. Metcalfe   The Effect of Phase Equilibria on
   L. Yarborough   the CO_2 Displacement Mechanism
                    Soc. Pet. Eng. J. (1979) p. 242

5. L.X. Nghiem   Compositional Modelling with a
   D.^{k} Fong   Equation of State
   K. Aziz   SPE 9306 (1980)

6. M.P. Leach   Compositional Model Studies - CO_2 Oil-
   W.F. Yelling   Displacement
                    Soc. Pet. Eng. J. (Feb. 1981), p. 89

THE USE OF SLIM TUBE DISPLACEMENT EXPERIMENTS IN THE ASSESSMENT OF MISCIBLE GAS PROJECTS

BERNARD J. SKILLERNE DE BRISTOWE

British Petroleum Company Limited

ABSTRACT

Slim tube displacement experiments can be employed to optimise dynamic miscible displacement projects with respect to both pressure and composition. By analogy with the minimum dynamic miscibility pressure concept a minimum dynamic miscibility composition may be defined which may be used to assess the suitability of alternative injection gasses.

Details are given of the way in which the slim tube displacement experiments are performed and are operated with automated data acquisition. Examples are provided of the phenomena observed during the course of the experiments which may be used to assess the nature of the displacement process. The results are discussed in relation to supercritical extraction phenomena which provides an explanation for the residual oil phase left behind in the slim tube at the end of the experiment.

INTRODUCTION

During the last decade a profusion of data has been generated illustrating the phenomena associated with dynamic miscible displacement. A short perusal of the literature quickly shows that the displacement gas used is almost always carbon dioxide and that most often the oil is a West Texas, Permian Basin Crude. Despite this restriction in composition little consensus of opinion has been reached as to what suite of laboratory experiments must be performed in order to evaluate a given field project (1) except that they will be extensive, time consuming and consequently extremely expensive to perform. For an operator whose interests lie far from West Texas and who needs to evaluate many potential prospects it is imperative that the maximum amount of information is generated in the most economical way possible. To do this, the experiments performed should be designed in such a way that they will form a rational basis for investment decisions.

Of the many experiments proposed, the slim tube displacement experiment offers the closest analogue to the processes that occur within the reservoir while at the same time it can be performed sufficiently rapidly for repeated experiments to be performed under varying operating conditions. In using the method the assumption is made that the processes which result in dynamic miscible displacements are independent of the nature of the host porous matrix and depend exclusively on the composition and physical properties of the fluids involved and on the temperature and pressure at which the displacement takes place. The displacement is confined within the walls of a

narrow tube so that the flow is essentially one dimensional. It can
thus provide no information regarding the gross fluid movements within
a reservoir which are subject to hydrodynamic instabilities such as
viscous fingering and gravity segregation. Its major use therefore
is in the initial assessment of a given project where the compatability
of the injected gas with the reservoir oil is sought. Nevertheless, since
a variety of evidence now exists to suggest that the petrophysical
properties of the rock are of secondary importance in determining the
displacement efficiency (2) it may be used to provide valuable
qualitative information about the elementary displacement mechanism.
The key to this is in identifying when the dynamic miscible displacement
process is operating. The purpose of this paper is to show that once
this can be done the process can be optimised with respect to either
the pressure or the composition of the injected gas.

THE SLIM TUBE DISPLACEMENT EXPERIMENT

In order to improve the efficiency with which the slim tube
experiments can be performed while making the best use of the
information available the data gathering part of the experiment
has been automated. The general experimental arrangement is shown
schematically in Figure 1 and a block diagram of the computer
controlled data acquisition system in Figure 2.

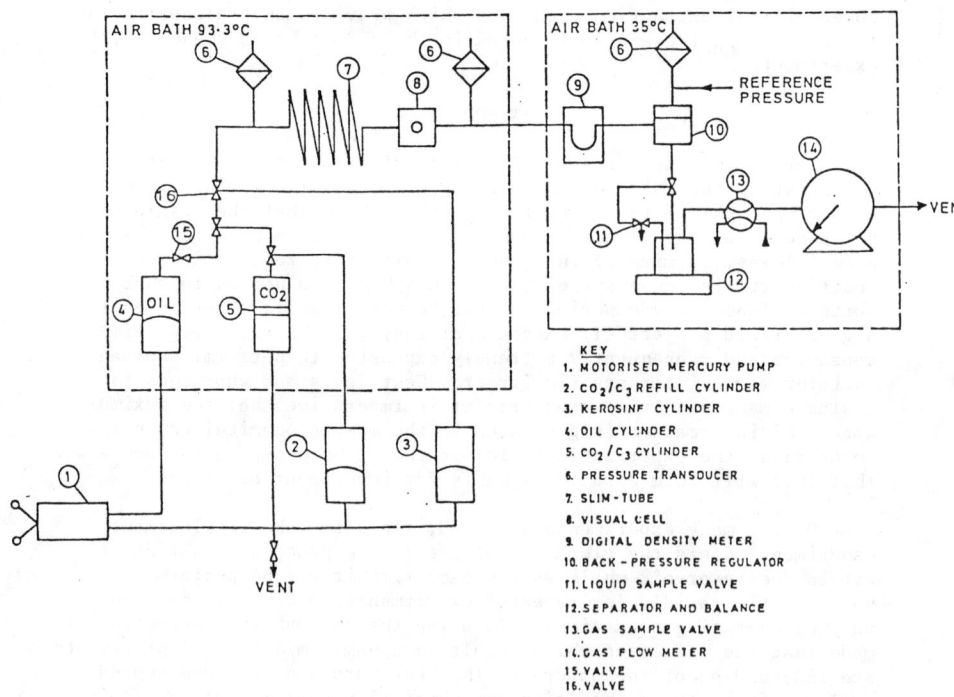

Figure 1. Schematic Diagram of the Slim Tube Apparatus.

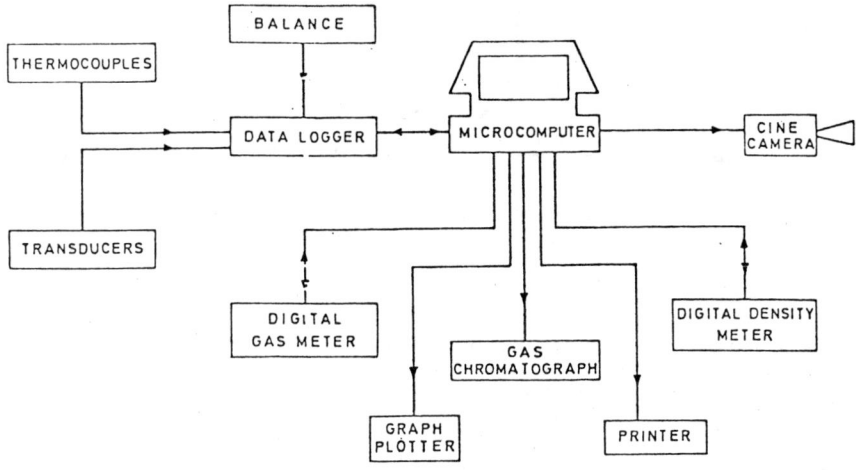

Figure 2. Microcomputer Control of Slim Tube Experiments.

Many different designs for slim tube apparatus have been reported
in the literature. These have been recently reviewed by Orr et al (1).
The present apparatus consists of a motorised Ruska pump which can
displace mercury into one or other of two sample vessels (4) and (5).
The single phase fluids contained in these vessels then pass sequentially through the slim tube (7), the windowed viewing cell (8) and
the digital density meter (9) to a back pressure regulator (10).
The regulator maintains the outlet pressure at a predetermined value
set by a gas reservoir vessel containing nitrogen. As the effluent
fluids pass through the back pressure regulator, they are separated
into a gas and liquid phase. The liquid is collected in a perspex
vessel placed on the pan of a digital electronic balance (12). The
volume of liquid in this separator can be determined independently
by observing the height of oil in the separator using a cathetometer.
Although the density of the oil changes after gas breakthrough the
liquid production falls off rapidly so that the assumption that the
density remains constant only introduces a small error into the
volume of oil recovered. The gas phase passes out of the separator,
through a gas sample valve (13) and a digital wet gas meter (14) and
is vented to atmosphere.

The slim tube is constructed from API Schedule 40 stainless
steel tubing and after packing was coiled to form a square section
helix and is mounted horizontally. The properties of the slim tube
are summarized in Table 1. The packing is held in place by stainless
steel sinters pressed into the ends of the tube. Specially designed
shut off valves which are shown in Figure 3 are fitted to each end
of the tube. These are equipped with bleed valves which facilitate
cleaning and allow the pore volume of the tube to be determined by
weighing the column before and after it had been filled with
distilled water at the temperature and pressure at which the experiments are to be carried out. Thus

$$V_p\ (p,\ T)\ =\ \frac{m_2\ -\ m_1}{\rho_w(p,\ T)}$$

where

$$\rho_w(p,\ T)\ =\ \frac{\displaystyle\sum_{i\ =\ 0}^{5}\ a_i T^i}{(1\ =\ bT)\ [\,1\ +\displaystyle\sum_{k=0}^{5}\ \sum_{j=1}^{3}\ x\ k_j\ T^k(p-p\theta)^j\,]}$$

is the equation of state for water of Kell and Whalley (3 and 4). In this way allowance can be made for the dilatation of the tube and the compression of the packing.

Table 1.   Properties of the Slim Tube

Internal Diameter	9.25 mm
Length	12.19m
Pore Volume (25°C, 0.1MPa)	290.4cm^3
Absolute Permeability	9.6(μm)^2
Packing	Lead Glass spheres
Porosity	36%
Diameter range of packing	ea 80% pass 0.115mm to 0.180mm

The pressure drop across the slim tube was measured by measuring the inlet and exit pressures using Bell and Howell scaled reference strain gauge pressure transducers. These were connected using the special fittings shown in Figure 4,

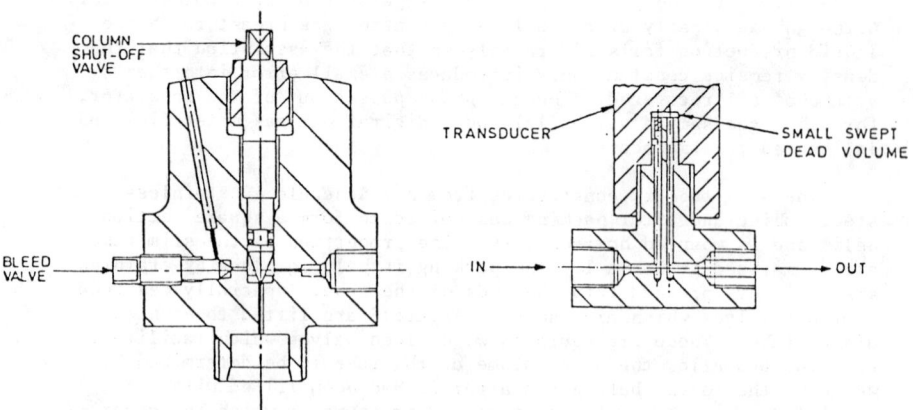

Figure 3.   Special Valves   Figure 4.   Transducer fittings

which minimise the dead volume while allowing it to be swept
by the flowing stream. This arrangement was adopted because
the differential pressure transducers available on the market
have large unswept dead volumes which may vary with pressure.
Furthermore, diffusion of the injection gas into the transducer
can lead to spurious mixing and mobilisation of dead volume oil
which cannot be accounted for and complicates subsequent interpretation. Prevention of this can only be achieved at the
expense of sensivity and the transducers must be very carefully
calibrated with respect to the excitation voltage and temperature
as well as the response of the diaphragm.

The pressure at which the displacement is performed, that
is the pressure at the displacement front, is best approximated
by the inlet pressure to the tube. The pressure is controlled,
however, at the outlet by the back pressure regulator. The flow
rate must therefore be kept low if a substantial pressure change
at the displacement front is not to take place during the displacement. This change may be regarded as an uncertainty in the
displacement pressure. The control of the gas dome-loaded back
pressure regulator depends upon the pressure in the nitrogen
vessel. Since $\frac{\delta p}{P} \simeq \frac{\delta T}{T}$ thermostating the pressure reference
effectively eliminates drifts. In this way the outlet pressure
in the tube can readily be kept within ±50kPa of the set point.

On exit from the slim tube the fluids flow through a windowed
cell which is depicted in Figure 5. It consists of two sapphire
windows held within a stainless steel body. The windows are held
ca 0.1mm apart by two PTFE "D" shaped inserts which reduce the
swept volume. This arrangement produces a large area flat field
which is photographed by a pulsed cine camera mounted outside the
oven. Time lapse photographs are taken at ca 0.001 injected pore
volume intervals so that a complete record is kept of the phases
flowing during the experiment. The utility of this when the data
is analysed after the experiment is over cannot be exaggerated.

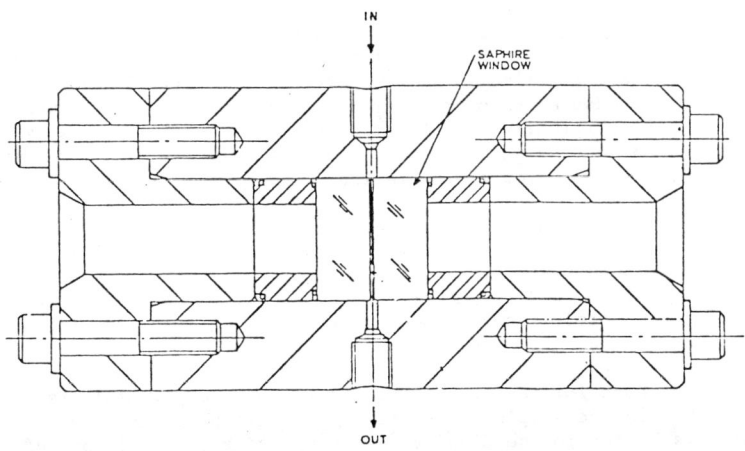

Figure 5. Windowed Cell.

From the windowed cell the fluids pass to an Anton Paar high pressure digital density meter which is housed in an auxillary air thermostat bath. This instrument allows a continuous record to be kept of the density of the fluids flowing through it. It is thus a very sensitive indicator of changes in composition and of the appearance of a low density "gas" phase within the oil and can thus be used to detect breakthrough.

The most important of the experimental parameters are the volume of gas injected and the volume of oil recovered. These are both normalised with respect to the pore volume of the slim tube. So that the displacement may be used as an indication of the process taking place within the reservoir, both of these measurements are referred to the inlet and exit faces of the slim tube under reservoir conditions. The volume of fluid entering the slim tube is obtained, if the flow rate from the pump is assumed to be constant, by

$$V_i(t) = \frac{\sum \dfrac{q_{Hg}(p_{in}, T_{amb})\, \rho_{Hg}(p_{in}, T_{amb})\, \Delta t}{\rho_{Hg}(p_{in}, T_{ov})} - \delta V_{in}}{V_p(p_{in}, T_{ov})}$$

where δV_{in} is the dead volume on the inlet side between valve (15) and the slim tube. This is arranged to be as small as possible. The summation is taken from the time $t=0$ at which the flow is changed from crude oil to displacing gas.

The volume of oil recovered at time t at the outlet from the slim tube is given by

$$V_r(t) = \frac{\left\{\dfrac{m(t) - m(t=0)}{\rho_o(p_{amb}, T_{amb})}\right\} B_o - \delta V_{in}(p_{in}, T_{ov}) - F_g(t)\, \delta V_{out}(p_{out}, T_{ov})}{V_p(p_{in}, T_{ov})}$$

The final term in the dividend makes allowance for the dead volume between the outlet of the slim tube and the separator. Up to the time of gas breakthrough this section of tubing is completely filled with oil before and after a time increment in the flow and consequently does not affect the measured recovery. From the time of gas breakthrough onwards the oil originally in the outlet dead volume is progressively drained and replaced by gas. This progressive drainage can be allowed for if the fractional flow of gas is known.

Before breakthrough $F_g(t) = 0$ and at the end of the run when no liquid is being recovered $F_g(t) = 1$. At intermediate times $F_g(t)$ is experimentally inaccessible. In principle it could be obtained from a flash calculation if the composition and densities of the flowing stream were adequately known. However, if only ultimate recoveries are required this complication is unnecessary.

The procedure for performing the displacement experiments is as follows. Oil in the sample vessel is made single phase and the apparatus is heated to the displacement temperature. Oil is pumped through the slim tube at a high rate, ca $200cm^3h^{-1}$ to miscibly displace the kerosine left in the tube after cleaning. After about

one pore volume has been injected the rate is reduced to that at
which the displacement is to be performed and injection is continued
until the pressure drop and the GOR are both constant. During this
period the base of the gas cylinder is connected to that of the oil
cylinder so that they come to the same pressure. When all is steady,
the oil cylinder is closed and the gas cylinder is opened to the
tube. A signal is fed to the microcomputer that the displacement
has commenced. All of the transducers are fed via suitable interfaces to the microcomputer which scans all of the input channels at
pre-arranged time intervals. Simple calculations such as converting
the millivolt transducer inputs to pressures and temperatures and
time averaging noisy signals are performed in real time. This is
most readily performed if a $Z80$ microprocessor is used which has a
programmable interrupt. All of the computed data is stored on a
floppy disc and all of the sequences of the measurement and control
are synchronised by a real-time clock. A summary report showing the
variation in the major variables as a function of the injected pore
volume is listed on the line printer as the experiment progresses.
At the end of the experiment any further computations are performed, and
the results are tabulated and output to a digital X-Y plotter. The
tube is then cleaned by flushing with solvents and finally with
kerosine in preparation for the next experiment.

PHENOMENA OBSERVED DURING SLIM TUBE DISPLACEMENTS

Displacement experiments may be performed with either composition
or pressure as the independant variable depending upon which parameter may most readily be optimised within a given project. These
two approaches are illustrated with reference to the two oils whose
properties are listed in Table 2. Oil A is from the Egmanton

Table 2. Properties of the oils studied.

		Oil A	Oil B
Composition (massfractions)	N_2	0.0001	0.0000
	CO_2	0.0012	0.0305
	C_1	0.0061	0.0535
	C_2	0.0028	0.0080
	C_3	0.0079	0.0079
	nC_4	0.0100	0.0066
	iC_4	0.0033	0.0022
	nC_5	0.0094	0.0055
	iC_5	0.0067	0.0040
	C_6	0.0187	0.0107
	C_7	0.0342	0.0217
	C_8	0.0466	0.0294
	C_9	0.0355	0.0308
	C_9+	0.8180	0.7892
Bubble point pressure		3.27MPa at 43.3°C	23.54MPa at 93.3°C
$\langle M \rangle$			0.377 mol kg^{-1}

reservoir and was studied in the early part of BP's East Midlands Additional Oil Project where a number of small, highly depleted reservoirs were evaluated as EOR candidates (5). This reservoir was over-pressured by water injection and it was of interest to know the extent to which it could be depressured while allowing the process to operate. The only source of gas available was pure CO_2 formed as a biproduct of ammonia production. Oil B is from a partially depleted reservoir close to a source of associated gas containing a substantial quantity of CO_2. It was therefore of interest to determine what enrichment of the CO_2 by the intermediate hydrocarbons might be required for a dynamic miscible displacement process to operate.

For oil A a series of displacement experiments were carried out at different displacement pressures. Two sets of experiments were performed with and without formation water being present at connate saturation within the slim tube. The results for these are shown in Figs. 6 and 7 where curve (a) is the result obtained when connate water was present in each case.

In Figure 6 the ultimate recovery, approximated by the recovery when 1.2 pure volumes had been injected is plotted, whereas in Figure 7 the recovery is that obtained at gas breakthrough. The curves are

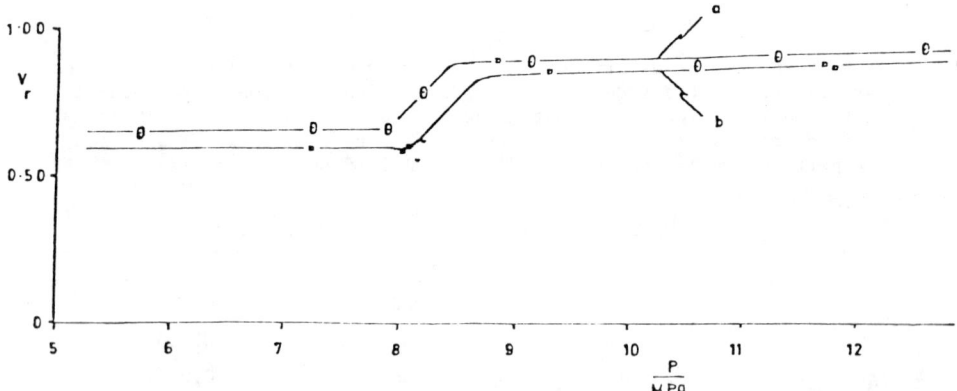

Figure 6. Ultimate recovery as a function of pressure for oil A.

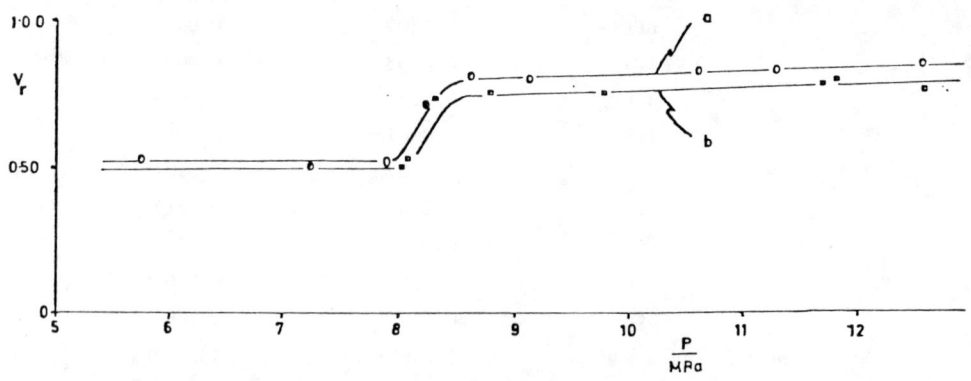

Figure 7. Recovery at breakthrough for oil A

typically sigmoid and there is seen to be a slightly better recovery
when connate water is present than in its absence. This is thought
to be due to the oil being a non-wetting phase when water is present
but the wetting phase when it is absent. The pressure at which the
sudden increase in recovery is observed is independent of the point
on the recovery curve at which measurements are made. However, it
is generally more convenient to use the value at ca $V_i = 1.2$ since

$$\frac{dV_r}{dV_i} \approx 0$$

The point of gas breakthrough was determined in these experiments by
observing the change in the GOR and by monitoring the pH of a small
flask containing distilled water through which the effluent gas was
passed. As the associated gas contains no CO_2 or H_2S a sudden decrease
in pH is observed as CO_2 first emerges from the tube as shown in Fig. 8.

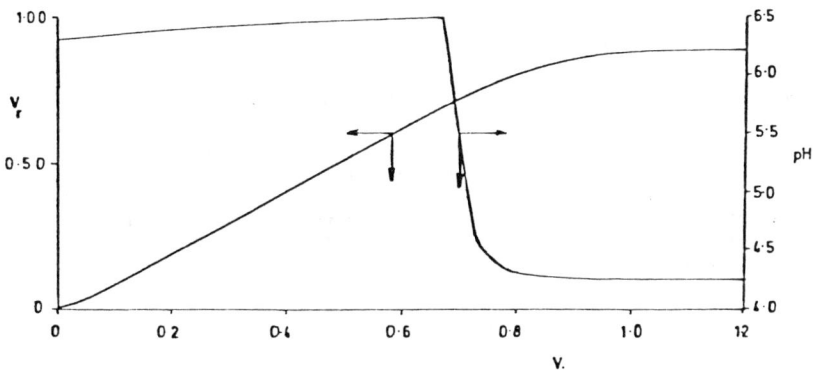

Figure 8. Recovery and pH variation during a slim tube
displacement of oil A.

The sequence of events observed in the visual cell was similar
to that described by Henry and Metcalf (6) with the exception that
a heavy phase was never observed. At pressures below 8MPa breakthrough
was accompanied by the appearance of a colourless CO_2 rich phase as
bubbles within the oil rich phase. At 8.5 MPa a progressive sequence
of lightening in the colour of the oil rich phase was observed but
with the presence of a colourless CO_2 rich phase. Bubbles of this
colourless phase were seen to accompany the oil up to ca 10.5 MPa.

In order to determine the effect of the operational variables
on the recoveries observed a series of measurements were performed
at 12MPa, where $\underline{dV_r}$ is small, in which the displacement rate was varied.
       dp
No difference was observed within experimental error as shown in Fig. 9.

Recently, we have been able to compare the recoveries with
those obtained using a vertical 2m long column 2.5cm in diameter used
by IFP who are now participating in the Egmanton CO_2 project. At
pressures above ca 9MPa very good agreement exists between the results
obtained from the two pieces of apparatus. The break in slope and
minimum dynamic miscibility pressures are likewise in good agreement.
However, at pressures below 8MPa where an immiscible gas displacement
is taking place the vertical column consistently gives higher

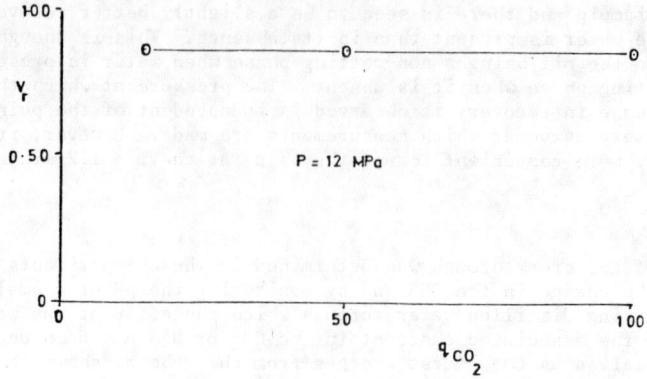

Figure 9.   Dependance of Recovery on Flow Rate at 12MPa.

recoveries than those obtained from the horizontally coiled tube.   As
the displacement in the vertical column is gravity stabilised this is
taken to indicate that hydrodynamic instabilities may be present in
the flow in the slim tube even though it is of small diameter.

For oil B, the effect of progressively increasing the mole
fraction of propane in the displacement gas (mixtures of CO_2 with
propane) is shown in Fig. 10.

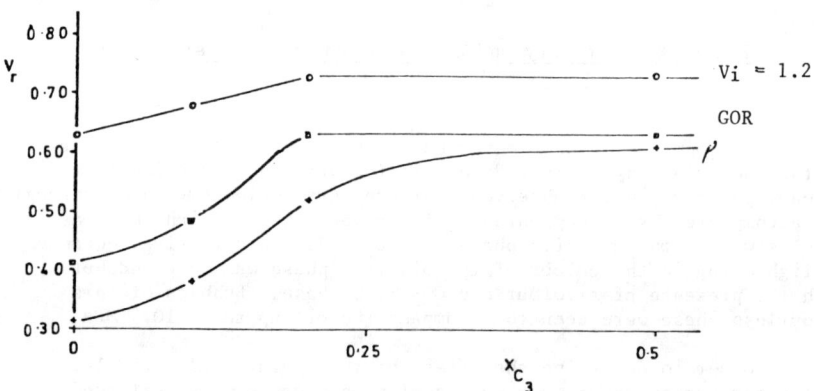

Figure 10.   Recovery as a function of gas composition for Oil B.

With $Xc_3 \geqslant 0.2$ the displacements show all of the characteristics
referred to previously which are observed when a dynamic miscible
process is taking place.   With $Xc_3 < 0.2$ the displacements are typically
immiscible in character.   The diagram shows a break in slope at
$Xc_3 = 0.2$ which is similar to that seen on the $Vr(p)$ diagrams.   This
point is referred to as the minimum dynamic miscibility composition
by analogy with the minimum dynamic miscibility pressure.

It has been observed that when the dynamic miscible process is operating ultimate recovery is reached by Vi = 1.2. To illustrate this $\frac{dVr}{dVi}$ is plotted as a function of X_3 in Figure 11.

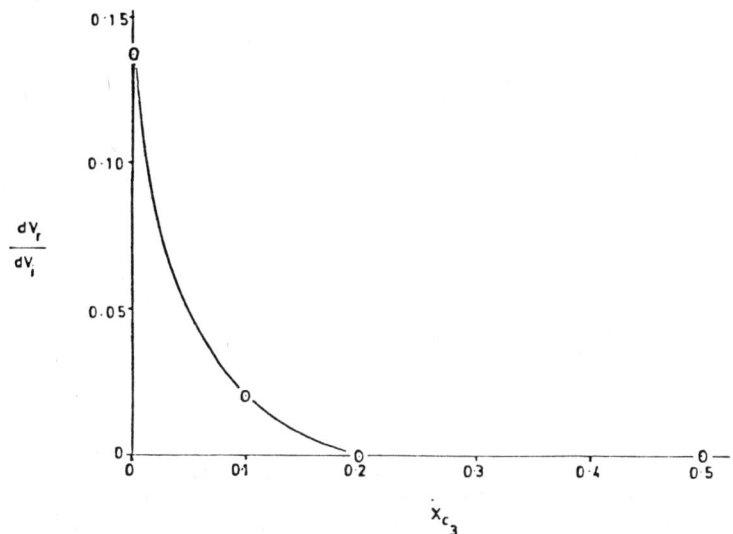

Figure 11. The variation of $\frac{dVr}{dVi}$ with composition

Since Oil B contains an appreciable amount of CO_2, gas break-through could not be detected by monitoring the pH as with Oil A. Figures 12 and 13 show the variation observed in the total stream density and the gas-oil ratio during an experiment. It was found that the density measurement gave a much more sensitive indication of the first change in composition than did the GOR. Furthermore if gas bubbles of low density are entrained within the oil the recorded density becomes very erratic. When a dynamic miscible process is operating the density varies fairly smoothly as shown in Figure 12 . This also serves to show that the composition of the transition zone may be more complex than is usually depicted.

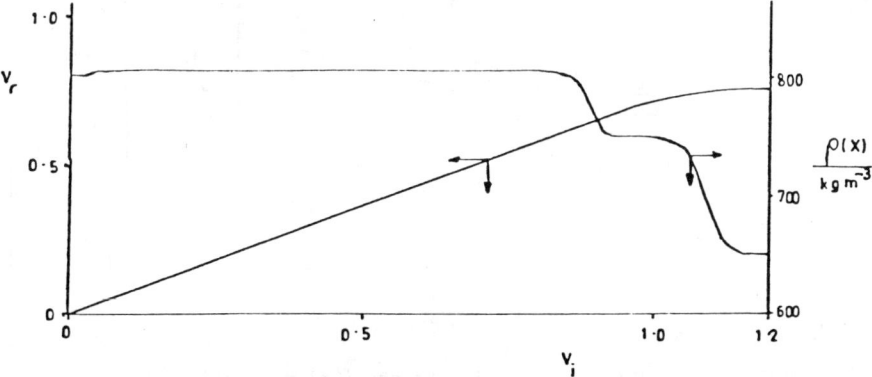

Figure 12. Recovery and density as functions of Vi at Xc_3 = 0.5

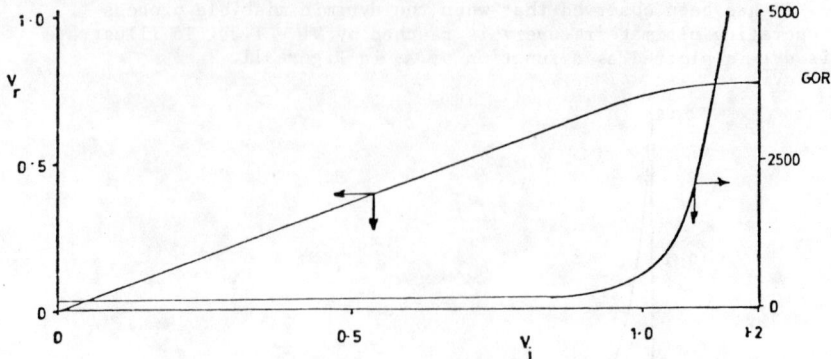

Figure 13. Recovery and GOR as functions of Vi at Xc_3 = 0.5

DISCUSSION

The break in slope of the Vr(p) and Vr(x) functions is now known
to be a function of the composition of the displacing fluid, the
composition of the oil and the displacement temperature (7). Recently,
Johnson and Pollin (8) have shown that the minimum dynamic miscibility
pressure for CO_2 displacing a series of pure n-alkanes is almost
exactly given by the critical pressure of the binary mixture at the
temperature at which the displacement is carried out. If equilibrium
exists within the displacement tube the sudden increase in recovery
which is observed must be largely a result of the increased solvent
powers of the displacement gas within the critical region. The
increased solubility is a result of the large deviations from ideality
which occur within the critical region. As criticality is reached
large changes are observed to occur in many physical properties as
shown in Figure 14 for pure CO_2. The Vr(p) curve for Oil A has been
superimposed upon this. At the same time as the solubility increases

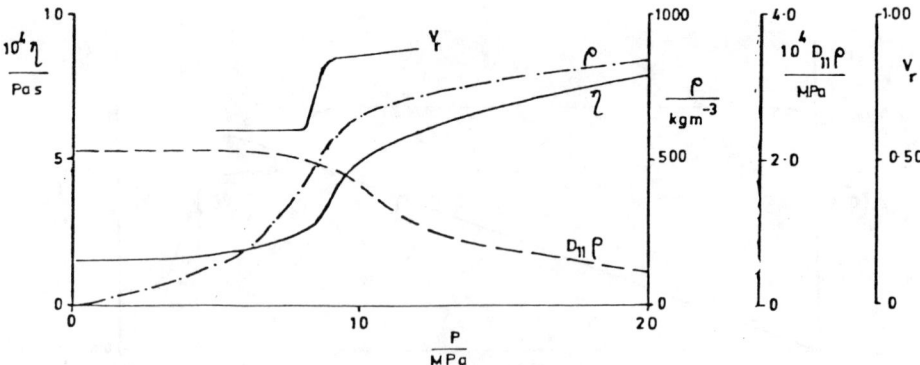

Figure 14. The variation in the physical properties of CO_2
in the critical region.

479

within the critical region other physical properties change in such
a way as to favour the displacement. Thus the density and viscosity
both increase while the product of the density and self-diffusion
coefficient decreases in agreement with kinetic theory. Although
the self diffusion coefficient decreases its value   still remains
considerably higher than that found in normal liquids enabling
trapped oil to be more readily contacted. It is not known at the
present time whether these changes in physical properties will
effect convective dispersion other than through the mutual diffusion
coefficient which appears to follow the behaviour of the self
diffusion coefficient in the critical region.

The solubility of a given compound depends upon the nature of
the intermolecular interactions between it and the solvent as is
reflected in the phase diagram. At the present time phase diagrams
have been determined for mixtures of CO_2 with a range of n-alkanes
and a few simple cycloalkane and aromatic hydrocarbons (9). p(T)
sections for the n-alkanes are shown in Figure 15. It can be seen
from this diagram that for the lower homologues the critical line
is separated into two branches, one of V-L critical point connecting
the critical points of the pure end members and the other of L-L
critical points which terminates at a critical end point. For the

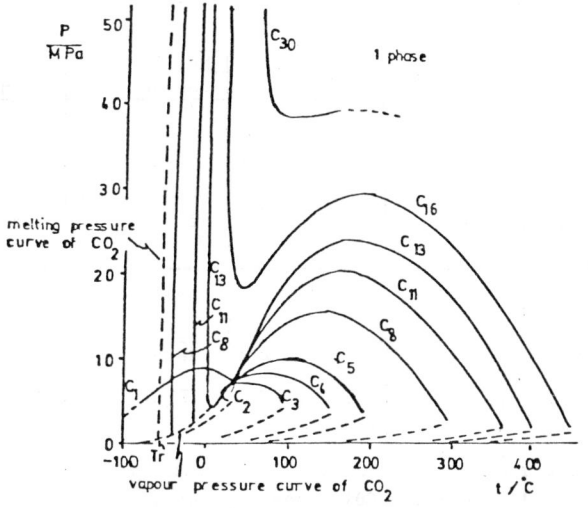

Figure 15.   Critical Lines for CO_2 + n-alkane mixtures.

higher homologues the critical line is continuous but may fold back
upon itself resulting in gas-gas immiscibility of the first kind.
The critical pressure increases with the carbon number at a given
temperature. Thus for extraction at a given temperature and pressure
some of the homologues will have their critical regions within the
range of pressures considered, while others will require a much
higher pressure and yet others, if the temperature is sufficiently
low, can never be brought into the critical region by increase in
pressure alone. This will account at least in part for a residual
"heavy" oil being left behind after the displacement and for the
small but persistant increase in recovery that is observed after
the minimum dynamic miscibility pressure has been reached. A

similar suite of curves exist for the aromatic and naphthenic compounds. The displacement of these curves from one another suggests a degree of selectivity in the extraction process and it is interesting to speculate that the paraffin, naphthene, aromatic distribution within the oils recovered may change as the pressure changes.

Although the minimum dynamic miscibility pressure appears to be largely governed by the equilibrium thermodynamic properties of the fluids the absolute recovery will depend both on the proportion of unextractable components in the oil and on the hydrodynamics of the displacement. It does not seem reasonable therefore to quote a fixed recovery which must be reached before it can be concluded that the multiple contact mass transfer mechanism is operating. Deduction of the process mechanism must thus be made on the basis of many different observations rather than a single one. An attempt to summarise the more general characteristics of the different process types is given in table 3.

TABLE 3   KEY TO PROCESS IDENTIFICATION

PROPERTY	Process Type			
	Immiscible	Low IFT	Multiple Contact	First Contact
Recovery at Vi = 1.2	low	intermediate	high	high
$\frac{dVr}{dVi}$ at Vi = 1.2	high	low	zero	zero
Rate Dependance	large	small	none	none
Breakthrough	early	intermediate	late	late
Density Change at breakthrough	Becomes very erratic	erratic at times	smooth	smooth
Sight Glass Observations at breakthrough	Colourless bubbles in dark oil	Colourless bubbles in lighter coloured oil	Dark to light colour change in oil with occasional colourless bubbles	Progressive lightening in colour of the oil without gas bubbles

If a series of displacements are carried out at differing compositions or pressures as in the example given above several of these mechanisms will be observed to operate. Considerable refinement of this scheme is required particularly in relation to the different types of displacement which may take place in the L-L and L-V regions.

CONCLUSIONS

It has been shown above that slim tube displacement experiments may be
used to optimise individual projects with respect to both pressure and
composition. A considerable amount of information may be gathered in
the course of the studies upon which the dominent mechanism which operates
in a given pressure or composition may be deduced.

ACKNOWLEDGEMENT

The author wishes to thank G.J.J. Williams, A.G. Steven,
A. Booth, C.G. Osborne, D.J. Thomas, S. Takhar, S. Bahal and
C. Liang from whose work the contents of this paper have been
drawn.

NOMENCLATURE

Bo	Flash formation volume factor of oil
Fg	Volume fraction of gas phase
GOR	Gas oil ratio
i, j, k	Indices in equation of state
m_1	Mass of evacuated slim tube
m_2	Mass of water filled slim tube
$<M>$	Number average molar mass
p	Pressure
p^θ	Standard pressure (101.325 kPa)
q_{Hg}, CO_2	Volumetric flow rate of mercury or CO_2
Vi	Injected pore volume
Vp	Pore volume of slim tube
Vr	Recovered pore volume
δVin	Inlet dead volume
$\delta Vout$	Outlet dead volume
Xc_3	Mole fraction of propane in mixtures
T	Temperature
δT	Temperature movement
Δt	Time movement
αij	coefficient in equation of state

482

δp pressure movement

ρo density of oil

ρw density of water

ρHg density of mercury

REFERENCES

1.   Orr, F.M., Jnr. and Taber, J.J.
"Displacement of Oil by Carbon Dioxide"
Final Report DOE/ET/12082-9, 1980.

2.   Yellig, W.F.
"Carbon Dioxide Displacement of a West Texas Reservoir Oil"
SPE/DOE9785, 1981.

3.   Kell, G.S.
"Precise Representation of the Volume Properties of
Water at One Atmosphere"
J. Chem. Eng. Data 12, 1, 66-69

4.   Kell, G.S. and Whalley, E.
"The pVT properties of Water"
Phil Trans Roy Soc (Lond) 258A 565 (1965)

5.   Gair, D.J., Grist, D.M., and Mitchell, R.W.
"The East Midlands Additional Oil Project"
SPE 195 Presented at the 1980 European Offshore
Petroleum Conference and Exhibition.

6.   Henry, R.L., and Metcalfe, R.S.
"Multiple Phase Generation During CO_2 Flooding"
SPE 8812, presented at the First Joint SPE/DOE
Symposium in Enhanced Oil Recovery, Tulsa,
April 20-23 (1980).

7.   Holm, L.W., and Josendal, V.A.
"Effect of Oil Composition on Miscible-Type Displacement
by Carbon Dioxide".
SPE 8814, presented at the first joint SPE/DOE Symposium
in Enhanced Oil Recovery at Tulsa, April 20-23, 1980.

8.   Johnson, J.P., and Pollin, J.S.
"Measurement and Correllation of CO_2 Miscibility Pressures"
SPE/DOE 9790, presented at the 1981 SPE/DOE joint
Symposium in Enhanced Oil Recovery, Tulsa, April 5-8.

9.   Schneider, G.M.
"Physicochemical Principles of Extraction with
Supercritical Gases".
Angew. Chem. Int. Ed. Engl. 17 716-727 (1978)

NUCLEAR MEASUREMENTS OF FLUID SATURATION IN EOR FLOOD EXPERIMENTS

N. A. BAILEY, P. R. ROWLAND, D. P. ROBINSON

AEE Winfrith, Dorchester, Dorset DT2 8BH

ABSTRACT

This paper describes the nuclear measurement methods which have been selected for the determination of fluid saturation distributions within the cores of high pressure flood experiments to be carried out at Winfrith in a study of enhanced oil recovery processes. Such methods should be capable of producing the accurate measurements of saturations, including their spatial and temporal variations, which are required to obtain a better understanding of the EOR process occurring within cores. It is intended that such measurements will provide a wider range of information for the validation of computer programs. These EOR codes are to be used for the assessment of the feasibility of EOR processes in North Sea Fields. A range of possible measurement techniques has been studied and the basis of the selection of the preferred nucleonic methods is described. The methods selected are based on the use of deuterium to determine water saturations or hydrocarbon gas components utilising a gamma-neutron reaction, and the introduction of radioactive ferrocene as an additive to oil to measure oil saturation from gamma emission. The development work carried out to establish these nucleonic techniques is discussed in some detail, showing their clear potential, and the methods of application to high pressure core fluids are discussed. These techniques are currently being used in on-going flood experiments.

INTRODUCTION

A programme of EOR studies is being carried out at AEE Winfrith under contract to the UK Department of Energy and an important element of this programme is the experimental programme. The experimental programme, whose objectives are the testing of EOR processes under relevant conditions and the validation of codes, is centred around a number of high pressure flood rigs, the first of which is currently being constructed. These rigs have been designed to allow experiments to be carried out at reservoir pressure, temperature and flow rate using reconstituted reservoir oils, correct salinity brines and a variety of EOR fluids including CO_2 and surfactants. The displacement process itself will take place in sandstone cores up to 5m long. Key measurements which are required of such a rig are the analysis of the fluids produced from the outlet end of the core and the determination of fluid saturations within the core as a function of axial position along the core and time.

The selection, development and application of methods for such measurements of fluid saturation is the subject of this paper. A number of techniques have been used in the past for such measurements, and these are discussed below, but most are not applicable to high pressure core floods where the core has to be surrounded by a thick-walled steel pressure vessel. Various nuclear measurement techniques have, therefore, been considered as they are less influenced by the presence of the pressure vessel and considerable experience with them already existed at Winfrith.

It has been concluded that at least two phases need to be marked within an EOR displacement test in which three phases can occur. Two preferred techniques have, therefore, been selected for the measurement of oil and water phases which have been subjected to detailed development, supported by theoretical assessments, to determine their performance under representative conditions. These tests have shown that very satisfactory measurements can be made of fluid saturations within sandstone cores, and work has continued to consider in detail the application of these techniques to the High Pressure Flood Rigs. One of the problems considered is the need to avoid any partition of the radioactive tracer from the phase which is being marked.

In parallel with these high pressure flood experiments, some low pressure studies have started at Winfrith, including a programme of waterflood displacement of oil, and these nuclear techniques are being used in these experiments where some comparisons are possible with other measurement methods.

SELECTION OF MEASUREMENT TECHNIQUES FOR FLUID SATURATIONS WITHIN CORES

The measurement of fluid saturation within cores is a difficult task and a review of the published literature suggests that it has only been attempted infrequently. Such measurements can, however, contribute substantially to the development of the understanding of EOR processes and their quantitative evaluation. Various techniques have been proposed for the present programme for such measurements and these have been considered for use in high pressure floods. Physical sampling of fluids along the length of a core has some attractions, but capillary effects lead to the extraction of samples which are unrepresentative of local saturations. The low flow rates which would occur in the sample lines if the core were operated at reservoir velocities would present flushing problems leading to incorrect fluid analysis.

Some measurements have been attempted previously using non-intrusive techniques such as nuclear magnetic resonance (NMR) and microwaves. NMR has been used successfully for downhole applications by Brown and Gamson[1] and Nikias and Eyraud[2] who were able to distinguish between brine and high viscosity oils. NMR does, however, require a totally non-magnetic containment if the high frequency alternating magnetic fields are to penetrate the core and this is not readily achieved in high pressure laboratory floods. Microwave techniques have been very successfully used by Parsons[3] to determine brine saturations but once again a non metallic containment is required to avoid reflection of the microwave radiation and this reduces its applicability to high pressure experiments. Preference is being given, therefore, to nuclear techniques as they are non-intrusive and are far less sensitive to the effects of pressure containment around the core.

Nuclear Techniques

A number of nuclear techniques are potentially of use in the measurement of fluid saturations within cores. The first technique considered is labelling particular fluids with γ-emitting tracers. The concentration of a particular fluid can then be inferred from measurements of radiation as shown in Figure 1. The water phase can be labelled by using radioactive forms of its dissolved salts, but the oil phase needs to have a material added to it which is of a hydrocarbon type containing a radioactive element. The main problem with such a technique is that of handling significant volumes of continuously radiating fluids. The method is, however, viable for fluid saturation measurement and is simple, direct and well understood.

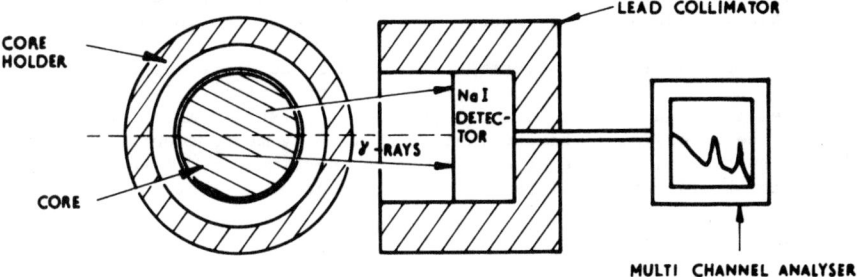

CORE HOLDER

CORE

γ-RAYS

NaI DETEC-TOR

LEAD COLLIMATOR

MULTI CHANNEL ANALYSER

FIG.I GAMMA TRACER METHOD

The alternative techniques rely on the response of fluids within the core to bombardment of γ-rays or neutrons from an external source which can be turned off when not required. This avoids the problem of handling radioactive fluids. One method of this type which is being investigated is neutron activation. The core is bombarded with neutrons and the resultant γ-radiation is detected and analysed using gamma spectroscopy to give a measurement of the relative abundance of the elements within the core.

A further alternative centres around the use of deuterium which emits a neutron when bombarded with high energy γ-rays. The energy threshold for the emission of a neutron from deuterium in such a reaction is 2.23MeV. Deuterium concentration could then be determined from neutron flux. Many fluid components could be labelled by substituting hydrogen atoms by deuterium. Supplies of heavy water, for example, are readily available at Winfrith and this can be added to H_2O for marking the brine phase. A range of deutero-carbons can also be synthesised from D_2O. The simplest substitute hydrocarbon to produce would be deutero-methane from the reaction of heavy water with aluminium carbide.

Preliminary Screening

Preliminary tests with the radioactive tracer and γ-neutron techniques showed both of these methods to be fundamentally feasible, although each has its own difficulties. Acceptable count rates, proportional to concentration, could be obtained from reasonable activity levels for each method suggesting that further development of these techniques was worthwhile.

The absorption of neutrons was examined at an early stage as this was equivalent to the neutron decay logging techniques which have been used for a number of years with pulsed neutron sources downhole to measure local brine saturation (4-6), or variations of salinity during waterflooding(7). Neutron absorption in a laboratory arrangement was investigated using a 1.5µg californium source placed close to a sandpack with a porosity of 40% within a 50mm bore glass cylinder. The source was surrounded by polythene to ensure that the neutron flux entering the sandpack was predominantly thermal and cadmium foil was used around the polythene to provide collimation. The sandpack was saturated with brine over the lower half and oil in the upper half. The sections were separated by a thin polythene interface to maintain a sharp front in saturation and the transmitted neutrons were detected on a glass scintillator. The measurement is based on the neutron absorption cross section in chlorine as a measure of brine saturation as this is the dominant cross section as shown in

486

Table 1. On traversing past the interface, however, the measured count rate
fell by only 7% which was inadequate for detailed saturation measurements. It
was concluded therefore that this technique was not viable unless fluids have
their cross sections increased by the addition of a neutron absorber. The
difficulty in this case could then be the separation of the additive from the
phase it was marking and this method has not been explored further.

Tests were then carried out investigating whether the differing absorption cross
section to thermal energy neutrons of hydrogen and deuterium listed in Table 1
could be employed, as theoretical calculations suggested a modest difference in
absorption behaviour could be obtained. Preliminary measurements showed that
such differences could be detected but not on the timescale or resolution
required to obtain a sharp picture of the movement of a front.

Table 1
Absorption Cross Sections (barns) to 0.09eV Neutrons

Element	H	D	C	O	Na	Si	Fe	Cl
Cross Section	0.173	2.6×10^{-4}	1.8×10^{-3}	10^{-4}	0.289	0.083	1.32	16.69

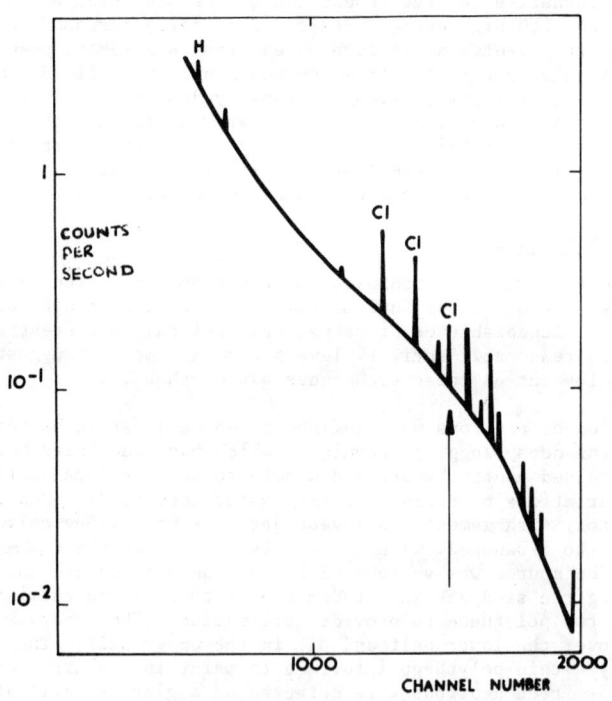

FIG. 2 TYPICAL NEUTRON CAPTURE GAMMA SPECTRUM WITHOUT PRESSURE VESSEL

A further technique which was investigated was the spectral analysis of gamma rays following activation by neutron irradiation. Such a method, covering both inelastic scattering and capture of neutrons, has been used by Hertzog[8] for downhole applications to determine the relative quantities of carbon, oxygen, silicon, calcium, iron, chlorine and hydrogen. Our experiments concentrated on the neutron capture process as the high energy neutrons needed for inelastic scattering led to shielding problems. The laboratory studies were carried out on a sandpack contained within a 50mm bore aluminium tube with a 1 µg californium source as a neutron generator. The initial test results (Figure 2) produced a spectrum showing quite clearly the peaks associated with chlorine. Chlorine concentrations could not, however be obtained to the required accuracy within an acceptable timescale. When the sandpack was placed inside an 18mm thick steel tube to simulate the high pressure rig situation, the spectrum was dominated by iron peaks, with hydrogen and chlorine peaks being reduced to an unacceptably low level compared with the background level produced by Compton scattering.

Because of these major difficulties encountered with neutron absorption and activation techniques it was decided that reliance should be placed on γ-tracer and γ-neutron techniques which were then subjected to further development.

GAMMA TRACER TECHNIQUES

Radio tracer techniques have to be restricted to γ-emitting isotopes as β and γ emissions will not penetrate the pressure vessel. As a gamma tracer technique depends on adding a radioactive material to a fluid, it is best achieved when the fluid normally contains materials which can be made radioactive. The possible choice of radioactive material is further restricted by the need for a sufficiently long half life (preferably greater than 1 month) and sufficiently high energy gamma rays (greater than 0.5MeV) to avoid excessive absorption in the pressure vessel around the core.

Labelling the brine phase is relatively straightforward, as the naturally occurring chlorides of sodium and caesium are both readily available in radioactive form and their characteristics are summarised in Table 2. The oil phase is more difficult as hydrogen and carbon do not have gamma emitting isotopes. A suitable isotope must therefore be wrapped in a hydrocarbon type molecule which will behave as an oil component. It should also have tightly bound electron orbitals and not be strongly electro-positive. This reduced the list of candidate isotopes to Sc-46, V-48, Fe-59 and Co-60. This list was further reduced to Fe-59 and Co-60 due to the vast range of organo-metallic chemistry associated with these elements, with a preference for Fe-59 as the very long half life of Co-60 (5.26 years) leads to decontamination and disposal problems. Ferrocene (Figure 3) was therefore selected as the most suitable material, as it contains iron which can be made radioactive, and is one of the simplest organo-metallic molecules, containing only carbon and hydrogen in addition to the iron. Its molecular structure suggests that it should behave as a heavy liquid oil fraction. The characteristics of Fe-59 are also included in Table 2.

Table 2 - Candidate Isotopes for Labelling Fluids

ISOTOPE	Na-22	Cs-137	Fe-59
HALF LIFE	2.6 years	30 years	45 days
ENERGIES (MeV)	0.51, 1.28	0.66	1.10, 1.29

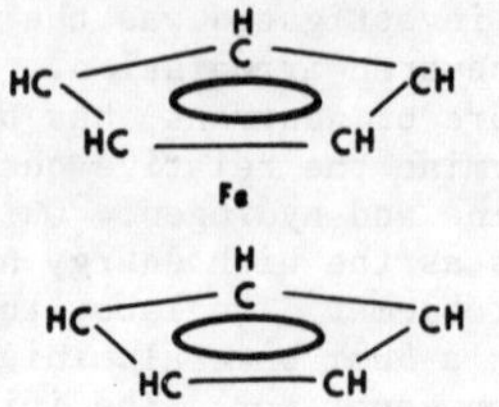

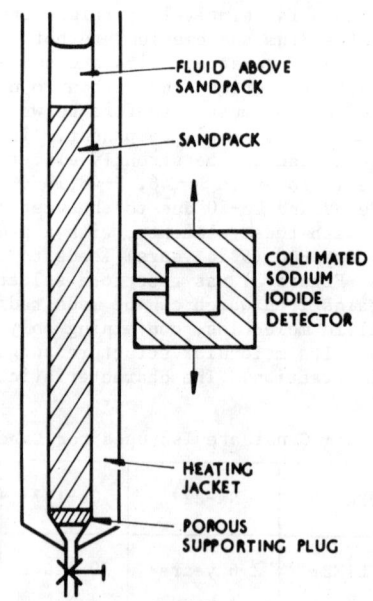

FIG. 3   FERROCENE MOLECULE.

Preliminary Tests

The first series of tests were carried out on a 1cm diameter pure silica sandpack, set up in special glass apparatus as shown in Figure 4 which allowed
visual examination of the displacements at the same time as operating at
temperatures of around 90°C. The sandpack was 25cm long and the displacement
tests were carried out with 3% brine, labelled with Cs-137 at an activity
level of 20μCi/ml, and medicinal paraffin. For visual purposes the active brine
could be marked with sodium flourosceine and the oil phase by Sudan Red dye.
A collimated sodium iodide detector was used to measure the radiation using
10cm thickness of lead and a slit width of 3mm. Calibration was carried out
in situ both in the empty tube and in the sandpack at 100% brine saturations.
The latter gave count rates of 3000 per minute. Using a count period of
1 minute this gives a standard error level of ±1.8%.

FLUID ABOVE
SANDPACK

SANDPACK

COLLIMATED
SODIUM
IODIDE
DETECTOR

HEATING
JACKET

POROUS
SUPPORTING PLUG

FIG. 4 APPARATUS FOR GAMMA TRACER
DEVELOPMENT TESTS.

The experimental programme on this apparatus included:-

i   downward displacement of inactive brine with active brine (miscible)

ii   downward immiscible displacement of active brine by oil (paraffin)

During the first test a record was kept of the brine level above the sandpack so that the sandpack porosity could be determined both from the velocities of the front, determined by radiation measurement and visual observation of the measured brine level as well as the activity levels in and above the sandpack.

A typical scan of the sandpack for the first experiment is shown in Figure 5 where the characteristic dispersion of the front in miscible displacement can be seen. The porosities calculated (38%) agreed closely, the agreement lying within the experimental uncertainty (±2%). When the active brine was displaced with oil (Figure 6) similar measurements were made and these indicated a frontal oil saturation of 85%. This was confirmed by direct measurement of the paraffin injected into the core.

The results were considered promising as it was clear that accurate measurements were being obtained and development of the technique was transferred to sandstone cores. At this point the first difficulty with this technique appeared which was adsorption of the tracer within the sandstone.

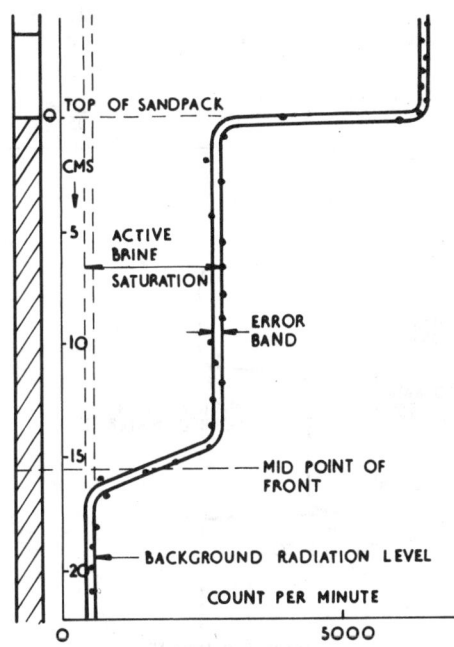

FIG.5 DOWNWARD DISPLACEMENT OF UNLABELLED BRINE BY LABELLED BRINE

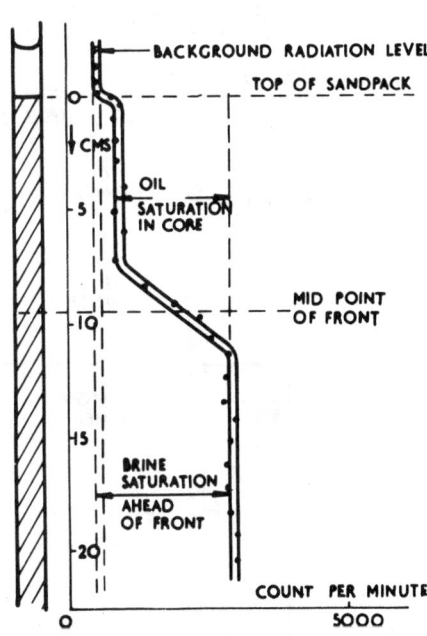

FIG.6 DOWNWARD DISPLACEMENT OF LABELLED BRINE BY OIL.

490

Adsorption

When radioactive salt tracers are dissolved in brine they form ions and the metallic components adsorb onto the surface of the rock and particularly onto clay structures within sandstones. This is very noticeable when a low concentration radioactive brine is added to dry sandstone with a substantial fraction of the measured activity eventually coming from the surface of the rock. A series of adsorption tests was, therefore, set up to investigate this behaviour using the sandstones to be used in the high pressure flood experiments. These sandstones, Clashach (200md permeability, 13% porosity) and Rosebrae (1250md permeability, 24% porosity), are Triassic quarry material and have only modest clay contents.

In these experiments a rock sample was mounted in a Table Tube (Figure 7) where knowing how much solution should exist within the porous material, the amount of adsorbed isotope could be determined. The salt concentrations were systematically varied and the results for the Clashach and Rosebrae are shown in Table 3.

Table 3 - Adsorption of Tracer on Sandstone

Condition	Adsorbed Activity / Dissolved Activity in Pores
Clashach	
a Solution of 6% NaCl + 1% CsCl + Cs-137 tracer	1.3
b Solution of 6% NaCl + 6% CsCl + Cs-137 tracer	0.53
c NaOH added to (b) to pH = 11	0.47
d Solution of 6% NaCl + Na-22 tracer	0.20
e Solution of 6% NaCl + 5% MgCl$_2$ + 5% CaCl$_2$ + Na-22 tracer	0.12
Rosebrae	
f Solution of 10% NaCl + 10% CsCl + Cs-137 tracer	0.10
g Solution of 10% NaCl + 10% CsCl + Na-22 tracer	0.05

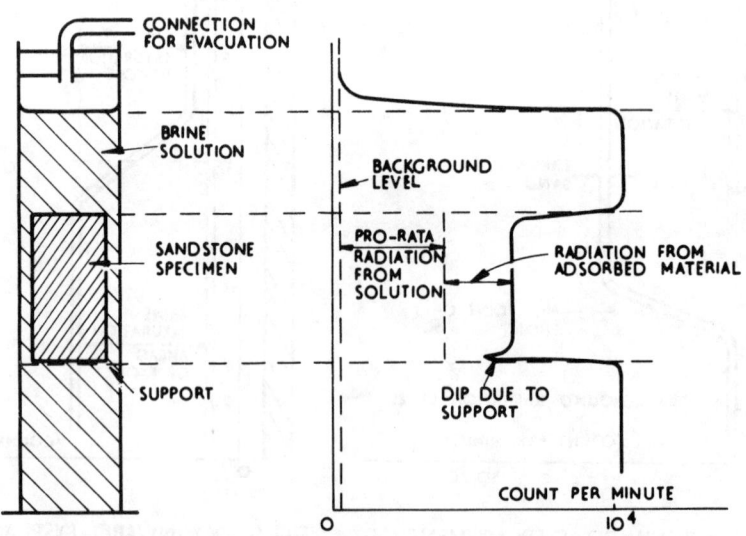

FIG. 7   APPARATUS FOR ADSORPTION MEASUREMENT AND TYPICAL GAMMA SCAN

In all cases a return to inactive brine resulted in the activity in the sample falling to zero. It can be concluded, therefore, that the adsorbed fraction decreased with increasing tracer salt concentrations, that adsorption was reversible, that the addition of divalent salts or increased pH had little effect on adsorption, and that caesium was adsorbed in preference to sodium. The tests were carried out with radiotracer concentrations of 10µCi/ml.

The most representative conditions were with brine made of 6% NaCl using a sodium-22 tracer. Under these conditions the radiation from adsorbed isotope was only 20% of that from the isotope dissoloved in the brine. It might be possible to overcome the adsorption problem by pre-saturating the core with active brine before saturating it with oil. If this quantity of tracer then remained adsorbed in the core during a flood experiment, it could be treated as an additional radiation background and measurements could still be made of water saturation albeit at a reduced accuracy. This would require that the rock remained totally water wet throughout and it is more likely that some of the adsorbed material will be displaced during a flood experiment making the interpretation of radiation measurements extremely difficult. Some tests were made in which the sandstone was preflushed with inactive brine in an attempt to saturate adsorption sites. Some improvement resulted from this but the inactive and active ions exchange during the subsequent active displacement.

Adsorption of water-borne tracers does, therefore, severely limit the power of this technique to measure water saturations, so consideration was switched to the labelling of the oil phase with a non-polar, non-ionising material, which should not adsorb and which would allow measurement of the oil phase saturation.

Labelling the Oil Phase

Ferrocene has been selected as a suitable organo-metallic material for use as an oil tracer as it satisfies the various criteria discussed above and it is readily available in inactive form. To convert it to its radioactive form requires either irradiation in a reactor or synthesis of ferrocene from radioactive iron via the reaction of ferrous chloride with cyclopentadiene using potassium hydroxide as a condensing agent. Investigations of the latter have shown that synthesis is possible, but the main emphasis in the development has been associated with irradiation which is expected to be more economic. Trial irradiations have been carried out on the PLUTO and DIDO reactors at Harwell for periods of up to 23 days to give a specific activity of about 0.4 mCi/gm. At the end of such an irradiation about 30% of the activity is in ferrocene with the remainder in a material which is insoluble in hydrocarbons. It is believed that this breakdown is induced by the Szilard-Chalmers reaction(9) which results in the iron being ejected from the molecule under irradiation, allowing the iron to oxidise in the surrounding atmosphere. The two radioactive components can readily be separated by dissolution, filtration and recrystalisation. A typical yield from a standard irradiation would be about 3gm of radioactive ferrocene containing about 1.25mCi of activity. When dissolved in 5 litres of oil to give a specific activity of 250µCi/litre, the statistical uncertainty in the saturation measurement on a representative core geometry will be less than ±3% on a 100% saturation. This accuracy is based on a 9mm slit width, a 10 minute counting time, and is derived from preliminary measurements in a representative geometry.

Such a concentration is far below the solubility limit of ferrocene in oil, which was found to be about 2.5% at 20°C and appreciably higher at higher temperature. Several tests have been carried out to examine the partition behaviour of ferrocene between oil and brine by dissolving it at a concentration of about 0.1% in oil and then contacting the mixture with brine at 100°C for several days. The amount of ferrocene transferred to the water phase was then determined colorimetrically or by radiation measurement. Both methods gave similar results with a partition coefficient less than 10^{-5} for neutral or alkaline brine but with an increased value of 10^{-3} for a strongly acid brine (pH=2). As the solubility of ferrocene in brine is so low, this compound will travel with and mark the oil. As quite modest quantities are required to mark the oil, it appears to provide a very suitable method for saturation measurements.

GAMMA-NEUTRON INTERACTION

If ferrocene is to be used for labelling the oil phase, and γ-emitters are of limited use for labelling the water phase, then an alternative method is required for the measurement of water phase saturations. The gamma-neutron reaction with heavy water has been selected to provide such a method. Such a reaction can occur in any element if it is bombarded by gammas with energies exceeding the binding energy, but the second lowest binding energy occurs in deuterium (2.23 MeV). Thus measurement of water saturations can be achieved by using heavy water and bombarding the core with gamma rays with an energy in excess of 2.23MeV. The neutron flux produced is proportional to water saturation. The only element with a lower binding energy is Beryllium-9, which will not be present in the flood experiments. The only convenient gamma source with a suitable energy level is sodium-24, but this unfortunately has a short half life (15 hrs) which means that special arrangements have to be made for the delivery of sources. An alternative would be to use a γ-beam generator but existing commercial devices would be too powerful for this application.

This technique has been considered here as a means of marking the water phase, as D_2O will also have a negligible effect on the physical and chemical characteristics of the water phase. Deuterium could also be used to determine the local saturation of a particular hydrocarbon as deutero-carbons can be synthesised. The simplest would be the production of deutero-methane from the reaction between D_2O and aluminium carbide, but other methods such as the Fischer Tropsch synthesis are available to synthesise higher deuterocarbons.

Development Tests

Development tests on the gamma-neutron method have been carried out on the apparatus shown in Figure 8. This apparatus consisted of a 2" diameter sandpack contained inside an aluminium tube with a heavy water saturation at one end separated from a light water saturation by a thin polythene film to maintain a sharp saturation front. The sandpack was set up on its own or within a 18mm thick steel vessel representing the pressure vessel surrounding the core in a High Pressure Flood experiment. Most of the circumference of the sandpack was surrounded by a polythene block containing BF_3 or He_3 neutron detectors. The polythene is required to thermalise and reflect neutrons before entering the neutron detectors. Optimisation studies showed that 16mm of polythene was required between core and counters to achieve optimum thermalisation and 50mm beyond the counters for reflection. The remainder of the circumference is taken up by the source and its collimator. The sealed source utilised had an activity of about 100 mCi and was mounted inside a lead collimator system with an adjustable slit allowing a gamma ray beam the full width of the sandpack. The slit design limited the beam to a few millimetres along the length of the sandpack, so that axial variations in saturations can be adequately resolved. In the γ-tracer techniques described earlier the detector was collimated, as it is not possible to collimate a source in the fluid.

The neutrons detectors were connected to a multi channel analyser initially which allowed the spectra to be investigated in detail and compared with those from pure neutron sources, but later tests have used more traditional Harwell 6000 apparatus counting all pulses above a particular threshold level which was set to reject noise.

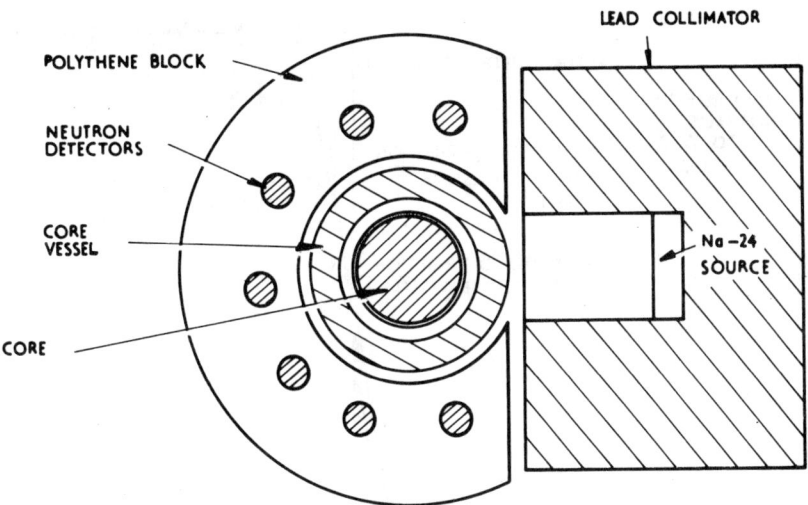

FIG.8   APPARATUS FOR GAMMA - NEUTRON INTERACTION DEVELOPMENT TESTS.

The development experiments were concentrated on saturation measurements through a saturation discontinuity and have allowed the systematic investigation of shielding thickness in the collimator and slit width. Initial tests were carried out using 50mm thick lead for the collimator, but it became clear that this was insufficient as a significant flux was occurring away from the slit position. It was found necessary to increase the lead thickness to 125mm to reduce this background gamma flux to an acceptable level. The weight of the collimator system at about 500kg is one of the factors to be taken into account in the design of traversing equipment.

The results of scanning through an interface with a thick walled collimator are shown in Figure 9-11. These figures show quite clearly that the step front in saturation has been smeared into a measured S-shape curve where saturations in the range 10%-90% appear over a length of about 1cm. Theoretical calculations have been carried out to determine the uncollided gamma flux distribution along the centre line of the sandpack using the relationship

$$\emptyset u(x) = \emptyset_o \varepsilon^{-\Sigma(\mu_i t_i)}/4\pi(\Sigma t_i)^2 \tag{1}$$

where $\emptyset u$ is the uncollided flux, $\emptyset_o$ the source flux, and μ_i the total linear attentuation coefficient for the different materials each of thickness t_i in the path of the flux, which is a function of x. From such calculations it is possible to predict the measured profile with an equation of the form

$$P_m(x) = \int_{-\infty}^{+\infty} f(x-x^1)P_t(x^1)dx^1 \tag{2}$$

where P_m and P_t are the measured and true profiles respectively and f is a function of the collimator and test section geometry which can be derived analytically or from calibration experiments. Any measured profile can, therefore be deconvoluted to give the true profile by means of the relationship

$$P_t(x) = \int_{-\infty}^{+\infty} g(x-x^1) P_m(x^1)dx^1 \tag{3}$$

where the Fourier transform of g is the reciprocal of the Fourier transform of f.

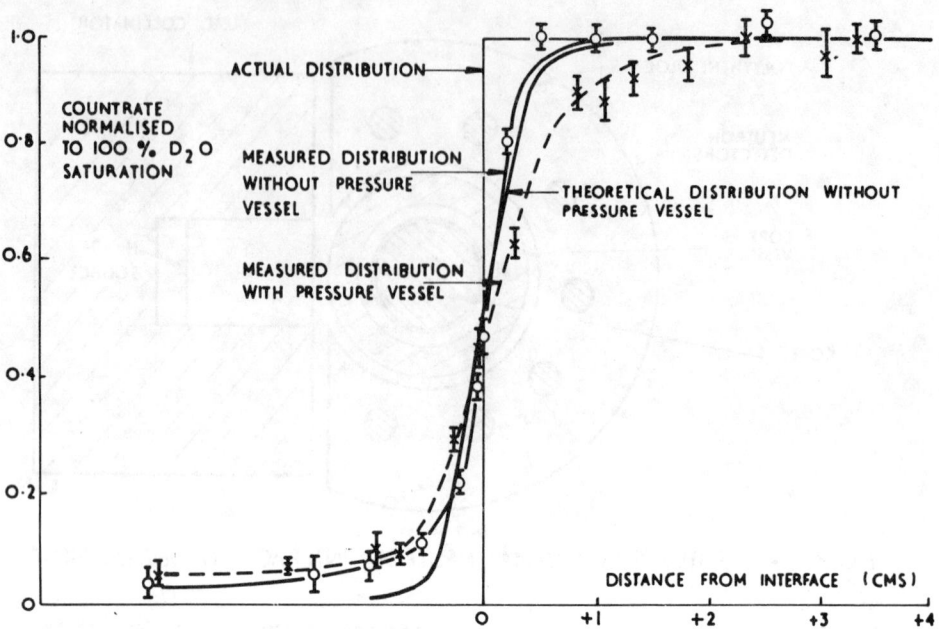

FIG. 9   MEASURED PROFILES FOR 0-100 % D$_2$O STEP IN SATURATION.
COLLIMATOR SLIT WIDTH = 4 mm

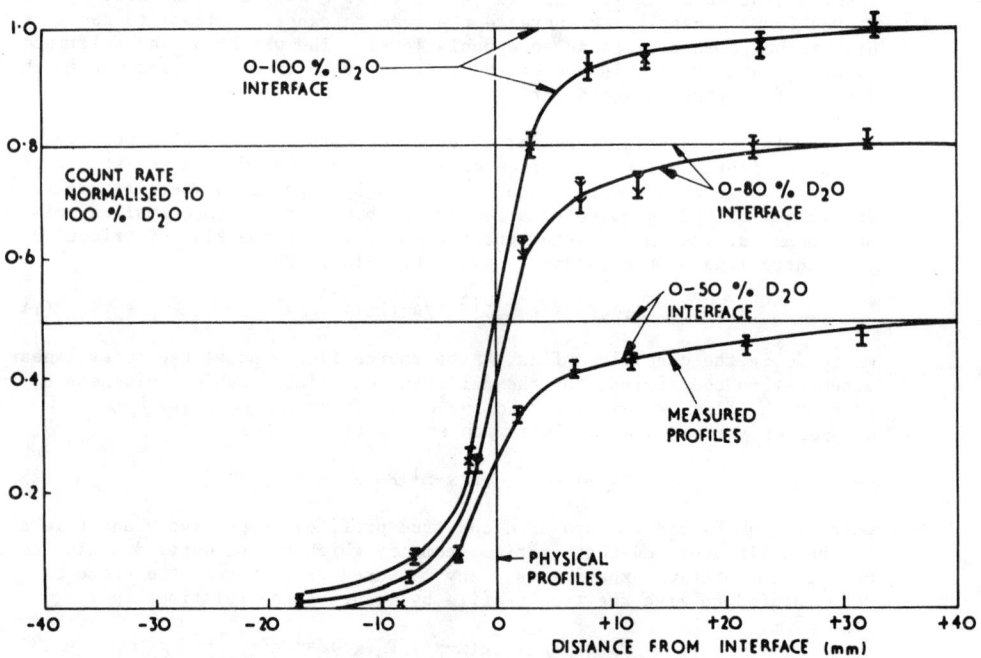

FIG.10 MEASURED SATURATION PROFILES FOR VARIOUS D$_2$O SATURATIONS
SLIT WIDTH = 4 mm

A profile derived from equation (2) for a sandpack without a pressure vessel
has been included in Figure 9. Close agreement can be observed for most of
the profile. The discrepancies at the left hand side of the profile are
probably induced by uncertainties in background level and the fact that some
collided gamma rays are scattered with energy levels in excess of 2.23MeV and
can produce neutrons from deuterium. The predicted width of the front between
10% and 90% saturation levels is 0.8cm which compares satisfactorily with the
measured values of about 1cm.

The signal is also shown in these results to be directly proportional to
deuterium concentration. Increasing the slit width in the collimator increases
the count rate and, hence, improves the counting statistics, but at the cost of
increasing the smearing of the saturation profile.

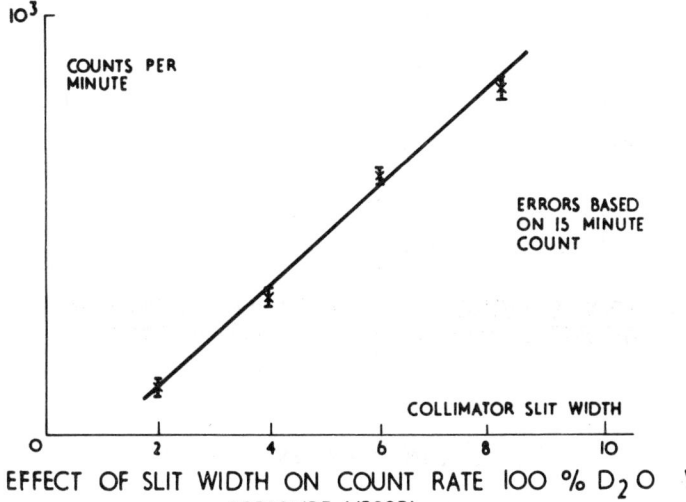

FIG.11   EFFECT OF SLIT WIDTH ON COUNT RATE 100 % D_2O WITH
PRESSURE VESSEL.

Tests have also been carried out on a sandpack with only a short length of light
water saturation preceded and followed by heavy water. The results shown in
Figure 12 confirm the ability of this technique to determine the length of such
a short slug of fluid. In Figures 9 to 12 data are included both with and without a pressure vessel simulation. It is observed that the pressure vessel leads
to some attenuation of the neutron flux and further smearing of the saturation
profile.

These results confirm that the γ-n interrogation of deuterium provides a method
which gives a good quantitative measurement of the saturation level of
deuterium bearing species as a function of axial position and time. It is
proposed to use this method for such measurements on high pressure flood
experiments.

A further merit of this measurement technique is that it can be used to determine
the dispersion coefficient for the core. At the start of an experiment the core
is flooded with normal brine prior to oil flooding to set up the starting
conditions for a displacement experiment. The brine is then displaced by heavy
water brine. This idealised first contact miscible displacement with negligible
density or viscosity differences, results in the development of a classical
dispersion front and by measuring this profile using the gamma-neutron technique
the dispersion coefficient for the core can be derived.

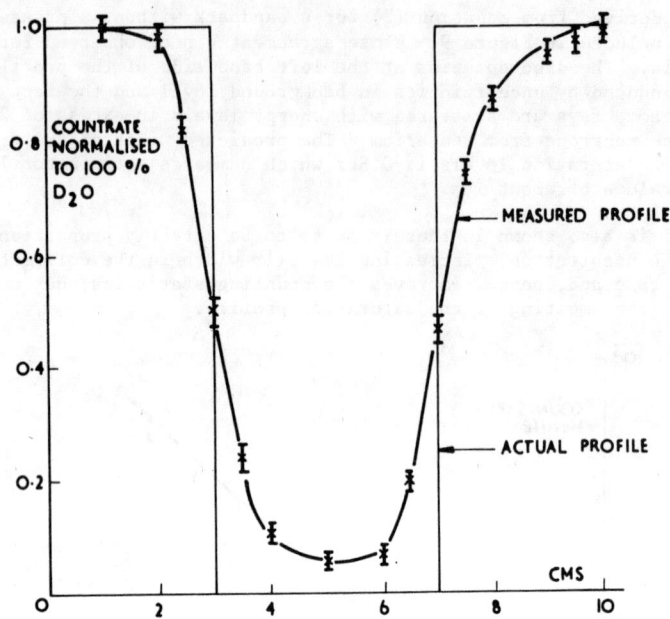

FIG. 12 PROFILE ALONG D$_2$O SATURATED SANDPACK WITH 4 CM LENGTH
SATURATED WITH H$_2$O. WITH PRESSURE VESSEL. SLIT WIDTH 8 mm

APPLICATION TO HIGH PRESSURE FLOOD TESTS

As a result of the development tests described above it has been decided to use radioactive ferrocene to label the oil phase in the High Pressure Flood Tests and heavy water to label the brine phase using a γ-n reaction interrogation method. It is intended that the measurement system should be capable of measuring saturations over the complete length of cores up to 5 metres long, traversing along cores to obtain the axial saturation distribution at intervals which can be as small as 1cm. At each position where measurements are required, counting will be carried out for approximately 10 minutes to give sufficiently large counts to ensure reasonable accuracy. Such a period is generally satisfactory for most conditions including a shock front travelling at 0.3m/day which might be typical of North Sea conditions. In the measurement period such a front will have moved only 2mm, which is small in relation to the slit width Although higher rates may have some experimental interest, the errors involved would be mitigated by front smearing induced by capillary forces in immiscible processes or dispersion in miscible processes. If necessary the counting time could be reduced by increasing the ferrocene concentration and using a higher activity sodium 24 source.

It should be noted that these two methods cannot be used simultaneously as the high level of gamma radiation occurring when the sodium 24 source is in use would swamp the radiation coming from the ferrocene. This restriction to one measurement at a time leads, however, to simplification of the apparatus to be used on the flood rigs as only one collimator is required. The apparatus will be similar to that used for development tests (Figure 8) with He^3 neutron detectors mounted in the polythene blocks for the γ-n method.

Detectors are not placed directly below the core so that the measuring apparatus can be removed readily from the core. When measurement is required of γ-radiation from the ferrocene, the sodium source will be removed and replaced, within the same collimator, by a sodium iodide crystal detector. Sodium iodide was selected in preference to a Ge(Li) detector to eliminate the need for cooling the crystal to cryogenic temperatures, and becuase of its higher sensitivity.

The detectors, fed from separate EHT supplies for neutron and sodium iodide detectors, will have their outputs taken to a common multi channel analyser together with the outputs from separate detectors which will be used to measure background radiation levels simultaneously. In this way the background level can be automatically removed from the measured signal. The multichannel analyser will be integrated into a minicomputer system which will control the flood rigs and carry out data processing. The detector and analyser system has to be calibrated on a representative geometry at 100% saturations of brine and oil.

The first high pressure flood rig is currently being constructed but these measurement techniques are now in use in low pressure studies of oil displacement by waterflooding. These experiments are being carried out using brine and tetradecane in the same sandstone materials which are to be used for the high pressure experiments, and encompass both displacement tests and steady state relative permeability measurements.

Conclusions

It can be concluded that two viable measurement techniques have been established for the measurement of fluid saturations within the cores of high pressure flood experiments. Labelling the oil phase with radioactive ferrocene, which can be produced synthetically from active iron or from reactor irradiation, allows the oil phase saturation to be measured. A convenient method for marking the water phase with heavy water and interrogating it via a gamma-neutron interaction to measure the water phase saturation has been evaluated. This method also has promise for measuring the concentration of a particular component in a hydrocarbon gas or liquid system by synthesising deutero-carbon additives.

Reasonable levels of source activity and conventional counting techniques allow these saturations to be measured within about 10 minutes with a statistical uncertainty of less than ±3% at the 100% saturation level and at the 25% saturation level the result would typically be 0.25 ±0.02. These techniques allow axial and temporal variations in fluid saturations to be determined as only about 1cm of the core length is viewed at any one time and the detectors can be traversed along the core. The location of any sharp front in saturation can be accurately determined and although the measurement introduces some smoothing of sharp fronts, it is possible to convert the measured profile back into a more accurate saturation profile by a deconvolution process.

The new techniques are to be used on the high pressure flood experiments to be carried out at Winfrith and they are already in use there on low pressure experiments.

Acknowledgement

This work has been supported by a contract from the UK Department of Energy.

REFERENCES

1   BROWN, R J S and GAMSON, B W: "Nuclear Magnetism Logging", Petroleum Trans.
     (1960) 219, 199-207.

2   NIKIAS, P A and EYRAUD, L E: "Some Examples of Nuclear Magnetism Logging in
     Three San Joaquin Valley Oil Fields" JPT (Jan 1963) 23-27.

3   PARSONS, R W: Microwave Attenuation - A New Tool for Monitoring Saturation
     in Laboratory Flooding Experiments. SPEJ (Aug 1975) 302-309

4   CLAVIER, C. HOYLE W. and MEUNIER, D: "Quantitative Interpretation of
     Thermal Neutron Decay Time Logs: Part 1. Fundamentals and Techniques".
     JPT (June 1971) 23 743-755.

5   WAHL, J S et al: "The Thermal Neutron Decay Time Log". SPEJ (Dec 1970)
     365-379.

6   RICHARDSON, J E et al: "Methods for Determining Residual Oil with Pulsed
     Neutron Capture Logs". JPT (May 1973) 593-603.

7   YOUNGBLOOD W E: "The Application of Pulsed Neutron Decay Time Logs to
     Monitor Waterfloods with Changing Salinity". JPT (June 1980) 987-963.

8   HERTZOG, R C: "Laboratory and Field Evaluation of an Inelastic Neutron
     Scattering and Capture Gamma Ray Spectrometry Tool". SPEJ (October 1980).
     327-340

9   OVERMAN, T and CLARK, H M: "Radioisotope Techniques". McGraw-Hill
     New York (1960) 378.

CHARACTERIZATION OF EOR POLYMERS AS TO SIZE IN SOLUTION

ROY DIETZ

*Division of Materials Applications, National Physical Laboratory,
Teddington, Middlesex, TW11 0LW UK*

ABSTRACT

The potential is explored of extending to the characterization of
EOR polymers the conventional electrical sensing zone (ESZ) technique.
Theoretical expectations and experimental factors are discussed. The method is
tested with well-characterized fractions of polyacrylamide and some biopolymer
samples for which characteristics relevant to EOR use are known.   Only the
largest solution species can be detected, but those species are significant in
determining technological properties. For polyacrylamides the ESZ response
correlated with hydrodynamic volume. For biopolymers there were correlations
with screen factor and with viscosity at concentrations of relevance in EOR.
The method offers promise for monitoring solutions rapidly for microgel.

INTRODUCTION

The size of solution species is relevant to the possible use of polymers in
EOR in two main respects.   Solution species must be large enough to give a high
viscosity at low concentrations for mobility control, but not so large that they
block pores in the substrate and give rise to poor injectivity.   A simple means
of characterizing polymers as to size in solution would aid the production of
improved polymers for mobility control, and could also serve as a means of
monitoring solutions for adequate injectivity.   The number and size of the
largest solution species present control injectivity; estimation of that part of
the size distribution gives rise to great difficulties for such conventional
techniques as gel permeation chromatography, ultracentrifugation and fractionation followed by independent characterization by light-scattering photometry or
viscometry.

It is clear from experiments with filters of controlled pore size (1) (2),
from determinations of radii of gyration by light-scattering photometry (3) and
from estimates of hydrodynamic volume from viscometry (4) that the size range of
relevance is some 0.5-1.0 μm.   In conventional particulate metrology suspensions
of particles of that size can be characterized by the electrical sensing zone
(ESZ) technique.   This paper describes an exploratory investigation into the
applicability of that technique to dissolved polymer.   Well-characterized
samples are used to relate the ESZ response to molecular characteristics for
polyacrylamide fractions and to rheological characteristics for biopolymers

DISSOLVED POLYMER IN AN ELECTRICAL SENSING ZONE

Principle and range of the ESZ method

In the electrical sensing zone method (Figure 1) a dilute suspension in a salt solution is made to flow through a small orifice (diameter ~ 100 μm). Coulostatic circuitry is used to maintain a constant current between electrodes placed on opposite sides of the aperture. When a suspended particle passes through the aperture, the electrolytic resistance is increased by the equivalent of the volume of electrolyte displaced.

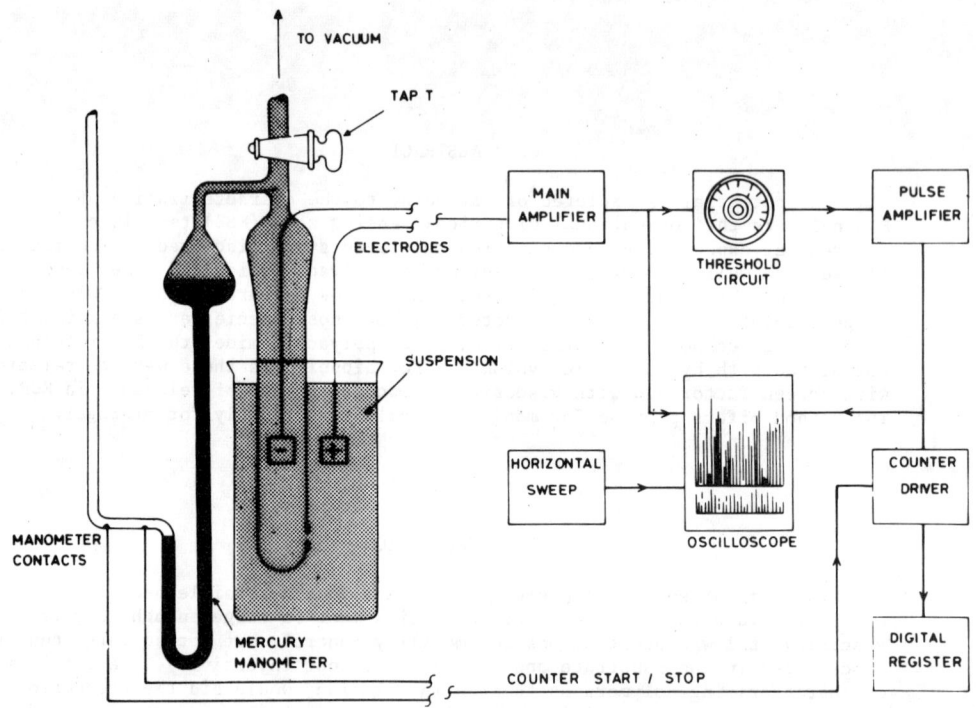

Figure 1. Schematic of the ESZ method

The constant current is maintained by a voltage pulse of magnitude proportional, to a good approximation, to the size of the suspended particle. Commercial instruments have discriminator circuits capable of counting and sizing the voltage pulses. The voltage scale can be calibrated in experiments with suspensions of particles of known size. Routine operation is possible with particles greater than some 1 μm in diameter; with precautions to reduce noise, operation with insulating particles of diameter 0.4 μm is feasible.

Polymer structure in solution

Application of the ESZ method to the sizing of dissolved polymer molecules introduces some special features connected with the nature of the solution species. Candidate polymers for EOR use fall into two classes of molecular structure (Figure 2). The synthetic polymers (polyacrylamide, poly(vinyl-pyrrolidone)) can be modelled as flexible chains. Most biopolymers have structures approximating more closely to rigid rods, at least near ambient temperature; for xanthan there is evidence (5) of a structural transition at temperatures above 60 °C.

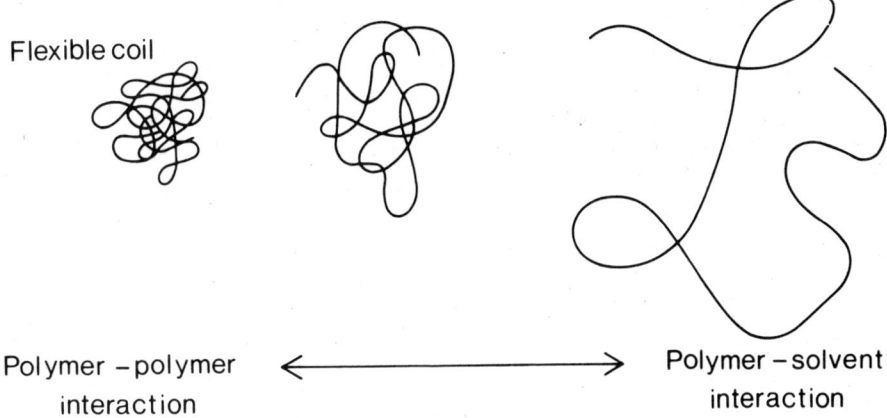

Figure 2. Polymer structure in solution

    Flexible chain polymer molecules pervade much larger volumes in solution than the volume of the polymer molecule itself. Within that pervaded volume, most of the solvent molecules are close to several polymer chain segments. It turns out that most of the solvent within that pervaded volume is incapable of independent hydrodynamic flow. The hydrodynamic properties of dilute solutions of flexible chain molecules can be treated (6) as those of suspensions of spheres, consisting largely of bound solvent. The effective volume of those spheres, for a monodisperse polymer of molar mass M in solution of concentration c and with an effective volume fraction ϕ is

$$V_h = \frac{\phi M}{N_o c}$$

where N_o is Avogadro's constant. For a suspension of spheres, the volume fraction is related to viscosity by

$$\eta = \eta_o(1 + 2.5\,\phi + \text{higher terms})$$

where η_o is the solvent viscosity. The measurable limiting viscosity number $[\eta]$ can be related to V_h

$$[\eta] = \underset{c \to o}{\text{Limit}} \frac{\eta - \eta_o}{\eta_o c} = \underset{c \to o}{\text{Limit}} \left[\frac{2.5\,N_o V_h}{M} \right]$$

so that for the low concentrations used in ESZ determinations, the volume of the equivalent sphere can be taken as

$$V_h = \frac{[\eta]M}{2.5\,N_o} \qquad\qquad 1$$

Real polymer samples are heterogeneous in molar mass and therefore in hydrodynamic volume. For a typical anionic polyacrylamide (molar mass 4000 kg/mol; 20% hydrolysed) the average hydrodynamic radius calculated from V_h is some 80 nm in 0.5 M NaCl. There is evidence (7) that much larger molecular aggregates may be present. For a xanthan sample purified by centrifugation, the hydrodynamic properties are consistent (8) with rigid rod molecules, of length some 0.6 μm and diameter some 2 nm. Again there is evidence of much larger solution species (9), particularly in unpurified commercial material (10).

Polymer 'particles' in the ESZ

Passage of a solution of a flexible chain polymer through the ESZ aperture corresponds to the passage of spheres of solvent of much reduced mobility. Such a 'particle' will be electrolytically conducting, but the resistance should be higher than that of the same volume of solvent. The salt-containing polymer gels used in electrochemistry to provide electrolytic conduction with minimal ion transport are a relevant analogy. It follows that the ESZ signal of a dissolved flexible chain polymer molecule will be smaller than that of an insulating particle, such as a polymer in latex suspension, of the same size. Biopolymer molecules in rigid rod conformation include smaller quantities of solvent, so that the discrepancy may be smaller, but in general dissolved polymer is to some extent 'transparent' in an ESZ.

There is also the question of molecular dynamics. Brownian motion causes the segment density, molecular shape and effective size of a flexible chain molecule to change continuously. Polymer properties are described in terms of average molecular dimensions, where the averaging is both over time for a given molecule and over the population of molecules of given chain length. Those fluctuations will be reflected in dispersion of the ESZ signals; an ideal polymer sample containing only one molecular species should give signals over a discrete range of apparent size, with a peak at the pulse height corresponding to the most probable size.

EXPERIMENTAL

Materials

Polyacrylamide fractions were produced by the controlled addition of ethanol to dilute (0.007 g cm^{-3}) solutions of commercial polymers, non-ionic and anionic, in water. Fractions were characterized by capillary viscometry (FICA Autoviscometer) and by light-scattering photometry (chromatix low-angle photometer) in the solvent (0.01g/cm^3 NaCl) used for the ESZ experiments. Biopolymers were gifts from Dr I G Meldrum (BP Research, Sunbury) and Dr I W Sutherland (University of Edinburgh).

Polymer solutions were prepared in 'Isoton' a proprietary (Coulter Electronics) saline solution, (ca 0.01g/cm^3) supplied for haemacytometry. The effect of electrolyte concentration was studied by using more concentrated saline solution (0.04 g/cm^3) filtered through Millipore membranes of pore size 0.1 μm.

ESZ measurement requires that only single particles traverse the aperture. The maximum counting rate of the instrument used limits the suspension concentration to 10^7-10^8 particles/cm^3. For a molar mass of 10^4 kg/mol, that number concentration corresponds to a mass concentration of 10^{-10}-10^{-9}g/cm^3. Solutions within that concentration range were prepared by successive volumetric dilution of parent solutions of polymers of concentration ca 10^{-3}g/cm^3; those parent solutions were prepared gravimetrically and with gentle magnetic stirring overnight.

ESZ technique

The size range of insulating particles to which the ESZ method is conventionally applied extends down to only 1 µm; smaller signals become obscured by background noise. In order to extend the useful range, detailed attention was given to earthing and shielding. The instrument [Coulter ZB] was housed in a Faraday cage, and the mains supply was routed through an isolating transformer. With these precautions the background count was acceptably low at a pulse height corresponding to an insulating particle of diameter ca 0.4 µm. A polymer latex suspension of that diameter was used to calibrate the size scale. The aperture was of nominal diameter 30 µm and the volume of liquid passed through was constant at approximately 0.05 cm^3.

Polymer solutions were analysed in the manual mode in order to avoid possible artifacts of automatic subdivision of the size range. Each analysis was preceded by a background count. With the polymer solutions counts were made by reducing the size threshold stepwise throughout the range for which the polymer count exceeded the background by a factor of ten or more. Duplicate determinations were made in all cases; counts were reproducible to within a few per cent except at the extremes of the range.

RESULTS AND DISCUSSION

ESZ analyses of the polymer solutions are presented as integral distributions of the number of particles per gram of polymer of apparent size greater than the abscissa values, which relate to the calibration with insulating particles. Smooth curves were drawn through 20 points representing counts throughout the size range. Since ESZ transparency is presumably a function of polymer-solvent interaction, and therefore of the chemical structure of the polymer, comparisons are made only between results for polymers of similar structure.

ESZ response versus molecular properties; polyacrylamide

The conjecture that the ESZ response of polymer samples of similar chemical structure should correlate with hydrodynamic volume as defined by equation 1 was tested in experiments with seven polyacrylamide fractions. Four fractions of nonionic polymer and three of anionic polymer were characterized by viscometry and light-scattering photometry; the mass-average molar masses ranged from 3400 to 18200 kg/mol. A monodisperse sample of molar mass 10000 kg/mol contains 6×10^{16} particles/gram. In the ESZ experiments [Figures 3 and 4] the number of particles per gram sensed before the signal to background ratio fell below 10 never exceeded 10^{15}. It follows that only a few percent of the particles present were sensed, even allowing for the polydispersity of the fractions. In this estimate no account was taken of possible adsorption of polymer on the glass surfaces. The shape of the measured distributions gave further evidence that only a fraction of the solution species present were detectable; thus most of the curves were rising steeply at the lowest accessible size.

The hydrodynamic volumes of the fractions calculated from viscometry and light-scattering photometry are average values and refer approximately to the most abundant species present, which are clearly not detected by the ESZ method, even for the fraction of largest molecular size. The experimental results do not provide, therefore, a critical test of the supposed correlation with hydrodynamic volume. There would be a correlation with the small part of the distribution measurable only if the complete size distributions of the fractions chanced to be of similar shape. Figure 3 shows that larger particles were detectable for the nonionic fraction of largest average hydrodynamic volume throughout the ESZ size range.

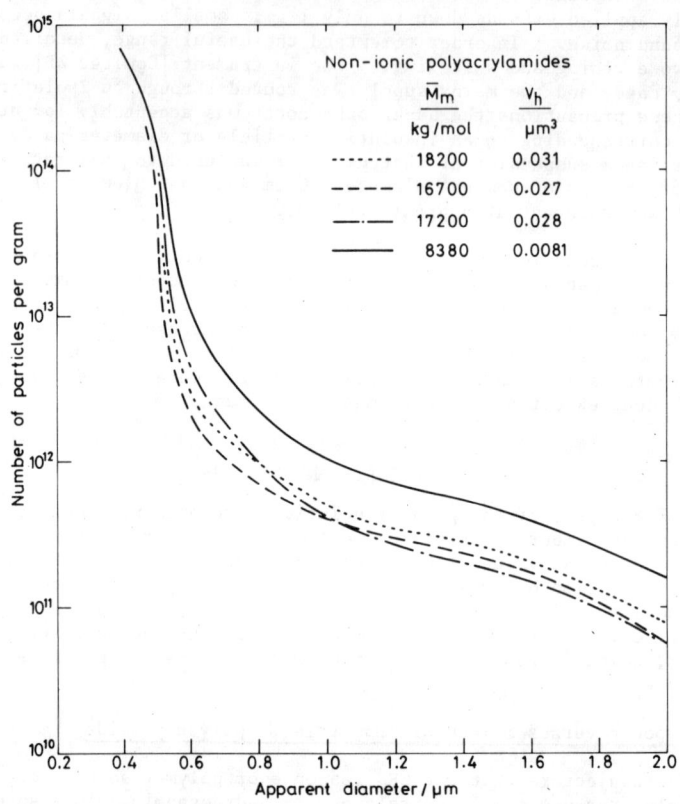

Figure 3. ESZ response of nonionic polyacrylamide fractions

For the anionic fractions (Figure 4) the relation is less clear, but in the ESZ size range below 0.6 μm the recorded distributions follow the order of average hydrodynamic volume. The effect of added electrolyte is also consistent with a relation between the ESZ response and hydrodynamic volume. In more concentrated electrolyte (0.04 g/cm^3) results for non-ionic polyacrylamides were little changed, but the number of particles sensed for an anionic polymer fell sharply. It is well known (11) that the radius of gyration and hydrodynamic radius of anionic polyacrylamides fall with increasing electrolyte concentration because the effect of repulsion between carboxyl groups along the chain is reduced.

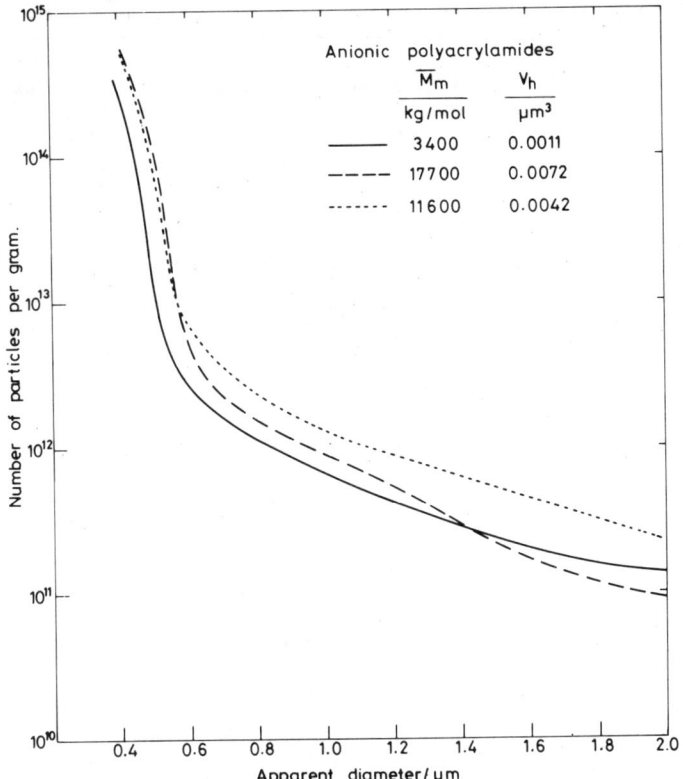

Figure 4.   ESZ response of anionic polyacrylamide fractions

In summary, the results for well-characterized polyacrylamides are not inconsistent with hydrodynamic volume as the molecular characteristic determining the ESZ response.   Only qualitative evidence can be offered since the fraction of particles sensed is small, even for fractions of large molar mass.   The ESZ technique in the conventional form employed here is sensitive only to the largest particles present, which may well be molecular aggregates.   It is worth noting that such aggregates are thought (12) (13) to be important in determining properties of water-soluble polymers.

ESZ response versus rheological properties; biopolymers

The relation between ESZ response and molecular size is more difficult to investigate with biopolymers since fractions of different molecular size but similar chemical structure are not readily accessible, at least at high molar mass.   Attempts were made to fractionate xanthan by controlled precipitation above 60 °C and by preparative gel permeation chromatography but without success.   Instead the ESZ response of whole biopolymers was related to rheological properties relevant to use in EOR.   Such information was available for a series of experimental biopolymers prepared under contract (OT/F/443) to the Department of Energy at the University of Edinburgh; a commercial xanthan (Keltrol) was included for comparison.   Since their thermal stability (14) makes scleroglucans possible candidates for use under North Sea reservoir conditions, a separate comparison was made between three samples.

Table 1. Rheological properties of biopolymers at 3×10^{-3} g/cm^3 in salt water

Sample	η at $1s^{-1}$	Screen factor 1 volume	Screen factor 10 volumes	Size of filter/μm that retains
Keltrol	2630	26.5	20	3
7824	2290	9.64	11.28	0.3
1.15	<790	1.5	1.8	0.22
9.4	1096	7.38	9.36	0.45-0.3

ESZ results for biopolymers (Figure 5) were consistent with those of poly-acrylamides in that only a small fraction of the particles present were sensed, assuming a plausible molar mass (5000 kg/mol). Similarly the integral distri-bution of apparent size did not reach a limiting lower plateau. Within those limitations, however, there were clear correlations with properties of signifi-cance for EOR use; those properties were measured at the University of Edinburgh at a concentration (0.003 g/cm^3) much higher than that used in ESZ analysis.

More large particles were detected (Figure 5) in solutions of the commercial xanthan, Keltrol, than for the remaining polymers; of the samples tested only Keltrol fails to pass in solution through a 1 μm filter. The viscosity of

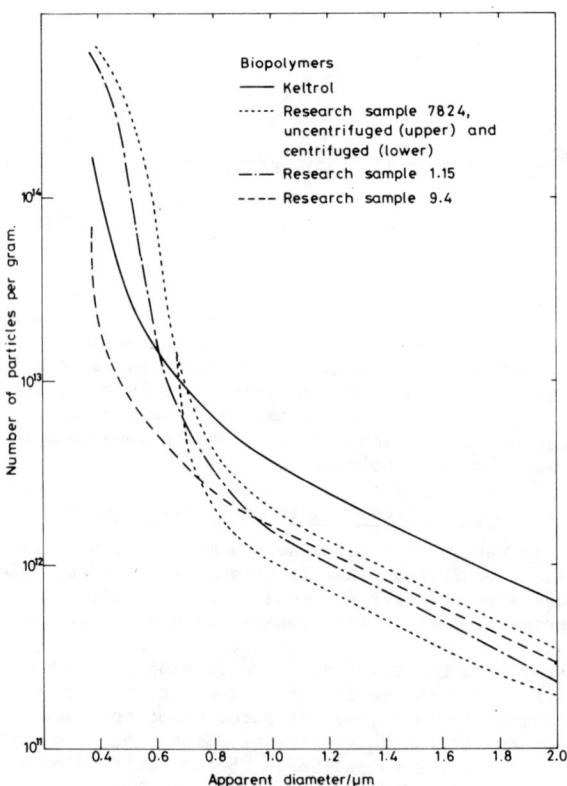

Figure 5.   ESZ response of biopolymers

solutions at a shear rate of $1s^{-1}$ follows the order of the number of particles of
apparent size ≥ 1 μm, but does not follow that of the number of apparent size
≥ 0.4 μm.   The screen factor, after 1 and 10 volumes, follows the former order.
Figure 5 also includes the effect of centrifugation (40 krpm) upon the ESZ
response of one sample (7824);   there was no significant change at apparent
sizes below 0.7 μm, but a clear decrease in count above that size.   Since
centrifugation selectively removes larger particles, the result links directly
the ESZ signal to size in solution.

      Three scleroglucans were available for test (Figure 6);   Actigum CS11 and
L21 were commercial materials and research sample E was from the University of
Edinburgh.   We have no rheological data for the set, but on the basis of the
correlations established above (Figure 5) one might predict that the screen
factor and viscosity at $1s^{-1}$ would be higher for L21 than both CS11 and sclero-
glucan E, and that the last two would have similar properties.

      As with the synthetic polymers studied, the ESZ response of biopolymers
reflects the few largest solution species present.   In many biopolymer solutions those species are deformable molecular aggregates termed microgel.   The
correlation between the ESZ response at a concentration such that interparticle
interactions can be neglected and rheological properties of the semi-dilute
regime in which interparticle interaction is strong suggests that molecular
aggregates can affect significantly rheological properties under EOR conditions.

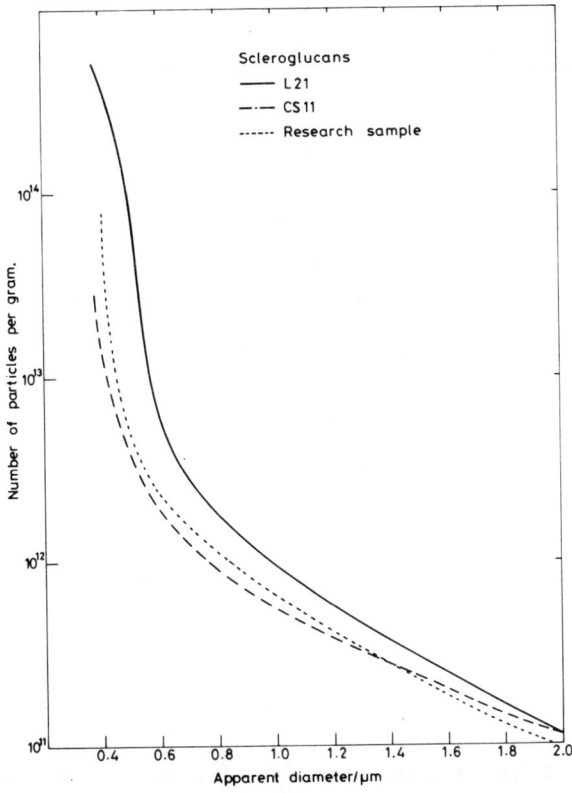

Figure 6.   ESZ response of scleroglucans

Microgel is also significant (10) (11) in determining injectivity of polymer solutions in porous media. Microgel is sufficiently deformable to pass through small pores at the high shear rate near the injection point, but can block pores at lower shear rates further into the substrate. A recently suggested (15) method for estimating microgel is based upon filtration at a very low shear rate. The ESZ approach has promise as a more rapid and convenient technique.

Future developments

With some modifications of the very conventional apparatus used, the ESZ technique could well yield more information of relevance to polymer use in EOR. Controlled variation of flow rate through the orifice is desirable; the shear rate in the conventional apparatus is high ($\sim 10^3 s^{-1}$) and polymer solution species must be deformed in the orifice. The effect of shear rate upon pulse size would help to characterize polymer deformation in a shear field. Since most of the solution species are below the accessible size range in the present work, it is superficially attractive to reduce the aperture size in order to enhance the signal. The true sizes are larger than the size scale calibrated with insulating particles, however, and there may be advantage in working at slow flow rates in wide apertures and with different electrolyte concentrations. Again the effect of using 'hydrodynamic focussing' to ensure passage through the centre of the orifice where the electrical field is symmetrical needs to be investigated.

SUMMARY

When used in the mode conventional for particle sizing, but with improved earthing and shielding, the ESZ technique can sense the largest solution species of water-soluble polymers. The response is related to the hydrodynamic volume of polacrylamides, and to rheological properties of biopolymers. The method has promise in monitoring solutions of candidate EOR polymers for microgel and, with some instrumental modification, in more fundamental studies of polymer solutions.

NOMENCLATURE

c	concentration (mass/volume) of polymer
M	molar mass
$\bar{M}_m$	mass-average molar mass
N_o	Avogadro's constant
V_h	equivalent hydrodynamic volume of a polymer molecule
η	viscosity of solution
η_o	viscosity of solvent
$[\eta]$	limiting viscosity number
ϕ	volume fraction of polymer in solution

ACKNOWLEDGEMENTS

Dr I.W. Sutherland (University of Edinburgh) kindly provided research samples and rheological data. Dr I.G. Meldrum (BP Research) presented some commercial samples. Molar mass characterization and fractionation was the work of M.A. Francis, and Mrs C.M.L. Atkinson carried out the ESZ determinations. The work was carried out under contract (OT/F/524) to the Department of Energy.

REFERENCES

1. SMITH, F.W.; "The Behaviour of Partially Hydrolysed Polyacrylamide Solutions in Porous Media" J. Pet. Tech. (Feb 1970), 148-156

2. SZABO, M.T.; "Molecular and Microscopic Interpretation of the Flow of Hydrolyzed Polyacrylamide Solution Through Porous Media" SPE 4028, presented at SPE 47th Fall Conference, San Antonio (1972)

3. KLEIN, J and CONRAD, K-D.; "Characterization of Poly(acrylamide) in Solution". Makromol. Chem. (1980), 181, 227-240

4. UNSAL, E., DUDA, J.L. and KLAUS, E.; "Comparison of Solution Properties of Mobility Control Polymers" in JOHANSEN, R.L. and BERG, R. (Eds) "Chemistry of Oil Recovery" ACS Washington (1978), 141-170

5. HOLZWARTH, G.; "Conformation of the Extracellular Polysaccharide of *Xanthanomas campestris*". Biochemistry (1976) 15 4333-4339

6. TANFORD, C.; "Physical Chemistry of Macromolecules". Wiley (1961) 333-344

7. BOYADJIAN, R., SEYTRE, G., BERTICAT, P. and VALLET, G. 'Caracterisation physico-chimique de Polyacrylamides utilises comme Agents Floculants'. Euro. Polym. J. (1975) 12 401-407

8. RINAUDO, M. and MILAS, M. "Polyelectrolyte Behaviour of a Polysaccharide from *Xanthanomas campestris*" Biopolymers (1978) 17 2663-2678

9. SOUTHWICK, J.G., LEE, H., JAMIESON, A.M. and BLACKWELL, J. "Self-association of Xanthan in Aqueous Solvent Systems" Carbohydrate Res. (1980) 84 287-295

10. KOHLER, N. and CHAUVETEAU, G.; "Polysaccharide Plugging Behaviour in Porous Media: Preferential Use of Fermentation Broth". SPE 7425. Paper presented at SPE 53rd Fall Conference, Houston (1978)

11. MACWILLIAMS, D.C., ROGERS, J.H. and WEST, T.J.; "Water-soluble Polymers in Petroleum Recovery" in BIKALES, N.M. (Ed). "Water-soluble Polymers" Plenum (1973) 106-124

12. DUNLOP, E.H. and COX, L.R.; "Influence of Molecular Aggregates on Drag Reduction". Phys. Fluids (1977) 20 S203-S213

13. WOLFF, C. "On the Real Molecular Weight of Polyethylene Oxide of High Molecular Weight in Water". Canad. J. Chem. Eng. (1980) 58 634-636

14. DAVISON, P. and MENTZER, E.; "Polymer Flooding in North Sea Oil Reservoirs" SPE 9030. Paper presented at SPE 55th Fall Conference Dallas (1980)

15. CHAUVETEAU, G and KOHLER, N.; "Influence of Microgels in Xanthan Polysaccharide Solutions on their Flow through Porous Media". SPE 9295 Paper presented at SPE 55th Fall Conference, Dallas (1980)

VISUALISATION OF THE BEHAVIOUR OF EOR REAGENTS IN DISPLACEMENTS IN POROUS MEDIA

ERIC G. MAHERS, ROBERT J. WRIGHT, RICHARD A. DAWE

Department of Mineral Resources Engineering,
Imperial College, London SW7 2BP

ABSTRACT

Micromodels have been successfully employed to observe directly displacement processes, and have assisted in understanding the physics of the complex fluid phenomena involved in Enhanced Oil Recovery. Both miscible and surfactant displacement sequences are reported here.

The models have been produced by etching the pores into nylon, from which replicas in epoxy resin have been made. Computer graphics and microfilm facilities have been used to produce accurately drafted network photomasks.

INTRODUCTION

The mathematical description and prediction of fluid flow behaviour has been much assisted by direct observation. Although this is not possible in real porous media, models can be made in transparent materials which permit direct observations within the pores of fluid interactions, displacements and entrapments. These models may be monolayer packs of glass beads, as used by Chatenever (1) and Egbogah (2), or two dimensional etched networks.

Mattax and Kyte (3), Michaels et al (4), Davis et al (5) and Wardlaw (6) have used etched glass models. Mattax's network comprised of a rectangular array of straight channels of similar width but varying length. He used this model to study the mechanism of water flooding, with regard to relative permeabilities and wettability. He described the distribution of the fluids and the effect of wettability on areal sweep efficiency, but did not extend this work to cover EOR mechanisms. Michaels et al used the same micromodel to analyse how changes in surface wetting, by the injection of aqueous hexylamine, might mobilise entrapped oil. They concluded that the observed stimulation of oil production was the result of transient changes in wettability. Davis et al made use of a commercial overlay shading medium to produce an irregular, random design. They used this model to qualitatively demonstrate the displacement of oil and water by the microemulsions used in the various Maraflood processes. A film is available from Marathon Oil Company showing the displacements of oil and water by micellar solutions specially formulated for selected U.S. crudes.

Wardlaw employed a heterogeneous, rectangular network with varying pore width. He describes the effect of pore throat size ratio on displacement efficiency, and drainage-imbibition cycles. Although he recognised the importance of pore connectivity and throat sizes on displacement mechanisms, he had not fully explored these factors in his networks.

The pores produced by etching into glass have been V-shaped in cross section, larger in width than those found in common reservoir rocks, usually very shallow, and with high surface roughness. This type of channel topology does not correctly scale capillary pressure effects. Special network patterns of realistic pore sizes have not, as far as we know, been incorporated into any model designs to study experimentally their effects on displacement phenomena.

The objective of this work is to understand the microscopic mechanics of low interfacial tension and miscible enhanced oil recovery processes by using micromodels. In our experiments we are exploring:

Pore network geometry and its effect on capillary pressure, displacement and entrapment, by varying pore shapes and throat sizes, and the connectivity of the pores, which, with the size distribution, defines the degree of freedom of route.

The scale and type of network heterogeneity, and the mechanisms by which areas of bypassed oil can be contacted and mobilised through diffusion or reduced interfacial tension.

Diffusion and mass transfer phenomena and the phase behaviour of ternary systems. Various alcohols can be used to simulate the different types of miscibility; i.e. preferentially oil or water soluble, and first or multiple contact.

CONCEPT OF THE MICROMODEL

The micromodel networks to be described are two-dimensional and it is therefore pertinent to discuss first the validity of results from these experiments. Any two pores within a real porous medium may be connected through a number of routes in three-dimensional space. If these pathways were rotated about a line between these two pores such that they lay wholly within a plane, a high porosity, two dimensional network would evolve. Although two-dimensional networks cannot allow bicontinua in the manner that three-dimensional models can, where the pathways can intertwine (e.g. in the manner of a double helix), simultaneous parallel flow is still permitted. This concept of high porosity, high coordination number networks lies at the heart of our micromodel designs.

NETWORK DESIGN

As already indicated, most networks previously employed have been of random or simple design. In our early work a photoreduced 'Letratone' texture was used to obtain networks which were homogeneous overall, but had variable pore structures on the microscopic scale. The pore 'necks' were of 10-30 microns in width with the pore 'bodies' being 50 to 100 microns. Figure 1 shows the etched network used. In an attempt to simulate more closely the heterogeneous microstructure of natural media, we have developed models with layered structures, figures 2 through 4, and some with additional serial variations, such as shown in figure 5. The latter type, with their high degrees of freedom, were designed to yield fairly realistic, and approximately predictable, relative permeability and capillary pressure functions.

A doublet network, similar to figure 6, was designed to demonstrate the effects of two pore sizes in parallel. Initially, it was drafted by hand and then photographed to create the photomask for etching.

Accurately drafted networks have now been produced using computer graphics and microfilm facilities; this method also enables the pore parameters to be easily varied. A unit cell is composed which is repeated to build up the model. Figures 6 through 9 have been produced in this manner; where the pores are in white.

The doublet network of figure 7 has abrupt changes in pore width, which contrast with the 45 degree divergence in the earlier design (figure 6). Figure 8 illustrates the ability to vary the pore parameters.

Although figure 9 is a regular array, an attempt has been made to create a smooth variation in channel width, with the walls comprised of arcs of circles. The pore throat to body ratio is 1:5. Development of this type of network is under way to investigate the effects of channel angularity.

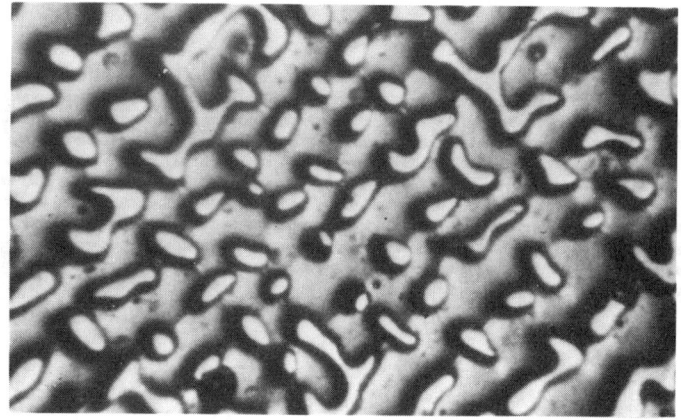

Figure 1   Letratone etched network

Figure 2   Parallel channel design

514

Figure 3 Parallel channel model with obstructions

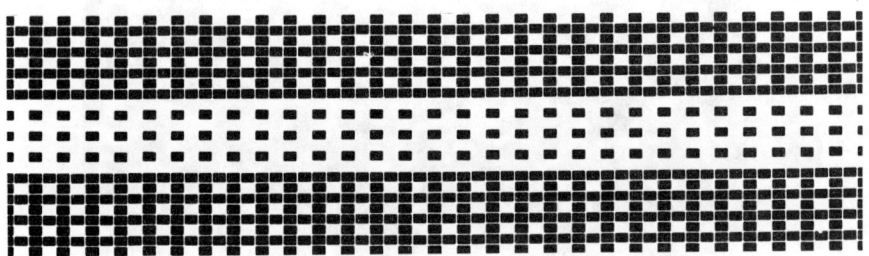

Figure 4 Network with high permeability central streak

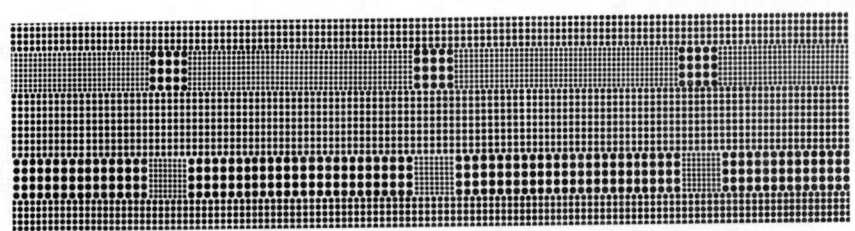

Figure 5a Parallel design with additional serial variations

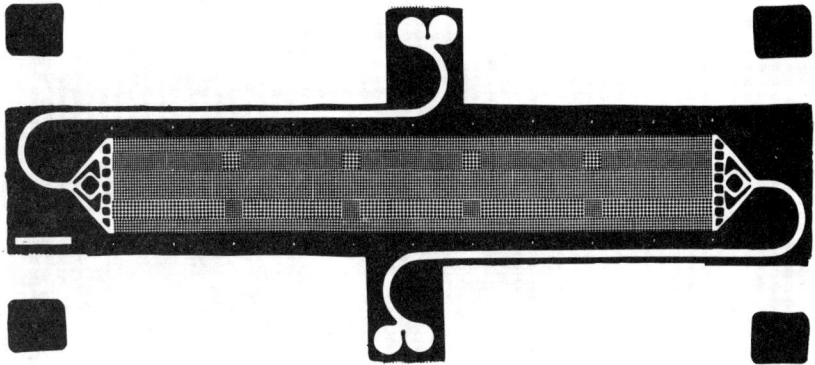

Figure 5b Serial model showing inlet ports

6a   6b

Figure 6 Pore doublet networks (computer drawn)

Figure 7 Doublet network with abrupt pore necks (computer drawn)

Figure 8   Variation of pore parameters

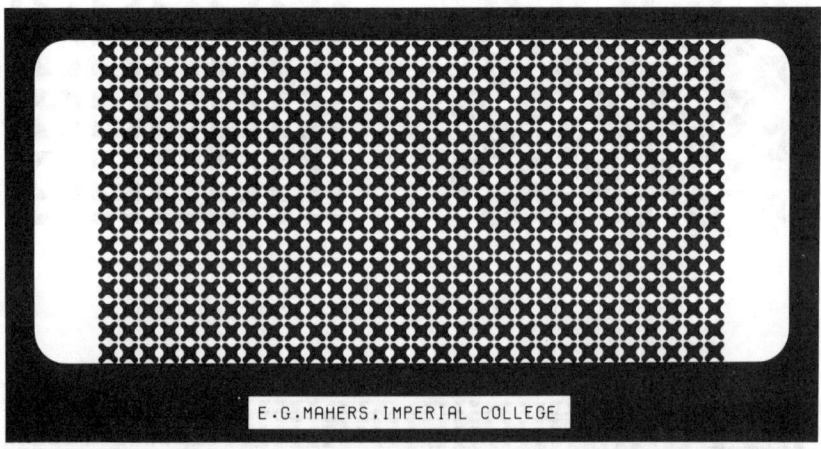

Figure 9   Regular network of curved channels (computer drawn)

CONSTRUCTION OF THE MICROMODEL

The micromodels are produced by etching the pores into nylon film. This is a process commonly used for making printing plates, e.g. using BASF Nyloprint, and has also been used by Bonnet and Lenormand (7). Such photoetching methods enable greater control over pore geometry than chemical etching, and can easily be done in the laboratory, without involving the hazards of hydrofluoric acid.

Detailed procedure

The procedure is illustrated in figure 10. The photographic negative of the network is placed over the nylon and placed under an ultra violet light source. An exposure time of about one hour was found to be adequate when the source (Philips HPW 125W F/70/2) was 200 mm from the negative, with a 10 mm diameter aperture placed midway in between. For deep pores it appears to be necessary to have a small air gap between the negative and the nylon, but the reason for this is not yet fully understood. The unexposed regions are then etched away by washing in a mixture of 90% (by volume) methylated spirits and 10% water at 35°C, for about 30 minutes. A turbulent stream of liquid is maintained across the surface of the nylon by circulation through a centrifugal pump. When etching is complete, the model is dried with warm air and exposed to ultra violet radiation for a further five to ten minutes to set the pore surfaces.

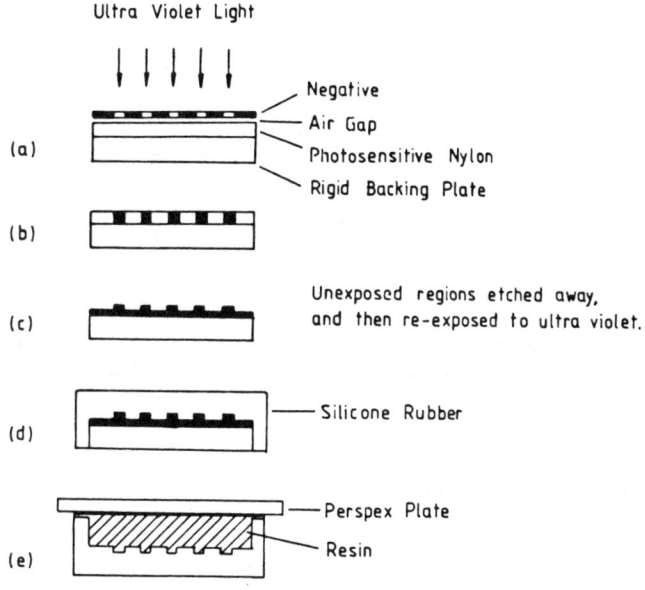

Figure 10   Etching procedure

In our early work we performed flow experiments directly with the nylon etchings, however these were found to suffer from significant sorption of dyes and solvents. Consequently we used the etchings as bases for preparing silicone rubber moulds (Hopkin and William Silastic 3110 RTV and Dow Corning Catalyst 1), from which rigid, non-absorbent epoxy resin replicas are cast, which accurately reproduce the microstructure of the nylon model. Araldite MY753, MY951 hardener, is a suitable resin for this purpose. We now have a relief structure on which a top needs to be secured, to form a two-dimensional pore network. To preserve uniform wettability, epoxy resin film has been produced in the laboratory, which can be sealed onto the epoxy cast by one of several methods:

1. The pores are filled with wax and then resin poured on top. The wax was removed by heating the model and injecting hot fluid, e.g. kerosene. This is a delicate operation with which we have had only limited success, however a similar method was reported by Bonnet and Lenormand (7).

2. The resin film is sealed on by external pressure, using screw clamps as shown in figure 11. This allows the model to be dismantled and cleaned easily. However, the seal is not always perfect and the film tends to depress into the pores, resulting in an unknown, and variable pore depth. The seal can be improved by placing the assembled model in an oven at $65\,^{\circ}C$ for one or two hours. This creates a weak adhesive seal at resin-resin contacts through increased surface interaction and plastic deformation.

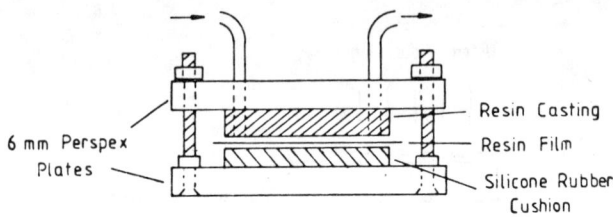

6 mm Perspex Plates

Resin Casting
Resin Film
Silicone Rubber Cushion

Figure 11   Micromodel seal by external pressure

3. The best method discovered to date is to coat the perimeter of the model with epoxy, and gently press the resin film on top. This is to prevent leakage while a solution of resin in methylated spirits is injected into the pores. After two to three hours a thin, wetting resin film will have been deposited. Injection of pure methylated spirits removes any excess resin leaving a bonding film in the tiny crevices; especially between the top of the pore walls and the sealing film. This method maintains constant pore geometry over a series of experiments, and allows short working distance microscope objectives to be used. Surface flatness is improved by using a thick resin film or block.

TOPOLOGY OF THE MICROMODEL PORES

Interfacial curvature, and therefore capillary pressure, is governed by the pore shape (the angle of divergence), contact angle and pore dimensions. In order to ensure that the capillary pressure and viscous resistance are controlled by the dimensions within the plane of the network, it is necessary for the depth of the pores to be constant and of the same order as the width. For widths greater than 50 microns, the depth tends to be constant at about 50 microns. Controlled pore widths at least as small as 20 microns are possible. Figure 12 shows the square shape of the pores in contrast to the V-shape of glass etchings.

1 mm

Figure 12   Shape of the pores of a resin cast of the early doublet model

DISPLACEMENT STUDIES

Displacements within the micromodels were observed through a microscope and recorded in colour on videotape or still photographs. The floods were carried out at low flow rates of less than 50 mm/hr (four ft/day) by means of a variable rate syringe pump. The fluids were introduced into the models through the valve arrangement shown in figure 13. This eliminated dispersion in the entry tube, thereby ensuring injection of uncontaminated fluid.

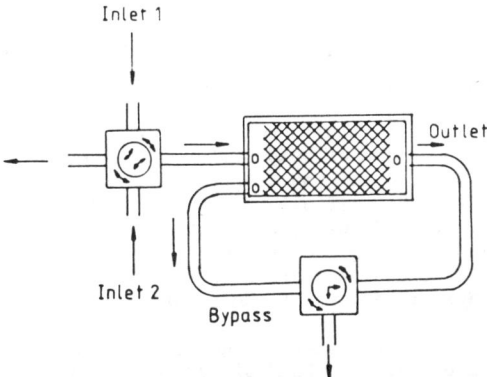

Figure 13   Valve arrangement for the injection of fluids into the micromodels

Use of dyes

Dyes were used to aid the visualisation of the fluids. Alternate injection of dyed and undyed, but otherwise identical, fluids highlighted the flow pathways, clearly showing the stagnant regions (figure 18). The distribution of this stagnant fluid is an important factor influencing the efficiency of an EOR flood.

The water-soluble dyes used in these experiments were Methylene Blue, Nigrosine Black and ICI Lissamines. The oils were dyed with ICI Waxolines. These dyes are, however, surface-active and their effect on wettability and interfacial phenomena should be taken into account when analysing results.

Mobilisation of oil

The mechanics of oil ganglia mobilisation in straight capillaries of varying cross sectional area have been described by Arriola et al (8,9). Under these conditions an EOR chemical can contact the downstream interface only through a wetting film surrounding the ganglion, i.e. a contact angle of 180 degrees measured through the globule. In a two-dimensional network this contact can be achieved through neighbouring pores. These parallel routes also affect the viscous pressure drop across the ganglion, thereby determining the reduction in interfacial tension required to mobilise the drop immiscibly. This discrepancy between single pore and network studies has been discussed by Stegemeier (10), and is effectively demonstrated by the pore doublet models.

Simulation of miscible processes

Alcohols were employed to simulate miscible displacements. The wide variety of alcohols available permits a spectrum of single and multiple contact miscible systems to be studied. For instance, low molecular weight alcohols can be used to model carbon dioxide injection. The partitioning of carbon dioxide between the aqueous and oleic phases is a function of pressure, and is reflected in the choice of ternary systems. This has been discussed by Stegemeier (10), Totonji and Farouq Ali (11), and Orr and Taber (12).

Examples of displacement processes

Figures 14 through 18 illustrate fluid displacement and distribution in the Letratone and doublet (hand drafted) models. Red and blue dyes were used in these experiments which, although clearly distinguishable in the original colour slides, appear as only slightly differing tones in these photographs. The interfaces are, however, well defined through light scattering at refractive index discontinuities.

We are also investigating techniques to quantitatively study fluid concentrations as functions of both position and time. Absorption of light by the dyes can be exploited to show dispersion and diffusion. It should be noted that because the diffusion and mass transfer of the dyes may be different to those of the fluids, and that the dyes may suffer dispersion through adsorption within the system, corrections need to be made to any measurements. It is therefore more desirable to utilise methods which exploit the refractive index properties of the liquids, e.g. interferometry.

Dynamic recordings of displacement sequences have been made on Sony U-matic videotape, and are held by the Professor of Petroleum Engineering, Imperial College.

In the following photographs flow is from left to right.

├──────────────┤ 1 mm

Figure 14 High tension, immiscible displacement of non-wetting fluid (dark tone) by a wetting phase of equal vicosity, in the Letratone model.

├──────────┤ 100 µm

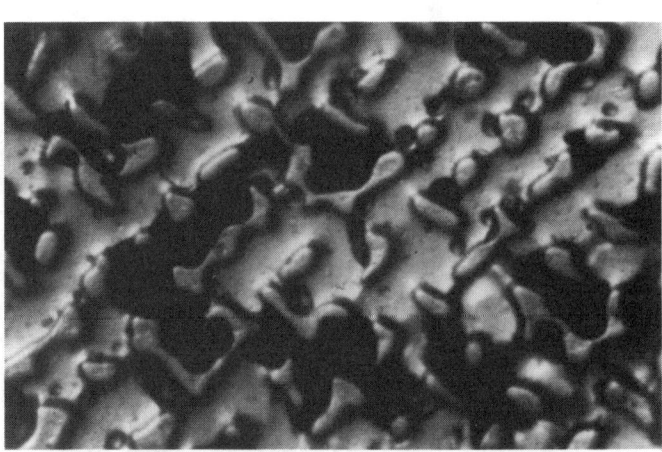

Figure 15 Residual non-wetting phase (dark areas) behind the flood front of figure 14.

1 mm

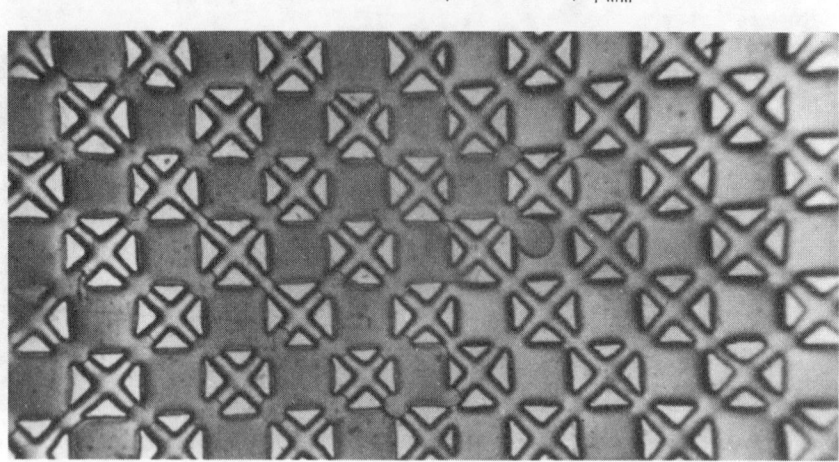

Figure 16   Displacement of water by Carnation oil, of viscosity 0.019 Pa.s, in the doublet network. The oil clearly preferred the large pores and entrapped the water within the smaller channels.

100 µm

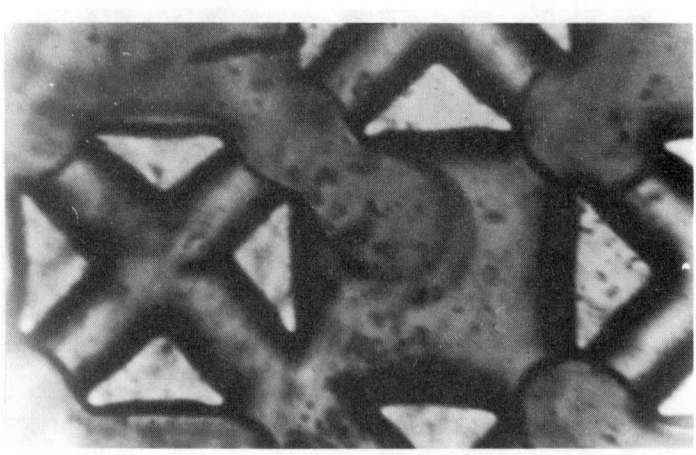

Figure 17   Illustration of a Haine's jump as the oil entered a large pore. The fluid interface moved during the exposure time.

├───────────┤ 1 mm

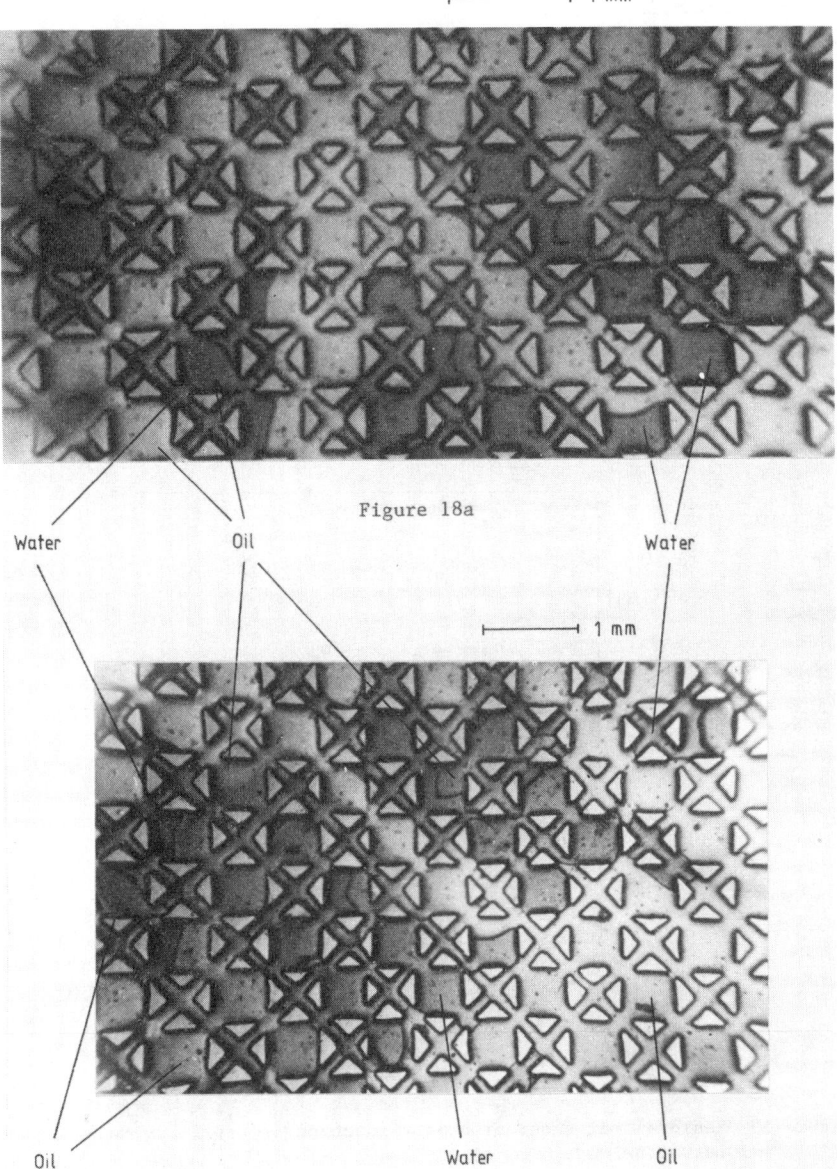

Figure 18a

Water   Oil   Water

├───────┤ 1 mm

Oil   Water   Oil

Figure 18b

Figure 18   Doublet model, initially fully saturated with Carnation oil, which
            was partially displaced by water, followed by a dyed oil flood.
            Subsequent injection of undyed oil highlighted the stagnant
            regions.

Quantitative results

The Letratone micromodels were used to study low tension displacements. As well as purely visual results, quantitative measurements were obtained from photographs in conjunction with flow rate data; viscosity and interfacial tension parameters were determined separately. Figure 19 shows the observed residual oil (as a percentage of the total pore volume) as a function of Capillary Number, for both high and low tension (petroleum sulphonate / kerosene) displacements. The Capillary Number was increased in steps by adjustment of the flow rate, and the remaining volume of oil measured. Some lower residuals were produced by Capillary Number 'shock' (sudden reduction in interfacial tension) at a surfactant displacement front, where microemulsions were often formed.

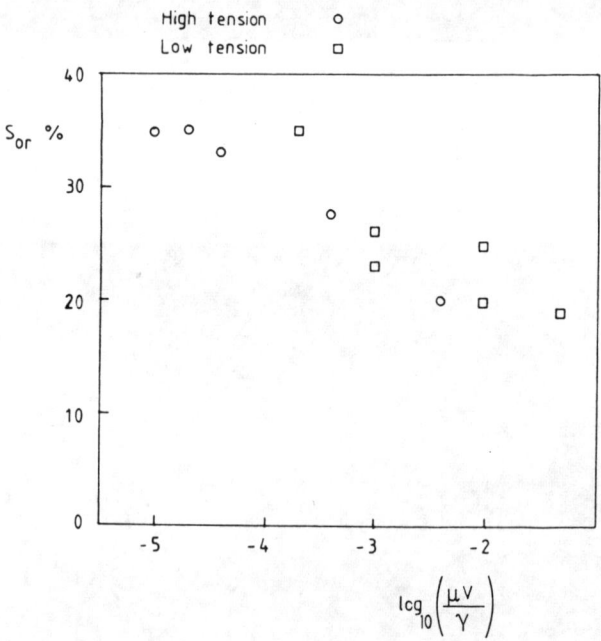

Figure 19   Residual oil saturation as a function of Capillary Number,   in   the Letratone model

Figure 20 illustrates the measured residual ganglion length distributions (taken in the mean flow direction). Clearly, blobs of oil longer than 500 microns are mobilised by increasing the Capillary Number within this range, while the stable globules at high Capillary Number are of the same order of size as the individual pores.

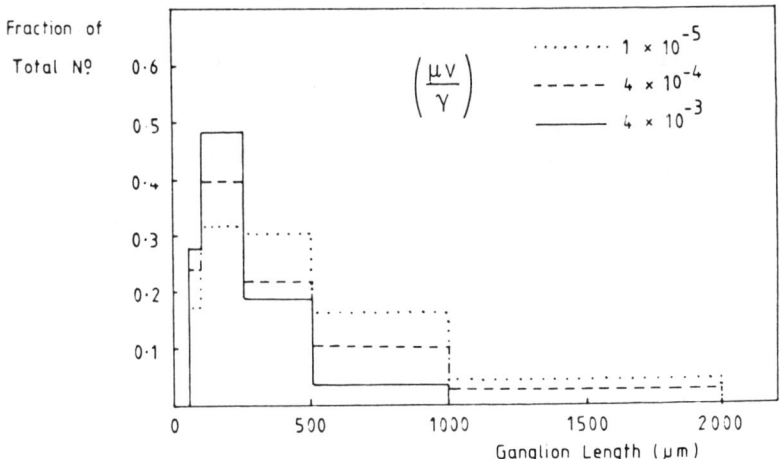

Figure 20 Size distributions of residual oil ganglia for three different values of Capillary Number.

CONCLUSIONS

Micromodelling techniques have been developed to gain insight into the physics of reservoir miscible and surfactant displacement processes by observations of alcohol and low interfacial tension systems at ambient temperature and pressure.

FUTURE WORK

We intend to develop the design of networks to relate micromodel displacement results to real reservoir rocks through pore size distribution and connectivity.

Holographic interferometry is currently being employed to investigate the role that diffusion can play in the recovery of oil entrapped by small scale heterogeneities.

Techniques for producing castings in glass of the nylon etchings are being examined to create more strongly water-wet models.

ACKNOWLEDGEMENT

The authors would like to thank the Department of Energy and the Science and Engineering Research Council for their support of this research, Professor Colin Wall for his encouragement, and Mr Martin Hughes for technical advice.

NOMENCLATURE

V = superficial (average interstitial) velocity, m/s
γ = interfacial tension, N/m
μ = viscosity, Pa.s
S_{or} = residual oil saturation
The Capillary Number is defined by:
$Nc = \mu V / \gamma$

REFERENCES

1. CHATENEVER, A., and CALHOUN, J.C.; "Visual Examinations of Fluid Behaviour in Porous Media - Part 1", Trans., AIME (1952) 195, 149-156

2. EGBOGAH, E.O., and DAWE, R.A.; "Microvisual Studies of Size Distribution of Oil Droplets in Porous Media", Bull. Can. Pet. Geol. (June 1980) 28, 200-210

3. MATTAX, C.C., and KYTE, J.R.; "Ever See a Water Flood?", Oil and Gas Journal (Oct 1961) 59, 115-128

4. MICHAELS, A.S., STANCELL, A., and PORTER, M.C.; "Effect of Chromatographic Transpart in Hexylamine on Displacement of Oil by Water in Porous Media", Soc. Pet. Eng. J. (Sept 1964) 4, 231-239; Trans., AIME, 231

5. DAVIS, J.A., and JONES, S.C.; "Displacement Mechanisms of Micellar Solutions", J. Pet. Tech. (Dec 1968) 20, 1415-1428; Trans., AIME, 243

6. WARDLAW, N.C.; "The Effects of Pore Stucture on Displacement Efficiency in Reservoir Rocks and in Glass Micromodels", SPE/DOE 1st Joint Symp. on EOR, Tulsa, Olklahoma (April 1980) 346-352; SPE paper 8843

7. BONNET, J., and LENORMAND, R.; "Constructing Micromodels for the Study of Multiphase Flow in Porous Media", Revue de L'Inst. Franc. du Pet. (1977) 42, 477-480

8. ARRIOLA, A., WILLHITE, G.P., and GREEN, D.W.; "Trapping of Oil Drops in a Noncircular Pore Throat", SPE paper 9404, SPE Annual Fall Meeting, Dallas (Sept 1980)

9. ARRIOLA, A., WILLHITE, G.P., and GREEN, D.W.; "Mobilization of an Oil Drop Trapped in a Noncircular Pore Throat upon Contact with Surfactants", SPE paper 9405, SPE Annual Fall Meeting, Dallas (Sept 1980)

10. STEGEMEIER, G.L.; "Mechanisms of Entrapment and Mobilization of Oil in Porous media", Improved Oil Recovery by Surfactant and Polymer Flooding, ed. Shah, D.O., and Schechter, R.S., Academic Press, Inc., New York (1977) 55-91; 81st Nat. Meeting AICHE, Kansas City (April 1976)

11. TOTONJI, A.H.M., and FAROUQ ALI, S.M.; "Solvent Flooding Displacement Efficiency in Relation to Ternary Phase Behaviour", Soc. Pet. Eng.J. (April 1972) 12, 89-95

12. ORR, F.M., and TABER, J.J.; "Displacements of Oil by Carbon Dioxide", Annual Report, U.S. DOE/MC/03260-4 (1980)

THE INTERPLAY BETWEEN RESEARCH AND FIELD OPERATIONS IN THE DEVELOPMENT OF THERMAL RECOVERY METHODS

J. OFFERINGA, R. BARTHEL and J. WEIJDEMA

Koninklijke Shell Explorative en Produktie Laboratorium, Rijswijk, The Netherlands (Shell Research B.V.)

ABSTRACT

The role of research in the development of thermal recovery processes is discussed, viz: steam drive, steam soak, hot-water drive and in-situ combustion. The importance of feedback from field experience to research is pointed out.

Five periods are distinguished: 1. Early and mid-fifties when mainly laboratory work was carried out. 2. Late fifties and early sixties when the processes were tested in field pilot projects. 3. Mid-sixties to early seventies when large steam-soak projects were started and research experiments were carried out in large physical models. 4. Mid-seventies to early eighties when the number of steam projects has been increasing fast. The design of these projects is being carried out with the aid of numerical simulators. 5. The present, when new techniques and applications for thermal methods are under investigation.

It appears from this historical survey that in particular the interplay between research and the field is stimulating for new developments.

INTRODUCTION

Present worldwide recovery of oil by EOR methods is estimated to amount to some 600 000 bbl/d. More than 80 per cent of this production is by thermal methods of which steam drive and steam soak take the major part. Usually thermal methods have been applied so far in reservoirs containing medium to heavy oil or tar.

Thermal methods are the oldest EOR methods. They have been developed during the past thirty years, partly in the laboratory and partly in the field.

The object of this paper is to show where research and field operations have stimulated each other in developing these processes or where, occasionally, one or the other came to a dead end. The paper is therefore presented in the form of a historical review. It is unavoidable that some of the material of older reviews such as, for instance, by RAMEY(1) (1967) and HARMSEN(2) (1971) on steam and hot-water injection and by DIETZ(3) (1970) on in-situ combustion will be repeated. However, we believe it fulfils a useful purpose. Firstly, because these reviews are now 10-14 years old and thus only cover half of the period of interest and, secondly, because we tend much more to discuss in retrospect the role of research.

If we sometimes seem to overemphasise Shell's role in the history of thermal recovery, this is not done intentionally but is mainly due to easier accessibility to reports and to the people who made part of this history. We are aware of the fact that similar developments to those within Shell research have also taken place in other companies. If this explanation is not fully satisfactory, it may even stimulate some petroleum engineer with a vocation for historian to write an "objective" history on the development of thermal processes.

In the following we do not present descriptions of the basic thermal processes. For those who are not familiar with these ample literature references are given in this paper.

EARLY AND MID-FIFTIES

In an internal Shell Report of 1951 entitled "Higher ultimate recovery by heating the reservoir", it is stated that: "The idea of heating an oil reservoir in order to decrease viscosity and consequently increase recovery is not new. Already in 1917 . . ." Then follows a long list of papers and patents containing suggestions for heating methods. A considerable number of these ideas date from before the second world war(1). Even some early field applications are mentioned. In retrospect, however, these tests should be considered to be isolated events having no follow-up. The actual rise of the thermal recovery methods did not occur before the early fifties. From then on, a continuous flow of papers on research investigations and field experiences has been maintained, which has not yet stopped.

It is interesting to note that in the internal Shell report mentioned above (part of which was later published (4)) some exotic heating methods were suggested which also nowadays regularly appear in the literature, such as application of sonic waves, electromagnetic radiation, electric conduction between electrodes in wells, injection of oxygen (instead of air) and even letting down an atomic bomb into a well. Although these methods were rejected (mainly on economic grounds) it still seems recommendable to re-evaluate these techniques regularly with changing economic conditions, new technical developments or even for deviant oil formations.

It is, furthermore, surprising to realise that in this early period attention was mainly directed to in-situ combustion. The reasons for this are that air injection was considered to be easier than, for instance, steam injection and that it requires less fuel at the surface.

In-situ combustion

The first field trials on in-situ combustion were carried out by Sinclair Research and by Magnolia Petroleum Corporation in the U.S. Sinclair (5) experimented with injection of air/fuel gas mixtures in a shallow oil sand. Ignition was achieved by means of a gas burner. They demonstrated the propagation of a combustion wave which left a clear burnt sand. Supporting laboratory experiments demonstrated that the oil in the formation could be moved by the front edge of the heat wave. Magnolia first tried out ignition of underground combustion in a three-well test using an electric heater (6) and then carried out a successful in-situ combustion drive in a 30 acre inverted five spot (7).

These early field trials stimulated research in various companies. In laboratory tube experimentation process variables were determined, such as minimum air flux for self-sustained combustion, fuel availability dependent on oil type, and frontal advance rate in relation to air rate (8).

An extensive study of combustion processes in oil sands led to a description of the drive mechanism of the process (9). In experiments in glass tubes oil-bank formation was made visible. Thermal analysis, using a pack of oil sand fluxed with air, showed that the reaction between oxygen and oil proceeds in two major steps; in a lower temperature regime (below 300°C) the oil molecules lose comparatively more hydrogen than carbon in reaction with the oxygen, leaving a carbon-rich coke-like residue which is subsequently burnt off in a temperature range of 300 to 500°C.

Hot-water injection

The method that was, at least at Shell in that period, considered as second best is hot-water injection. Therefore a feasibility study was carried out on the applicability of this process in the Schoonebeek field in the· Netherlands. This field, which had once been called a play-ground for petroleum engineers specialised in thermal recovery, contains oil of 25°API with an initial viscosity of 180 cP. It consists of two parts: the high-pressure (70 bar) waterdrive area and the low-pressure (10 bar) solution gas drive area. The total STOIIP is 1.2×10^9 bbl. Hot water was intended to be injected into the waterdrive area.

As it was realised that the temperature distributions in the water and the oil zone are the determining parameters of the process, an analytical model was developed to determine these temperature distributions. This model, still known as the Lauwerier model (10), mainly consists of a procedure for calculating heat losses in the formations over- and underlying the hot-water zone. It enables the average water and the average oil temperature to be calculated at a certain time and thus the average viscosities and the mobility ratio. The recovery as a function of time can then be approximated from this varying mobility ratio.

Furthermore, scaled laboratory experiments were carried out to confirm the model for an actual situation with a hot-water tongue underrunning the oil. From economical calculations it appeared that the process could be profitable provided cheap fuel is available (4).

Steam injection

In the meantime also, a start had been made with tube experiments on the steam-drive process (11). It appeared in these one-dimensional experiments that a steam zone with a stable front develops, in front of which the temperature decreases gradually to the initial temperature. Further important observations were the low residual oil saturation in the steam zone, the relative independence of the process to oil viscosity and sand permeability and again the dominating effect of heat losses to over- and underlying formations.

LATE FIFTIES AND EARLY SIXTIES

After the preceding period of (mainly) laboratory investigations, a period had now come of active field testing of the studied processes. All three processes were tested by Shell in the Schoonebeek field: Hot-water drive in the high-pressure water-drive area and in-situ combustion and steam drive in the low pressure solution gas-drive area. In the heavy-oil fields of the Bolivar Coast in Venezuela (10-15°API), steam drive and in-situ combustion were tested in the Mene Grande field and the Tia Juana field.

In-situ combustion

In designing an in-situ combustion test at that time, ignition was considered a matter of particular concern. Ignition procedures using powerful well heaters to heat the formation round the injector to combustion wave temperatures (some 300°C or more) (5,6), or by use of reactive chemicals (12,13) all had their specific drawbacks, often necessitating well repairs and repeated trials (3). At Schoonebeek, for instance, ignition was achieved in the three injectors of the triple seven-spot by squeezing-in concentrated nitric acid as a strong oxidiser. In one of the injectors, however, an explosion occurred, tearing up a tubular section.

As in the South Belridge field test spontaneous ignition had quite unexpectedly occurred after prolonged air injection (4), interest in this manner of starting in-situ combustion led to the development of a predictive method for spontaneous ignition, based on low-temperature oxidation rates of oil sands (15). Predictions for the Tia Juana test site in Venezuela led to the decision to rely on spontaneous ignition, which actually occurred some five weeks after air injection.

Later, on basis of the predictive model, it was formulated that by heating the formation round the well bore gently and gradually to only some 100°C with a low-powered well heater or by steaming the injector, a smooth well-controlled ignition could be achieved after one or two days' air injection (16,17). By steaming the injector a smooth ignition was obtained in the Tia Juana wet in-situ combustion test in 1965, discussed below.

Although much experience had been gained with both in-situ combustion pilots at Schoonebeek and Tia Juana, they were considered economically unsuccessful. Oil returns were low, mainly due to rapid up-dip channelling of the air.

In the meantime a theoretical one-dimensional model study had been made on the tolerance of a dry combustion wave to water injected simultaneously with the air (18). The water evaporates to steam in the hot burnt-out zone and the steam carries the heat downstream through the combustion zone. The effect of such heat recuperation is two-fold: the steam stabilises the propagation of combustion by preheating the oil and, most important, the growing steam zone effectively sweeps the mobile oil far ahead of the combustion wave. As a result, less air is needed for the combustion-drive process, so that compressor capacity can be reduced, thus economising on investment and running costs. The optimum tolerable water-injection rate relative to the air-injection rate was found to be the one at which the evaporation front moves steadily and closely behind the combustion wave.

Based on the relatively high oxidation rates of oil in the low-temperature range found in combustion kinetic studies (19,20,21) a series of laboratory tube experiments was carried out at 40 bar, in which the water-injection rate was deliberately increased to far above the quoted tolerance limit (22). It appeared that as the water under these conditions enters the combustion zone and evaporates, it suppresses the combustion temperature to near-saturated steam temperatures, with the result that the less reactive coke (final oxidation step) remains unburnt. It was shown that, in spite of partial quenching of the combustion, a steady progress of the heat wave is ensured, while achieving a further important economisation on the air requirements.

The experimental results fitted a simple theory which showed that in partially quenched combustion the speed of the combustion wave is no longer governed by the air-injection rate (as in dry and "normal" wet combustion) but by the water-injection rate and that the air requirement per unit formation volume decreases with increasing water-injection rate. Later, more refined theories on "superwet" combustion were developed (23,24) which were reviewed in Ref. (3). Seemingly different experimental observations by others (25) appeared to fit in with these theories (26).

On the basis of their experience with wet and superwet combustion in one-dimensional models, Shell decided at an early stage to try out wet combustion in the field.

In 1962, the triple seven-spot dry combustion test at Schoonebeek was converted into a wet combustion project (3). The test was terminated in 1965 after a severe production decline had set in, presumably caused by an overall formation plugging, while in addition several injectors and producers suffered from corrosion. Although the test was not considered an economic success, it had shown that while injecting water simultaneously with the air, more than three times as much tertiary oil could be produced per unit volume of air than during the dry combustion phase.

In Venezuela, a wet combustion drive test was carried out in a seven-spot in the Tia Juana field, in the period 1965 to 1968 (3). Compared to Schoonebeek, where conditions were rather in the range of 'normal' wet combustion, water/air injection ratios were chosen to be twice as large, to reach the range of partially quenched combustion of the laboratory experiments. Temperature profiles determined in observation wells and in producers remained below 200°C. Though speculative, this might be taken as an indication that the combustion has perhaps indeed proceeded in the partially quenched mode.

The test made a small net profit. Although much uncertainty arose concerning how much oil should be attributed to the effect of the wet combustion drive (3), probably at least twice as much tertiary oil had been produced per unit volume of air as compared with the Schoonebeek test.

Steam injection

The most important event of this period is probably the discovery of the steam-soak process in 1959 in the Mene Grande field (27). The process was discovered rather accidentally when, during the planned steam drive, steam eruptions around an injector made it necessary to relieve the reservoir pressure by backflowing the injector. It appeared then that the well continued to flow at a rate of more than 100 bbl/d oil at a relatively low watercut, whereas surrounding producers had been pumped before steam injection at oil rates varying from 3 to 10 bbl/d.

Further testing was carried out in the Tia Juana field, the favourable results of which prompted a large scale project. In the meantime also, Shell Oil tested the process successfully in its Yorba Linda field in California (28) which was the start of the large-scale applications of the process in the Californian heavy oil fields.

In retrospect, this accidental discovery has been mentioned as being more or less inevitable (29). Others consider "the observation of a phenomenon, the realisation of its value and the initiative to apply it" to be less obvious (30). An interesting question in this respect is: Why has this process been discovered in the field and not been proposed by research? Afterwards, the idea to reduce the pressure drop around a production well by heating seems rather obvious. Some simple calculations could have demonstrated that this effect would last a reasonable time. A possible answer to the question why this idea has never been proposed is that, even if anybody had the idea, he probably would have rejected it himself because he would expect to produce mainly water. Even in recent literature different explanations (31,32) are given for the low water cut during the production phase.

The steam-drive projects in Schoonebeek (33) (4 injectors and 8 producers) and in Tia Juana (34) (7 injectors and 24 producers) both proved to be a technical success. Additional oil recoveries from the test areas were estimated to be 38% and 21% STOIIP for Schoonebeek and Tia Juana respectively. In both projects it appeared that owing to gravity the steam flows only through the upper part of the formation. The lateral flow patterns in both projects were far from symmetric. This phenomenon can partly be accredited to the dips of the reservoirs (6.5° in Schoonebeek and 3° in Tia Juana) but is also due to heterogeneous sand developments.

For correct interpretation of such projects it is important to know the volumetric development of the steam zone. At that time only theories describing the development of a one-dimensional steam zone were available. The thickness of the steam zone which determines the vertical sweep efficiency, had to be derived from field observations. These thicknesses were estimated to be between 7 and 11 m.

It followed from the analyses that oil was not only displaced by free steam but also by condensation water, in Tia Juana even in equal amounts.

The Schoonebeek project was extended to adjacent areas. In the meantime it had become apparent from the steam-soak tests in the Tia Juana field that a combination of reservoir compaction and steam soak could maintain primary recovery for a considerable period. It was therefore decided for the near future to discontinue steam-drive activities in the Bolivar Coast fields.

Hot-water injection

Although the hot water injection test in the Schoonebeek field (2 injectors and 7 producers) proved to be reasonably successful, it appeared that the process is much more complicated than was initially envisaged (35). Water breakthrough occurred much earlier than expected. Later studies based on model experiments showed that the process is intrinsically subject to lateral instability and that the hot water tends to concentrate in a few tongues. Nevertheless, the performance in the Schoonebeek field was attractive enough to extend the project over a considerable part of the high pressure water-drive area. In some parts of the reservoir hot-water injection is still being carried out. This is a quite exceptional situation since most early pilot projects were discontinued because of poor areal sweep efficiency. An explanation for the acceptable performance of Schoonebeek could be its relatively low initial oil viscosity of 180 cP (180 mPa.s).

MID-SIXTIES TO EARLY SEVENTIES

This period shows a spectacular increase in the number of steam-soak projects at the cost of the other thermal processes. Within a period of ten years the production of heavy oil due to this process increased in California to about 130 000 b/d and in Venezuela to an even higher level. Although by its nature only a stimulation process, it enables at relatively low costs production from undepleted reservoirs which would otherwise only produce at very low rates.

Theoretical (36,37) and experimental (38) studies were carried out to investigate the performance of the process and to define optimal injection and production schemes. The effect of a number of parameters had to be investigated, among which slug size, cycle length, number of cycles, soaking time, etc. Although at that time the complete set of flow- and heat transport equations could not yet be solved, the simplified equations were already being solved numerically with the aid of the computer. The most advanced models did not predict the performance purely based on physical input parameters but had to be matched to actual well performance.

Answers to important questions concerning whether a particular reservoir at its particular stage of depletion was suitable for steam soaking still very often had to be found by field trials.

So far we have only mentioned the Californian and Venezuelan Bolivar Coast heavy oil fields as targets for thermal recovery but not the even more important heavy oil sands of Canada and the Orinoco belt in Venezuela.

In this period attention in Venezuela was mainly directed to the Bolivar Coast rather than to the Orinoco belt. In Canada, however, the first pilots were started as early as 1957 (39).

Amoco started a field programme in which dry and wet combustion were tested. Also, reversed combustion was tried out. To obtain injectivity, fracturing of the formation appeared to be necessary. As a follow-up to many years of testing, a relatively large in-situ combustion test is currently being carried out in co-operation with the AOSTRA.

A special condition exists in the site of AOSTRA and Shell Canada where the oil zone is underlain by a zone with a high water saturation which is more permeable than the oil sand. No fracturing is required to inject steam in this zone. From the field-testing programme which started in 1963, a cyclic steam-injection recovery scheme has been developed which is at present being tested in a seven-7-spot pattern. Physical model experiments with vacuum models were carried out to investigate the performance of this project (40).

In all field trials on the various thermal processes it had appeared that, however well the process might be understood in a one-dimensional tube experiment, this understanding did not guarantee reliable predictions for an actual field performance. Areal distribution of the injected fluid and fluid segregation due to density difference play a major role. To study these phenomena, three-dimensional model experiments were carried out which were scaled to actual reservoir conditions.

Two different types of models were applied: high-pressure models and vacuum models. Not only was it necessary to develop all kinds of new laboratory techniques but also to derive scaling rules. In the tests usually field symmetry elements were simulated consisting of a few injection and production wells.

In the earlier high-pressure models, temperatures and pressures were equal to (which is essential for ISC experiments) or approaching those in the field. Typical dimensions for these sand-filled models were 3 m x 1.5 m x 0.15 m. To enable maintenance of the high pressures, the models were contained in bulky pressure vessels. To simulate dip effects, the vessels could be placed in a tilted position.

In high-pressure model experiments on wet in-situ combustion, using a medium viscosity (Schoonebeek type) oil, surprising observations were made (3). As expected, the injected air rapidly moved to the top of the formation, driving a combustion spearhead to the production wells, finally resulting in a tilted coke deposit (41). Temperature observations showed, however, that at several spots and at several moments combustion in the burnt-out zone revived. This was an indication that oil was being driven upwards by the growing underlying water tongue. Also, it was observed that the combustion heat made itself felt to near the bottom of the sand pack. Thus, an effective recovery mechanism was recognised by which a considerable amount of the oil is driven upwards by the invading water tongue into the hot regime near the top of the formation where the oil becomes much more mobile and is easily driven toward the producers. A similar effective flow regime has been discovered earlier (42) for a water drive in a reservoir having a high mobility streak along the top.

Occurrence of this specific flow regime would explain why model experiments, run at a given water-injection rate but with air-injection rates differing by a factor of 4, showed practically identical recovery curves ending at an ultimate recovery of more than 1.5 higher than that obtained in a comparable plain water drive.

It would be interesting to experiment with modern computer simulators and assess how far the air injection rate can be reduced without affecting these favourable recovery results. Furthermore, it might be investigated to what extent the sketched recovery mechanism would remain operative in reservoirs containing heavier oil, e.g. as in Tia Juana, with greater mobility contrasts to water and air.

In retrospect, it is noted that the concept of partially quenched combustion, being a marked step forward in combustion control, has been developed on basis of the frontal drive laboratory experiments and corresponding theories. These research activities have led, however, to an effective recovery process in which a mechanism other than partially quenched combustion also plays a role.

In steam-drive model experiments particular attention was paid to the development of the steam zone. Also, the utilisation of the heat injected was studied. From these experiments an analytical theory (43) could be derived which predicts the shape and the growth of the steam zone by making use of the concept of a pseudo-mobility ratio between oil and steam.

In experiments directed to the Schoonebeek field trials (33) it was observed that steam followed the downdip draining hot condensate and thus broke through in a downdip well. This phenomenon had also been observed in the field. Agreement between field- and model observations was satisfactory.

Vacuum models to predict reservoir performance under steam drive were in particular applied by Shell Dev. Co. (44). In vacuum models a rigid structure is obtained by imposing a vacuum on a packed bed of glass beads confined between two plastic sheats. Although these models were initially developed for low temperatures, they could also be applied for low temperature steam at the sacrifice of correct matching of the ratio between latent and sensible heat. The advantages are that they are much simpler to use and safer.

Information from model experiments was thus obtained for complicated field projects such as for example Mt. Poso (44), Midway Sunset, Yorba Linda and Peace River (40).

MID-SEVENTIES TO EARLY EIGHTIES

In the last ten years it has become more and more clear that the competition between the three thermal processes has been won by steam. After the oil crisis of 1973, the demand for heavy oil has increased and in particular in the U.S. many existing steam projects have been extended and new projects started. Many reservoirs which have already been producing for 10 to 15 years under steam soak are now being converted to steam drive combined with steam soak.

To give an order of magnitude of these projects (45): Many of them are producing in the range of 1000 to 5000 b/d. Large projects are Getty's Kern River Project with a production due to steam injection of 52,000 b/d, Shell's Mount Poso project with 20,000 b/d and Texaco's San Ardo project with 22,000 b/d.

Another important large steam drive project is being carried out by Maraven in Venezuela in a nearly depleted part of the Tia Juana field (46). Although a large project on a commercial scale with a production rate of about 20,000 b/d, it is still considered to be of an exploratory nature for the wider application of steam drive in the Bolivar Coast heavy-oil fields.

For comparison, we mention some of the largest ongoing in-situ combustion projects: The Rumanian project in the Suplacu de Barcau field (47) with an oil production of about 6500 b/d, Getty's project in the Bellevue field in Louisiana with about 2800 b/d and Mobil's project in the South Belridge field with 1900 b/d. Most of these projects are technically considered to be successful and profitable. This means that although steam is by far the most successful thermal method, in-situ combustion should certainly not be considered obsolete.

The main reservoir-engineering problem in designing new projects is the lack of simple and reliable performance prediction methods. Until the early seventies, a selection had to be made from extrapolation of pilot projects or more or less similar other projects, time-consuming scaled model experiments in the laboratory and a few simplified analytical models.

In the field of steam drive, a number of analytical methods for computing the volume and the shape of the steam zone have been derived over the years. The common basis of these various models is a procedure to account for the heat losses from the steam zone to over- and underlying formations. Furthermore, an attempt is made to disconnect the coupled heat transport equations (conduction and convection) and those for fluid flow (water, oil and steam). In one case this is done by neglecting fluid flow completely (Marx and Langenheim (48)), in another case by taking fluid flow partially into account to arrive at a better analysis of the heat distribution in front of the steam zone (Mandl and Volek (49)). A severe limitation of both approaches is that they describe only the development of a one-dimensional steam zone and do not predict its thickness. The method of Neuman (50) predicts areal and vertical development, assuming this is only determined by heat conduction and convection. Van Lookeren (43) assumes that the shape of the steam zone is determined by gravity and viscous forces but has to simplify the heat distribution. Nevertheless, some of his predictions check quite well within a defined range of applicability with observations from laboratory experiments.

With the aid of these analytical models it is possible to approximate the volume of oil displaced by steam. No straightforward methods existed to predict the volume of oil displaced by the hot condensate; neither did methods to take areal effects into account. This could only be done with the scaled model experiments and since the mid-seventies with a numerical simulator.

A lot has already been said in literature on the advantages and disadvantages of physical and numerical models (51). The consensus nowadays is, more or less, that physical experiments should provide the physical insight, and that the quantitative effect of any physical parameter could be investigated with the aid of the numerical simulator. This means that geometrically scaled model experiments as carried out in the late sixties and early seventies, will no longer be carried out in the future, since geometrical effects can much more easily be studied with a numerical simulator than with a physical model.

The main development in thermal simulators, or in particular those simulators that can handle hot water and steam is connected with the increasing capacity of the computers: increasing speed and memory space. These factors enable the study of more details by means of the application of more grid blocks and acceptable runtimes made possible by replacing the older explicit methods by implicit methods.

Numerical simulators for the in-situ combustion process are not new either: One of the first mentioned in the literature dates from as early as 1965 (52). However, owing to the complexity of the process, their development is far less advanced than in the case of steam models. One of the major problems in simulating field performance is the fact that the essential phenomena occurring in the combustion zone of a few metres thick have to be represented for practical reasons in grid blocks with sizes in the range of 10 m and more.

PRESENT DEVELOPMENTS

The developments which are at present taking place in the field of thermal recovery can be grouped in the following way:
   a. follow-up methods for ongoing steam projects.
   b. new thermal methods for the recovery of heavy oil.
   c. search for new targets for thermal methods.
   d. improvement of existing and development of new equipment.

Each of these groups is discussed briefly:

a. The need for follow-up processes is felt in particular in steam-drive projects which have been in progress for a couple of years, in which steam breakthrough has already taken place and where the oil-steam ratio is declining.

In reservoirs with medium viscosity oil (e.g. Schoonebeek) water injection or (if present) a strong aquifer may cause collapsing of the steam zone when steam injection is discontinued. In this way, relatively cold oil may be pushed into the hot formation. As the heat stored in the reservoir is utilised very efficiently, in this way, high oil/steam ratios may be obtained. It is clear that this process is not suitable for heavy-oil reservoirs. Nevertheless, water injection, with or without caustics, needs further consideration.

A promising method to improve the sweep efficiency of steam drive seems to be the application of blocking agents (53,54), such as foams to divert steam into unswept areas. This method is under active study at various companies and institutes, both in the field and in the laboratory.

b. As already mentioned above, re-evaluation of heating heavy oil reservoirs by electromagnetic radiation or electric conduction regularly occurs. The major economic drawback of these methods is the low thermal efficiency inherent in the generation of electricity.

Also, the interest in the injection of oxygen (instead of air) for in-situ combustion has been revived (55,56). The potential advantage of oxygen would lie in suppression of early breakthrough of large quantities of hot combustion gases in the production wells.

Methods which may become of interest with increasing oil prices are combinations of mining techniques and thermal methods (57) to increase recovery and reduce heat losses.

c. With a very few exceptions, all thermal projects have been carried out in reservoirs containing heavy oil. Light-oil reservoirs were not considered because water is generally considered to be a cheaper driving fluid. If the residual oil remaining after water drive is considered to be a target, thermal methods should also be taken into account. Although not all proven to the same degree, injection of steam and hot water as well as in-situ combustion may be considered. Steam distillation (or strip) drive in light-oil reservoirs is a process which has been technically proven both in the field (58) and in the laboratory (59). Residual oil saturations in the steamed-out zone are in the range of 3 to 8%. The economic weak points of the process are its high initial investment and high operation costs (fuel). Further study on the economic viability of the process seems necessary.

The technical feasibility of in-situ combustion in a watered-out light oil reservoir has already been demonstrated in the mid-sixties by Amoco in the Sloss field in Nebraska (60).

A potential process for high-pressure reservoirs might be derived from the property of hydrocarbons to dissolve in water at near-critical conditions. In practice, this means that the pressure should be above 200 bar and the temperature above $300^{\circ}C$. This means that this process is, anyhow, limited to deep reservoirs. Russian investigators (61) claim to have obtained high recoveries in tube experiments. Which crudes are suitable candidates for this process needs further investigation, as well as the economic viability of the process. Technical limitations can also be caused by the design of the deep injection wells.

A completely different type of target where thermal methods seem promising are fissured limestone reservoirs. These reservoirs often consist of very low permeable matrix blocks containing nearly all the oil and a highly permeable fissure (or fracture) system (62). Drive processes have a very low recovery because the oil in the matrix blocks is bypassed by the drive fluid (either water or gas). Water imbibition does not occur or is very weak because the rock is oil-wet or neutral-wet. Gravity drainage is often hampered by capillary forces, the low permeability of the matrix rock and sometimes by the high viscosity of the oil.

The effect of heat can be manifold (63,64): expulsion of oil from the matrix blocks due to swelling and gas development within the matrix blocks (oil vapours and steam), improvement of gravity drainage and countercurrent imbibition due to viscosity reduction and reduction of the capillary retention. Both heavy and light oils come into consideration.

Heat can be supplied by either injection of steam or hot water or by in-situ combustion. An important requirement for these processes to be effective is that the average fissure spacing should not be too wide to enable sufficient heat penetration into the matrix blocks.

d. Although in the field it is very often an area of serious problems, engineering of thermal projects has not been discussed so far in this paper. We will briefly touch on some of the major problem areas encountered with steam injection and some of the developments taking place in this field. These major problem areas are: water treatment, boiler design (efficiency, H_2S emission, resistivity to feedwater and fuel), thermal well completions, production of sour gas due to thermal cracking and production of oil-water emulsions.

These problems cannot be considered separately: water treatment and boiler design are fully interwoven and are furthermore determined by the properties of the available water.

With the generation of steam downhole which is, at present, being actively investigated (65,66), a number of these problems is circumvented, such as boiler efficiency, H_2S emission and well completion.

Mechanical problems in hot wells which increase with the depth of the wells are tackled by testing various insulations and high-temperature packers (67). The problems do not seem to be solved easily.

CONCLUDING REMARKS

In the early period much was expected of the in-situ combustion process with hot-water drive in second and steam drive in third place. At present, much more oil is produced by steam than by the two other processes. Steam soak, although not a drive process, has produced most of the "thermal" oil in the world. This process was discovered in the field and not proposed by research.

Looking at the way the three processes were developed, it appears that hot-water drive followed the sequence: desk study - laboratory experiments - pilot test. In the case of steam drive, laboratory experiments preceded the detailed desk study. In-situ combustion research was prompted by very early pilot tests.

From the above observations, one tends to conclude that all three phases (desk study, laboratory experiments and pilot test) are essential in the development of the process, while the sequence seems to be of less importance. On the other hand, it is very important really that all phases have been passed through. Research without field testing may lead to sterile hobbyism and field testing without detailed preceding and following interpretation studies does not produce more than an abundance of poorly understood data.

REFERENCES

1. RAMEY, H.J.;
    "A current review of oil recovery by steam injection",
    Proc. 7th World Petrol. Congr. Vol. 3, pp. 471-476.
2. HARMSEN, G.J.;
    "Oil recovery by hot-water and steam injection",
    Proc. 8th World Petrol. Congr. Vol. 3, pp. 243-251.
3. DIETZ, D.N.;
    "Wet underground combustion. State of the art",
    J. Pet. Tech. (May 1970) 605.
4. VAN HEININGEN, J. and SCHWARZ, N.;
    "Recovery increase by thermal drive",
    Proc. 4th World Petrol. Congr. Section II, pp. 299-311.
5. GRANT, B.F. and SZASZ, S.E.;
    "Development of an underground heat wave for oil recovery",
    Trans. AIME (1954) 201 108.
6. KUNH, C.S. and KOCH, R.L.;
    "In-situ combustion - newest method of increasing oil recovery",
    Oil and Gas J. (Aug. 10, 1953) 52, No. 14, p. 92.
7. MOSS, J.T., WHITE, P.D. and McNIEL, J.S.;
    "In-situ combustion process - results of a five-well field experiment in
    Southern Oklahoma",
    Trans. AIME (1959) 216, 55.
8. WEINAUG, C.F. et al., editors;
    "Thermal recovery processes",
    Petrol. Trans. Reprint Series No. 7, Soc. of Petrol. Engineers AIME 1964.
9. TADEMA, H.J.;
    "Mechanism of oil production by underground combustion",
    Proc. 5th World Petrol. Congr. (1959) Sec. II, paper 22,279.
10. LAUWERIER, H.A.;
    "The transport of heat in an oil layer caused by the injection of hot
    fluid",
    Appl. Sci. Res., Section A, Vol. 5, pp. 145-150.
11. SCHENK, L.;
    "Steam drive - Results of laboratory experiments and first field tests in
    Mene Grande, Venezuela",
    Symposium on Thermal Recovery Methods, Caracas 1965.
12. TADEMA, H.J. and QUANT, J.Th.;
    "Process for igniting hydrocarbon materials present within oil bearing
    formations",
    U.S. Patent 2,863,510 (Dec. 9, 1958).
13. TADEMA, H.J. and QUANT, J.Th.;
    "Recovery of oil by combustion in-situ",
    Dutch Patent 87145 (Jan. 15, 1958).
14. GATES, C.F. and RAMEY, H.J.,Jr.;
    "Field results of South Belridge thermal recovery experiments",
    Trans. AIME (1958) 213 236.
15. TADEMA, H.J. and WEIJDEMA, J.;
    "Spontaneous ignition of oil sands",
    Oil & Gas J. (Dec. 14, 1970), 70-80.
16. WEIJDEMA, J. and ZELDENRUST, H.;
    "Formation ignition with moderate preheating",
    Dutch Patent Applic. 297100 (Oct. 1963); Venezuelan patent specific.
    14629.
17. STRANGE, L.K.;
    "Ignition: key phase in combustion recovery",
    Petrol. Engin., November 1964, p. 105, and December 1964, p. 97.

18. DE HAAN, H.J.;
    "In-situ steam generation. The case of frontal displacement",
    (November 1961, unpublished).
19. WEIJDEMA, J.,;
    "Zur Oxydationskinetik Kohlenwasserstoffe in poroesen Media in Bezug aug
    underirdische Verbrennung",
    Erdol und Kohle, Erdgas, Petrochemie (Sept. 1968) 21 520.
    (Determination of the oxidation kinetics of the in-situ combustion
    process).
20. BURGER, J.G. and SAHUQUET, B.C.;
    "Chemical aspects of in-situ combustion. Heat of combustion and kinetics",
    Soc. Petr. Eng. J., (October 1972), 410-422.
21. FASSIHI, M.R., BRIGHAM, W.E. and RAMEY, H.J.,Jr.;
    "The reaction kinetics of in-situ combustion",
    Paper SPE 9454, presented at Dallas (Tex.), September 21-24, 1980.
22. DIETZ, D.N. and WEIJDEMA, J.;
    "Wet and partially quenched combustion",
    J. Pet. Tech. (April 1968) 20 411.
23. BECKERS, H.L. and HARMSEN, G.J.;
    "The effect of water injection on sustained combustion in a porous
    medium",
    Soc. Pet. Eng. J. (June 1970) 145-163.
24. BASKIR, E., BECKERS, H.L., DIETZ, D.N., TER HAAR, L.G.J. and
    KRUIZINGA, J.H.;
    Shell Research (unpublished).
25. PARRISH, D.R. and CRAIG, F.F.,Jr.;
    "Laboratory study of a combination of forward combustion and waterflooding
    - the COFCAW process",
    J. Pet. Tech. (June 1969) 753-761.
26. HARMSEN, G.J.;
    "A note on COFCAW",
    J.Pet. Tech. (July 1969) 801.
27. DE HAAN, H.J. and VAN LOOKEREN. J.;
    "Early results of the first large-scale steam soak project in the Tia
    Juana field, Western Venezuela",
    J. Pet. Tech. (Jan. 1969), pp. 101-110.
28. STOKES, D.D. and DOSCHER, T.M.;
    "Shell makes a success of steam flood at Yorba Linda",
    Oil and Gas J. (Sept. 2, 1974), pp. 71-78.
29. RAMEY, H.J.;
    "Thermal recovery - A troublesome neophyte",
    Interview in J. Pet. Tech. (Jan. 1969), pp. 7-8.
30. DIETZ, D.N.;
    "Letter to the Editor",
    J. Pet. Tech., July 1969, p. 862.
31. PRATS, M.;
    "A current appraisal of thermal recovery",
    SPE 7044.
32. COATS, K.H., RAMESH, A.B. and WINESTOCK, A.G.;
    "Numerical modelling of thermal reservoir behavior",
    Proc. of the Canada-Venezuela Oil Sands Symposium 1977, pp. 399-410.
33. VAN DIJK, C.;
    "Steam drive project in the Schoonebeek field, the Netherlands",
    J. Pet. Tech. (March 1968), pp. 295-302.
34. DE HAAN, H.J. and SCHENK, L.;
    "Performance analysis of a major steam drive project in the Tia Juana
    field, Western Venezuela",
    J. Pet. Tech. (Jan. 1969), pp. 111-119.

540

35. DIETZ, D.N.;
    "Hot-water drive",
    Proc. seventh World Petrol. Congr. Vol. 3, p 451.
36. BOBERG, T.C. and LANTZ, R.B.;
    "Calculation of the production rate of a thermally stimulated well",
    J. Pet. Tech. (Dec. 1966), pp. 1613-1623.
37. OFFERINGA, J.;
    "A mathematical model of cyclic steam injection",
    Proc. 8th World Petrol. Congr., Vol. 3, pp. 227-234.
38. NIKO, H. and TROOST, P.J.P.M.;
    "Experimental investigations of the steam-soak process in a depletion-type
    reservoir",
    SPE 2978.
39. NICHOLLS, J.H. and LUHNING, R.W.;
    "Heavy oil sand in-situ pilot plants in Alberta (Past and Present)",
    Proc. of the Canada-Venezuela Oil Sands Symposium 1977, pp. 527-538.
40. PRATS, M,;
    "Peace River steam drive scaled model experiments".
    Proc. of the Canada-Venezuela Oil Sands Symposium 1977, pp. 346-363.
41. PRATS, M., JONES, R.F. and TRUIT, N.E.;
    "In-situ combustion away from thin horizontal gas channels",
    Soc. Pet. Eng. J. (March 1968), 18.
42. VAN DAALEN, F., VAN DOMSELAAR, H.R. and HOOYKAAS, H.;
    "Method of producing liquid hydrocarbons from a subsurface formation",
    U.K. Patent specification 1,112,956 (April 7, 1966).
43. VAN LOOKEREN, J.;
    "Calculation methods for linear and radial steam flow in oil reservoirs",
    SPE 6788.
44. STEGEMEIER, G.L., LAUMBACH, D.D. and VOLEK, C.L.;
    "Representing steam processes with vacuum models.
    SPE 6787.
45. MATHENY, S.L.;
    "EOR methods help ultimate recovery",
    Oil and Gas Journ. (March 31, 1980), pp. 79-124.
46. VAN DER KNAAP, W.;
    "M-6 steam drive process. Preliminary results of a large scale field
    test",
    SPE 9452.
47. PETCOVICI, V.;
    The experience of the Rumanian petroleum engineers with thermal recovery",
    Congreso Panamericano de Ingenieria del Petroleo, Mexico 1979.
48. MARX, J.W. and LANGENHEIM, R.N.;
    "Reservoir heating by hot fluid injection",
    Trans. AIME (1959) pp. 312-315.
49. MANDL, G. and VOLEK, C.W.;
    "Heat and mass transport in steam drive processes",
    Soc. Pet. Eng. J. (March 1969), pp. 59-79.
50. NEUMAN, C.H.;
    "A mathematical model of the steam drive process - Applications",
    SPE 4757.
51. FAROUQ, ALI,S.M. and REDFORD, D.A.;
    "Physical modelling of in-situ recovery methods for oil sands",
    Proc. of the Canada-Venezuela Oil Sands Symposium 1977, pp. 319-326.
52. GOTTFRIED, B.S.;
    "A mathematical model of thermal oil recovery in linear systems",
    J. Pet. Tech. (Sept. 1965), 196-210.
53. DOSCHER, T.M. and HAMMERSHAIMB, E.C.;
    "Field demonstration of steam drive with ancillary materials",
    SPE/DOE 9777.

54. ESON, R.L. and FITCH, J.P.;
    "North Kern front field steam drive with ancillary materials",
    SPE/DOE 9778.
55. ANON.;
    "ARCO wants to test oxygen for in-situ",
    Enhanc. Recov. Week, November 10, 1980.
56. PUSCH, G.;
    "Tertiärölgewinnungsverfahren – Die Untertageverbrennung mit Sauerstoff
    kombiniert mit Wasserinjektion (ISCOWI)",
    Erdöl und Kohle-Erdgas-Petrochemie (Jan. 1977) 30, 13–25.
57. BETC-STAFF;
    "Technical constraints limiting application of enhanced oil recovery
    techniques to petroleum production in the United States",
    DOE/BETC/RI-80/4, May 1980.
58. KONOPNICKI, E.F., TRAVERSE, E.F., BROWN, A. and DEIBERT, A.D.;
    "Design and evaluation of the Shiells Canyon field steam distillation
    drive project",
    SPE 7086.
59. HAGOORT, J., LEIJNSE, A. and VAN POELGEEST, F.;
    "Steam-strip drive: A potential tertiary recovery process",
    SPE 5570.
60. PARRISH, D.R., POLLOCK, C.B., NESS, N.L. and CRAIG, F.F.;
    "A tertiary COFCAW pilot test in the Sloss Field, Nebraska",
    J. Pet. Tech. (June 1974), pp. 667–675.
61. CHEKALJUK, E.B. et al.;
    "Method of recovering oil from an oil bearing bed",
    British Patent Appl. No. 15256/71.
62. REISS, L.H.;
    "The reservoir engineering aspects of fractured formations",
    IFP, Editions TECHNIP, Paris 1980.
63. SAHUQUET, B.C. and FERRIER, J.J.;
    "Steam drive pilot in a fractured carbonated reservoir Lacq Superieur
    field",
    SPE 9453.
64. DE VRIES, A.S.;
    "Aspects of enhanced recovery in densely fissured carbonate reservoirs
    containing heavy oil",
    Congreso Panamericano de Ingenieria del Petroleo, Mexico 1977.
65. WRIGHT, D.D. and BINSLEY, R.L.;
    "Feasibility evaluation of a downhole steam generator",
    SPE/DOE 9775.
66. FOX, R.L., DONALDSON, A.B. and MULAC, A.J.;
    "Development of technology for downhole steam production",
    SPE/DOE 9776.
67. JOHNSON, D.R. and FOX, R.L.;
    "Examination of techniques for thermally efficient delivery of steam to
    deep reservoirs",
    SPE 8820.

U.S. DEPARTMENT OF ENERGY R&D ON DOWNHOLE STEAM GENERATOR FOR THE RECOVERY OF HEAVY OIL

RONALD L. FOX

Sandia National Laboratories

J. J. STOSUR

U.S. Department of Energy

ABSTRACT

The energy loss associated with delivering steam from surface generators to the reservoir is one of the factors that has limited most commercial steaming operations to relatively shallow oil bearing formations (about 1000 feet). The Sandia National Laboratories under contract to the U. S. Department of Energy has initiated an ambitious program for the development and field testing of a downhole steam generator. The advantages are impressive: exceptionally high overall thermal efficiency; good potential for alleviating air pollution from generating steam at the surface, and; significant economic benefits from accelerated oil recovery due to the introduction of combustion products along with steam.

Two designs are being developed: a low pressure and a high pressure steam generator. The low pressure combustion design transfers energy to water through a heat exchanger, thus enabling the combustion process to be conducted at a pressure less the the injection pressure; a high pressure combustion design mixes the combustion gases directly with water, resulting in the injection of steam and combustion gases into the reservoir.

Field testing of a high pressure combustion generator was carried out in a shallow reservoir (275 meters) to determine if the system was compatable with field conditions, if recovery with this device resulted in modifications to the reservoir or produced crude, and to assess the injection of combustion gases into the formation as a method of reducing air pollution associated with steam injection. Follow on field tests have examined the performance of the device for downhole operations in reservoirs below 700 meters.

INTRODUCTION

THE U. S. Department of Energy initiated development of tools for production of steam at the oil producing formation in 1978. The technical implementation of the development and testing program is being carried out by the Sandia National Laboratories as part of the Department of Energy's Project DEEP STEAM. The DEEP STEAM project incompasses development of methods for application of conventional steam drive to deep reservoirs as well as the downhole steam generator program. The project provides for the conception, feasibility analysis, laboratory testing, and field testing of methods for downhole production of steam.

Two concepts for downhole production of steam for drive operations have been selected for comparative development. The two designs differ in method of transferring heat from hot combustion gases to produce steam. A low pressure combustion design transfers energy to water through a heat exchanger thus enabling the combustion process to be conducted at a pressure less than the injection pressure; a high pressure combustion design mixes the combustion gases directly with water, resulting in the injection of steam and combustion gases into the reservoir.

The high pressure combustion design has been utilized in a series of field experiments, and this paper presents the status of downhole steam generator technology as revealed in these experiments.

DOWNHOLE STEAM GENERATOR PROGRAM

The development of technology for downhole steam production has been carried out at the Sandia National Laboratories and under contract with Rockwell International, Rocketdyne Division, and with Foster-Miller Associates. The design and testing of a low pressure combustion generator has been pursued at Rocketdyne. A high pressure combustion downhole steam generator is under development at Foster-Miller. The Sandia National Laboratory is investigating a high pressure generator which differs from the Foster-Miller design in the method for obtaining clean combustion at pressures required for steam injection operations. The field tests which have occurred to date have utilized the Sandia Systems. The field testing has included these experiments:

Test A: Intermediate term test of equipment on surface with injection into shallow reservoir.
Test B: Installation and recovery of the generator from a deep well.
Test C: Intermediate term test of equipment in downhole operations in a deep reservoir.

The conditions for Test A were for 3 to 4 months of continuous operation utilizing oil field water and utilities with injection of the generator effluent into a 270 m deep reservoir. Test A was performed in the Kern River Field of California in cooperation with Chevron, USA during January-May, 1980. The conditions for Test B were to install and retrieve a downhole generator below 700 m in a standard oil field casing with a mechanically set packer below the device. Test B was performed near Lovington, New Mexico, in cooperation with ARCO Oil and Gas during September 1980. The conditions for Test C were for 3 to 4 months of continuous downhole operation in a reservoir at a depth greater than 700 m. The results of these tests are given in the remainder of this paper.

SHALLOW WELL OPERATIONS

A. Steam Generator. The direct contact steam/generation concept was chosen for the preliminary field test. A commercially available steam generator unit was sought to minimize the development time required to perform a test. Vapor Energy Co. of Grand Prairie, Texas, produced a direct contact generator of slender cylindrical geometry. These units had been previously operated at pressures up to 100 psig. A specially designed unit which could approximate dimensions for downhole operation was procured for 5×10^6 btu/hr at 3000 psi with an outside diameter of 6.5 inches. The design is illustrated in Figure 1 and is designated "before." In this design propane vapor is brought into the air stream in either one or both inlets indicated in the figure. The propane and air are mixed as they travel down the channel. Ignition is by spark plug and occurs at the area expansion. Water is passed up an annulus outside the combustion region and is entrained into the combustion gases. As this mixture passes down the inside of the steam generator, vaporization of the water occurs resulting in a mixture of steam and combustion products (mainly CO_2 and N_2).

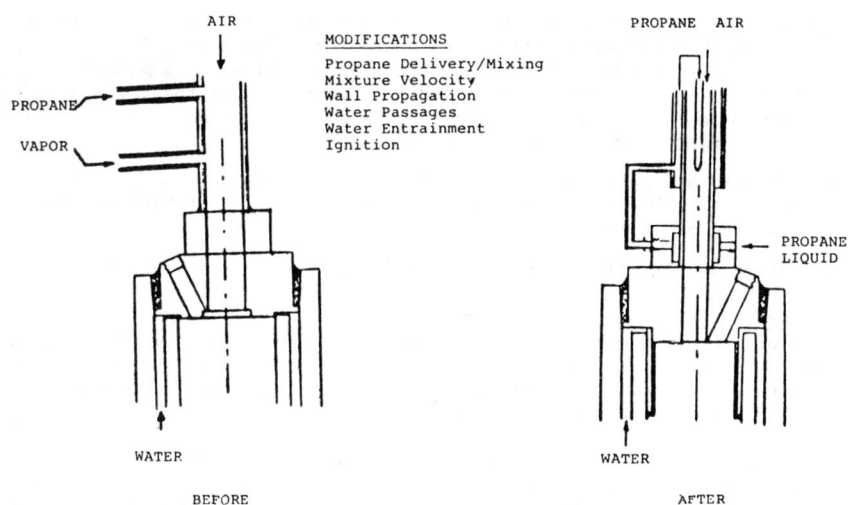

Figure 1. Direct Contact Steam Generator Before and
After Modification

This design was tested at Sandia at pressures of 100 psig or lower. The pressure level was limited by the vapor pressure of propane at the existing ambient
temperature. Several shortcomings of this system for the field test application were noted. These problems were resolved by the following modifications:

1. In order to provide propane vapor at pressures above 400 psia, liquid
propane is first pumped to pressure. Then the propane liquid is partially
vaporized by heat transfer from two sources: 1) propane is circulated through
the flange which attaches the mixer section to the combustor section (the
flange is heated by conduction from the flame zone); and 2) propane is then
circulated through a jacket around the air line (the air is warm because of
compression heat). These two zones are sufficient to cause partial vaporization of the propane. The resulting flow then passes down a tube and is
injected into the air stream through four holes which are 0.030" dia. The
holes are normal to the air flow to achieve penetration and improved mixing
of the two streams. All surfaces in the mixing section are smooth to eliminate
stagnation regions (flame holders). The wall cooling has the effect of reducing temperatures in the boundary layer which eliminates flame propagation up
the walls.

2. The diameter of the mixing section was decreased from 1.5 inches to 1.27
inches. This has the effect of increasing the flow velocity by 40% which
decreased the possibility of burn-back.

3. The water passage was modified to include a sleeve which separates the
water and combustion zone to a point at which combustion is essentially complete. Two beneficial effects arise from this modification: 1) greater ease
of ignition, and 2) better combustion efficiency. An intermediate step in
this modification was for the original fabricator (Vapor Energy) to install
a sleeve for this purpose, with Sandia's concurrence. However, after approximately 20 hrs operation with this modification, the sleeve was severely
damaged. This was apparently due to the sleeve not being fully wetted by
water. Hence, the final modification included a lip on the end of the sleeve
to restrict water flow and insure that the annulus remained fully wetted.

Approximately 30 slots were machined through this lip to allow water passage, even after thermal expansion during operation. Additionally, the water passage gap at the top of the combustor was increased from 0.040" to 0.100" so that it would not close upon thermal expansion. Further, the area of contact between water and the top cap was increased by extending the area expansion further into the combustor.

4. Instead of spark ignition, the final design utilized a flow plug for ignition. This modification facilitated ease of starting at high pressure levels without use of an exotic power source.

The above modifications are the basis of a patent application filed by DOE.

B. Kern River Field Test Results. The DEEP STEAM test started with the injection of steam and combustion products into the reservoir on February 6, 1980, after one week of operations to calibrate equipment and to instruct personnel in the operation of the system. The test was scheduled for and was conducted on a 24-hour, 7-days a week basis.[1]

After the start up on February 6 the test was run for 109 days to May 15. During that period the test was interrupted 51 times: 8 for power and water failures which were supplied by others, 12 were deliverate shutdowns, and the remaining 31 were represented by problems with the fuel supply, computer, instruments and people. The only major down period was from March 7 through March 20 and was caused by a detonation being propagated upstream through the air line due to oil in the line. The run time excluding this particular problem was 80% of the total test time.

The shallow reservoir field test at Kern River demonstrated the field performance of the generator system and that neglectable material corrosion was encountered with the low sulfur LPG fuel. The combustion gases moved rapidly through the reservoir, reaching production wells in the 2.5 acre 5-spot pattern within 18 hrs. The oil produced did not exhibit any special emulsions due to presence of the combustion gases. The effect of injecting the combustion gases into the reservoir reduced the environmental impact due to atmospheric exhaust of NO_x and SO_x.

TEST OF GENERATOR INSTALLATION AND RETRIEVAL

The downhole generator requires supplies of fuel, water, and oxidizer to produce steam at the sand face. Multiple strings must be run to supply the generator with these fluids. A test to determine proceedure for installation of the device below 700 meters was performed in late September 1980.[2] The generator utilized in this test was 15 meters long and designed for insertion with a 18 cm diameter casing. The multi-string supply consisted of two jointed tubulars, two small diameter continuous tubulars, and an electrical control cable.

A mechanically set Baker HB-1 packer was located below the generator. The packer was set after the generator was in place. The multistring delivery system was pressure tested before setting of the packer, after setting, and after release of the packer. The test demonstrated that the integrity of the system was maintained at all stages of insertion and retrieval.

DEEP DOWNHOLE OPERATION

The generator design which was tested in the shallow test in the Kern River Field has been replaced by a design which operated on liquid fuels (diesel No. 2). The liquid fuel design was developed in order to have wider application by using more abundant fuels.

A site for testing of the generator in a reservoir below 700 m was selected in the Wilmington Field in California. The test was carried out in cooperation with the City of Long Beach, California, and the Long Beach Oil Development Co. A new injection well was drilled for the test while existing production wells were utilized to form a five-spot 5-acre pattern. The injection well was directionally drilled at a 36 degree angle, the total length of the well is 830 meters.

The generator was inserted on June 19, 1981. The continuity of supply lines was tested and the responces of the reservoir to gas injection were studied after insertion of the device. The generator was ignited on June 22, 1981. The computerized ignition was achieved without incident, and continuous operation of the generator has proceeded ignition. The arrival of combustion gases at production wells was observed on June 26. The intermediate term test is scheduled for completion during September 1981. A longer term follow on test may be performed in the same location for a total operation time of one year.

CONCLUSIONS

The DOE program for development of a downhole steam generator for recovery of heavy oil from deep reservoirs has proceeded through a series of successful field tests. The ability of the device to eliminate heat losses and reduce environmental impact of steam drive has been demonstrated. The operational characteristics of the device in deep downhole operations are being evaluated in current field testing.

REFERENCES

1. Mulac, A. J., et al, Project DEEP STEAM Preliminary Field Test, Sandia National Laboratories Report SAND80-2843, April 1981

2. Mulac, A. J., et al, Multiple String Demonstration Test for Project DEEP STEAM, Sandia National Laboratories Report SAND80-2872, April 1981

STEAM DRIVE – THE SUCCESSFUL ENHANCED OIL
RECOVERY TECHNOLOGY

TODD M. DOSCHER and FARHAD GHASSEMI

Department of Petroleum Engineering, University of Southern California,
Los Angeles, California 90007

I. ABSTRACT

Continued work with physically scaled models of the steam drive process has confirmed earlier conclusions that the process should be viewed as being comprised of two distinct components: the heating of the oil at the interface between the oil column and the overriding steam, and then the displacement of the heated oil by an exceptionally high velocity gas (steam vapor) drive.

The studies described herein were conducted to validate certain hypotheses that result from the appreciation of the foregoing mechanism by which the steam drive operates. In general, all these hypotheses have been proven.

The oil steam ratio, when communication between injection and producing wells is established early, is not dependent upon reservoir thickness when recovering moderately viscous crude oils. The oil steam ratios in a wide range of field operations, in keeping with this conclusion, are shown to fall within a very narrow band of values.

Perhaps the most novel conclusion arising from this study, intimated by the work of others in the past, is that the steam drive is a powerful process for recovering high gravity crudes, even waterflood residuals if high injection rates of high quality steam can be achieved. The substitution of inert gas for some of the steam in a mature steam drive can signficantly increase the thermal efficiency but it is not certain, at this time, that the economic efficiency would thereby be increased. Finally, the observed effect of the viscosity of the crude at steam temperature on the efficiency of the process threatens the possibility of using the steam drive for unassisted recovery of truly viscous crudes and bitumens.

II. INTRODUCTION

The steam drive was first attempted in the Mene Grande field in Venezuela in the late 50's where its failure gave rise to the use of cyclic steam stimulation for accelerating the recovery of crude oil from reservoirs containig viscous crudes. However, during the subsequent decade experiments with the steam drive in the San Joaquin Valley of California and in the Athabasca bituminous sands of Canada proved that it was a viable scheme for recovering viscous crudes and certain bitumens.

Today, production in California as a result of both cyclic steam injection and steam drive operations is approximately 400,000 barrels of oil a day, or 40% of the state's total. Recovery efficiency in some mature operations is already well over 50% of the original oil in place and is projected to approach 70%, a value that is achieved in only a few reservoirs that are exploited by conventional technology. Cumulative recovery from reservoirs subjected to steam injection in California is now approaching 1.5 billion barrels of oil.

The steam drive has been intensively studied in the Department of Petroleum Engineering of the University of Southern California using physically scaled models. The results of the studies have corroborated a number of conclusions that had been empirically reached by studies of field operations and in addition have provided further insight into the mechanisms by which the steam drive functions[1,2,3,4]. One of the most important conclusions from these studies is that the efficiency of the steam drive is a function of the viscosity of the crude oil at steam temperature. This in turn leads to the further conclusions that the steam drive may not be as efficient as may have been surmised for the recovery of extremely viscous bitumens and may be the ultimate recovery scheme for the recovery of high gravity crude oils.

In the classic analytical derivation of the way in which a steam drive functions, attention is focussed on the development of a steam zone, which occupies the entire cross section of the reservoir, and from which oil is assumed to be depleted to some naturally determined residual oil saturation. The fact is, however, that the pressure required to frontally displace a viscous oil bank at an appreciable (economic) rate can rarely be applied in a real reservoir.

Reported field results have demonstrated that steam does not frontally displace heavy oil in a steam drive operation. Injected steam initially enters the formation through a depleted or wet interval, a fracture, or, in unconsolidated formations, a fluidized interval. In a successful steam drive in reservoirs containing viscous crude, the steam penetrates to the producing well quite early in the life of the operation. Without this, the influx of oil into the producing well would still be limited by the high viscosity of the reservoir crude. Even if the initial steam entry is not through a depleted zone at the top of the oil section, then the steam soon migrates to the top if there is any significant vertical permeability at all. Depletion of the oil to a very low saturation occurs in the interval through which the steam flows, and with time the depletion extends downwards[5,6]. In the scaled physical model studies, this vertical extension of the steam zone can be followed in some detail. The heated crude at the interface between the steam and the oil column is stripped off or dragged along by the flowing steam. This stratified flow of steam is of course not unexpected in view of its very low density in comparison to the density of the reservoir fluids. However, the unexpected result is that the flowing steam is capable of driving the oil saturation down to such low levels.

The interfacial tension of oil against saturated steam has been verified in our laboratories to be little different from that oil against water, and therefore it appears that the only parameter to which the low residual can be attributed is the high velocity of the gas (steam vapor). It might be noted that the injection of 500 barrels of steam per day into a reservoir at an averge pressure of 200 psi. is equivalent (neglecting condensation) to the injection of 6 MM SCFD of an ideal gas. The resulting velocities on 2.5 to 6 acre spacing (the usual spacings for steam drive with injection rates of 500 barrels per day) when coursing through a 25 foot depleted zone are very high; of the order of 100 feet per day. Of course, a significant fraction of the injected steam condenses, but offsetting this somewhat is the fact that the production pressure is less than the injection pressure and therefore the uncondensed steam will expand and further increase its velocity through the reservoir.

III. THE OIL STEAM RATIO - THE MEASURE OF SUCCESS

The amount of oil produced per unit quantity of steam injected is the most important criteria for judging the success of a steam injection project. This is so because the amount of energy used to generate the steam in even the most successful projects is a substantial fraction of the produced oil. The value of the fuel is overwhelmingly the largest single component of the cost of producing oil by steam injection.

During the 60's Marx and Langenheim[7] published a procedure for calculating the growth of a steam zone as a function of steam injection rate, time of injection, reservoir dimensions, and the thermal properties of the reservoir and the cap and base rocks. Mandl and Volek[8] later undertook a somewhat more detailed analysis of the mass and thermal balances at the condensation front and developed a somewhat improved analytical technique for estimating the growth of the steam zone. Subsequently Myhill and Stegemeier[9] codified Mandl and Volek's analysis to permit ready calculation of the oil steam ratio as a function of a variety of operating conditions and well spacings.

It is important to note that the oil steam ratios calculated by these methods are for uniform sands (although allowance can be made for unproductive but permeable layers disseminated through the reservoir sand). Also, it is assumed at the outset of these calculations that the injection rate of steam is established or assumed since the analysis is strictly a thermal one and does not treat the questions related to fluid flow. Eventually, numerical methods[10] were developed for predicting the performance of a steam drive. It soon became obvious that for the presumed reservoir conditions in most heavy oil reservoirs did not permit the injection of steam at the rates observed in field operations. In order to overcome this deficiency, an extraordinary high compressibility was assigned to the reservoir. The important advantage of the numerical analysis over the analytical one is the ability of the former to include more than one layer and thus permit the development of the obvious gravity override of the steam. By making suitable adjustments in the input parameters to get a match of the performance of particular operations, the resulting numerical analysis can be presumed to be a generalized simulation of the steam drive process.

Both the analytical and numerical models of the steam drive process predict a significant effect of reservoir thickness on the oil steam ratio, see Figures 1 and 2. This comes about from the fact that when frontal displacement occurs, the heat loss at the cap and base rock is independent of the reservoir thickness. Hence, the thicker the reservoir, the greater the fraction of the injected heat that is captured within the reservoir.

Recent reviews of the performance of steam drive operations indicate that a much lower range of oil steam ratios occur in field operations than would be expected from the results of these calculations. The average of 7 current and completed steam drive operations in thick sands, in excess of 70 feet and approaching 200 feet, reported in a recent study[11] is only 0.22 with a standard deviation of only 0.06. On the other hand, the average for four steam drives in thinner sands, ranging from 18 feet to 50 feet[12, 13, 14, 15] is still 0.20, ranging from 0.15 to 0.25. Further, the recovery efficiency from two of these relatively thin reservoirs, viz., Slocum and San Joaquin have been reported to have approached 80%.

This anomalous behavior of the oil steam ratio combined with the observed override of steam suggested that the performance of steam drives, at

552

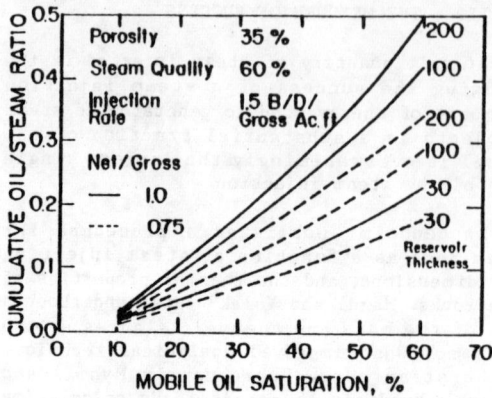

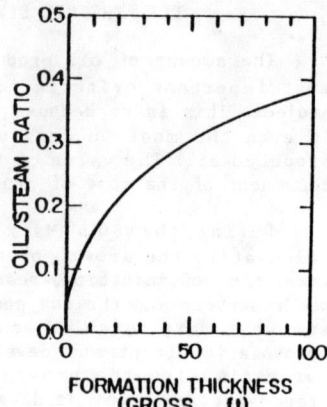

Fig.1. Effect of Oil Sat.,Reservoir Thickness
and Net/Gross Ratio on O/S Ratio

Fig.2. Effect of Res.
Thickness on Oil/
Steam Ratio

least when dealing with viscous crudes, is not correctly estimated by an
analysis that assumes frontal displacement. The early work on the physical
model experiments[1,16] indicated that the production of oil occurs at a
(vertically) moving boundary between the overriding steam zone and the oil
column. Correlative with this performance is the existence of the following
occurrences:

1. There is an optimum steam injection rate for a given spacing between
injection and production wells. This phenomenon had been observed in Kern
River operations sometime earlier[17].

2. The oil steam ratio is virtually linearly related to the quality of
the injected steam. This result is implicit in the Mandl Volek analysis. Hot
water is virtually ineffective in recovering any signficant quantity of
viscous crudes.

3. The oil steam ratio is a function of the viscosity of the viscous
crudes at (the average) steam temperature. For moderately viscous crudes
the ratio appears to be a function of the half power of the inverse of the
steam temperature viscosity, Figure 3.

This observation has serious implications for the applicability of the
steam drive process to highly viscous bitumens for which a very high steam
temperature would be required. It would appear that the resulting oil steam
ratio would fall to uneconomic values (below 0.12?) when the produced crude
is used as fuel for generating steam. The fall in the oil steam ratio would
be further exacerbated by the lower fraction of the latent heat and the
lower specific volume of the steam at high temperatures (pressures).

4. The permeability of the formation, above a darcy, has virtually no
effect on the oil steam ratio. Detailed investigations have not been
conducted at lower permeabilities, but there is a suggestion that very low
permeabilities have a depressing effect on the oil steam ratio.

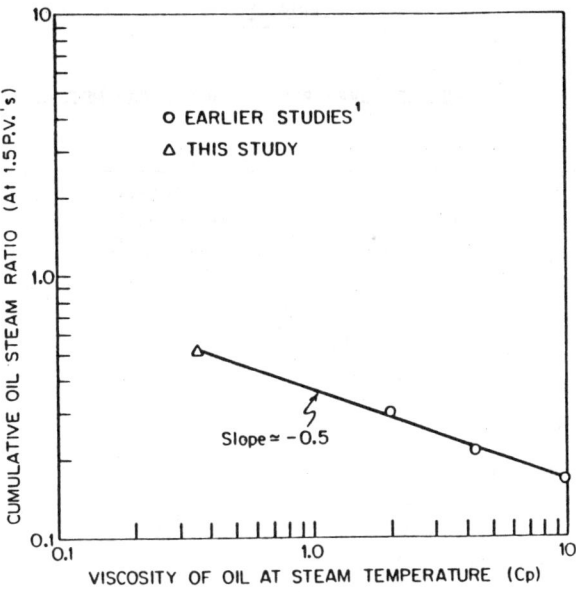

Fig. 3. Oil/Steam Ratio as a Function of
Reservoir Fluid Viscosity

5. Diurnal injection of steam, as would occur if steam was generated by solar devices, does not seriously affect the oil steam ratio as long as the total daily rate is the same as in continuous injection[3].

The current study was therefore designed to investigate just why the oil steam ratio appears to be so less in actual operations than would be predicted by analytical and numerical models, and in addition, why the oil steam ratios appear to fall within a narrow band of values for reservoirs having a wide range of thicknesses. Further, because of the increasing value of the oil steam ratio with a decrease in oil viscosity, it was decided to investigate just how high an oil steam ratio could be achieved in the displacement of high gravity crudes, particularly at residual oil (to water flood) saturations.

Finally, because the steam drive process appears to be comprised of two mechanisms: firstly, the heating of the oil at the steam oil interface, and secondly the displacement of the oil from that interface by the flow of the steam vapor; the substitution of nitrogen for some of the steam - that part used for displacement alone and not for heating - was investigated.

IV. THE EFFECT OF RESERVOIR THICKNESS

The experimental procedures and techniques used in carrying out the physically scaled model experiments have been thoroughly detailed in earlier studies[18,1,16]. Tables 1 summrizes the prototype and model parameters for the experiments conducted with viscous oils in this study.

TABLE 1

PROTOTYPE AND CORRESPONDING MODEL PARAMETERS

Parameter	Prototype Lloydminster Reservoir	Model
Initial Res. Temp., °F	75	35
Pressure, psia Injection	300	8.5
Production	50	5.0
Sand Thickness, ft	27	0.354
Permeability, Darcy	1	550
Porosity, %	33	33
Steam Injection Temp., °F	417	186
Initial Gas Saturation, %	2	2
Oil Gravity, °API	14	11.8
Steam Quality, %	75	27.1
Scaling Factor	1	76.9
Flow Rate, (P-B/D, M-cc/min)	1000	186
Time (P-year, M-minute)	1	85.7
Pattern Area (P-acre, M-ft^2)	5	4.6
Distance of Injector to Producer, ft	330	4.29
Thermal conductivity, BTU/hr.ft.°F		
a) overburdon	1.09	1.13
b) underburden	1.1	1.0
Oil Viscosity, cp		
a) reservoir temp.	100,000	2000
b) steam temp.	28	3.8

Figure 4 shows the results obtained for a 26 foot thick prototype on 5 acre spacing (or for a 13 foot thick reseroir on 2.5 acre spacing at half the indicated injection rate, or for a 52 foot reservoir on 10 acre spacing at double the indicated injection rate). A comparison of this data with the results obtained earlier for the 70 foot prototype, Figure 5, indicates that the optimum stem injection rate is in the same range for both models; it is not dependent on the thickness of the formation. Figure 6 compares the performance of the optimum steam injection rate in the two models. (The initial, quite pronounced difference in the two runs is due to the fact that in the later work with the thinner reservoir the initial few hundreths of a pore volume of oil that was produced was not attributed to the steam drive.) Earlier conclusions based on field observation to the effect that the optimum rate is a function of the acre feet in the pattern area was probably due to the fact that all the patterns had the same thicknes.

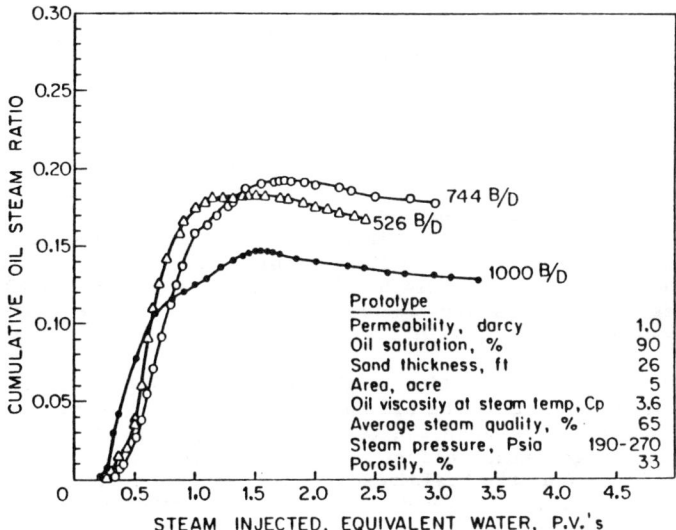

Fig. 4. Effect of Injection Rate on Oil/Steam Ratio

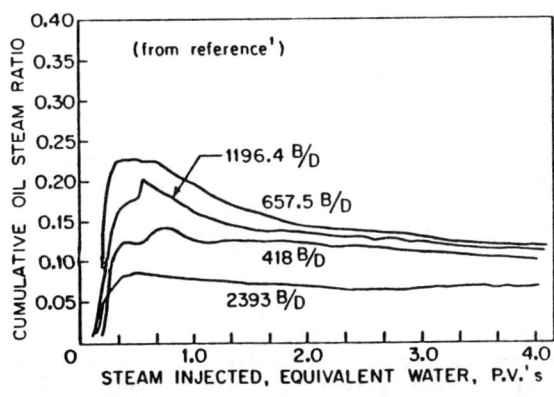

Fig. 5. Effect of Injection Rate on Oil/Steam Ratio

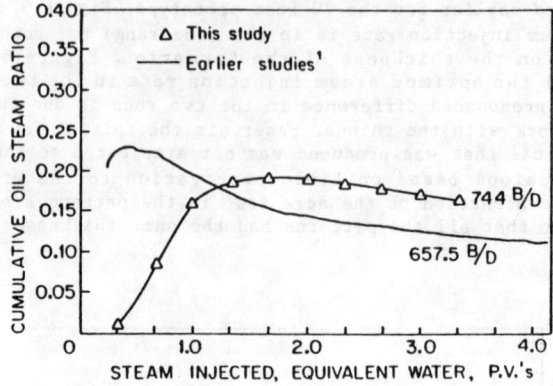

Fig. 6. Comparison of Oil/Steam Ratio for Thick and Thin Sand

The steam quality used in the 70 foot prototype runs averaged 45% whereas the steam quality for the 26 foot runs averaged 65%. Calculations based on frontal drive theory indicate the oil steam ratio in the first five years should be somewhat higher for the thick reservoir than that observed in this work, and just slightly higher for the thin reservoir. Discounting the effect of steam quality, the results for the two reservoirs would be virtually identical.

Given that the description of the steam drive presented earlier, viz., that the steam overrides the oil column and gradually strips the hot oil from the interface between the latter and itself, there is no reason to anticipate a significant effect of reservoir thickness on the process. Indeed, this is what the field results indicate. The almost constant oil steam ratio observed in the wide range of steam driven reservoirs reported above is in agreement with the range of oil steam ratio, 0.15 to 0.20 observed in the physically scaled model runs with viscous crudes. A simple mathematical formulation of the displacement of oil at the moving boundary has been developed which leads to the prediction of just such a range of values for the oil steam ratio[16].

V. EFFECT OF FLUID VISCOSITY

Figure 7 shows the results for the 26 foot prototype reservoir when the prototype reservoir fluid has a viscosity at steam temperature of only 0.4 centipoise, slightly greater than water would have at the same temperature. A comparison of this production history with that of the more viscous reservoir fluids quickly shows that displacement of the low viscosity fluid by steam results in a much higher oil steam ratio.

Figure 3 shows the oil steam ratio as a function of viscosity of the oil at steam temperature. The data point for the three highest viscosity fluids were obtained in the earlier studies in a 70 foot prototype, and the data point for the lowest viscosity fluid is the one described in this study adjusted downwards because of the higher quality of the steam that was used. It is apparent that for the same oil saturation, the oil steam ratio continues to increase as the viscosity of the reservoir fluid decreases.

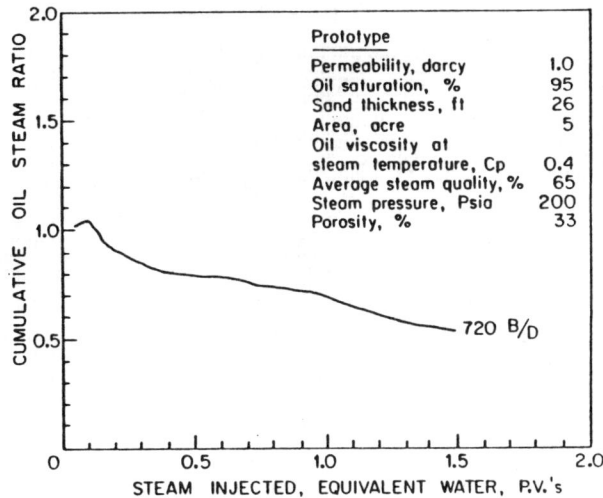

Fig. 7. Oil/Steam Ratio History for Low Viscosity Fluid

A comparison of the steam (and therefore temperature) distribution in the reservoir for the displacement of a viscous fluid and one with a low viscosity is quite informative. Figures 8 and 9 show the temperature distribution for the displacement of the heavy oil and the mobile oil, respectively, after the injection of an amount of steam generated from 0.5 pore volume of liquid water. Figures 10 and 11 portray the temperature distribution after the injection of 1.5 pore volumes.

◄── **FLOW DIRECTION**

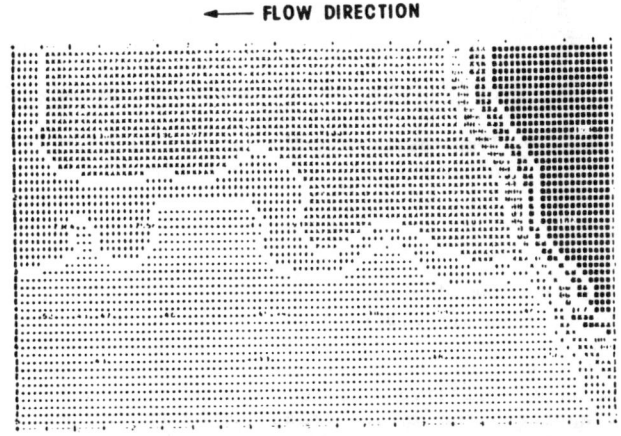

Fig. 8. Temperature Distribution for the Heavy Oil
After 0.51 P.V.'s of Steam Injected

◄——— **FLOW DIRECTION**

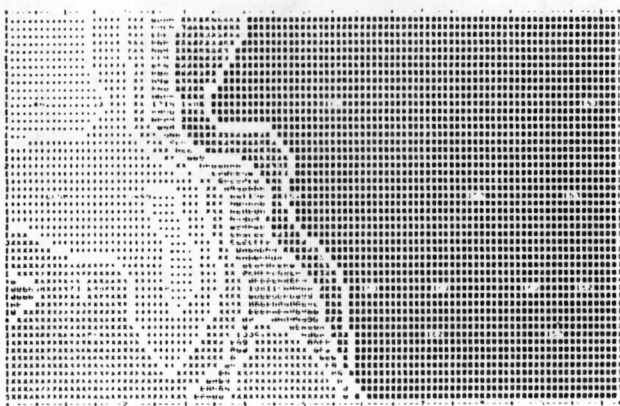

Fig. 9. Temperature Distribution for the Light Oil
After 0.5 P.V.'s of Steam Injected

◄——— **FLOW DIRECTION**

Fig. 10. Temperature Distribution for the Heavy Oil
After 1.5 P.V.'s of Steam Injected

It should be noted that with a decrease in viscosity of the reservoir
fluid the temperature distribution indicates the mechanism is gradually
taking on the aspects of a frontal displacement. This again should be
anticipated since with decreasing velocity the available pressure can indeed
displace the bank of reservoir fluid. The oil steam ratio should now be
expected to approach the values predicted by the frontal displacement
analyses, and indeed it does. An earlier numerical study also indicated that
the override of steam decreased significantly when injecting steam into a
reservoir that had been water flooded to a resdiual oil saturation[19].

◄——— FLOW DIRECTION

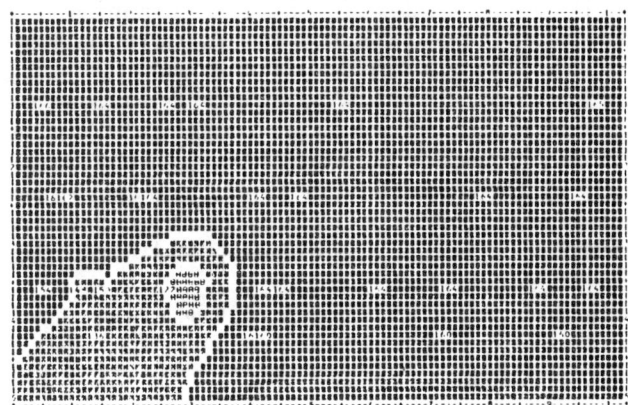

Fig. 11. Temperature Distribution for the Light Oil
After 1.5 P.V.'s of Steam Injected

VI. STEAM DRIVE OF A RESIDUAL OIL SATURATION

The relationship of the oil steam ratio to oil viscosity, Figure 3, indicates that a steam drive in a reservoir having a 33 percent porosity and saturated with water will produce the latter at a reservoir water/steam injected ratio of 0.7 after the injection of 1.0 pore volume of steam. Even higher, if the quality of the injected steam at the sand face is above 65% and the pressure is less than the prototype value of approximately 250 psi used in our studies.

If the reservoir is not 100% saturated with water, but contains a residual saturation of a low viscosity crude oil which is displaced more or less in proportion to its saturation in the reservoir; then the resulting oil steam ratio would be anticipated to be $(0.7s_{or})$.

For a residual saturation of 0.25, or greater, the resulting oil steam ratio would be 0.18, or greater; as high or higher than the oil steam ratios experienced in the steam drive of heavy oils. The results of a model experiment with a residual saturation of 22% of a prototype crude oil having a viscosity of 0.15 centipoise at a steam temperature of 401°F. are shown in Figure 12. The oil steam ratio is 0.21 after the recovery of 32% of the residual oil, and the oil steam ratio is still 0.14 after the recovery of 50% of the oil in place.

It is important to note that reaching such high oil steam ratios is dependent both on a suitably high steam injection rate and a sufficiently high porosity. However, the results do not appear to be dependent on distillation effects, as had been suggested by previous workers studying the recovery of residual oil by a steam drive[20]. Further, the mere fact that a high oil steam ratio is realized is not sufficient to indicate that an economic steam drive operation is feasible. Certainly, an economic operation would be at hand if no additional well drilling costs are encountered in implementing the drive; however, if many new wells had to be drilled in relatively thin sands (resulting in a high capital investment per recovered barrel), the advantages of high oil steam ratios might be overcome.

560

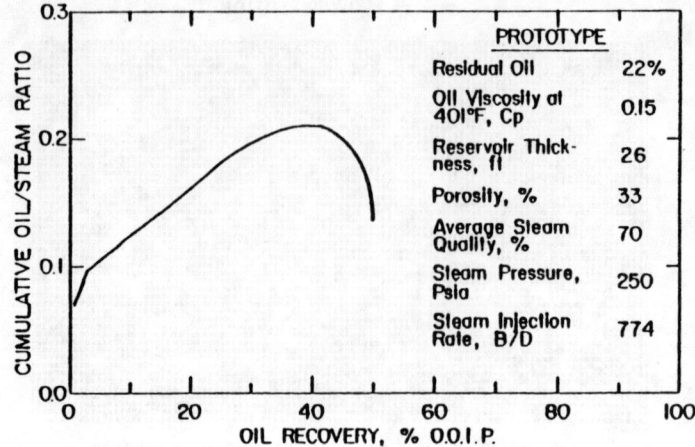

Fig. 12. Cum. Oil/Steam Ratio history for Waterflood Residal Oil

VII. EFFECT OF CO-INJECTION OF NITROGEN AND STEAM

Figure 13 compares the performance of the steam drive of a viscous oil with with that in virtually two identical reservoir situations in which the injection of steam and nitrogen was substituted for the injection of steam alone after one pore volume of steam had alredy been injected. In Run 36 steam and nitrogen were simultaneously injected, and in Run 37 slugs of nitrogen were alternated with the steam.

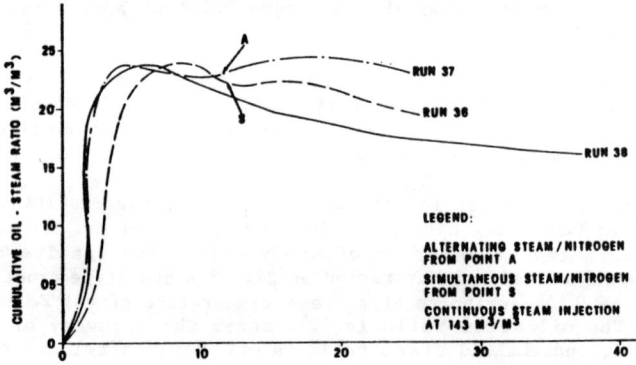

Fig. 13. Nitrogen as a Steam Additive

Compring the results of Run 36 and 37 with the control, Run 38, it is apparent that oil production is being maintained even though steam injection is curtailed as a result of the ancilliary effect of the injected inert gas. These results clearly show that in a steam drive operation the steam is playing the multiple role already described.

At this time, it is not clear that there would be a marked economic gain in the substitution of an inert gas for some of the steam in a mture steam drive because of the unit cost of compressed, inert gaases. However, this is not likely to be true in large instllations at this time, and may not be true at all in the future as the cost of energy continues to escalate. There is a far larger component of energy costs in the unit cost of steam than there is the unit cost of inert gas.

VII. CONCLUSIONS

The continued study of the steam drive in physically scaled models, and correlative observations made on reported steam drive operations in the field lead to the following conclusions:

1. The oil steam ratio in reservoirs which contain a moderately viscous crude oil, and in which steam overrides the oil column will substantially be the same regardless of reservoir thickness. The most probable range for the oil steam ratio will be 0.15 to 0.20. This conclusion presumes that adequate injectivity and communication between injection and production wells has been secured. The recovery efficiency itself will decrease as the reservoir thickness increases much beyond 50 or 60 feet.

2. Because of the higher steam temperatures required to achieve sufficient mobility of reservoir crude, the oil steam ratio in reservoirs which contain very viscous crudes are not likely to permit economic recovery unless a fuel cheaper than the crude oil itself is used for generating steam.

3. The oil steam ratio increases markedly as the viscosity of the reservoir crude decreases. This is due to the more efficient stripping of the heated fluid at the oil steam interface. As the viscosity of the reservoir fluids decrease to that of water, the displacement gradually conerts to a frontal advance and the efficiency of displacement of the reservoir fluids increases still further. The increase in the oil steam ratio is sufficient to indicate that in numerous situations waterflood residual oil saturations can be economically recovered by a steam drive. (An extension of the same argument may indicate that many reservoirs containing high gravity crudes may be more efficiently exploited by a steam drive than by a water flood.

4. An inert gas such as nitrogen can be substituted for a signficant fraction of the steam, that would otherwise be injected into a mature steam flood, sto maintain the production of oil and achieve a higher oil/steam ratio. The economic advantage of such a substitution will become more signficant as the value of the crude oil itself increases.

5. A simple conceptual model for the steam drive which comprises the overlay of the steam, the heating of the oil at the resulting oil steam interface, and the displacement of the heated oil by the gas drive accounts for virtually all of the observations made on steam drive operations in the field and in the laboratory physically scaled models.

VIII. REFERENCES

1. Doscher, T. M., et. al.; "Scaled Physical Models of the Steam Drive Process", Annual Report, Contract EY-76-S-03-0113 36 PA, United States Department of Energy, Oakland, California

2. Doscher, T. M., and Haung, W.; "Steam Drive Performance Judged from Use of Physical Models", Oil and Gas J. (Oct 1979) 52

3. Doscher, T. M., Ghassemi, F., and Omoregie, O. S.; "The Anticipated Effect of Diurnal Injection on Steam Drive Efficiency", Paper SPE 8885 presented at SPE 50th California Regional Meeting, Los Angeles, April 9-11, 1980

4. Ree, S. W., and Doscher, T. M.; "A Method for predicting Oil Recovery by Steamflooding Including the Effect of Distillation and Gravity Override", Soc. Pet. Eng. J. (Aug 1980) 249-266

5. Blevins, T. R ., and Billingsley, R. H.; "The Ten Pattern Steam Flood, Kern River Field", J. Pet. Tech. (Dec 1975) 1505-1514

6. Neuman, C. H.; "A Mathematical Model of Steam Drive Process Applications", Paper SPE 4757 Presented at SPE 45th Annual Meeting , Held in Ventura, April 2-4, 1975

7. Marx, J. W., and Langenheim, K. H.; "Reservoir Heating by Hot Fluid Injection", Trans., AIME, 216, 312-315

8. Mandel, G., and Volek, C. W.; "Heat and Mass Transport in Steam Drive Process", Soc. Pet. Eng. J. (Mar 1969) 59-79

9. Myhill, N. A., and Stegemeier, G. L.; "Steam-Drive Correlation and Prediction", J. Pet. Tech. (Feb 1978) 173-182

10. Gomma, E. E.; "Correlations for Predicting Oil Recovery by Steamflood", J. Pet. Tech. (Feb 1980) 325-332

11. Farouq Ali, S. M., and Meldau, R. F.; "Current Steamflood Technology", J. Pet. Tech. (Oct 1979) 1332-1342

12. Hall, A. L., and Bowman, R. W.; "Operation and Performance of the Slocum Thermal Recovery Project", J. Pet. Tech. (Apr 1973) 402-408

13. Blevins, T. R., Aseltine, R. J., and Kirk, R. S.; "Analysis of a Steam Drive Project, Inglewood Field, California", J. Pet. Tech. (Sept 1969) 1141-1150

14. Wooten. R. W.; "Case History of a Successfull Steamflood Project-Loco Field", Paper SPE 7548 Presented at 53rd SPE Annual Meeting, Houston, Oct 1-3, 1978

15. Greaser, G. R., and Shore, R. A.; "Steamflood Performance in the Kern River Field", Paper SPE 8834 , Presented at First Joint SPE/DOE Symposium, Tulsa, April 20-23, 1980

16. Doscher, T. M., and Ghassemi, F.; "The Effect of Reservoir Thickness and low Viscosity Fluid on The Steam Drag Process", Paper SPE 9897, Presented at The California Regional Meeting, Bakersfield, March 25-26, 1981

17. Brusell, C. G., and Pittman, G. M.; "Performance of Steam Displacement in the Kern River Field", J. Pet. Tech. (Aug 1975) 997-1004

18. Stegemeir, G. L., Laumbach, D. D., and Volek, C. W.; "Representing Steam Process with Vacuum Models", Paper SPE 6787, Presented at 52nd SPE Annual Fall Metting, Denver, Oct 9-12, 1977

19. Aydelotte, S. R., and Ramesh, A. B.; "Economic Feasibility of Steam Drive in Light Oil Reservoirs", Presented at 5th Annual DOE Symposium, Tulsa, August 22-24, 1979

20. Hagoot, J. Leijnse, A., and Van Poelgeest, F.; "Steam Strip Drive: A Potential Tertiary Recovery", J. Pet. Tech. (Dec 1976) 1409-1419

DOWNHOLE STEAM GENERATION USING A PULSED BURNER

D. A. CHESTERS, C. J. CLARK and F. A. RIDDIFORD

BP Research Centre, Chertsey Road, Sunbury

ABSTRACT

The recovery of viscous crude oils using downhole steam generation provides a
significant improvement in thermal efficiency over surface steam generation.
It is also likely to be an economically viable solution for oil recovery at
depths greater than 300m. In addition, by discharging combustion products into
the formation, the technique is virtually pollution free and provides a source of
CO_2 to assist viscosity reduction. The operation of a continuously fired,
high intensity burner within the confines of an oil well, however, presents a
number of technical problems, the most intractable of which relates to the
selection of materials to operate at high burner temperatures.

In order to obviate these difficulties BP has been developing a system based
upon a pulsed burner. The unit operates in a "quasi" detonation mode and
achieves the overall required combustion intensity as a series of high intensity,
short duration pulses. Steam is produced by atomising water into the high
velocity combustion products, the quantity of steam being governed by ignition
frequency and mixture flow rate. The feasibility of this system has been
demonstrated on a 75 m test well.

A high pressure test rig, designed to simulate oil well conditions at greater
depths, has been built to test pulsed burners operating on both gaseous and
liquid fuels. To date, operating experience on this scale has been obtained
only on gaseous fuels. Under pulsating conditions, the temperatures on the
combustion chamber walls are close to that of saturated steam at the operating
pressure.

Current developments are being directed towards liquid fuel operation, with the
ultimate objective of proving a system on residual fuel.

INTRODUCTION

Downhole steam generation for the recovery of viscous crude oils offers a number
of potential advantages over conventional surface methods. In particular, it is
predicted from numerical simulation of the downhole steam generation process
that the combined injection of steam and combustion derived CO_2, which assists
in viscosity reduction, will significantly increase recovery rates compared to
conventional steam drive (1). As a result of this interaction, it is also
predicted that the operating cost per barrel of recovered crude oil is
independent of depth up to ~1500 m (1). Since surface and well thermal losses
are eliminated, the technique is also more thermally efficient, an advantage
that becomes progressively more significant as deeper reservoirs are probed and
higher pressures encountered. Downhole steam generation therefore appears to be
an attractive solution to thermal recovery in deep reservoirs.

The advantages of downhole steam generation were recognised a number of years ago when the US Department of Energy set up the research programme "Project Deep Steam". This project has resulted in the development of indirect (low pressure combustion (2)) and direct (high pressure combustion (3)) methods of downhole steam generation. In the indirect method, steam is generated within a downhole heat exchanger and the combustion gases are ducted back up the well casing. Pressures are, therefore, only those required to maintain the flow of reactants and products within the system. In the direct method, however, water is flashed directly to steam by the combustion gases and the mixture injected into the formation. Combustion therefore takes place at reservoir pressure. Both systems have reached an advanced state of development and have been tested downwell (4,5). The operation of high output continuous burners within the confines of an oil well, however, presents a number of technical problems. In particular, the very high heat fluxes encountered within the combustion chamber coupled with the degree of mixing required to achieve high combustion intensities necessitate careful attention to cooling and fuel/air mixing. The problems are accentuated in high pressure applications where the technology literally enters the space age; advanced rocket engines using exotic fuels rarely exceed pressures of 70 bar. The design of a continuous burner and associated downhole hardware is therefore complex and requires the use of highly specialised materials. Field experience gained to date has already highlighted the importance of these particular areas (6).

In order to overcome the potential problems of a continuously fired downhole burner, BP have developed a downhole steam generator that operates in a pulsed combustion mode. This paper deals with the current state of the development and discusses:-

(a) the principles of operation and general construction of a pulsed downhole steam generator;

(b) operational experience of a methane fired downhole steam generator;

(c) current progress in the development of a liquid fuel fired burner.

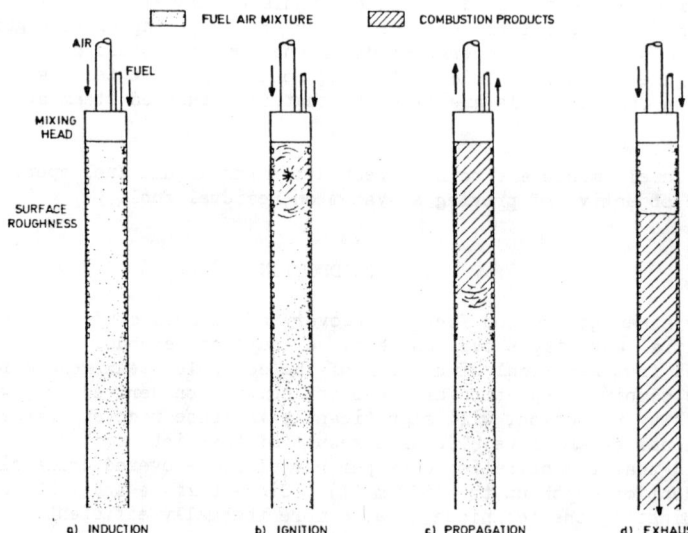

Figure 1. Schematic Diagram of the Individual Phases in the Firing Cycle of a Pulsed Burner.

DESCRIPTION OF A PULSED DOWNHOLE STEAM GENERATOR

The essential difference between the BP downhole generator and those being developed elsewhere is that it operates in a pulsed mode rather than in a continuous mode.

Within the pulsed burner, combustion occurs by the repetitive firing of discrete volumes of fuel/air mixture. The individual phases in the firing cycle are shown schematically in Figure 1.

At the start of the firing cycle (Figure 1a), fuel and air are introduced from separate supply lines at one end of a tube roughened over part of its length. The tube is filled over its entire length with fuel/air mixture. On ignition (Figure 1b) flames propagate both upstream and downstream. The pressure rise within the combustion tube modulates the reactant flow and the upstream propagating flame is extinguished at the mixing head (Figure 1c). As a result of turbulence induced acceleration, the flame within the roughened tube reaches a velocity that is about 25% of the detonation velocity of the fuel/air mixture (~ 500 ms^{-1}). This corresponds to a pressure ratio across the flame front of 2.5. The flame and associated shock structure propagate as a coupled system or "quasi" detonation, the velocity of which is determined primarily by the roughness of the tube walls. A detailed description of the propagation mechanism is given in Ref 7. In the final phase of the cycle (Figure 1d) combustion gases are exhausted from the burner at high velocity. Since their residence time within the burner is short, little heat is transferred to the combustion tube walls; for the greater part of the cycle the tube is filled with relatively low temperature gases.

The power output, and hence potential steam output of a pulsed burner, is determined by the frequency with which each discrete volume of reactants, that is the volume of the combustion tube, is ignited. Power output is simply adjusted by varying fuel/air flow rate and ignition frequency. The upper limit to power output is theoretically determined by the maximum propagation velocity that can be achieved in a roughened tube. In practice, however, power output is limited by a realistic frequency and fuel/air flow rate. For burners presently being designed, an output of 10 MW (~ 2000 bbl steam/day*) at a depth of 1500 m is believed to be achievable. An operating pressure of 70 bar has been selected as a target for the initial phase of burner development. A specification for such a burner is given in Table 1.

Table 1 Typical Specification for a 10 MW Pulsed Burner Operating at 70 bar

Principal Burner Characteristics	
Diameter of combustion tube	89 mm
Overall diameter of generator	127 mm
Length of combustion tube	1.3 m
Overall length of generator	4.0 m
Heat output range	0 - 10 MW
Ignition Frequency	0 - 5 HZ

*
This is defined as the number of barrels of water converted to steam.

A comparison of the combustion intensities that can be achieved in continuously fired high intensity combustion systems is given in Table II. Examination of these data shows that the combustion in a pulsed burner is only exceeded by the most exotic systems.

Table II Comparison of Typical Combustion Intensities Achieved in High Intensity Combustion Systems.

Combustion System	Operating Pressure (bar)	Combustion Intensity MW/m^3
Steam boiler	1	1 - 3
Gas turbine	4	370
Ram jet	5	750
Rocket engine	35	10^5
Pulsed burner	70	$1.3 \ 10^3$

Although high combustion intensities may be achieved using a pulsed burner, the fact that flame stabilisation is not required and that the entire combustion tube is available to mix fuel and air prior to combustion, permits significant simplification in the design of the mixing head and combustion tube. In addition, because of the low wall temperatures that result from a pulsed mode of operation, the burner requires only simple cooling and may be constructed from conventional materials. That low wall temperatures are indeed measured during normal operation is demonstrated in Figure 2a. These data were obtained from thermocouples set in the wall of a burner operating at a pressure of 7 bar. It is seen that the wall temperature is highest nearest the mixing head and that it drops to well below the saturation steam temperature some distance downstream. If, however, the burner changes to a continuous mode of operation, the change in wall temperature is dramatic. The effect is shown in Figure 2b. The short duration plateau, followed by a rapid temperature rise, indicates a change to film boiling within the water jacket. Continuous operation under these conditions would result, after a few minutes operation, in total failure of the unit. In the prototype burner, instrumentation is included to sense such an event and initiate shut down.

The overall design of the entire downhole steam generator is shown schematically in Figure 3. Fuel and air are supplied to the burner through a dual string system, with the ignition and instrumentation lines strapped to the air line. The upper part of the assembly comprises instrumentation and ignition packages which provide, respectively, a continuous monitor of combustion performance and ignition control. Fuel and air are injected into the combustion tube and the ignition cycle is initiated, as already discussed. Water is sprayed directly into the combustion chamber via an annular water jacket. The water is flashed directly to steam and the resulting mixture of steam and combustion products is injected into the formation. A high pressure packer prevents the escape of steam up the annulus.

Small scale experimental burners of this general design have been successfully operated on gaseous fuel (methane and hydrogen) in a high pressure test rig. Limited field work has also demonstrated that the system represents a practical means of downhole steam generation. Burners are currently being developed to use liquid fuels (ranging from middle distillate to residual fuels) to meet a variety of operational requirements. Each of these aspects of the development is now separately discussed.

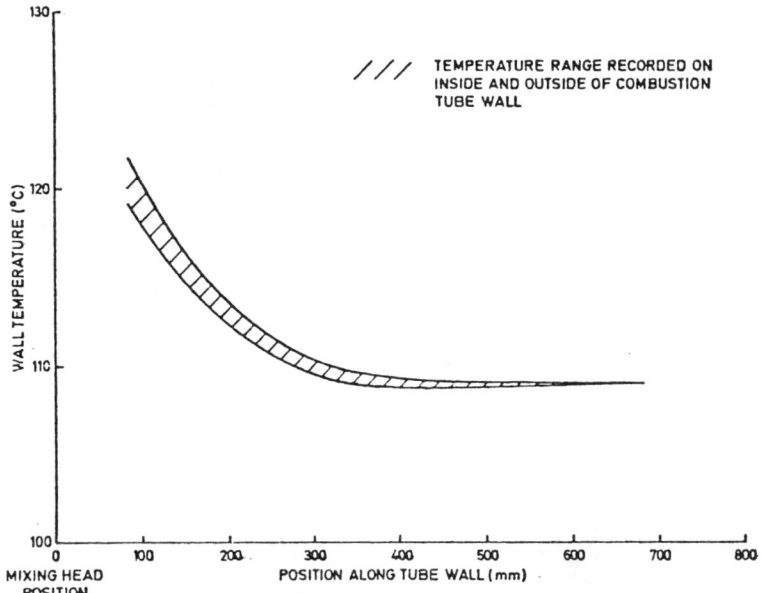

Figure 2a Temperature profile along the combustion tube of a prototype 0.25 MW pulsed burner during operation at 7 bar.

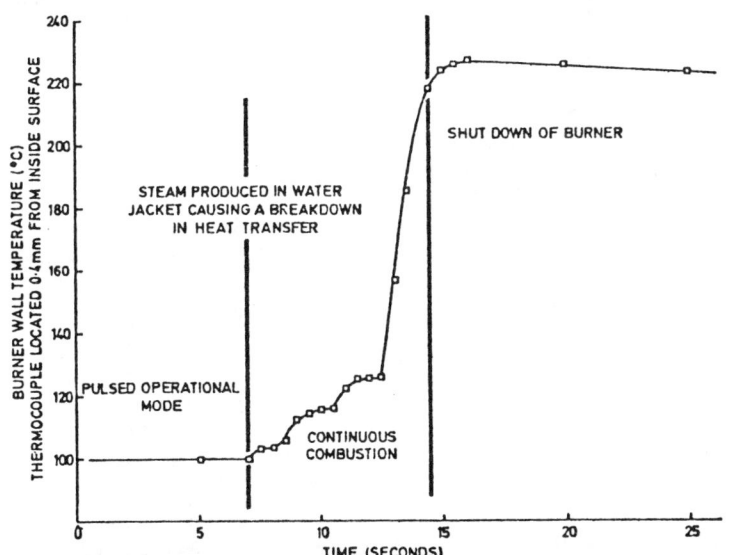

Figure 2b Variation of combustion tube wall temperature with time following a change to continuous operation at 2 bar.

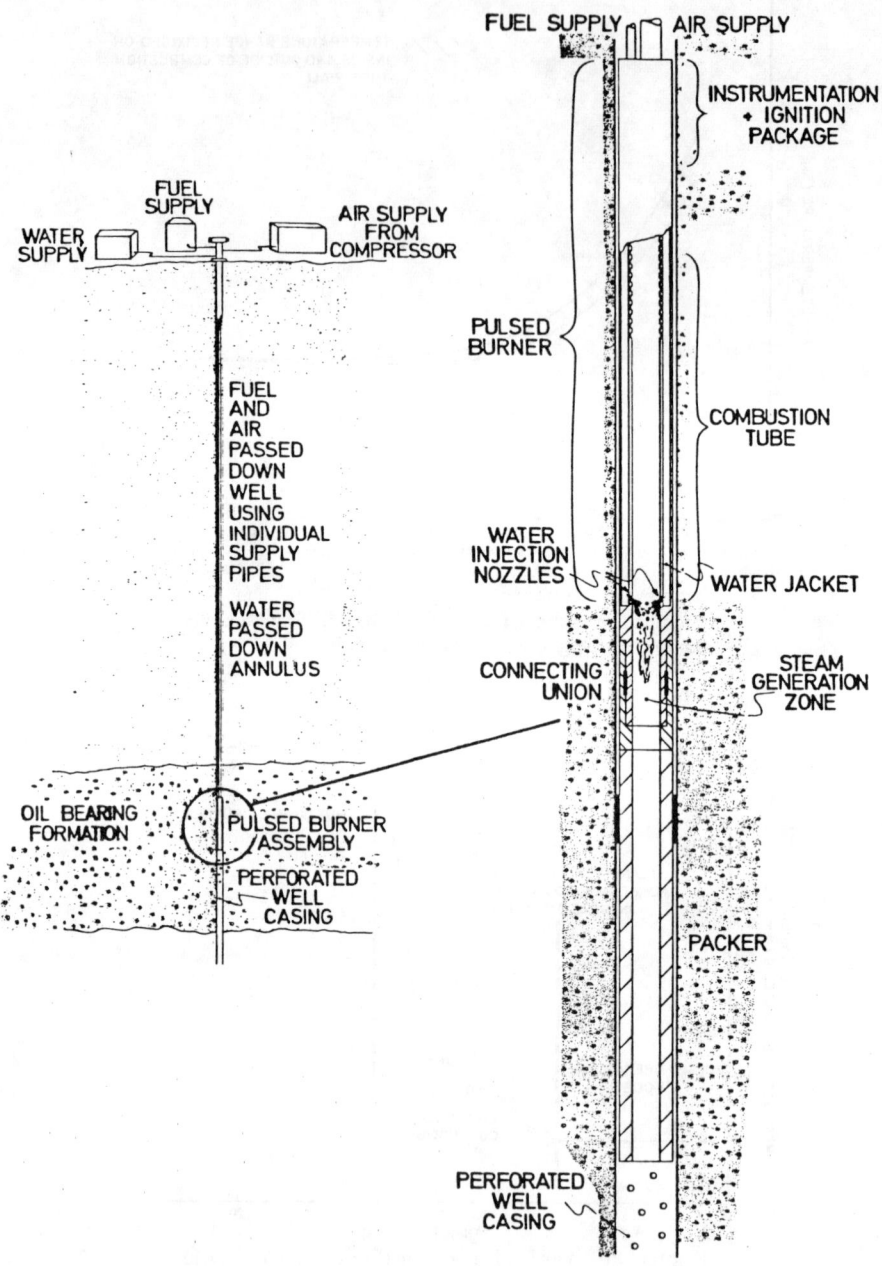

Figure 3 - Schematic diagram of a prototype Pulsed Downhole Steam Generator

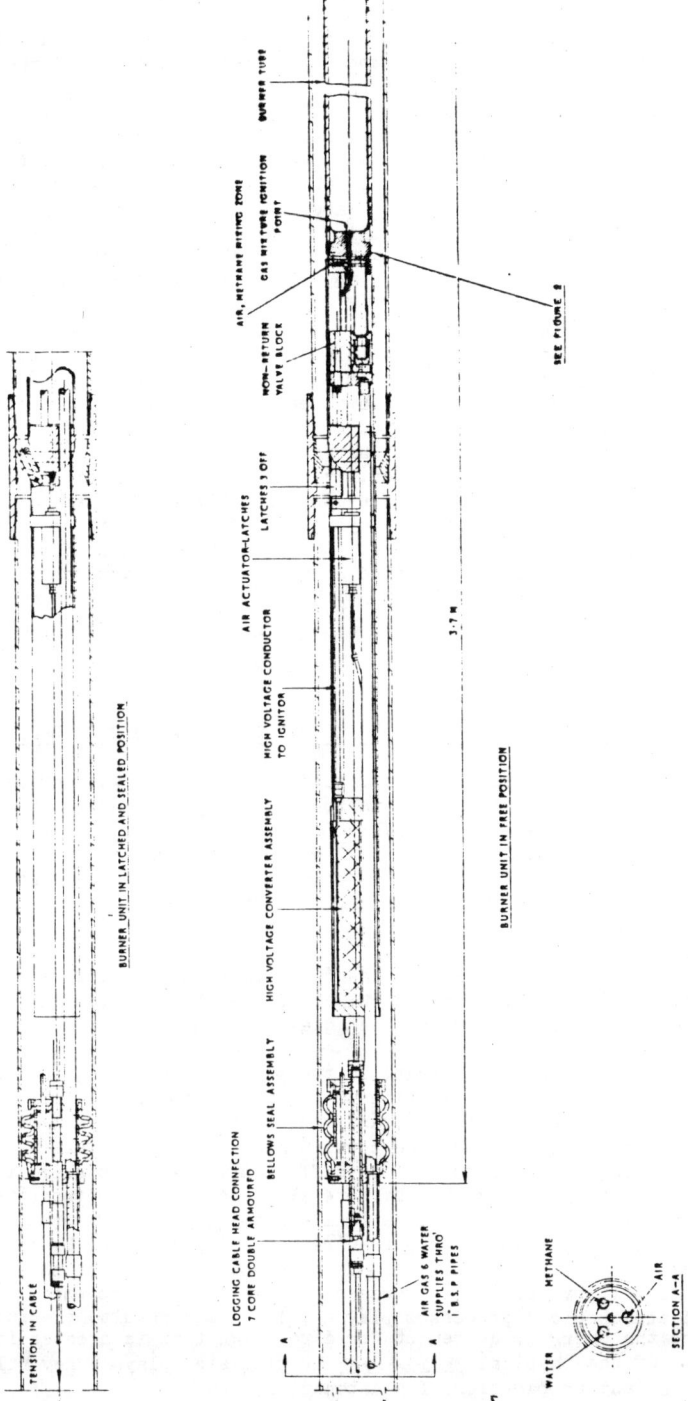

Figure 4 General arrangement diagram of the 0.5 MW burner used in field trials.

OPERATIONAL EXPERIENCE OF A DOWNHOLE GAS FIRED GENERATOR

A series of experiments was carried out in a water well drilled into Bunter Sandstone and located in the Midlands' gas field. The aim of the work was to prove the operation of a pulsed burner in a downhole environment. The burner was designed to operate on locally available methane at depths of up to 75m. The test burner had a maximum diameter of 0.13m, a length of 1.0 m and a combustion tube volume of 0.007 m^3. The firing frequency was 0.5 - 2 Hz, giving heat output in the range 0.125 - 0.5 MW. A general arrangement drawing of the test burner is given in Figure 4. Water, fuel and air were fed in separate lines through a high pressure bellows seal. Ignition was achieved by means of a high energy (10 J) aeroengine igniter situated a short distance downstream of the mixing head; in this design the mixing head is located under the packer.

During the early stages of the trial, difficulties were experienced with the ignition system; short circuits within the high tension units and leads resulted in the test being discontinued after a few hours' operation. Tests were resumed with an improved ignition system which enabled continuous operation for 24 hour periods at a depth of 75 m. Although the burners operated for only a limited period, the work did demonstrate that downhole operation of a pulsed burner is technically possible. In addition, it highlighted a number of deficiencies in the system that required further development. In particular, it was clear that the design of a reliable ignition system is of critical importance to the successful operation of the burner. A closer integration of the burner and packer assembly was also recognised as being an important requirement for downhole burners.

CURRENT PROGRESS IN THE DEVELOPMENT OF A LIQUID FUEL BURNER

Initial development work on the pulsed downhole burner was carried out using hydrogen or methane as fuel. Owing to economic/political constraints, gas supplies may not always be available for downhole steam generation. It is considered essential, therefore, that a downhole burner should be capable of operating on a range of fuels. For this reason, the development has been extended to liquid fuels with the ultimate aim of operating on residual fuel or produced crude.

The operation on a liquid fuel presents a number of difficulties for both pulsed and continuous burners. For the pulsed burner, the greatest difficulty is associated with the initiation of a "quasi" detonation.

It is generally agreed (8) that, in two phase detonations, fuel droplets are broken up by secondary atomisation to produce a combustible mixture of fuel micromist (micron size droplets) and hot oxidiser in the wake of each parent droplet. The large surface area between micromist and oxidiser enables chemical reaction to occur at a rate that is sufficiently rapid to support the incident shock front. A similar mechanism generates the "quasi" detonation. In order to initiate a direct detonation in a fuel/air mixture, a high energy ignition source is required, the precise energy requirement being dictated by the overall induction period of the reaction. In a gaseous system the induction period is governed solely by chemical processes, whilst in a heterogeneous system both physical and chemical processes play a role. As a result, the induction period and hence the energy requirement for direct ignition is greater for heterogeneous systems. Of the physical parameters, droplet size plays a key role in the initiation and propagation of a detonation (9).

The operation of a pulsed burner on liquid fuels has therefore necessitated the development of two key areas of burner operation, namely:

(a) Fuel atomisation and mixing.

(b) High energy ignition systems.

To date, a burner has been successfully fired at atmospheric pressure on liquid kerosene. Within the burner, atomization is effected by a ring of twin fluid atomising nozzles using combustion air as the atomising medium. Detonation is initiated by a high energy source generating approximately 250 J at each ignition event. This system operates on an energy augmentation principle triggered by a plasma plug. The latter has overcome the problem of igniter reliability, discussed previously. The plasma plug is a recent development [10] and operates in a surface discharge fashion, the main discharge taking place within the confines of a small cavity.

In this mode of operation the breakdown voltage is lower than that required for a conventional air gap and, furthermore, exhibits a smaller pressure dependence. Trigger voltages are maintained at an acceptable level even at pressures of 70 bar. The plasma plug has the additional advantages of a variable output, an enclosed electrode assembly and a self-cleaning action. These features have enabled plasma plugs to be operated reliably in laboratory test rigs at high pressure.

Future work will be directed towards developing burners to operate on fuels of lower volatility.

CONCLUSIONS

1. A burner that operates in a pulsed mode, rather than in a continuous mode, offers a number of advantages for downhole steam generation. These include:-

(a) low burner temperature;

(b) relatively simple construction from conventional materials;

(c) high combustion intensities;

(d) high turndown ratio

2. A steam generator incorporating the principle of pulsed combustion has been successfully demonstrated downhole.

3. A reliable ignition system is an essential requirement for a burner operating in a pulsed continuous mode.

4. An atmospheric pulsed burner has been successfully operated on liquid kerosine using a high energy ignition source.

ACKNOWLEDGEMENT

Permission to publish this paper has been granted by the British Petroleum Company Limited.

The authors acknowledge the support given by the European Economic Community.

REFERENCES

1. BADER, B.E. and FOX, R.L.; "The Potential of Downhole Steam Generation to the Recovery of Heavy Oils", UNITAR 1 First International Conference on the Future of Heavy Crude and Tar Sands, Edmonton, Alberta 4 - 12 June 1980.

2. WRIGHT, D.E. and BINSLEY, R.L.; "Feasibility Evaluation of a Downhole Steam Generator"; SPE/DOE 9776, Paper presented to the Second Joint Symposium on Enhanced Oil Recovery of the Society of Petroleum Engineers, Tulsa, Oklahoma, 5 - 8 April 1981.

3. MULAC, A.J., et al; "Project Deep Steam - Preliminary Field Test Bakersfield, California, SAND 80 - 2843.

4. "Downhole unit said to be ready for sale"; Enhanced Recovery Week, May 11, 1981.

5. "Test slated for downhole steam generator"; Oil and Gas Journal, March 30, 1981.

6. JOHNSON, D.R., et al; Project Deep Steam Quarterly Report, July 1 - September 30, 1980.

7. LEE, J.H.S. and MOEN, I.O.; "The Mechanism of Transition from Deflagration to Detonation in Vapour Cloud Explosions", Prog. Energy Combust. Sci., (1980) $\underline{6}$, 359 - 389.

8. DABORA, E.K. and WEINBERGER, L.P.; "Present Status of Detonations in Two-phase Systems"; Acta Astronautica (1974), $\underline{1}$, 361 - 372.

9. LU, P.L., SLAGG, N. and FISHBURN, B.D.; "Relation of Chemical and Physical Processes in Two-phase Detonations", Acta Astronautica (1979), $\underline{6}$, 815 - 826.

10. ASIK, J.R., PIATKOLSKI, P., FOUCHER, M.J. and RADO, W.G.; "Design of a Plasma Jet Ignition System for Automotive Application "(No 770355), Society of Automotive Engineers, International Automotive Engineering Congress and Exposition Cobo Hall, 28th February 1978.

HOT CAUSTIC FLOODING

R. JANSSEN-VAN ROSMALEN and F. Th. HESSELINK

*Koninklijke Shell Explorative en Produktie Laboratorium,
Rijswijk, The Netherlands (Shell Research B.V.)*

SUMMARY

A chemical recovery method used as a follow-up to or in combination with thermal methods is restricted to processes that are not too sensitive to elevated temperatures and temperature gradients. Caustic flooding is one of the few processes which can meet these requirements. Three different recovery mechanisms can be created by varying the salt content and/or adding extra chemicals to the caustic slug. The caustic process at high salinities is regarded to offer the best prospects for "hot caustic", since the process does not require extra chemicals (e.g. polymer) for mobility control.

In our studies on caustic floods, laboratory experiments have shown that an increase in temperature can be favourable for in-situ emulsification and, once an emulsion is formed, can drastically influence the flow properties of the emulsion. For a caustic flood following a hot water drive, a minimum shear rate was found to be necessary for emulsification.

In the caustic flooding experiments where a mobile emulsion was formed, oil recoveries superior to those of a comparable hot-water flood have been obtained. In one-dimensional packs of reservoir sand a hot caustic flood recovered 16-18% PV additional oil after a hot water drive. The sweep-improving effects of a sufficiently mobile caustic emulsion are expected to give a substantially higher additional recovery in a real three-dimensional flood.

1. INTRODUCTION

Caustic flooding is an enhanced oil recovery method which is based on the principle that organic acids, naturally occurring in some crude oils, can react with the alkali of the caustic injected. This chemical reaction leads to the formation of surfactants at the oil-water interface, resulting in a decrease in interfacial tension between the oil and the water phase and in in-situ emulsification when a caustic solution of suitable alkalinity is injected into an oil-bearing formation.

A basic requirement for the caustic recovery method is that the oil contains a sufficient amount of natural acids. In this connection an acid number of about 1 mg or higher is desirable (1) (2) but active caustic systems have also been reported at lower acid numbers (2). This activity is, among other things, related to the percentage of natural acids having a sufficiently high molecular weight to be effective in reducing the oil-water interfacial tension.

Several mechanisms have been proposed to describe the effect of caustic injection on the recovery of oil. Johnson (3) has given an overview of the different mechanisms involved in caustic flooding, such as emulsification and entrapment (4), wettability reversal (2) (5) (6) and emulsification and entrainment (2). Recently Mayer et al (7) have summarised data on field tests, and related the results to caustic concentrations and salt contents of the floods. Their status report shows that different recovery mechanisms are dominating in the various caustic flooding field trials.

In the present paper the effect of an increase in temperature on the caustic flooding process will be discussed. A method of this kind is indicated in Ref. 8. The application of "hot caustic" may be considered for the following reasons. In some cases a hot caustic injection may yield increased oil recoveries, while at reservoir temperature the injection would not lead to a better performance in comparison with a plain water drive. It should be noted that injection of hot caustic into an ambient temperature reservoir may require preheating of the reservoir, since the thermal front will move through the reservoir at a much slower rate than the caustic front. In other cases the reservoir may already have been preheated by a steam drive or a hot water drive. A chemical recovery method to follow up these thermal methods is restricted to processes that are not too sensitive to high temperatures and temperature gradients as occur in preheated zones. Such conditions adversely affect a surfactant or polymer flood. The prospects offered by a foam or caustic process are more promising under conditions as described above.

The scope of the present study is to provide an indication of:
- which caustic recovery mechanism can best be chosen at elevated temperatures.
- how the parameters that are important for a caustic process are influenced by an increase in temperature.
- the oil recovery when applying hot caustic flooding.

The investigation of caustic flooding in conjunction with a thermal project is especially relevant, since the high-acid-number crudes suitable for caustic flooding are usually low-gravity crudes (1) for which thermal EOR processes are often considered. The preceding steam flood may even have increased the acid number of the crude.

2. THE CAUSTIC PROCESS

2.1. Principal recovery mechanisms
The interfacial tension between the oil and the water phase as well as the type of emulsion formed during caustic flooding have been found to be dependent on the salinity of the caustic injected (2) (4). This means that the recovery mechanism is likewise affected by salinity. Reisberg and Doscher (9) have reported that caustic solutions used in conjunction with surfactants were effective in increasing oil displacement. The combination of an alkaline solution with a co-surfactant was also mentioned for the formation of a micro-emulsion to be injected into a reservoir (10). Research on the application of polymers in caustic flooding to provide mobility control is in progress (7) (11) (12).

In order to catalogue the variety of alternatives for the caustic process, we would propose to classify the caustic recovery mechanisms (see Fig. 1) in parallel with the various mechanisms identified in surfactant flooding, as follows:
- The under-optimum system.
  At low salinity of the caustic injected a non-viscous oil-in-water emulsion is formed. This system is often formed at sodium ion

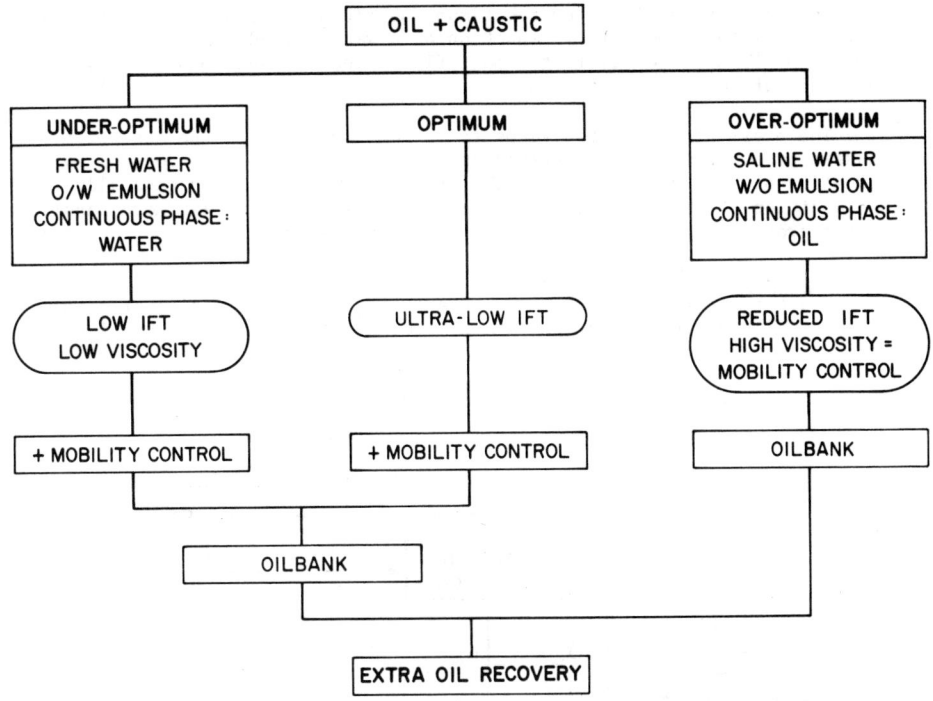

FIG.1: CAUSTIC RECOVERY MECHANISMS

concentrations lower than 0.5 mol/l. Oil/water interfacial tensions (IFT) can be very low ($< 10^{-2}$ mN/m). Oil recovery can be increased either by entrapment of emulsified oil drops in the pore throats, reducing water mobility (4), or by very low interfacial tension, leading to reduced residual oil saturation through a favourable capillary number (13). In the first case an improvement in sweep efficiency is achieved. In general, however, the latter mechanism is more common; it may lead to oil bank formation in cases where interfacial tensions are sufficiently low. Mobility control (polymer, foam) is required for stable displacement of such an oilbank.

The optimum system.

By optimisation of the amount of salt and/or the addition of surfactants (9), interfacial tensions may become ultra-low, so that capillary-trapped oil is mobilised and an oilbank is formed. This system also requires mobility control.

The over-optimum system.

At high salinities a viscous water-in-oil emulsion is formed (2). Interfacial tensions are reduced somewhat. Once an emulsion bank has formed, caustic fingers through the emulsion bank are converted into emulsion, sealing off the finger. Extra oil recovery is obtained by improved mobility, leading to a better areal and vertical sweep efficiency in the reservoir. In addition, the increased viscosity of the viscous emulsion drive, combined with the reduced IFT, may (at least in the laboratory) cause such an increase in the capillary number that residual oil becomes mobilised, leading to oilbank formation ahead of the emulsion drive.

An application of the over-optimum caustic system, not shown in Fig. 1, is the possible permeability reduction of a water-flooded zone by emulsion formation. This would divert the main flow direction and consequently improve the sweep efficiency.

The over-optimum system seems to be most suitable in the case of a hot caustic process, since this system provides its own mobility control by the formation of a viscous emulsion. The other two systems would generally require the addition of extra chemicals, which would make the process more sensitive to temperature, and also more complicated and costly.

2.2. Parameters for the over-optimum system

The parameters that are of importance for caustic flooding are schematically given in Fig. 2. When a caustic solution is injected into a

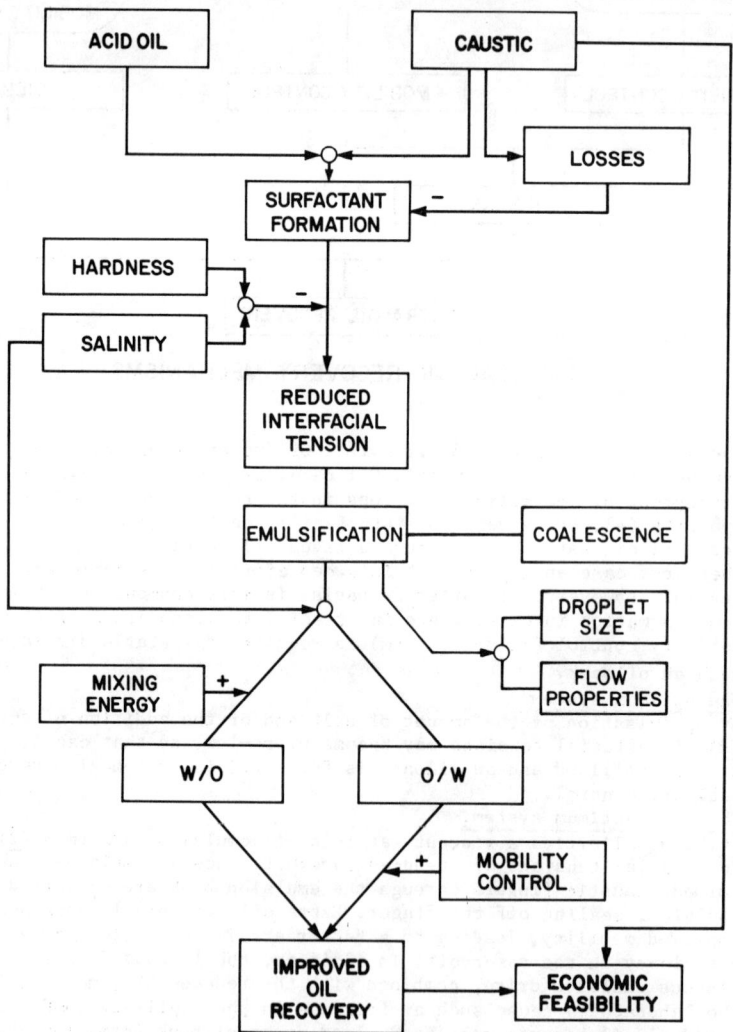

FIG. 2: INTERACTION-DIAGRAM FOR CAUSTIC FLOODING

reservoir, part of the caustic is used to form surfactants, and part of it is depleted by interaction with rock and reservoir water. This interaction is highly dependent on type of rock, pH and composition of the caustic solution, reservoir salinity and temperature (14). If depletion is excessive, it may accordingly retard or even prevent the onset of increased oil production (12).

The surfactants formed in situ, will reduce oil-water interfacial tension. Since the water phase is saline, this reduction is not drastic. Divalent ions (Ca^{++}, Mg^{++}) may be detrimental to the process, leading to the formation of less interfacially active soaps (see Fig. 3). Calcium is much more detrimental than magnesium in this respect. The calcium concentration in the water should therefore be controlled by the addition of chemicals to the flood water. For instance, addition of soda ash (Na_2CO_3) to the alkaline water causes most of the calcium ions originally present to precipitate as calcium carbonate.

As a result of reduced interfacial tensions, emulsification is promoted under the action of interfacial tension gradients (Marangoni effects which may result in turbulence at the oil/water interface) and/or by an extra mixing effect (see Fig. 2) generated by external shear forces. In general, no extra mixing energy is required to form an oil-in-water emulsion, since in that case interfacial tensions are much lower. Wettability reversal of the rock to oil-wet (2) may play a role in the formation of a stable water-in-oil emulsion. Flow behaviour of the emulsion, in relation to droplet size is an important aspect for the overall flooding behaviour.

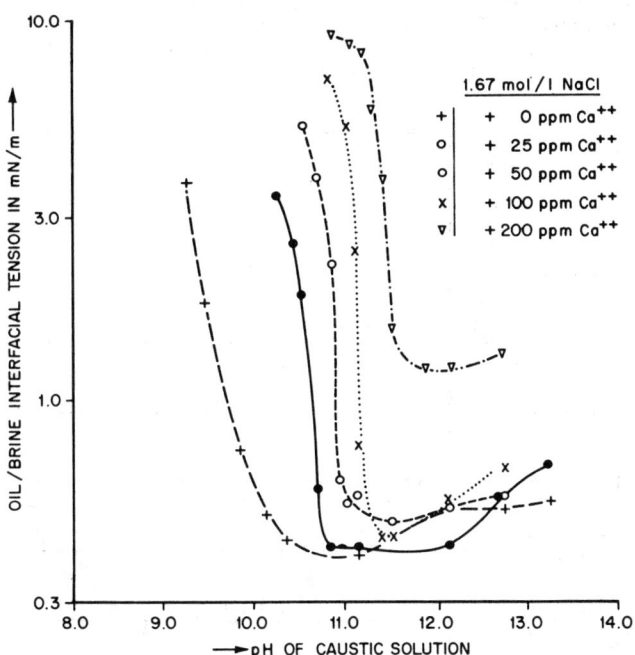

FIG. 3: ILLUSTRATION OF INTERFACIAL TENSIONS AS A FUNCTION OF pH AT DIFFERENT CALCIUM CON-CENTRATIONS (Acid number of crude used : 3.0)

Besides the emulsification phenomena, coalescence of the emulsion also contributes to the performance of a caustic flood. Since an emulsion is a thermodynamically unstable system, phase separation occurs through coalescence of the droplets. The rate of coalescence is dependent on temperature, flow rate, interfacial properties and viscosity. In an effective caustic process the coalescence rate is balanced by the emulsification rate.

Extra oil recovery by improved sweep efficiency and oil bank formation in relation to the caustic required for an effective process will determine the economic feasibility of a caustic flooding process.

2.3. The effect of an increase in temperature

An increase in temperature will have a considerable impact on the ease of emulsification (the interaction of Marangoni effects and external shear forces), the coalescence rate, the flow behaviour of the emulsion and the total caustic losses (see Fig. 2).

Emulsification and coalescence

The reduced oil viscosity at higher temperatures promotes the diffusion of organic acids to the oil/water interface. Thus the acid–caustic reaction is accelerated and the mixing process due to Marangoni effects is promoted. As a result, water droplets become more readily dispersed in the oil. In some cases, however, as will be shown in section 3, extra mixing energy is needed to further break up the water droplets to achieve a better emulsion stability. This energy is to some extent provided by the flow in the porous medium. If velocity gradients and ensuing shear forces are large enough, interfacial forces are no longer able to keep fluid particles intact, and they are broken up into smaller droplets. The theory of deformation and break-up of a droplet in a flow field was first formulated by Taylor (16) and modified by e.g. Karam et al. (17).

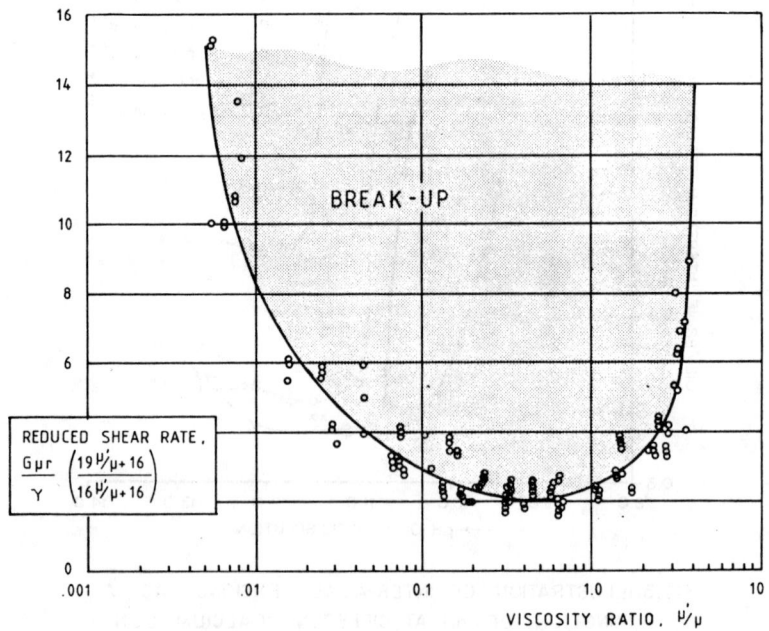

FIG. 4 : BREAK-UP OF DROPLETS IN A SIMPLE SHEAR FIELD

Taylor showed that the drop behaviour only depends on the two dimensionless parameters μ'/μ and $Gr\mu/\gamma$, where μ' is the viscosity of the discontinuous phase, μ of the continuous phase, G the shear rate, r the diameter of the droplet and γ the interfacial tension. In accordance with this, Karam et al. have plotted their results of droplet break-up of different fluids as a function of these two dimensionless parameters (see Fig. 4). Irrespective of the system studied, a single curve should be obtained by such a dimensionless plot. Droplet break-up occurs above this curve.

In connection with the caustic process the following main conclusions can be drawn from this plot. If the interfacial tension is increased, for example by divalent ions present in the reservoir water, the shear rate in the porous medium needs to be increased by the same factor to cause similar droplet break-up (see Fig. 4). It is therefore very important that most of the divalent ions are precipitated using a suitable buffer solution. Secondly, the break-up of a liquid droplet occurs readily when the viscosity ratio between the water and oil phase (μ'/μ) is of the order of 0.2 to 1 (see Fig. 4). This implies that not only the acid number, but also the viscosity of the oil under consideration is an important criterion for the caustic process. By the application of heat the difference in viscosity between the oil and the water phase becomes less pronounced, resulting in easier emulsification.

In a good caustic process a balance exists between emulsification and coalescence. Since an increase in temperature will enhance the coalescence rate, the temperature can affect the performance of a caustic flood in this way.

Flow behaviour of the emulsion

The temperature effect on the flow behaviour of the emulsion is important in the following respect. The apparent viscosity of the emulsion should preferably be somewhat higher than that of the oil phase in order to provide mobility control, and it can be much higher in the case of diversion of the main flow from a water-flooded area (see also section 2.1). The strong dependence of apparent viscosity on temperature will be shown in section 3.

Caustic loss

Caustic in the form of sodium hydroxide has been shown to strongly interact with the rock at elevated temperatures (14). On the positive side, the dissolution interaction generates in situ water-soluble silicates which may have a beneficial effect on oil recovery (15). In contrast to these high losses resulting from sodium hydroxide, we found considerably lower consumptions when using carbonate buffer solutions. This is probably due to the lower initial pH values required when a buffer solution is applied instead of sodium hydroxide.

3. EXPERIMENTAL RESULTS AND DISCUSSION

The crude used for the caustic flooding and emulsification experiments was characterised by an acid number of 3.6 and a viscosity of 170 mPa.s(=cP) at reservoir temperature (42°C). The reservoir brine contained 76115 ppm TDS, of which 3200 ppm Ca^{++} and 960 ppm Mg^{++}.

Optimum emulsion stability was found at a pH of 9.5 of the caustic solution. The caustic solution was a carbonate/bicarbonate buffer with sodium chloride added (total Na^+ content: 1.25 mol/l). The effectiveness of such a caustic solution in recovering the oil has been studied in Bentheim sandstone cores (permeability: 1.5 μm^2) and in sandpacks of reservoir sand (permeability: 3-9 μm^2). In the case of sand packs more divalent ion exchange can be expected, by which emulsification could be hampered.

The caustic flooding experiments were carried out at high initial oil saturation at initiation of the caustic injection and at low oil saturations, i.e. after a water drive. The effect of temperature, shear rate and permeability on the emulsification process and the corresponding oil recovery has been studied. Shear rates have been calculated from (18):

$$G = 4u/\sqrt{(8k\Phi)} \tag{1}$$

where u is the flooding rate, k the permeability and Φ the porosity of the porous medium.

Since the flooding experiments in the porous media used represented one-dimensional floods, the oil recoveries observed only accounted for possible extra oil recovery because of slightly reduced residual oil saturation and/or oil bank formation (see Fig. 1) on account of the steep pressure gradients in the emulsion zone. A possible improvement in sweep efficiency for a real three-dimensional caustic flood can be expected when emulsion formation in combination with pressure build-up across the porous pack was observed.

3.1. Caustic flooding experiments at high initial oil saturations

Caustic flooding experiments were carried out at elevated temperature, $80^{\circ}C$, and for comparison at a reservoir temperature of $42^{\circ}C$. Caustic was injected after saturating the porous medium with synthetic reservoir brine and subsequent flooding with tank oil to irreducible water saturation.

The experiments in Bentheim sandstone (see Table I) were performed at a flooding rate of 1.06×10^{-5} m/s (= 3 ft/day). At a temperature of $42^{\circ}C$ (exp.1) a considerable pressure build-up over the core length was registered during the caustic flood, indicating the formation of a viscous emulsion. The low mobility of this emulsion may explain why the oil recovery in this caustic drive was not significantly improved over that of a plain water drive (compare exp. 1 and 2). In a caustic drive experiment at $80^{\circ}C$ (exp. 3) the pressure initially increased, pointing to in-situ emulsification, and afterwards decreased when most of the oil, followed by some emulsion, had been produced, indicating that the emulsion in this case did not have too low a mobility. The

TABLE 1 - DISPLACEMENT EXPERIMENTS IN POROUS PACKS

flooding rate: $1.06 * 10^{-5}$ m/s (= 3 ft/day), except exp. 7b: $0.35 * 10^{-5}$ m/s

Exp. No.	Caustic/water drive	Temp. [$^{\circ}C$]	Shear rate at wall	Emulsion formed	Initial oil saturation, S_{oi}	Oil recovered per cent of S_{oi}	
						After 1 PV	Final
Experiments in Bentheim sandstone cores (porosity: 0.2; permeability 1.5 μm^2, length: 30 cm)							
1	caustic dr.	42	27.4	YES	0.95	45	48
2	water dr.	42	27.4	-	0.95	39	45
3	caustic dr.	80	27.4	YES	0.78	82	83
4	water dr.	80	27.4	-	0.78	42	47
Experiments in sand packs (porosity:0.4 and permeability:9.5 μm^2,except exp.7b and 8; length:22 cm)							
5	caustic dr.	42	7.8	NO	0.79	40	48
6	water dr.	42	7.8	-	0.79	48	56
7a*	caustic dr.	80	7.8	YES	0.84	64	75
7b*	caustic dr.	80	4.2	YES	0.87	63	73
8*	water dr.	80	27.4	-	0.78	40	41

* exp. no. 7b and 8: permeability 4 μm^2; porosity 0.4.

emulsion obviously acted as a viscous phase, forcing the oil out of the porous pack. The oil recovery (83%) was far superior to that of the comparable water-drive experiment (47%, see exp. 4). This large difference in recovery and pressure behaviour between a caustic drive and a water drive at 80°C was confirmed in duplicate experiments.

In reservoir sand at 42°C the caustic drive showed no pressure build-up, and the oil recovery was even lower than in the corresponding water drive (see exp. 5 and 6 in Table I). At 80°C the drives were conducted at two different flooding rates: $0.35 * 10^{-5}$ and $1.06 * 10^{-5}$ m/s (1 ft/day and 3 ft/day, resp.). Although shear rates (see eq. 1) were different (cf. exp. 7a with exp. 7b, Table I), both cases showed some pressure build-up, pointing to the in-situ formation of an emulsion bank. The oil recoveries were in both cases equally high. The corresponding hot water drive (see exp. 8) yielded a 20% (of OIP) lower oil recovery. At higher permeabilities, however, less difference in recovery was found between the caustic and hot water drives, probably because of the effectiveness of the viscosity reduction of the oil (from 170 mPa.s (= cP) at 42°C to 27 mPa.s at 80°C) in porous media of high permeability.

These experiments in sandpacks indicate that at higher temperatures (80°C) mixing induced by Marangoni effects and also diffusion processes may be sufficient for the formation of a small emulsion bank, probably because of a large contact area at the interface between the oil and the caustic slug just after injection. So this process at high initial oil saturations was found not to be dominated by external shear forces.

3.2. Caustic flooding experiments after a (hot) water drive

The flooding experiments were performed at reservoir temperature (42°C) and at elevated temperatures. Crude and brine were injected in order to bring the porous system at connate water saturation and subsequently reduce it to a low oil saturation. The water drives prior to the caustic injection were performed at the same temperatures as the caustic flood.

The results of these tests are given in Table II. They show that in-situ emulsification can indeed occur at low oil saturations, and that this process is critically dependent on shear rate, in contrast to the process at high initial oil saturations. Both the flooding rate using one particular type of porous material, and the permeability times porosity of the core are important

TABLE II - CAUSTIC FLOODING EXPERIMENTS AFTER A HOT WATER DRIVE

Exp.	Permeability k [μm²]	Temp. [°C]	Flooding rate* x 10⁻⁵ m/s	Shear rate at wall	Emulsion formed	Oil saturation after water drive, S_{olp}	Oil recovered after 1 PV fraction of S_{olp}
Experiments in Bentheim sandstone cores (porosity: 0.20; length: 30 cm)							
1	1.5	88	3 x 0.35	27.4	YES	0.40	64
2	1.5	80	3 x 0.35	27.4	YES	0.50	51
3	1.5	80	0.35	9.1	NO	0.38	20
4	1.5	60	3 x 0.35	27.4	YES	0.40	44
5	1.5	42	3 x 0.35	27.4	YES	0.39	11
Experiments in sand packs (porosity: 0.40; length 22 cm, exp. 8 : 42 cm)							
6	9.1	80	3 x 0.35	7.6	NO	0.36	7
7	3.2	80	8 x 0.35	35.4	YES	0.41	41
8	3.0	80	6 x 0.35	27.1	YES	0.37	42
9	3.2	60	5 x 0.35	22.4	YES	0.50	10
10	3.2	60	8 x 0.35	35.8	YES	0.40	12

* $(0.35 \times 10^{-5}$ m/s = 1 ft/day).

582

parameters for emulsification (cf. exp. 2 and 3, and exp. 2 and 6). In the
experiments where no emulsion is formed, the additional oil recovery is not
higher than obtained by a prolonged water drive. Also in experiments at
sufficiently high shear rates where an emulsion is formed, but which were run
at lower temperatures, the additional oil recovery is poor (see exp. 5, 9 and
10). This is for the same reason as given for the experiments at high initial
oil saturations, which were carried out at relatively low temperatures, i.e.
the emulsion mobility is too low. The result is that at low temperatures most
of the emulsion formed remains in the core, and that the caustic solution
breaks through the emulsion bank.

    Improved caustic flood recoveries are obtained at higher temperatures,
where an emulsion bank of higher mobility is formed; see Table II. This is
supported by viscosity measurements. At 80°C the viscosity of the emulsion
produced in the effluents during the caustic flood was found to be 200 mPa.s,
in contrast with an emulsion viscosity of 2200 mPa.s at 60°C.

    As a second effect of temperature, the amount of emulsion formed during
flooding at one particular shear rate was found to decrease with increasing
temperature. This phenomenon may be caused by enhanced coalescence, which is
common for emulsions at higher temperatures.

    Under optimal conditions for the formation of a stable and at the same
time mobile emulsion bank, a sharp increase in oil saturation in front of the
emulsion bank could be observed in both Bentheim sandstone and reservoir sand.
This resulted in general in a decrease in water cut from about 100% to 60%, as
is illustrated in Fig. 5 (exp. 1). After the emulsion bank had been produced,
the oil production stopped completely. In Bentheim sandstone the oil recovery
was far superior to that of a comparable hot water drive (≈ 25% of OIP extra
oil recovery at 80°C). In reservoir sand the caustic flood performance at 80°C
was similar to that in Bentheim sandstone in terms of pressure behaviour and
oil bank formation, yielding an extra oil recovery after a waterflood of about
20% of OIP (≈ 18% PV), as illustrated in Fig. 6 (exp. 8). For comparison, the
production curve of a caustic flood at 42°C is given in the same figure.

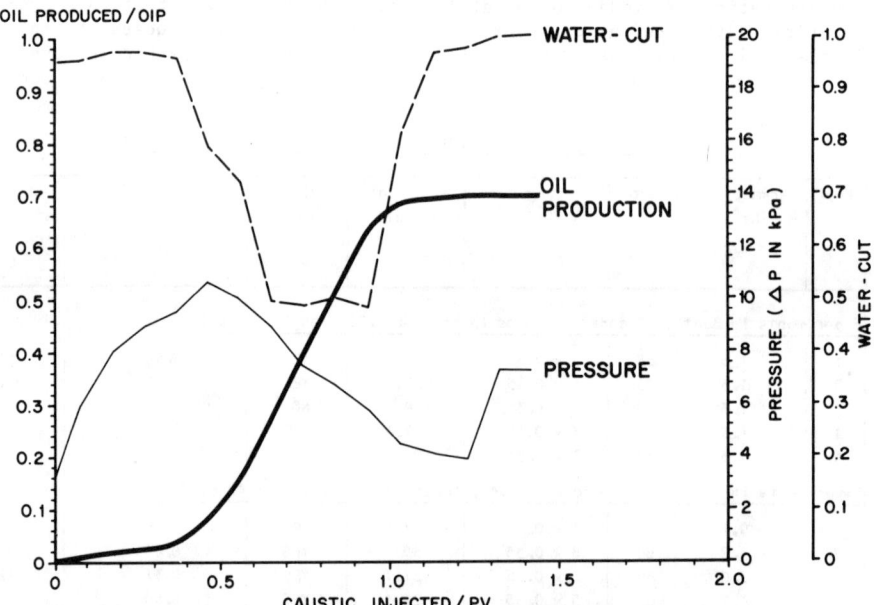

FIG.5:CAUSTIC FLOODING EXPERIMENT IN A BENTHEIM SANDSTONE CORE
FOLLOWING A WATERDRIVE.

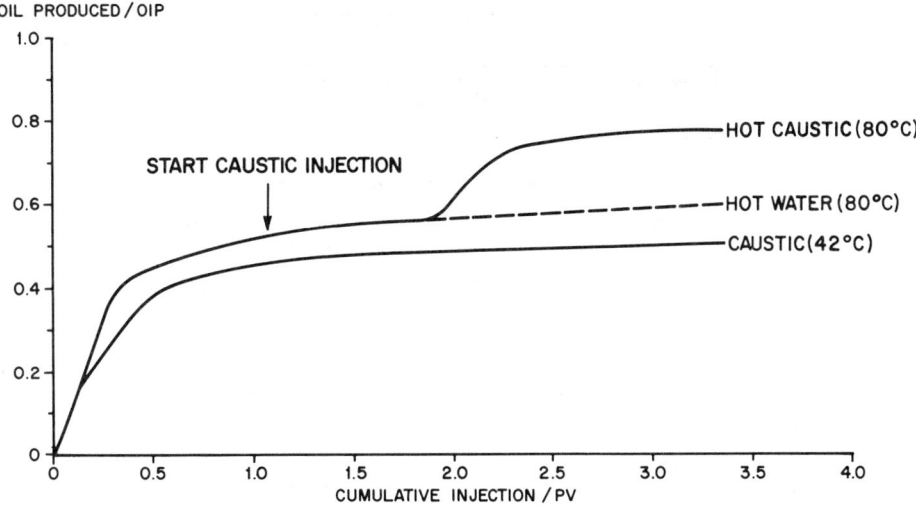

FIG.6:FLOODING EXPERIMENTS IN SANDPACKS.

4. COMPARISON OF THE EMULSIFICATION IN A COUETTE APPARATUS AND IN A POROUS MEDIUM

Since the emulsification process at low initial oil saturations was found to be dependent on shear rate, we decided to study emulsion formation at different shear rates in more detail, using a thermostatted Couette apparatus. This apparatus (Fig. 7) consists of two concentric cylinders, the inner one of which can rotate. The Couette apparatus is particularly suitable for this investigation, since the shear rate is almost constant throughout the small gap between the inner and outer cylinder. The emulsification behaviour of a caustic solution in contact with crude in the Couette apparatus (see Table III) has been compared with that in a porous medium at relatively low oil saturations (Table II).

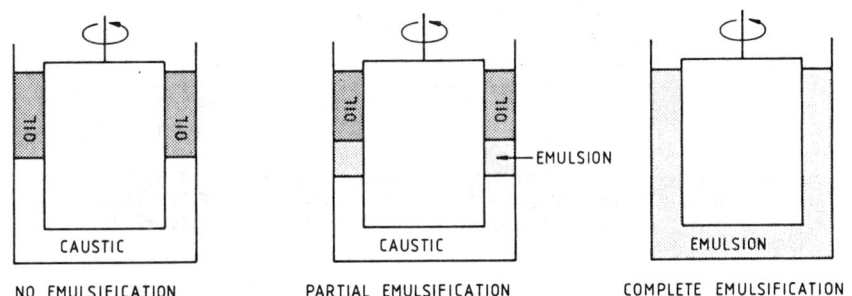

FIG.7:EMULSION FORMATION IN A COUETTE APPARATUS

The emulsions formed in the porous medium and in the Couette apparatus were characterised by droplet size analysis (using a HIAC 520 particle-size analyser) and by microphotography. In general, a fairly good similarity between the emulsions was observed. In both cases peaks were found at a droplet size of 2.8 and 4.3 μm (see Fig. 8).

584

TABLE III - EMULSIFICATION IN COUETTE APPARATUS

| Shear rate | Formation of W/O emulsion | | |
$[s^{-1}]$	$40^\circ C$	$60^\circ C$	$80^\circ C$
28	YES (16% free water)	YES (30% free water)	NO
38	YES	YES (5% free water)	NO
46	YES	YES	YES
68	YES	YES	YES
113	YES	YES	YES

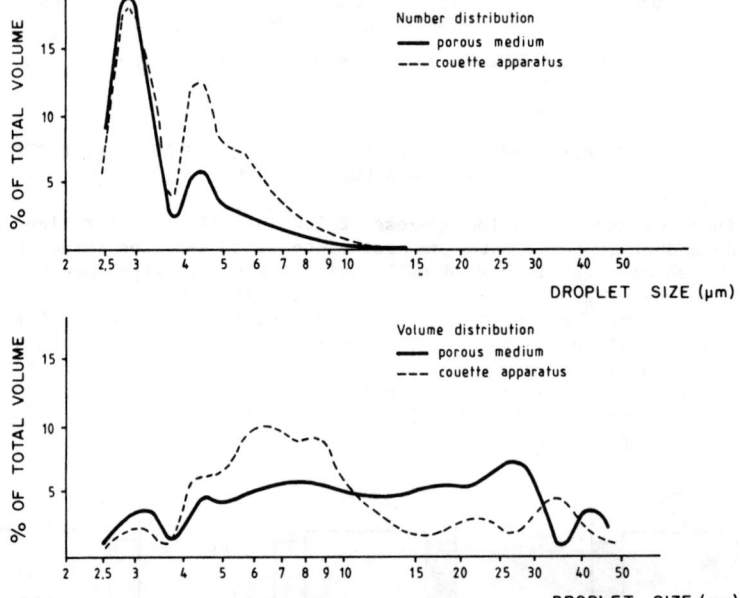

FIG. 8 : DROPLET SIZE DISTRIBUTIONS OF EMULSIONS FORMED AT 60°C IN
A POROUS MEDIUM (Exp. 9, Table II) AT A SHEAR RATE OF 22S⁻¹, AND
IN A COUETTE APPARATUS AT A SHEAR RATE OF 46S⁻¹ (Table III)

At the onset of the emulsification process in the Couette apparatus, the development of instabilities at the interface between the caustic solution and the oil was observed, resulting in the formation of relatively large droplets. These droplets only broke up further and formed a stable emulsion when the shear rate exceeded a critical value (see Fig. 4, section 2.3). Dependent on the shear rate applied, three situations could be distinguished (see Fig. 7): (i) the two phases remained completely separated, (ii) only part of the two phases was emulsified, (iii) complete emulsification.

The results of the Couette experiments performed at different temperatures and shear rates (see Table III) indicate that the higher the temperatures, the higher are the shear rates needed for emulsification. Parallel to this, it was found that in the core floods an increase in temperature at a fixed shear rate resulted in a decrease in the amount of emulsion produced (see section 3.2). This phenomenon can be attributed to reduced stability, which is in general observed with emulsions at higher temperatures, probably due to enhanced coalescence.

In the core flood tests (Table II) stable emulsions were formed when the shear rates exceeded about 20 s^{-1}, whereas in the Couette apparatus this threshold value was somewhat higher, i.e. about 40 s^{-1}. The lower threshold value in porous media may arise from the fact that, apart from simple shear flow, extra velocity changes involved in the flow through constrictions and dilatations also contribute to the emulsification process.

5. CONCLUSIONS

1. The caustic process at high salinities is considered to be most suitable in the case of a hot caustic drive, since this process provides its own mobility control by the formation of a viscous water-in-oil emulsion.
2. High temperatures can be favourable for the onset of in-situ emulsification.
3. Apparent viscosities of water-in-oil emulsions were found to be very temperature-dependent (in our specific case ≈ 200 mPa.s (= cP) at 80°C, ≈ 2200 mPa.s at 60°C).
4. In laboratory experiments on caustic flooding where an emulsion bank was formed, but where the emulsions were too viscous, generally no extra oil recovery was observed. This system may, however, be feasible in a water-flooded area to decrease the permeability and consequently divert the main flow direction to areas of higher oil saturation. This would imply injecting a small slug of caustic, followed by (hot) water or steam.
5. At low oil saturation, which implies a caustic flood after a water drive, the emulsification process was found to be critically dependent on shear rate (α flooding rate, permeability, porosity). The shear rate probably becomes less important if an oilbank has built up in front of the emulsion bank, since from that moment on a situation resembling caustic flooding at high initial oil saturation is reached.
6. An emulsion bank of sufficient mobility was effective in displacing the oil, and resulted at low initial oil saturation in the formation of an oilbank. In one-dimensional packs of reservoir sand extra oil recoveries of about 18% PV have been achieved as compared to water floods of the same temperatures.
7. Apart from a reduction in S_{or}, in a three-dimensional flood a sufficiently mobile caustic emulsion is expected to yield additional recovery because of improved sweep efficiency over that of a water drive.

REFERENCES

1. JENNINGS, H.Y., Jr.;
   "A study of caustic solution-crude oil interfacial tensions",
   Soc. Pet. Eng. J. (June 1975), 197.
2. COOKE, C.E., Jr., WILLIAMS, R.E. and KOLODZIE, P.A.;
   "Oil recovery by alkaline waterflooding",
   J.Pet. Tech. (Dec. 1974) 26 (12), 1365.
3. JOHNSON, C.E., Jr.;
   "Status of caustic and emulsion methods",
   J.Pet. Tech. (Jan. 1970), 85.

4. JENNINGS, H.Y., Jr., JOHNSON, C.E., Jr., and McAULIFFE, C.D.;
   "A caustic waterflooding process for heavy oils",
   J.Pet. Tech. (Dec. 1974), 1344.
5. WAGNER, O.R. and LEACH, R.O.;
   "Improving oil displacement by wettability adjustment",
   Petr. Trans. AIME (1959) 216, 65.
6. EHRLICH, R., HASIBA, H.H. and RAIMONDI, P.;
   "Alkaline waterflooding for wettability alteration – Evaluation a
   potential field application",
   J.Pet. Tech. (Dec. 1974) 1335.
7. MAYER, E.H., BERG, R.L., CARMICHAEL, J.D. and WEINBRANDT, R.M.;
   "Alkaline injection for EOR – A status report",
   SPE 8848, (April 1980).
8. SCHULZ, W.;
   "Verfahren zur Förderung von Erdöl",
   Deutsches Patentamt, Auslegeschrift 26.02.450, Bekanntmachungstag
   (1.6.1978).
9. REISBERG, J. and DOSCHER, T.M.;
   "Interfacial phenomena in crude oil–water systems",
   Producers Monthly (Nov. 1956) 21, no. 2, 43.
10. CHANG, H.L.;
    "Oil recovery by micro-emulsion injection",
    US patent 4,008,769 (Feb. 22, 1977).
11. SZABO, M.T.;
    "An evaluation of water–soluble polymers for secondary oil recovery –
    Part I",
    J.Pet. Tech. (May 1979) 553.
12. DE ZABALA, E.F., VISLOCKY, J.M., RUBIN, E. and RADKE, C.J.;
    "A chemical theory for linear alkaline flooding",
    SPE 8997 (May 1980).
13. MELROSE, J.C. and BRANDNER, C.F.;
    "Role of capillary forces in determining microscopic displacement efficiency
    for oil recovery by waterflooding",
    J.Can. Petr. Techn. (Oct. 1974) 1.
14. SYDANSK, R.D.;
    "Elevated temperature caustic sandstone interaction – implications for
    improving oil recovery",
    SPE-DOE 9810 (April 1981) 517.
15. CAMPBELL, T.C.;
    "A comparison of sodium orthosilicate and sodium hydroxide for alkaline
    waterflooding",
    SPE 6514, 47th Annual Calif. Reg. Meeting, Bakersfield CA.
    April 13–15, 1977.
16. TAYLOR, G.I.;
    "The information on emulsions in definable fields of flow",
    Proc. Roy. Soc. London (1934) 146A, 501.
17. KARAM, H.J. and BELLINGER, J.C.;
    "Deformation and break–up of liquid droplets in a simple shear field",
    IEC Fundamentals (1968) 7, no. 4, 577.
18. KOZENY, J.;
    "Über kapillare Leitung des Wassers im Boden",
    Berichte Wien Akad., 136–2A (1927) 271.

The authors would like to express their thanks to their colleagues in KSEPL.
Special thanks are due to Mr. J.C.Stekelenburg, who carried out the experimental
work.

ENHANCED OIL RECOVERY R&D
IN THE UNITED STATES AND IN THE U.S. DEPARTMENT OF ENERGY

J. J. GEORGE STOSUR

Office of Oil, Gas and Shale Technology, U.S. Department of Energy

ABSTRACT

The paper provides a general outline of the status of the enhanced oil recovery technology in the United States with emphasis on technical problems and the search for solutions. Upon this background, the U.S. Department of Energy's effort and the research priorities in enhanced oil recovery are described including the new comprehensive data collection system and analysis on several hundred field projects.

INTRODUCTION

There is universal agreement that enhanced oil recovery (EOR) presents one of the best options for liquid fuels production in the next two decades.

The U.S. resource target for EOR is very large; of the 450 billion barrels of oil that have been discovered to-date, only one-third, or 150 billion barrels will be produced through primary and secondary methods, that is through depletion and waterflooding. That leaves a target of over 300 billion barrels of oil, the location of which is known and in reservoirs which, though depleted are usually reasonably well defined and outlined.

Of the 450 billion barrels found to-date in U.S., 350 billion barrels are considered as light oil (generally with gravities above 25° APT or 0.91 g/cc). Even after waterflooding, 230 billion barrels of this light oil remain in the ground awaiting enhanced recovery technologies, and even larger percentage fraction of heavy oils remains unrecovered. While there is uncertainty as to exactly how much of this oil can be recovered by EOR processes, a range of from 18 to 52 billion barrels is reasonable, depending on technological successes and energy prices in the future. Here light oil accounts for between 12 and 33 billion barrels and heavy oil for 6 to 19 billion barrels. By way of comparison, 52 billion barrels is nearly twice U.S. proved reserves and equals the total output from 71 synfuel plants, each producing 100,000 barrels per day over a 20-year plant life.

While oil production through EOR can be brought on line far more quickly than synfuels, there are a number of constraints that must be resolved before this can happen. Some of the more prominent are technical constraints. The subject of this paper is research and development conducted at U.S. Department of Energy to mitigate the technical constraints in the application of EOR technologies.

The views expressed in this paper are those of the author, and they do not necessarily represent those of the U.S. Department of Energy.

EOR TECHNIQUES - AN HISTORICAL OVERVIEW

There are three generic groups of EOR processes: chemical, gas (miscible and immiscible) and thermal. Each has several variants, but the processes which are most widely applied are:

Chemical

- Polymer-augmented waterflooding. It relies on the addition of "thickening" agents to water in order to increase displacement efficiency by reducing mobility of the displacing fluid.

- Alkaline flooding. It is based on the addition of strong caustic substances to injection water in order to affect reduced surface tension between reservoir fluids, thus permitting easier fluid movement.

- Surfactant polymer flooding. Surface active agents are injected to displace oil by reduced surface tension which allows building an oil bank that is subsequently pushed by polymers and water.

Gas (miscible and immiscible)

- Hydrocarbon miscible. Miscibility is obtained by the injection of hydrocarbon gases which dissolve in oil, reduce viscosity and help the creation of an oil bank which can then be pushed to producing wells by water. Now that the cost of natural gas and LPG is high, the method is not much used.

- Carbon dioxide flooding. It is based on injection of CO_2 to strip the lighter components, swell the oil, partially mix with it and create an oil bank which can then be displaced by additional gas or water.

- Nonhydrocarbon gas drive. Inert gases can be used such as nitrogen or flue gas, primarily to add pressure to reservoir but also to attain miscibility (depending on pressure) and displace the oil in much the same way as CO_2 or natural gas.

Thermal processes

Thermal processes apply largely to the recovery of heavy oils which are too viscous in their natural state to flow freely. Here, advantage is taken of the exponential decline of oil viscosity as temperature rises when reservoir is heated with steam injected from the surface or generated in the reservoir by in situ combustion.

- Steam soak. The steam soak variety is typically used for stimulation to either accelerate or establish primary production in an otherwise unproductive heavy oil reservoir. Multi-cycle steam applications can be quite efficient under favorable conditions, but efficiency quickly falls.

- Steam drive. Steam is injected continously in one well, viscosity of oil is reduced until it becomes mobile and can be displaced or produced by gravity drainage in surrounding wells.

- Fireflooding. It uses energy derived from burning part of the oil in a reservoir to assist in the recovery of the remaining unburnt oil. The combustion is supported by injected air and often water to increase efficiency of the process.

Potential EOR technologies

- Microbial EOR. Microorganisms can be used to generate surfactants, to produce CO_2 in the reservoir and to otherwise change the composition of the oil for improved recovery. This method is largely in research stage, though a few field tests were performed.

- Combination mining. Several approaches have been proposed for the direct extraction of crude oil, including large diameter shafts from which horizontal or upwardly slanted wells are drilled for drainage.

- Steam drive in light oil. There is evidence that steam drive could hold promise in shallow light oil reservoirs where other methods failed.

- RF heating. It is based on beaming radio frequency energy into a heavy oil reservoir. The method is currently being field tested to determine its potential.

An historical summary of EOR projects in the U.S. based on biennial Oil and Gas Journal surveys (1) is shown in Table 1.

Table 1. United States EOR projects in perspective

Method	1970	1973	1975	1977	1979
Chemical					
Polymer Flood	14	9	14	21	22
Caustic Flood	0	2	1	3	6
Micellar/Polymer	5	7	13	22	14
Miscible Gas					
Hydrocarbon Miscible	21	12	15	15	9
Carbon Dioxide	1	6	9	14	17
Other Gases	0	1	1	6	8
Thermal					
Steam Drive	22	22	31	43	79
Cyclic Steam	31	42	54	56	54
In Situ Combustion	38	19	21	16	17
	132	120	159	196	226

Closer examination of the ten-year history of chemical EOR projects in U.S. shows: steady increase in the number of polymer projects since 1973; careful experimentation with caustic floods, though steadily increasing over time, and; steady increase in new starts of the micellar/polymer projects until 1977 and then a sharp decline in response to discouraging results.

The miscible gas projects show a gradual phasing out of hydrocarbon miscible projects due to rapidly increasing value of natural gas; steady increase in the number of carbon dioxide projects, which apparently replaced the increasingly costly natural gas, and; a recent surge of interest in nonhydrocarbon gases, even though most of the tests are very small.

. The thermal recovery projects are most numerous, reflecting the relative maturity of the technology and show steady increase in the number of steam drive projects with a sharp increase in 1979 (even sharper increase is expected for 1981); gradual increase, then leveling off and slight decline in 1979 of steam soak projects in favor of the more efficient steam drive projects, and; a sharp decline of in situ combustion projects after 1970 when euphoria over the new technology gave way to the somber reflection that the technology is a lot more difficult to apply than it appeared.

These observations underline the tendency by the private sector to prefer the lower-risk, proven technologies such as cyclic steam and steam drive and to avoid the less certain and the less predictable but advanced approaches such as micellar/polymer floods, due to the high degree of risk and poor performance predictability.

Equally interesting is the operators' own evaluations of their projects which was compiled from the Oil and Gas Journal survey (1), Table 2. Again, most of the projects judged technically and economically successful or promising are those in the more or less established and less risky processes which include thermal recovery and polymer flooding.

Table 2. Operators' own evaluation of project performance—March 1981

METHOD	Tech. and Econ. Success	Tech. but not Econ. Success	Promising	Terminated or Discouraging	Too Early to Tell	Total Eval.
Steam	57	6	26	5	19	113
In Situ Comb.	7	2	1	1	3	14
Polymer	10	–	4	3	5	22
Caustic	–	2	1	–	3	6
Micellar/Polymer	–	1	2	3	8	14
Carbon Dioxide	2	1	5	2	7	17
Other Gases	2	–	3	–	3	8
Total	77	11	42	14	48	194

The year 1981 will prove to be the year of a sharp increase of new EOR projects in the United States. To stimulate industry activity in EOR, a special incentive program was set forth by the Economic Regulatory Administration. Under this program an operator of a qualifying EOR project was permitted to realize world oil price for controlled oil, provided that the difference was invested in the project. There were several limits: a project could recover no more than 75% of qualifying costs with a limit of $20 million per project. The purpose was to ameliorate the high front-end costs associated with EOR. The response was overwhelming, with as many as 423 EOR projects by the time the program was concluded in March 1981, following total decontrol of crude oil prices.

The incentive program for EOR was designed to begin many EOR projects that would lead to rapid oil production and, as expected, most of the projects proposed by industry involved the use of current technology, those processes that are less risky, better understood and requiring smaller capital investment. Table 3 shows the types and number of projects in various categories.

Table 3. New Enhanced Oil Recovery projects due to Energy
Regulatory Administration's special incentive program

Oil Recovery Technique	Number of Projects
Miscible Fluid Displacement (Less CO_2)	13
CO_2 Miscible Fluid Displacement	95
Conventional Steam Drive Injection	93
Unconventional Steam Drive Injection	34
Microemulsion Flooding	37
In Situ Combustion	35
Polymer Augmented Waterflooding	48
Cyclic Steam Injection	16
Alkaline (Caustic) Flooding	32
Immiscible Non-hydrocarbon Gas Displacement	11
Enhanced Heavy Oil Recovery (Other Than Thermal)	3
Other Tertiary Enhanced Recovery Techniques	6
Total	423

The new EOR projects represent over $15 billion of private investment and are
expected to recover nearly 3 billion barrels of crude oil, but as much as 80%
of the total number use proven, less risky technology. The major promise of
EOR is, however, with the newer, high risk "advanced" processes that have
shown great promise in theory and in the laboratory, but have not fared well in
the transition to the field.

The large number of new EOR field tests provide a unique opportunity for gathering and analyzing actual field data in a compressed period of time. The 423
incentive projects include a sufficient number of advanced technology applications that, when measurement and analyses are linked with oil production data,
an excellent opportunity will present itself for scientific observation of their
performance. Accordingly, a few of the more advanced field tests will be
selected from among the cost-shared and incentive field tests for extensive pre-
and post-test observation, diagnostics and analysis. It is hoped that these
data coupled with the results from laboratory experiments and small nonproducing
mini-tests, will be used to greatly improve the fundamental understanding of how
and why the advanced EOR techniques behave as they do. The results should
contribute towards more effective prediction of various processes under different
reservoir conditions (2).

TECHNICAL CONSTRAINTS AND SEARCH FOR SOLUTIONS

The high risks that deter operators from the advanced technologies are the
principal focus of the goals of the DOE Enhanced Oil Recovery R&D program. R&D
priorities among the processes directly reflect the degree of risk associated
with each technology and the size of the potential of the respective processes.
Research priorities have been developed to reduce or eliminate certain constraints. These priorities for the chemical, miscible gas and thermal EOR
processes are reflected in the current R&D program, as follows:

Chemical

- Basic studies on mechanism of displacement of bypassed oil.
- Effect of surface and physical chemistry of microemulsions on displacement efficiency.
- Chemical degradation at high temperature, high connate water salinity and high clay content.
- More effective formulations of microemulsions and additives for mobility control.
- Fundamental R&D on rock/fluid interactions, which includes adsorption, wettability, ion exchange and formation damage.
- Quantitative determination of the effects of dispersion, relative permeability, apparent viscosity and inaccessible pore volume on mobility control under one-, two- and three-phase flow for the development of equations to be used for improving the precision of predictive reservoir simulators.

Gas Miscible Displacement (CO_2 Flooding)

A number of technical constraints to a wider application of gas miscible displacement are similar to those of chemical and even thermal recovery methods. One such generic problem is lack of adequate mobility control and the resultant low areal and vertical sweep efficiencies. Even when the displacing phase is fully miscible with crude oil under all conceivable reservoir conditions, an unfavorable mobility ratio leads to fingering which causes premature breakthrough and poor sweep efficiency. Other than the ever present problem of mobility control the R&D effort in miscible displacement processes includes:

- Fundamental studies on miscibility with reservoir oil, criteria for miscibility (single contact and multiple-contact miscibility).
- Formation damage in carbonate reservoirs due to CO_2 flooding.
- The effect of N_2 and other gases on phase behavior and displacement efficiency.
- Static and dynamic laboratory investigations of phase behavior of CO_2-crude oil systems.
- Studies of the effect of foams, polymers, graded-viscosity slugs and emulsifiers on the mobility ratio and displacement efficiency.
- Supply of natural CO_2 and its cost.

Thermal Recovery

- Development of downhole steam generation capability at depths exceeding 2500 feet with the triple objective of substantially reducing transmission heat losses, overcoming flue gas emission problems and increasing recovery efficiency due to the action of CO_2 with steam.

- Fundamental research on the effect of ancillary materials with steam on displacement and sweep efficiencies.

- Improved insulation of steam injection wells and the development of a metal-extrudable packer for high temperature wells.

- Development of techniques for tracing the position of the high temperature front in steam flooding and in situ combustion methods (surface mapping of thermal fronts).

- Field experimentation to determine the feasibility of bitumen recovery from tar sands using steam flooding, reverse in situ combustion and the combination of both methods.

One of the more interesting EOR related projects sponsored by the Department of Energy is the development of a downhole steam generator at Sandia National Laboratories. The project was started 3 years ago and has recently attracted wide publicity in trade journals and some independent work by private interests. The downhole steam generator has indeed unique potential: it offers exceptionally high overall thermal efficiency (not only due to the elimination of heat lost in the transmission of steam from the surface to the bottom of the bed, but also due to the elimination of heat rejected and lost up the stack of a conventional steam generator); it can help overcome air pollution by injecting combustion products with steam into oil reservoirs, and; its economic benefits may be superior to the currently used steam generators. A prototype has already been tested under simulated conditions at the surface and a real downhole test of the high pressure system will be conducted in Long Beach, California, in the summer and fall of 1981.

Another example of advanced technology brought to fruition and transferred to the private sector is in the area of faster, more efficient drilling. Sandia National Laboratory has developed a special bonding technique which permitted a new bit design with polycrystalline diamond cutters. One source quotes oil companies' reports that three synthetic-diamond bits can drill one 5000 ft.-deep well section in 112 hours, while in the past, such a section would use up as many as 13 conventional bits (3). Far more important that the difference in hardware cost is the saving in time; synthetic-diamond bits can cut drilling time by several days for a saving of up to $1 million in the more expensive offshore wells. Worldwide, 13 companies now fabricate the synthetic diamond bits, with total capacity of the U.S. companies in the range of 2,800 bits per year and several of them have order backlogs extending for a year or more.

A good example of the kind of long-term R&D pursued by the Department of Energy is microbial EOR (4). There are currently five contracts with universities to determine whether microbial cells can be used to selectively plug high-permeability layers and thus improve sweep efficiency; to examine the potential of using microorganisms for EOR by mechanisms other than selective plugging, such as in situ production of biosurfactants and biopolymers; to screen qualitatively and quantitatively for microbial organisms' ability to produce gases such as carbon dioxide, hydrogen and methane and; to determine the feasibility of bacteria to produce acids such as acetic acid, solvents such as alcohols and acetone as well as small molecules possessing surfactant properties. By most accounts microbial EOR is avant-garde, insufficiently pursued by private interests and yet, according to many scientists, offers exciting possibilities for brand new concepts in EOR.

CONCLUSIONS

Enhanced oil recovery has the highest probability of new-term impact on the production of liquid fuels, with lowest cost premium. However, a coordinated program of field experiments and supporting R&D is required to accelerate both the development of technology and early commercial implementation. To this end, the U.S. Department of Energy has established a highly organized effort through its Energy Technology Centers, National Laboratories and nearly 30 universities to systematically attack the complex technical problems that have inhibited industry application of the potentially efficient but expensive and risky technologies.

The U.S. Department of Energy R&D priorities reflect the new emphasis on high-risk, long-range, high potential EOR technologies. It is presumed that the industry will respond to the recent oil price decontrol with stepped-up development activities, including the economically marginal petroleum resources and increasing recovery efficiency. In all cases, the EOR program is closely coordinated with industry to avoid duplication of effort.

REFERENCES

1. SHANNON, L. Matheny, Jr.; "EOR Methods Help Ultimate Recovery", Oil and Gas Journal (Mar 31, 1981) 79-124.

2. "The DOE Light Oil Research-and-Development Program," Draft publication dated May 21, 1981.

3. "Synthetic Diamonds Shake-Up The Drill-Bit Market", Business Week, December 1, 1980, 98-99.

4. "Contracts for Field Projects and Supporting Research on Enhanced Oil Recovery and Improved Drilling Technology", 24, DOE/BETC-80/4, Sept 30, 1980.

AUTHOR INDEX